Achieving sustainable cultivation of sugarcane

Volume 2: Breeding, pests and diseases

It is widely recognised that agriculture is a significant contributor to global warming and climate change. Agriculture needs to reduce its environmental impact and adapt to current climate change whilst still feeding a growing population, i.e. become more 'climate-smart'. Burleigh Dodds Science Publishing is playing its part in achieving this by bringing together key research on making the production of the world's most important crops and livestock products more sustainable.

Based on extensive research, our publications specifically target the challenge of climate-smart agriculture. In this way we are using 'smart publishing' to help achieve climate-smart agriculture.

Burleigh Dodds Science Publishing is an independent and innovative publisher delivering high quality customer-focused agricultural science content in both print and online formats for the academic and research communities. Our aim is to build a foundation of knowledge on which researchers can build to meet the challenge of climate-smart agriculture.

For more information about Burleigh Dodds Science Publishing simply call us on +44 (0) 1223 839365, email info@bdspublishing.com or alternatively please visit our website at www.bdspublishing.com.

Related titles:

Achieving sustainable cultivation of sugarcane Volume 1:
Cultivation techniques, quality and sustainability
Print (ISBN 978-1-78676-144-6); Online (ISBN 978-1-78676-146-0, 978-1-78676-147-7)

Achieving sustainable cultivation of coffee
Print (ISBN 978-1-78676-152-1); Online (ISBN 978-1-78676-154-5, 978-1-78676-155-2)

Global tea science
Print (ISBN 978-1-78676-160-6); Online (ISBN 978-1-78676-162-0, 978-1-78676-163-7)

Chapters are available individually from our online bookshop: https://shop.bdspublishing.com

climate•SMART•publishing
IN AGRICULTURAL SCIENCE

BURLEIGH DODDS SERIES IN AGRICULTURAL SCIENCE

NUMBER 38

Achieving sustainable cultivation of sugarcane

Volume 2: Breeding, pests and diseases

Edited by Professor Philippe Rott
University of Florida, USA

burleigh dodds
SCIENCE PUBLISHING

Published by Burleigh Dodds Science Publishing Limited
82 High Street, Sawston, Cambridge CB22 3HJ, UK
www.bdspublishing.com

Burleigh Dodds Science Publishing, 1518 Walnut Street, Suite 900, Philadelphia, PA 19102-3406, USA

First published 2018 by Burleigh Dodds Science Publishing Limited

Library of Congress Control Number: 2017956488

British Library Cataloguing in Publication Data
A catalogue record for this book is available from the British Library

ISBN 978-1-78676-148-4 (print)
ISBN 978-1-78676-150-7 (online)
ISBN 978-1-78676-151-4 (online)
ISSN 2059-6936 (print)
ISSN 2059-6944 (online)

Typeset by Deanta Global Publishing Services, Chennai, India
Printed by Lightning Source

Contents

Series list

Title	Series number
Achieving sustainable cultivation of maize - Vol 1 From improved varieties to local applications Edited by: Dr Dave Watson, CGIAR Maize Research Program Manager, CIMMYT, Mexico	001
Achieving sustainable cultivation of maize - Vol 2 Cultivation techniques, pest and disease control Edited by: Dr Dave Watson, CGIAR Maize Research Program Manager, CIMMYT, Mexico	002
Achieving sustainable cultivation of rice - Vol 1 Breeding for higher yield and quality Edited by: Prof. Takuji Sasaki, Tokyo University of Agriculture, Japan	003
Achieving sustainable cultivation of rice - Vol 2 Cultivation, pest and disease management Edited by: Prof. Takuji Sasaki, Tokyo University of Agriculture, Japan	004
Achieving sustainable cultivation of wheat - Vol 1 Breeding, quality traits, pests and diseases Edited by: Prof. Peter Langridge, The University of Adelaide, Australia	005
Achieving sustainable cultivation of wheat - Vol 2 Cultivation techniques Edited by: Prof. Peter Langridge, The University of Adelaide, Australia	006
Achieving sustainable cultivation of tomatoes Edited by: Dr Autar Mattoo, USDA-ARS, USA & Prof. Avtar Handa, Purdue University, USA	007
Achieving sustainable production of milk - Vol 1 Milk composition, genetics and breeding Edited by: Dr Nico van Belzen, International Dairy Federation (IDF), Belgium	008
Achieving sustainable production of milk - Vol 2 Safety, quality and sustainability Edited by: Dr Nico van Belzen, International Dairy Federation (IDF), Belgium	009
Achieving sustainable production of milk - Vol 3 Dairy herd management and welfare Edited by: Prof. John Webster, University of Bristol, UK	010
Ensuring safety and quality in the production of beef - Vol 1 Safety Edited by: Prof. Gary Acuff, Texas A&M University, USA & Prof.James Dickson, Iowa State University, USA	011
Ensuring safety and quality in the production of beef - Vol 2 Quality Edited by: Prof. Michael Dikeman, Kansas State University, USA	012
Achieving sustainable production of poultry meat - Vol 1 Safety, quality and sustainability Edited by: Prof. Steven C. Ricke, University of Arkansas, USA	013
Achieving sustainable production of poultry meat - Vol 2 Breeding and nutrition Edited by: Prof. Todd Applegate, University of Georgia, USA	014
Achieving sustainable production of poultry meat - Vol 3 Health and welfare Edited by: Prof. Todd Applegate, University of Georgia, USA	015

Acknowledgements

We wish to acknowledge the following for their help in reviewing particular chapters:

- Dr Karen Aitken, Commonwealth Scientific and Industrial Research Organisation, Australia
- Dr Niranjan Baisakh, Louisiana State University, USA
- Dr Julien Beuzelin, University of Florida, USA
- Dr Graham Bonnett, Commonwealth Scientific and Industrial Research Organisation, Australia
- Dr Frikkie Botha, Sugar Research Australia, Australia
- Dr Kathy Braithwaite, Sugar Research Australia, Australia
- Dr Peta Campbell, South African Sugarcane research Institute, South Africa
- Dr Lisa A. Castlebury, United States Department of Agriculture, USA
- Dr Ronald Cherry, University of Florida, USA
- Dr Jean-Heinrich Daugrois, Centre de Coopération Internationale en Recherche Agronomique pour le Développement, France
- Dr Asha Dookun-Saumtally, Mauritius Sugarcane Industry Research Institute, Mauritius
- Dr Jean-Claude Girard, Montpellier, France
- Dr Jean-Yves Hoarau, Centre de Coopération Internationale en Recherche Agronomique pour le Développement, France
- Dr Hermann P. Hoffman, Universidade Federal de São Carlos, Brazil
- Dr Jeffrey Hoy, Louisiana State University, USA
- Dr Jonathan Jacobs, Institut de Recherche pour le Développement, France
- Dr Jeffrey B. Jones, University of Florida, USA
- Dr Anthony Kennedy, West Indies Central Sugar Cane Breeding Station, Barbados
- Dr David Kuhn, United States Department of Agriculture, USA
- Dr Ramon Leon, University of Florida, USA
- Dr Robert Magarey, Sugar Research Australia, Australia
- Dr Sharon McFarlane, South African Sugarcane Research Institute, South Africa
- Dr Martha Malapi-Wight, United States Department of Agriculture, USA
- Dr Clarissa Maroon-Lango, United States Department of Agriculture, USA
- Dr Marcelo Menossi, Universidade Estadual de Campinas, Brazil
- Dr Erik Mirkov, Texas A & M University, USA
- Dr Claudia Monteiro-Vitorello, University of São Paulo, Brazil
- Dr Deborah Neher, University of Vermont, USA
- Dr Albert Orgeron, Louisiana State University, USA
- Dr Charles Overstreet, Louisiana State University, USA
- Dr Roy Parfitt, Sugar Research Australia, Australia
- Dr Michel Peterschmitt, Centre de Coopération Internationale en Recherche Agronomique pour le Développement, France
- Dr George Piperidis, Sugar Research Australia, Australia
- Dr Richard N. Raid, University of Florida, USA
- Dr John J. Riascos, Centro de Investigacion de la Caña de Azucar de Colombia, Colombia
- Dr Takeo Sakaigaichi, NARO Kyushu Okinawa Agricultural Research Center, Japan
- Dr Salem Saumtally, Mauritius Sugarcane Industry Research Institute, Mauritius

- Dr Sandy Snyman, South African Sugarcane Research Institute, South Africa
- Dr Ramesh Sundar, Sugarcane Breeding Institute, India
- Audrey Thong-Chane, eRcane, France
- Dr Patricia Timper, United States Department of Agriculture, USA
- Dr Johnnie van den Berg, North-West University, South Africa
- Dr Derek Watt, South African Sugarcane Research Institute, South Africa
- Dr William White, United States Department of Agriculture, USA
- Dr Steve Wylie, Murdoch University, Australia
- Dr Marvellous Zhou, South African Sugarcane Research Institute, South Africa

Introduction

Sugarcane is the source of about three quarters of the world's sugar, and is grown widely in the tropics and sub-tropics. Despite rising demand, average yields have not increased significantly, partly because of continued vulnerability to pests and diseases. In addition, cultivation has been seen as damaging biodiversity and soil health with a negative effect on both yields and the environment.

These volumes on achieving sustainable cultivation of sugarcane summarise the wealth of research addressing these challenges. Volume 1 reviews cultivation techniques and sustainability issues. This volume, Volume 2 reviews how the challenges facing sugarcane production can be addressed through developments in breeding as well as better management of pests and diseases. The volume covers key advances in breeding, including conventional and molecular breeding techniques. It also summarises key advances in understanding bacterial, fungal and viral diseases of sugarcane and assesses best practices in integrated disease, pest and weed management.

Part 1 Plant physiology and breeding

The first part of the volume assesses the latest research on sugarcane genetics, physiology and genetic diversity, and how this is informing advances in conventional and molecular breeding techniques. The focus of Chapter 1 is on sugarcane genome sequencing and genetic mapping. The chapter explains the difficulties of sequencing and mapping the genome of sugarcane and strategies to overcome these difficulties. The chapter outlines progress on sugarcane sequencing, genetic mapping of simply inherited and complex traits. The chapter then focuses on the mapping of a gene controlling sugarcane brown rust resistance, *Bru*1, which exemplifies the potential of marker-assisted selection. Finally, the chapter discusses prospects for future research in sugarcane genome sequencing and genetic mapping.

Following on from the previous chapter's focus on sugarcane genomics, Chapter 2 examines advances in understanding sugarcane plant growth and physiology. Crop yield in sugarcane depends on how well the source leaves can supply the assimilates needed for the growth and filling of the harvestable plant. This is the source-sink relationship which remains poorly understood in sugarcane. The chapter discusses significant advances made in this area in the past few decades, as well as suggesting future trends in research such as high throughput genomic, metabolomic and phenotypic analyses that could allow refinement in crop and metabolic modelling, as well as improved breeding and production strategies.

Returning to the theme of the genetics of sugarcane, Chapter 3 examines the challenges of ensuring and exploiting the genetic diversity of sugarcane. The majority of cultivars in current commercial sugarcane breeding programs can be traced back to a few key interspecific hybrids that were developed during the early 1900s. Sugarcane breeders have expressed concerns about the narrow sampling of ancestral clones in modern sugarcane breeding programs, and this concern has prompted periodic attempts at so-called 'base-broadening' programs. The chapter provides an overview of sugarcane germplasm collections and then describes introgression-related research and breeding efforts focused on use of plant species belonging to the genera of *Saccharum*, *Erianthus* and others. It emphasizes the difficulties and challenges that need to be overcome in order to achieve successful outcomes from introgression breeding, and considers a few possible future directions for research.

Chapter 4 shifts the focus to advances in conventional sugarcane breeding programs. All sugarcane cultivars currently grown throughout the world arise from breeding programs which have used a reasonably similar approach sustained over many decades. This approach comprises a continuous pipeline of regular (usually annual) crossing among selected parent clones to produce large populations of seedling clones, followed by selection of these clones in successive stages of field trials for 9–12 years for important traits. The chapter outlines the history and structure of sugarcane breeding programs as a context for considering efforts to advance rates of progress. The chapter reviews studies conducted in the last 30 years to improve many specific aspects of operations in sugarcane breeding programs. The chapter describes emerging concerns about whether current rates of genetic gain in sugarcane are optimal, and suggests some avenues for faster gains.

Continuing the theme of sugarcane breeding and complementing the preceding chapter, Chapter 5 considers advances in marker-assisted breeding of sugarcane. Sugarcane breeding has, until very recently, been based solely on phenotype, and marker-assisted breeding of sugarcane remains in its infancy compared with row crops such as maize and soybean. A major reason for this is the complex genetics of sugarcane. This chapter reviews the uses of DNA marker technology in sugarcane for fingerprinting and diversity analysis. This is followed by a review of the development of linkage maps and initial trait/QTL mapping, including the *Bru*1 locus for resistance to brown rust (*Puccinia melanocephala*). Finally, the chapter reviews the use of newer next generation sequencing (NGS)-based technologies in sugarcane, including genome-wide association analysis (GWAS) and genomic selection (GS).

Concluding the first part of the volume, Chapter 6 focusses on genetic improvement of sugarcane by transgenic, intragenic and genome editing technologies. The chapter explores the role of tissue culture, gene transfer and selectable marker systems in genetically improving sugarcane. It includes a detailed case study on transgenic, intragenic and genome editing approaches to improve bioethanol production from sugarcane. The chapter also addresses the role of metabolic engineering in elevating sucrose yields and producing alternative sugars or value-added biomaterials.

Part 2 Progress in understanding and managing diseases, pests and weeds

The second part of the volume discusses progress in understanding bacterial, fungal and viral diseases and their management, followed by the management of insect and nematode pests as well as weeds. Chapter 7 focusses on the challenge of ensuring biosecurity in sugarcane cultivation. Biosecurity is an important factor in maintaining and improving yields. It can refer to the day-to-day management of established pests, diseases, and weeds, as well as preparing for unusual threats to the industry, and safeguarding agriculture and natural resources. The chapter focuses on the sugarcane industry's preparedness for exotic pathogens, describing techniques such as pest risk analysis, quarantine control and surveys. The chapter discusses emerging technologies and examines and compares current guidelines. Finally, the chapter presents a case study of the 2006 sugarcane smut incursion in Australia.

Continuing the theme of understanding diseases of sugarcane, Chapter 8 addresses viral metagenomics and sugarcane pathogens. Plant viral metagenomics has recently proved effective for studying the collection of plant viruses. The advent of metagenomics-based approaches has led to the discovery and characterization of new plant viruses and helped solving etiological enigmas. The chapter describes the plant virus metagenomics methods developed during the last decade before focusing on the use and application of these approaches in the field of sugarcane pathology. The chapter presents three recent studies that have used viral metagenomics-based approaches aimed at solving long-standing etiological enigmas, identifying known and potentially new sugarcane viruses in a sugarcane quarantine context, and assessing the diversity of a known sugarcane virus (*Sugarcane yellow leaf virus*). Finally, the chapter discusses the challenges and opportunities of using such approaches to address issues of plant pathology.

Moving from viral to fungal diseases of sugarcane, Chapter 9 discusses progress in understanding fungal diseases affecting sugarcane, with a specific focus on red rot, a disease of sugarcane stalks caused by the fungal pathogen *Colletotrichum falcatum*. Red rot has been recorded in 77 countries worldwide, but occurs most severely in south Asian countries, especially India, Nepal, Pakistan, Myanmar and Bangladesh, as well as Thailand and Vietnam. Developing red rot-resistant varieties has been the major management strategy to deal with the disease in most countries. The chapter reviews the economic impact of red rot, diagnosis, pathogen variability, application of molecular tools to understand pathogen biology, varietal resistance, screening methods and new management approaches including fungicide delivery.

Complementing the previous chapter's focus on fungal diseases of sugarcane, Chapter 10 focuses on sugarcane smut. Fungal diseases of sugarcane constitute a worldwide threat to cultivation. Smut is caused by the biotrophic fungus *Sporisorium scitamineum*. After offering a historical perspective on the disease, the chapter presents an overview of data gathered over the past few years using 'omics' techniques (genomics, transcriptomics, proteomics and metabolomics) in combination with functional studies performed to identify candidate genes involved in the interaction between the pathogen and the host plant. The chapter discusses studies aimed at deciphering the molecular cross-talking between sugarcane and *Sporisorium scitamineum* along with new trends and recurring issues in the study of smut disease.

Continuing the theme of fungal diseases, Chapter 11 considers progress in understanding sugarcane rusts. Rust fungi (Uredinales or Pucciniales) are a globally distributed order of obligate parasitic fungi occurring on vascular plants. The typical symptom is a 'rust-coloured' (often orange or yellow) pustule on a plant leaf. Economically, rust fungi cause enormous crop losses, including in sugarcane. The chapter summarizes current knowledge of sugarcane rusts in terms of the relevant background biology, pathogen dispersal, disease identification, environmental conditions favouring disease emergence, crop loss, host-plant resistance and control measures. The chapter focuses on the history, identification and control of three rust diseases of sugarcane: brown, orange and tawny rust.

Chapter 12 moves from fungi to consider progress in understanding viruses affecting sugarcane. There are currently around ten diseases of sugarcane known to be caused by viruses. In recent years, our understanding of these diseases has greatly benefitted from access to low cost sequencing technologies which have revolutionised virus discovery, improved our understanding of genetic variation in viral populations, improved diagnostic options and helped us to understand gene functions and identify gene targets for control. The chapter addresses the broad themes of virus discovery, genetic variation, diagnostic technologies and control. The chapter presents case studies describing the impact of two major viral diseases in Papua New Guinea and Indonesia.

Switching to bacterial diseases of sugarcane, Chapter 13 considers recent progress in understanding three major bacterial diseases: gumming, leaf scald and ratoon stunting. The chapter describes new insights provided by genomic and microscopy analyses of the bacterial pathogens *Xanthomonas vasicola* pv. *vasculorum*, *X. albilineans*, and *Leifsonia xyli* subsp. *xyli*, the causal agents of gumming, leaf scald, and ratoon stunting, respectively. The chapter addresses genomic-based hypotheses regarding adaptation of the three pathogens to sugarcane and the origin of their respective ancestors. It examines the strategies used by *X. albilineans* and *L. xyli* subsp. *xyli* to spread in sugarcane, including bacterial invasion of xylem vessels and non-vascular tissues. The chapter looks ahead to the development of further functional studies to improve the understanding of the interactions between sugarcane and bacterial pathogens, which are likely to contribute to better control of bacterial diseases of sugarcane.

Chapter 14 moves to consider the challenge of nematode pests of sugarcane by focussing on managing the soil biological community to improve soil health and reduce losses from nematode pests. Most sugarcane fields are infested with at least five species of plant-parasitic nematodes, with *Pratylenchus* and *Meloidogyne* considered the most damaging genera. The chapter describes the main nematode pests and the community of organisms with which they co-exist before analysing existing nematode control measures and their economic importance. The chapter then describes how a broader approach focusing on improving soil health can mitigate the impact of nematodes and ensure sustainable sugarcane production.

The theme of Chapter 15 is progress in understanding and managing insect pests affecting sugarcane. The chapter describes major achievements in pest management in sugarcane agroecosystems based on applied research conducted in different countries. The focus of the chapter is on two major groups of pests, stalk borers (Lepidoptera) and white grubs (Coleoptera). Chapter sections address determination of crop losses, biological control using parasitoids and entomopathogens, push-pull strategies with the use of companion (or service) plants, agricultural practices and their influence on pest population dynamics, insecticide use and biosecurity. The chapter also discusses new tools for pest detection and management such as modeling, remote sensing, and use of geographical information systems (GIS). The chapter looks ahead to future developments in integrated pest management of sugarcane.

Bringing together the insights of preceding chapters on particular diseases, Chapter 16 focuses on Integrated Disease Management (IDM) strategies in sugarcane cultivation. As has been noted, sugarcane diseases cause severe losses to sugar production around the world. More than a hundred bacterial, fungal, phytoplasma and viral diseases are present in sugarcane growing areas worldwide. Some diseases are present in most sugarcane growing regions while others are confined to specific countries. The chapter outlines and explores the various disease management strategies used in sugarcane cropping systems around the world and the potential for integrated disease management to minimize losses and maximize profitability. The chapter covers management of diseases by controlling the pathogens, management of diseases through the host, and management of diseases by controlling the environment, before looking ahead to future research trends in this field.

Complementing the preceding chapter's focus on integrated disease management, the volume's concluding chapter, Chapter 17, concentrates on the related challenges of Integrated Weed Management (IWM) in sugarcane cultivation. Weeds reduce sugarcane yield by competing for water, nutrients and light. They may also act as alternate hosts for disease and insect pests, increase production costs and lower land value. Implementation of integrated weed management in sugarcane is essential for reducing the negative impact of weeds to economically acceptable levels. Understanding of the critical period of weed control is important so that IWM strategies can be developed to maximize control measures. The chapter deals with an integrated approach involving different methods for effective weed management in sugarcane that can be adopted for sustainable production, involving cultural, mechanical and chemical weed control.

Part 1

Plant physiology and breeding

Chapter 1

Sugarcane genome sequencing and genetic mapping[1,2]

Xiping Yang and Ramkrishna Kandel, University of Florida (Gainesville), USA; Jian Song, University of Florida (Gainesville), USA and Dezhou University, China; Qian You, University of Florida (Gainesville), USA and Fujian Agriculture and Forestry University, China; Mengjuan Wang, Fujian Agriculture and Forestry University, China; Jianping Wang, University of Florida (Gainesville), USA and Fujian Agriculture and Forestry University, China

1 Introduction

Sugarcane (*Saccharum* spp.) contributes approximately 80% of the sugar to the world, greatly exceeding sugar beet as a source of sugar (Dahlquist, 2013). In addition to being a source of sugar, sugarcane is also an important bioenergy crop, with an energy output-to-input ratio of ethanol production from sugarcane five times higher than that of maize (Goldemberg, 2008; Waclawovsky et al., 2010). In 2013, the Food and Agriculture Organization (FAO) estimated that sugarcane had a worldwide gross production value of $81.5 billion (FAO, 2013). It was grown on ~27.1 million hectares with a worldwide harvest of 1.9 billion metric tons, higher than maize (1.0 billion metric tons), rice (741.0 million metric tons) and wheat (729.0 million metric tons) (FAO, 2014).

Breeding superior commercial cultivars is critical for maintaining sugarcane production, which will benefit from research in sugarcane genome sequencing and genetic mapping. These research areas focused on understanding sugarcane's genome structure, organization and inheritance patterns. They also aid in understanding genetic variations within sugarcane

1. Xiping Yang, Ramkrishna Kandel, Jian Song and Qian You contributed equally.
2. Corresponding author: Jianping Wang (wangjp@ufl.edu).

http://dx.doi.org/10.19103/AS.2017.0035.02

that control important agronomic traits. Discovery of genes or quantitative trait loci (QTL) of interest provides resources for breeding programmes. Genes could be pyramided or removed depending on the role played by the gene, via marker-assisted selection (MAS). Meanwhile, advances in sugarcane genome sequencing and genetic mapping are driving the identification of underlying causes of genetic variation by providing a 'road map' containing the locations of genes or markers. Therefore, research into sequencing and mapping the sugarcane genome could speed up the release of superior sugarcane cultivars by an efficient and effective selection process.

The sequencing of the first plant genome sequence of *Arabidopsis thaliana* (AGI, 2000) stimulated further genome sequencing and genomic-based research in crops. However, due to their relatively large genome size and complex genome structure, sequencing crops is much more difficult compared with model species. The genome of *A. thaliana* was generated by sequencing bacterial artificial chromosome (BAC) clones using Sanger sequencing technology (Feuillet et al., 2011), which relied on constructing a physical map. This process was tedious and time-consuming. In addition, Sanger sequencing has a low throughput and is relatively expensive (AGI, 2000; Feuillet et al., 2011). Subsequently the 'whole-genome shotgun' (WGS) strategy was developed and used to rapidly draft whole genome-sequences, including that of the first fruit crop, the grapevine (*Vitis vinifera*) (Jaillon et al., 2007), in which WGS using Sanger sequencing technology generated 6.2 million end reads representing an 8.4-fold coverage of the genome. In WGS, the target genome is fragmented for sequencing, and a consensus sequence is then assembled relying on computational algorithms. WGS does not require a physical map, making it much less time-consuming compared to a BAC-by-BAC strategy. The WGS strategy can be further enhanced by the addition of next-generation sequencing (NGS) technology, which can produce millions of sequencing reads using massive parallel sequencing systems. For example, using Illumina GAII system, 46.18 Gbp of high-quality short reads (<100 bp) were generated, representing 108.6-fold coverage of the entire watermelon genome (Guo et al., 2013). However, an extra challenge exists for WGS when it comes to sequencing large crop genomes, due to their high repetitive and transportable element DNA and/or ploidy levels (Feuillet et al., 2011). For example, the size of the maize genome is 2.3 Gbp and nearly 85% of the genome is composed of transposable elements (Schnable et al., 2009). To sequence the maize genome, a minimum tiling path of BAC clones were selected from an integrated physical and genetic map, and unique regions of BAC clones were further enriched for sequencing. Besides the large genome size and repetitive elements, the redundancy of the homo(eo)logous chromosome creates an additional difficulty for polyploid genome assembly.

Sequencing their corresponding diploid ancestor is one strategy used for polyploid genomes, such as in cotton (Gossypium) (Chen et al., 2007), Brassica (Wang et al., 2011) and peanut (Bertioli et al., 2016). Another strategy reported for sequencing polyploid (such as wheat) genomes is to isolate, sequence and assemble each chromosome (Marcussen et al., 2014). Recently, progress in third-generation sequencing (TGS) technology, which can generate significant long reads by reading single DNA molecules (Bleidorn, 2016), has dramatically alleviated the challenges of *de novo* genome assembly of complex genomes, such as sugarcane.

With the whole-genome reference available, genome re-sequencing can now be utilized effectively to discover sequence variations, construct genetic maps, identify QTL, estimate genetic diversity, understand genome evolution and do more (Huang et al., 2010; Li et al., 2009; Page et al., 2013; Smith et al., 2008; Varshney et al., 2009). This approach has long been suitable for species with a small genome, but has been a challenge for sugarcane, with its estimated genome size of up to 10 Gbp and a total of 100–130 chromosomes (D'Hont, 2005). To capture genome-wide sequence variations and develop genetic markers for species with complex genomes, it is feasible to sequence a subset of the genome after reducing the genome's complexity. Several methods that utilize NGS to detect sequence variations by sequencing a subset of genomes have previously been discussed (Davey et al., 2011). In this chapter, we review the difficulties associated with and strategies for sequencing and mapping sugarcane genomes, and highlight significant achievements in this area. In addition, we discuss future trends for this area of research.

2 Sequencing sugarcane genomes

2.1 Difficulties and strategies

Today's commercial sugarcane cultivars are derived from hybridization between the species *Saccharum officinarum* with high sugar content (SC) and the wild species *Saccharum spontaneum*. The commercial cultivars have a chromosome number of 100–130 with 70–80% of the chromosomes derived from the *S. officinarum* species, 10–20% from *S. spontaneum* and 5–17% from a recombination of these two species (D'Hont, 2005; D'Hont et al., 1996; Piperidis et al., 2001). The *S. officinarum* species is octoploid (2n = 8x = 80) and has no available diploid and tetraploid progenitors. *S. spontaneum* (2n = 4x-16x = 32-128), which shows excellent stress tolerance to sugarcane cultivars, varies widely in chromosome numbers (D'Hont et al., 1998; Roach, 1989). As a result of interspecific hybridization events, modern sugarcane cultivars combine the two polyploid ancestors' genomes. Consequently, the genome of modern sugarcane cultivars contains multiple homo(eo)logous chromosomes inherited from the two ancestors. The existence of the homo(eo)logous alleles is further complicated by close paralogs within genomes derived from duplication events. This may affect the accuracy of genome sequence assembly into haplotypes. For example, a functional gene in a sugarcane genotype can have more than ten different homo(eo)logous alleles per locus (Paterson et al., 2010; Souza et al., 2011) and maybe a few similar paralog alleles. Therefore, a strategy that can distinguish between different alleles and precisely assemble each allele and paralog is needed. Whole-genome sequencing for polyploid plant species has so far been hindered, especially for sugarcane lacking diploid ancestors.

The sequencing of the sugarcane genome is also limited by its large genome size, especially in modern sugarcane cultivars. The monoploid genome size of the French sugarcane cultivar R570 is estimated to be approximately 1 Gbp (D'Hont, 2005; Tomkins et al., 1999) in comparison with rice, whose genome size is ~400 Mbp (Goff et al., 2002; Yu et al., 2002). The average monoploid genome sizes of the two sugarcane ancestor species, *S. officinarum* and *S. spontaneum*, are 985 Mbp and 843 Mbp, respectively (Zhang et al., 2012). D'Hont and Glaszmann (2001) reported that the monoploid genome size of *S. spontaneum* is 750 Mbp, which is 80% of the size of the *S. officinarum* monoploid. The genome size of the monoploid in *S. officinarum* is around 930 Mbp, which is comparable to the gametic genome size of sorghum (730 Mbp) (Grivet and Arruda, 2002; Paterson et al., 2009). The genome size of the *Saccharum* species varies greatly. Genome size variation was investigated among 71 *Saccharum* spp. indicating genome sizes ranging from 7.50 to 8.55 Gbp for *S. officinarum*, and from 3.36 to 12.64 Gbp for *S. spontaneum* (Zhang et al., 2012). Overall, the genome sizes of sugarcane cultivars or of their ancestor genotypes are much larger than most sequenced crops (Table 1). In general, the monoploid sugarcane genome is quite large, as is the whole sugarcane genome size with its multiple haplotypes. Therefore, selection of a suitable genotype with a low ploidy level to sequence is helpful for reducing the genome's homo(eo)logous composition.

Table 1 Genome size of sequenced major crops

Crop name	Genome size (Mbp)	References
Sorghum	730	Paterson et al., 2009
Rice (*Oryza sativa* L. ssp. *japonica*)	420	Goff et al., 2002
Rice (*Oryza sativa* L. ssp. *indica*)	466	Yu et al., 2002
Maize	2300	Schnable et al., 2009
Potato	844	PGSC, 2011
Tomato	900	TGC, 2012
Soybean	1100	Schmutz et al., 2010
Wheat (*Aegilops tauschii*)	4360	Jia et al., 2013

Several sugarcane genome sequencing strategies have been proposed (Paterson et al., 2010; Souza et al., 2011). Initially, sequencing focused on the gene-rich or euchromatin regions, using a combination of BAC clone sequencing and WGS sequencing. The genetic map would further facilitate genome assembly (Souza et al., 2011). However, the existence of homo(eo)logs, paralogs, transposable elements or repetitive sequences complicates sugarcane genome assembly, particularly when NGS technology with short read outputs is applied to genome sequencing. The pure haploid genome was exploited to reduce the complexity of polyploids by reducing the ploidy level, thus minimizing the collapse of similar sequences caused by high levels of homo(eo)logs. Haploid sugarcane plants were successfully induced from anther culture using a range of genotypes from China, Hawaii and India (Fitch and Moore, 1996). The induced haploid plants had only half of chromosome number of the original parental clones subjected to anther culture. The whole process of genome sequencing can take advantage of a reduced ploidy level (Ming et al. Personal communication). With the advent of TGS technology, the much longer reads will greatly improve the accuracy of polyploid genome assembly.

2.2 Progress of genome sequencing

It has been a huge and daunting task to sequence the whole genome of sugarcane, owing to the challenges mentioned above. When the reference genome became publicly available for sorghum (Paterson et al., 2009), a diploid species closely related to sugarcane, scientists initially sampled sugarcane fragments for sequencing, which was then compared to the sorghum genome sequences. Microcolinearity between sugarcane and sorghum genomes was examined using 20 representative sugarcane BACs of cultivar R570 (Wang et al., 2010), which revealed extensive conservation of structure and sequence and a 95% sequence identity in the genic regions of the two species. Sequencing and comparison of eight BAC clones of modern sugarcane harbouring seven homo(eo)logous haplotypes revealed that the homo(eo)logous genes were highly conserved, and all of them were predicted as functional genes (Garsmeur et al., 2011). A brief profile of the sugarcane genome and the genomic composition of sugarcane were analysed by BAC-end sequences (BESs) of cultivar R570 (Kim et al., 2013). They found that out of 16 514 BESs longer than 200 bp, 13.5% were annotated as protein-coding sequences, 42.8% as repeated sequences and 11.7% did not match any sequences in public databases.

This representative BAC-based sequencing provided a glimpse of the composition and features of the sugarcane genome.

Due to species-specific features and chromosome rearrangements between species, creation of a reference genome for sugarcane itself remains critical. The French sugarcane cultivar R570 was selected as a priority for the Sugarcane Sequencing Initiative (SUGESI) (Souza et al., 2011). Following the strategy proposed by the SUGESI Consortium, BAC-based sequencing of sugarcane euchromatic regions was broadly applied (Garsmeur et al., 2011; Tomkins et al., 1999; Wang et al., 2010). A BAC library of the cultivar R570 was constructed containing 103 296 clones with a mean insert size of 130 Kbp, representing 14-fold monoploid genome coverage and 1.3-fold whole-genome coverage (Tomkins et al., 1999). This sugarcane BAC library was unable to distinguish between all homo(eo)logs and paralogs in the whole-genome scale (Grivet and Arruda, 2002; Paterson et al., 2010). Paterson et al. (2010) recommended a 'modified BAC-based approach' that would only assemble overlapping BAC regions without considering allelic variation of homo(eo)logs, which will generate a mosaic monoploid genome for sugarcane. Sequencing additional targeted regions should resolve the allelic variations between regions. Several groups around the world contributed to the BAC library sequencing. The French group led by D'Hont selected a core set of R570 BACs (5000 BACs) to generate a gene-rich part of a monoploid genome reference by utilizing whole-genome profiling technology provided by Keygene to analyse a set of 20 736 BACs from cultivar R570's sequence. This was funded through the International Consortium for Sugarcane Biotechnology (ICSB) (Garsmeur et al., 2015). In contribution to the French ICSB project in generating the monoploid genome, an Australian group led by Aitken selected more than 1000 R570 BAC clones for further sequencing with part of these BAC clones targeting QTL identified in Australian germplasm. A total of 31 Gbp of data with an average read length of 7282 bp was generated from the PacBio platform, which covered 73-fold of the complete genome sequence, representing approximately a 1000-fold of the monoploid genome (Aitken et al., 2015). To further improve the assembly of the R570 monoploid genome and annotation, recently, a new project led by Aitken in collaboration with scientists from Australia, France and the United States has been launched with the aim to generate the first sugarcane genome sequence of R570 by 2018/2019 (http://genome.jgi.doe.gov/SacoffStandDraft_FD/SacoffStandDraft_FD.info.html).

There have been several attempts to apply the WGS sequencing approach to sugarcane, which had been successfully applied in reference genome construction for almost all the species with a complete reference genome. However, no successful assembly was generated for sugarcane. For example, in Brazil, the new TruSeq Synthetic Long Read sequencing approach was reported to have sequenced a Brazilian sugarcane cultivar, SP80-3280. The overall sequences covered 4x–5x of the haploid genome (Riaño-Pachón et al., 2016; Riaño-Pachón and Mattiello, 2017). However, assembly of the sequence was hindered by the low coverage, resulting in an N50 of 8438-bp assembly. Nevertheless, 90% of gene space was covered, providing an important resource for the genic region studies of the genome. Recently, CP 96-1252, one of the top commercial sugarcane cultivars in Florida, USA, was subjected to WGS sequencing and 693 million paired-end reads were generated (Miller et al., 2017). Currently, the genome sequences are in the processes of assembly and annotation (Miller et al., 2017).

The mosaic monoploid reference genome is of great importance, since it provides a sketchy view of sugarcane genome structure. However, due to the heterozygosity of sugarcane and the existence of homo(eo)logous alleles, the collapsed monoploid reference genome will lack diverse allele information or haplotype details and further sequencing should be done for selected and interesting regions. A group in the United States and China led by Ming attempted to sequence sugarcane ancestor species including *S. spontaneum*, *S. officinarum* and *S. robustum* with a BAC-by-BAC sequencing approach (Souza et al., 2011) using TGS technology with efforts to resolve all the haplotypes. The genomes of these three species are expected to be publicly available in 2017 or 2018. With reduced complication from homo(eo)logs and reduction of heterozygocity, the genome sequence of ancestor species will greatly elucidate sugarcane chromosome organization and provide comprehensive genome references for the community.

Due to all the challenges associated with assembling sugarcane genomes, a reference genome for sugarcane is currently not available for the public to use. However, analysis of deep sequence data generated from target-enriched regions of the sugarcane genome makes it possible to investigate allelic variation in target regions, specifically in genic regions of sugarcane. The target-enrichment approach can select genomic regions of interest based on the sequence homology. All allelic variants within the targeted regions can be retrieved and deep sequenced. For example, allelic variations of parental accessions in a mapping population (IJ76-514 and Q165) were assessed from the probe-targeted genic region (Bundock et al., 2012) to develop markers for the mapping population. A variety of natural allelic variations including single nucleotide polymorphism (SNP), insertion deletion (InDel), copy number variation and present and absent variations among 12 representative genotypes were catalogued, using a solution-based target-enrichment sequencing approach (Song et al., 2016). Another gene-enrichment approach, methyl filtration with McrBC endonuclease digestion, was applied to the sugarcane genome (Grativol et al., 2014). Compared with unfiltered dataset, the methyl-filtered technology increased the coverage of gene regions by 134x and produced 1.5x more scaffolds and 1.7x more assembled sequences in length. Understanding the gene-rich regions is still important at this stage, since it is difficult to assemble the whole genome. Even more importantly, these sequence variations can be utilized for enriching the density of genetic maps and discovering the underlying causes of agronomic traits.

In addition to nuclear genome sequencing, chloroplast and mitochondria genomes in sugarcane were surveyed. Photosynthesis takes place in the chloroplasts and the chloroplast DNA of plants forms a circular double-stranded molecule with a genome size ranging from 120 to 160 Kbp (Asano et al., 2004). Chloroplast genomes of six sugarcane genotypes were retrieved by analysing low-coverage sequence data and the results revealed that *S. spontaneum* is less repetitive and more polymorphic than *S. officinarum*, with *S. hybrids* somewhere in between the two ancestor species (Berkman et al., 2014). Plant mitochondria, with a genome size ranging from 200 Kbp to 11.3 Mbp, play a key role in energy production and metabolism (Gualberto et al., 2014). PacBio RSII was applied in order to sequence an Asian cultivar's (Khon Kaen 3) mitochondrial genome, and two mitochondria circular chromosomes were assembled using the long read length technology (Shearman et al., 2016). Comparison with five commercial hybrids in the study supported the hypothesis that a species from New Guinea is the maternal origin of modern sugarcane cultivars.

2.3 Transcriptome sequencing

Before the sugarcane whole-genome and transcriptome reference sequences were fully assembled, many sugarcane genomic studies were focused on characterizing functional genes by surveying transcript sequences or transcriptome of certain tissues (Bower et al., 2005; Carson and Botha, 2000; Casu et al., 2003, 2004; Ma et al., 2004; Vettore et al., 2003). Transcriptome analysis has been an important tool for biological inquiry by providing high-throughput expression data and offers a useful means of accessing gene sequence information. In transcriptome, a collection of expressed sequence tags (ESTs) and their assembly provide a large number of candidate gene sequences that are involved in agronomic traits such as sucrose accumulation, stress tolerance, biomass yield, plant architecture, etc. (Kido et al., 2012; Souza et al., 2001). The availability of the large EST database enables association analysis of genes involved in specific traits in sugarcane as well as similarity assessment of sugarcane genes with those from other taxa (Ma et al., 2004). Comparative analysis based on the consensus sequences assembled from ESTs also enabled the identification of new putative genes (Vincentz et al., 2004). A group from South Africa began an EST collection for cultivar Nco376 (Carson and Botha, 2000). A huge number of ESTs have been produced and deposited in the EST database (Arruda, 2001; Bower et al., 2005; Casu et al., 2003, 2004; Ma et al., 2004; Vettore et al., 2003). The largest achievement in EST collection was the Brazilian ONSA consortium's Sugarcane EST Project (SUCEST) which contained 237 954 ESTs from 37 cDNA libraries derived from different organs and tissues sampled at different developmental stages (Simpson and Perez, 1998; Vettore et al., 2001; http://sucest.lad.ic.unicamp.br/en/). Cluster analysis of these SUCEST ESTs has revealed 43 141 sugarcane-assembled sequences (SASs) encoding putative transcripts (Vettore et al., 2003). After reassembling of SASs, 33 620 unique genes were identified, indicating that nearly 90% of the sugarcane expressed genes were tagged. Analysis of the SUCEST database allows a number of genes associated with specific pathways to be identified, such as those involved in protein metabolism, cellular communication and signal transduction, bioenergetics and stress responses. More than 3500 genes corresponding to signal transduction components were identified by taking advantage of the present SUCEST database (Souza et al., 2001).

Transcriptome analysis has been extensively used to investigate biological phenomena. Gene expression profiling analysis was carried out to identify genes related to traits or stresses. For example, cold-response genes were discovered by comparing the expression level of 1,536 ESTs of SP80-3280 between the cold-treated seedlings and the control using nylon filter arrays (Nogueira et al., 2003). Out of 34 cold-inducible genes, 20 were reported as novel genes (Nogueira et al., 2003). A specific stress response gene, an *S. spontaneum* homolog of an NOD26-like major intrinsic protein gene (SspNIP2), was selected and characterized through transcriptome analysis of hybrid CP72-1210 (cold susceptible) and *S. spontaneum* TUS05-05 (cold tolerant) (Park et al., 2015). A molecular model for how sugarcane responded to low temperature was proposed using the expression profile of cold-inducible genes from SP80-3280 combined with cold-related genes mined from SUCEST. These are involved in many different functions including signalling pathways, transcriptional regulation and other metabolic processes (Nogueira et al., 2003). This model will facilitate researchers to further characterize candidate genes and develop strategies for improving sugarcane cold tolerance. In another case, using cDNA microarray, transcriptional changes were studied in sugarcane roots with exogenous methyl jasmonate (MJ) treatment. The MJ-inducible transcripts in sugarcane roots have provided an indication of the biological processes triggered by the MJ treatment, which helps us understand transcriptional responses to pest and disease at the application of induced resistance (Bower et al., 2005).

Because sugarcane is a sugar and biofuel crop, the genes associated with cell wall metabolism and sugar accumulation are of critical interest to the sugarcane community

(Casu et al., 2003, 2007). The genes associated with sugar accumulation have been assessed by ESTs and microarray analysis to investigate the regulation of metabolic enzymes and sugar transporters in the sugarcane stem (Casu et al., 2003, 2004, 2007; Watt et al., 2005). The expression profiles of two genotypes IACSP04-065 and IACSP04-627 were analysed, contrasting them for lignin content and showing more than 2000 differentially expressed transcripts between the two genotypes, providing molecular information on lignin biosynthesis in sugarcane (Vicentini et al., 2015). In another study using the Affymetrix GeneChip array, 119 genes were identified as being differentially expressed at different developmental stages, and the members of cellulose synthase (CesA) and cellulose synthase-like (Csl) gene families were further characterized in this study (Casu et al., 2007). The transcriptome of sugarcane comprises transcripts from homo(eo)logous chromosomes, owing to the polyploid nature and interspecific hybridization origin of sugarcane cultivars. Since the sugarcane GeneChip array for transcript profiling was developed using the entire collection of sugarcane ESTs, it is a powerful resource for resolving the expression of different homo(eo)logous alleles (Casu et al., 2007). Using this array, Casu et al. (2007) dissected each homologous allele in gene families encoding functions in cell wall synthesis. This has the potential to increase our understanding of the influence of homo(eo)logous gene dosage on transcript composition.

The sugarcane ESTs provide an extensive sequence resource to develop genetic markers, such as SNPs and simple sequence repeats (SSRs), for genetic mapping and linkage or association analysis. In addition, the large number of ESTs have proven to be a valuable resource for investigating allelic variation within gene coding regions. For example, the SUCEST database can be utilized to develop SSR and SSR-derived restriction fragment length polymorphism (RFLP) for QTL mapping (Pinto et al., 2004; Silva and Bressiani, 2005). Differentially expressed candidate genes from maturing stems of sugarcane were used to investigate SNPs and their frequency, and each SNP locus was quantified across nine sugarcane genotypes. These results demonstrated that SNP markers can be used to distinguish genotypes in a mapping population with a copy number resolution for a specific genotyped locus (Cordeiro et al., 2006). Sequence polymorphism of three genes belonging to the alcohol dehydrogenases (Adh) family was investigated among four different cultivars using 217 transcripts recovered from approximately 250 000 sugarcane ESTs. A total of 37 SNPs inside three Adh genes were identified and eight InDels were observed (Grivet et al., 2003). These results demonstrated that the EST database served as a valuable source for sequence polymorphism discovery and genetic marker development (Grivet et al., 2003).

3 Genetic linkage mapping

Genetic linkage mapping is a technique that determines the linear order of genetic loci or markers in a chromosome based on chromosomal recombination frequency during meiosis. Developing a mapping population containing a significant number of recombination events for a large number of genetic loci is critical for successful linkage map construction. DNA markers representing genetic loci can be used to discern polymorphisms between parents, which can then be used to genotype the recombinant individuals in the population. Computer programmes are then used to analyse the genotypic data to generate genetic linkage maps. Genetic maps are not only an important tool for mapping genes or genomic regions controlling traits of interests, but also a fundamental tool for improving genome assembly, which is especially critical for species with a complex genome like sugarcane.

To date, genetic linkage mapping studies in sugarcane have primarily focused on studying pairing behaviours of chromosomes during meiosis, understanding QTL and structural differences between *S. officinarum* and *S. spontaneum* genomes and aligning homo(eo)logs within or between species. None of these studies attempted to capitalize on the availability of genetic map information on genome assembly in sugarcane. Although high-resolution genetic maps have been successfully integrated with sequence maps to achieve a high-quality reference genome in other species such as Eucalyptus (Bartholome et al., 2015), cotton (Zhao et al., 2012), Japanese eel (*Anguilla japonica*) (Kai et al., 2014), barley (Mascher et al., 2013) and physic nut (Wu et al., 2015), it is still in its infancy with respect to sugarcane. With its large and complex genome, extensive repetitive elements (Alkan et al., 2011) and numerous near-identical copies, which are difficult to separate into each homo(eo)logous chromosome by using NGS short reads, researchers have tended to assemble sugarcane sequences into a single collapsed contig (Mascher and Stein, 2014). Current methods of NGS produce contigs and scaffolds of limited length, leaving thousands of sequence gaps. This is why high-resolution linkage maps are useful, offering valuable tools for correctly placing scaffolds on the chromosomes, while correctly ordering SNP and InDel markers within genome assemblies. The process of bringing contigs into a linear order using SNPs and InDels generated from NGS methods is called anchoring (Lewin et al., 2009). The contigs are anchored into a pseudo-molecule based on the position of marker source sequences assembled on contigs and the locations corresponding to markers on a high-quality and high-resolution linkage map (Kai et al., 2014).

The first ever effort to integrate linkage maps with sequence maps to facilitate genome assembly is under way and conducted by Aitken et al. (2016) using genotyping data obtained from an SNP chip (Aitken et al., 2014). The team is anchoring scaffolds onto sugarcane chromosomes in accordance with the order of markers in the high-quality and high-resolution genetic map (Kai et al., 2014; Lewin et al., 2009). Coupling genetic map information with *de novo* assembly provides more information on the genome structure than either process in isolation. This is because genetic maps are accurate at the chromosomal scale, but inaccurate at fine scale due to infrequent recombination, while the opposite is true for *de novo* assembly (Fierst, 2015). Thus the challenges of genome assembly in sugarcane with a large number of small contigs can be alleviated by integrating genetic maps with whole-genome sequences.

3.1 Population and marker systems for genetic mapping

It is difficult to attempt genetic linkage mapping using mapping populations such as F2, recombinant inbred lines, inbred-backcross and doubled-haploid (DH) lines in *Saccharum* species because they are heterozygous, clonally propagated and suffer from inbreeding depression (Alwala et al., 2010). A segregated F1 population can be used as a pseudo-testcross population to construct linkage maps for one or both parents (Aitken et al., 2007; Edmé et al., 2006; Ming et al., 1998), as first proposed by Wu et al. (1992). Similarly, the progeny derived from selfing a clone can be used as a pseudo-F2 mapping population (Alwala and Kimbeng, 2010).

The key to genetic map construction is accurately identifying the genotypes at each marker locus of all individuals in the mapping population. Genotype calling in diploids is straightforward owing to disomic inheritance. Thus, with the advent of DNA markers, many genetic linkage maps have been constructed in diploids using traditional mapping populations. However, accurately identifying sequence variations and their dosages has been a challenge for polyploids, owing to a large number of possible genotypes at every single locus. This is because sugarcane's polysomic inheritance creates complex segregation ratios that are difficult to classify accurately. In addition, there are no genetic models to help interpret the complex segregation ratios. Difficulties of genetic linkage mapping in polysomic polyploids arise due to: i) a large number of segregating genotypes; ii) many possible modes of gamete formation due to the random disjunction of multivalent and different frequencies of multivalent formation and iii) difficulties in identifying different gametic genotypes by a single backcross to a tester (Fisher, 1949). In order to accurately genotype the genetic loci with a high level of confidence, Wu et al. (1992) proposed to use single-dose (SD) markers for genetic map construction in polyploid species, because the SD markers have a simple and distinct inheritance mode. Indeed, SD markers are present in a single copy and will produce a segregation ratio of 1:1 (presence: absence) or 3:1 in the progeny depending on their presence in either one or both parents of the mapping population, respectively, when the marker is dominant. In addition, SD markers can aid in discriminating auto- and allopolyploids, and in identifying preferential chromosome pairing (Alwala and Kimbeng, 2010). A complete preferential pairing of homologous chromosomes should yield equal frequencies of both repulsion- and coupling-phase linkages. Conversely, SD markers cannot be used to distinguish duplicated genomic regions for precise mapping, which is a hard task for diploid species mapping as well. In addition, it is difficult to detect repulsion-phase markers when homologous chromosomes are pairing completely randomly in a population of less than 750 individuals (Wu et al., 1992). The linkage phase of markers could be detected based on their segregation ratios in the progeny. The segregation ratio would be 1:1 (repulsion: coupling) for allopolyploids (disomic), while 0.25:1 for an autotetraploid or approaching 0:1 for even higher ploids (polysomic) (Qu and Hancock, 2001). The deviation of segregation ratios from 1:1 or 0:1 indicates partial preferential or random pairing of chromosomes (Wu et al., 1992). In addition, the SD markers make up the majority (~70%) of total polymorphic loci in the sugarcane genome, as revealed by linkage mapping studies (Aitken et al., 2005; Garcia et al., 2006; Silva et al., 1993). This means that applying SD markers for genetic mapping is not a huge sacrifice.

Although the use of double-dose (DD) or higher dose markers may provide better coverage of a genome, it may also pose a challenge to accurate discernment of segregation ratios. The segregation ratios for DD or higher dose markers are likely to be > 1:1 irrespective of disomic (allopolyploids) or polysomic inheritance. Markers are expected to have segregation ratios of 11:3, 13:1 and 69:1 in the case of an auto-octoploid biparental population for DD, triple-dose and quadruple-dose markers. At least 13, 41 and 209 progeny are required to observe one homozygous recessive individual, respectively (P = 0.05). This makes the detection of high-dosage markers unrealistic if the mapping population size is not large enough. However, multiple-dose markers can be used to re-assign linkage groups (LGs) as homologous groups (HGs), which adds extra challenges to constructing a genetic map of sugarcane. In addition, SD markers can only detect linkages in the coupling phase because detection of linkages in the repulsion phase requires extremely large populations (Wu et al., 1992).

Several kinds of DNA markers have been used to develop genetic linkage maps in sugarcane, such as amplified fragment length polymorphism (AFLP), RFLP, random amplified polymorphic DNA (RAPD), SSR, diversity array technology (DArT), sequence-related amplified polymorphism (SRAP), target region amplification polymorphism (TRAP) and SNP. Recently, Aitken et al. developed a 345k SNP chip on sugarcane and used it to screen a population of 367 sugarcane clones, of which half were selected parental clones from the Australian core breeding programme and half were commercial cultivars (Aitken et al., 2016). This SNP array will be of great value in speeding up sugarcane genotyping. SD markers have been used for genotyping the mapping populations in the majority of linkage mapping studies conducted so far (summarized in Table 2).

3.2 Linkage maps constructed in *S. spontaneum, S. officinarum, S. robustum* and hybrids

A linkage map is an important tool for dissecting the genomic constitution of an organism. The efforts made by various researchers in constructing linkage maps in *S. spontaneum, S. officinarum, S. robustum* and *Saccharum* spp. hybrids have been summarized in Table 2. The construction of linkage maps delineating DNA markers and genes controlling important agronomic traits will facilitate genome assembly, comparative genomics, QTL analysis and sugarcane cultivar improvement. However, highly saturated linkage maps are still not available for sugarcane. Although a large number of DNA markers have been identified, genome coverage (60% best so far) is still low because sugarcane has so many chromosomes. Nonetheless, with the advent of high-throughput genotyping platforms, there is a future prospect of generating linkage maps with saturated genome coverage in the future. The status of genetic map construction and the features of *S. spontaneum, S. officinarum, S. robustum* and *Saccharum* spp. hybrids are further discussed below.

3.2.1 Genetic maps of *Saccharum spontaneum*

Saccharum spontaneum is a major potential source for the enhancement of sugarcane cultivar biomass production and therefore sparked interest among early sugarcane geneticists eager to understand the crop's genomic organization, constitution and inheritance. The first ever genetic linkage map constructed for *S. spontaneum* 'SES 208' (2n = 64) used a mapping population comprising 88 progeny derived from a cross between 'SES 208' and 'ADP 85-0068', a DH developed by anther culture of 'SES 208' and using SD RFLP and RAPD markers (Al-Janabi et al., 1993). This was further improved and later saturated with more markers (da Silva et al., 1995; Silva et al., 1993). However, none of these studies detected single repulsion-phase markers, which suggested that *S. spontaneum* displays polysomic inheritance. This is further supported by other later studies (Edmé et al. 2006; Ming et al., 2002). Ming et al. (1998) developed linkage maps using RFLPs in two interspecific F1 populations derived from crosses between *S. officinarum* 'Green German' (2n = 97-117) × *S. spontaneum* IND 81-146 (2n = 52-56) and *S. officinarum* 'Muntok Java (2n = 140) × *S. spontaneum* 'PIN 84-1 (2n = 96). Later, Ming et al. (2002) further refined the linkage map of *S. spontaneum* by developing a consensus map of *Saccharum* species using a sorghum linkage map as a template. Edmé et al. (2006) also studied the genome structure, which revealed similarities in some sections of *S. officinarum* and *S. spontaneum* genomes for potential recombination, as well as differences in terms of segregation distortion, chromosomal fusion, translocations and rearrangements. Segregation distortion depends on the relatedness or divergence of parents (Tanksley and Nelson, 1996) and could be caused by 2n and n transmission of gametes to the progeny from *S. officinarum* and *S. spontaneum*, respectively (Alwala et al., 2008), or by artificial selection of traits;

Table 2 Genetic linkage mapping efforts in sugarcane

Mapped species	Mapping population	Parent lines	Mapped parent	Marker system used	Map length (cM)	References
S. spontaneum	Ss × Ss	SES208 × ADP85-0068	SES208 (2n = 64)	RAPD	1500	Aljanabi et al., 1993
	Ss × Ss	SES208 × ADP85-0068	SES208 (2n = 64)	RFLP, RAPD	3300	da Silva et al. 1993, 1995
	So × Ss	Green German × IND-146	IND81-146 (2n = 52-56)	RFLP	2172	Ming et al., 1998
	So × Ss	Green German × IND-146	IND81-146 (2n = 52-56)	SSR	614	Edmé et al., 2006
	So × Ss	PIN84-1 × Muntok Java	PIN84-1 (2n = 64)	RFLP	1395	Ming et al., 1998
	So × Ss	LA Striped × SES 147B	SES 147B (2n = 64)	AFLP, SRAP, TRAP	1491	Alwala et al., 2008
S. officinarum	So × Sr	LA Purple × Molokai 5829	LA Purple (2n = 80)	RAPD	1152	Mudge et al., 1996
	So × Sr	LA Purple × Molokai 5829	LA Purple (2n = 80)	RAPD, RFLP, AFLP	1881	Guimaraes et al., 1999
	So × cultivar	IJ76-514 × Q165	IJ76-514 (2n = 80)	AFLP, SSR, RAPD	4906	Aitken et al., 2007
	So × Ss	LA Striped × SES 147B	LA Striped (2n = 80)	AFLP, SRAP, TRAP	1732	Alwala et al., 2008
S. robustum	So × Sr	LA Purple × Molokai 5829	Molokai 5829 (2n = 80)	RFLP	NA	Ming et al., 1998
	So × Sr	LA Purple × Molokai 5829	Molokai 5829 (2n = 80)	RFLP, RAPD, AFLP	1189	Guimaraes et al., 1999
Hybrids/commercial cultivars	Hybrid × Hybrid	PIN84-1 × Muntok Java	Muntok Java (2n = 140)	RFLP	1472	Ming et al., 1998

Mapped species	Mapping population	Parent lines	Mapped parent	Marker system used	Map length (cM)	References
Hybrid × Hybrid	Green German × IND81-146	Green German (2n = 97-117)	RFLP	2466	Ming et al., 1998	Hybrid × Hybrid
Green German × IND81-146	Green German (2n = 97-117)	SSR	1180	Edmé et al., 2006	Selfed population	R570
R570 (2n = 107-117)	RFLP	2008	Grivet et al., 1996	Selfed population	R570	R570 (2n = 107-117)
RFLP	5894	Hoarau et al., 2001	Selfed population	R570	R570 (2n = 107-117)	RGA
	Rossi et al., 2003	So × cultivar	IJ76-514 × Q165	Q165 (2n = 115)	AFLP, SSR, RAPD	9053
Aitken et al., 2005	Cultivar × Cultivar	Q117 × MQ77-340	Q117	AFLP, SSR	3167	Reffay et al., 2005
Cultivar × Cultivar	MQ77-340	3582	Hybrid × Ss	Green German × IND81-146	Green German (2n = 97-117)	
SSR	1180	Edmé et al., 2006	Cultivar × Cultivar	R570 × MQ76-53	R570	RFLP, SSR, AFLP
3144	Raboin et al., 2006	Cultivar × Cultivar	R570 × MQ76-53	MQ77-340	RFLP, SSR, AFLP	4329
Raboin et al., 2006	Cultivar × Cultivar	SP80-180 × SP80-4966	Integrated map	RFLP, SSR, AFLP	2602	Garcia et al., 2006
Cultivar × Cultivar	SP80-180 × SP80-4967	Integrated map	EST-AFLP/RFLP	6261	Oliveira et al., 2007	Selfed population
LCP85-384	LCP85-384	AFLP, SSR, TRAP	5617	Andru et al., 2011	Selfed population	LCP85-384
LCP85-384	AFLP, SSR, TRAP	7406	Liu et al., 2016	Hybrid × Cultivar	IAC66-6 × TUC717	Integrated map

(Continued)

Table 2 (*Continued*)

Mapped species	Mapping population	Parent lines	Mapped parent	Marker system used	Map length (cM)	References
AFLP, EST-SSR, retrotranspon-based markers	4843	Palhares et al., 2012	So × Cultivar	IJ76-514 × Q165	Q165	DArT, SNP, AFLP, SSR
9774	Aitken et al., 2014	*Erianthus/Saccharum* hybrid × Cultivar	YC96-40 × CP84-1198	YC96-40	AFLP, SSR	1209
Chen et al., 2015	*Erianthus/ Saccharum* hybrid × Cultivar	YC96-40 × CP84-1199	CP84-1198	AFLP, SSR	2283	Chen et al., 2015
Erianthus/Saccharum hybrid × Cultivar	YCE01-116 × NJ57-416	YCE01-116	AFLP, SSR	973	Chen et al., 2015	*Erianthus/ Saccharum* hybrid × Cultivar
YCE01-116 × NJ57-416	NJ57-416	AFLP, SSR	1955	Chen et al., 2015	Cultivar × Cultivar	SP71-1088 × H57-5028
Integrated map	SNP, SSR, TRAP	3268	Balsalobre et al., 2017	Hybrid × Cultivar	CP95-1039 × CP88-1762	CP95-1039
SNP	4224	Yang et al., 2017	Hybrid × Cultivar	CP95-1039 × CP88-1763	CP88-1762	SNP
4373	Yang et al., 2017					

Ss = *S. spontaneum*, So = *S. officinarum*, Sr = *S. robustum*.

this has practical significance in introgression breeding (Cui et al., 2015). The information on the location of structural differences between the two genomes will facilitate mapping QTL for traits of agronomic importance in sugarcane.

The information on recombination, segregation distortion, chromosomal locations and chromosomal rearrangements in *S. spontaneum* could be useful for interpreting the genome structure variations of *Saccharum* spp., and also perhaps for dissecting its genomic constitution. This will facilitate the transfer of biomass and resistance-related alleles into sugarcane cultivars.

3.2.2 Genetic maps of *Saccharum officinarum*

Saccharum officinarum, a major contributor to modern sugarcane cultivars, has a high sucrose content. It was used in backcross in breeding programmes called nobilization to increase the sucrose content of hybrids after initial crosses with *S. spontaneum* (Edmé et al., 2005; Roach, 1989). Thus it was imperative to understand its genome structure and chromosomal inheritance in modern sugarcane cultivars. The first ever study to determine the chromosomal inheritance of *S. officinarum* was conducted by Al-Janabi et al. (1994) in 44 progeny from a cross between meiotically regular (strict bivalent pairing) *S. officinarum* 'LA Purple' (2n = 80) and *S. robustum* 'Mol 5829' (2n = 80) (Table 2). The linkage analysis of 55 SD markers revealed both repulsion- and coupling-phase linkages, suggesting that *S. officinarum* does not behave like *S. spontaneum*. *S. officinarum* displayed a mixture of disomic and polysomic chromosomal inheritance. Mudge et al. (1996) and Guimares et al. (1999) further saturated the map and confirmed sugarcane's incomplete polysomal inheritance, as did many other later studies. Additional maps constructed with different clones, such as 'LA Striped' (Alwala et al., 2008) and 'IJ76-514' (Aitken et al., 2007) exhibited repulsion-phase linkages which suggested that preferential pairings among homo(eo)logs exist in *S. officinarum*.

3.2.3 Genetic maps of *Saccharum robustum*

The first ever linkage analysis of *S. robustum* was conducted by Al-Janabi et al. (1994) in 'Mol 5829'. Because repulsion-phase linkage was detected using as few as 44 progeny in *S. robustum*, the possibility of complete polysomic inheritance in *S. robustum* could easily be ruled out. Ming et al. (1998) and Guimares et al. (1999) attempted to develop a denser linkage map of *S. robustum* with more markers. However, neither complete preferential pairing nor complete random pairing of homologous chromosomes were observed, which suggested that *S. robustum* genomes are neither fully polysomic nor disomic during meiosis (just like *S. officinarum*); it can be deduced from this that *S. robustum* is an immediate progenitor of *S. officinarum*.

3.2.4 Genetic maps of sugarcane hybrids and commercial cultivars

With the agronomic importance of commercial cultivars (interspecific hybrids), significant effort has been invested in constructing genetic maps of sugarcane hybrid cultivars. This effort is critical for further mapping the loci that control important agronomic traits, as well as for understanding the transmission genetics of *S. spontaneum* (autooctoploid) and *S. officinarum* (allooctoploid) genomes after hybridization.

Ming et al. (1998) developed genetic maps of two clones, namely 'Green German' (2n = 97-117) and 'Muntok Java' (2n = 140), which were hybrids, though were called *S. officinarum*. The maps were compared to a high-density sorghum linkage map and revealed a high level of collinearity between the two species except for two chromosomal fusions and rearrangements. For the first time, this study provided a base of valuable knowledge for applying the sorghum genome as the reference for genomic studies of sugarcane. Two popular sugarcane cultivars have been used extensively for linkage mapping studies: 'R570' (2n = 115), a French cultivar with 80% of its chromosomes from officinarum, 10% from *S. spontaneum* and 10% recombinant chromosomes (D'Hont et al., 1996), and 'Q165' (2n = 110), an Australian cultivar containing 82–83 (75%) of the chromosomes from the *officinarum* genome, 15–17 (15%) from the *S. spontaneum* genome and 11–12 (11%) recombinant chromosomes from the two genomes. The first attempt to construct a linkage map in a selfed population of 'R570' was undertaken by Grivet et al. (1996) and further improved by Hoarau et al. (2001), Rossi et al. (2003) and Raboin et al. (2006). Complete polysomic inheritance was ruled out by the studies, as several LGs harboured markers in repulsion phase. No significant structural differences were detected between *S. officinarum* and *S. spontaneum* genomes, which could be due to the low marker coverage or relative independent segregation between the two genomes. Aitken et al. (2014) developed a comprehensive linkage map using 227 F1 progeny from a cross between the cultivar 'Q165' and a *S. officinarum* accession 'IJ76-514'. The linkage map for 'Q165' consisted of 2267 markers distributed across 160 LGs, and spanned a total length of 9774 cM, the most comprehensive genetic linkage map of sugarcane so far. The linkage study detected a similar level of repulsion (19%) in 'Q165' to that of 'R570' (21%) discovered in Raboin et al. (2006). The limited preferential pairing of chromosomes suggested that these cultivars did not display complete disomic or polysomic inheritance. The linkage map for 'Q165' was partially aligned to the 'R570' genetic map developed by Rossi et al. (2003) and Raboin et al. (2006) using common SSR markers. Importantly, except for HG5, each HG in the 'Q165' map was aligned to a single HG in the 'R570' map.

Genotyping-by-sequencing (GBS) has recently been used for linkage map construction on two commercial Brazilian cultivars (Balsalobre et al. 2017). A total of 223 LGs were generated from 993 markers and 48 LGs were further assembled into 18 HGs. Yang et al. (2017) used GBS and constructed high-density genetic maps for sugarcane hybrid/cultivar, containing 2453 markers forming 162 LGs with a total map length of 4224.4 cM for CP95-1039 and 2154 markers forming 142 LGs with a total map length of 4373.2 cM for CP88-1762. Based on sequence information of mapped markers, ~70% LGs were assembled into HGs. Use of high-throughput genotyping has cut down the time required to develop large numbers of SNPs and InDels for linkage mapping and thus has helped improve genome coverage. Construction of sugarcane linkage maps using reduced representation GBS will enable rapid construction of high-density genetic maps in sugarcane, which will significantly advance candidate gene analysis of economically important traits.

4 Mapping simply inherited and complex traits

4.1 Mapping simply inherited traits

Simply inherited traits are also referred to as qualitative or Mendelian traits, such as the seed and pod colour of pea plants in Gregor Mendel's work (Mendel, 1866). They are controlled by one or a few genes, and none of those is significantly influenced by the environment (Levin and Schaffer, 2013). Unlike complex traits, simply inherited traits typically segregate following Mendelian ratios, and can be scored qualitatively (Kuhl et al., 2011). Due to its large and odd number of chromosomes, as well as its polysomic inheritance, simply inherited traits are rarely found in sugarcane. So far, only a couple of simply inherited traits have been identified, such as resistance to brown rust (Daugrois et al., 1996) and stalk colour (Raboin et al., 2006).

A linkage or genetic map serves as a 'road map', with markers providing location information. This is a prerequisite for mapping any trait. However, when no genetic map is available, near-isogenic lines and bulked segregant analysis can be used to tag molecular markers for many simple traits of interest in various plant species, such as tomato (Thomas et al., 1995; Young and Tanksley, 1989), maize (Yang et al., 2004) and bean (Blair et al., 2006). Due to high inbreeding depression, obtaining inbred lines in sugarcane is not realistic, so F1 or first progeny are used widely in the construction of genetic maps for mapping simply inherited traits.

The earliest report on a simply inherited trait in sugarcane was on sugarcane's resistance to brown rust. Several molecular markers were associated with a major sugarcane brown rust-resistant gene, which was identified by using the self-pollinated progeny of the French cultivar 'R570'. The segregation of brown rust resistance was observed in a clear-cut 3:1 (resistant: susceptible) ratio (Daugrois et al., 1996). This single dominant resistance gene was linked with an RFLP marker, revealed by probe CDSR29, at 10 cM. In 2000, the rust-resistant gene (*Bru*1) was identified in a co-segregating group, consisting of 19 RFLP markers, of the R570 map, while the marker CDSR29 was closest to the *Bru*1 gene (Asnaghi et al., 2000). In 2004, eight AFLP markers were found surrounding *Bru*1 within an interval of 10 cM (Asnaghi et al., 2004). The two closest AFLP markers were located on either side of the gene with 1.9 and 2.2 cM, respectively. In 2008, a high-resolution map was developed covering two molecular markers located at 0.28 cM and 0.14 cM on each side of *Bru*1, which was co-segregating with 13 markers (Le Cunff et al., 2008). These fine mapping results provided reliable markers for employing MAS in the breeding programmes.

More recently, through use of a panel of 380 modern cultivars, 22 markers linked to *Bru*1 within an interval of 8.2 cM have been identified (Costet et al., 2012a). Of these linked markers, R12H16 and 9O20-F4 co-segregated completely with *Bru*1 and were completely absent in susceptible accessions. Genotyping *Bru*1A using the 9O20-F4 marker requires enzyme restriction and gel electrophoresis, which is time-consuming and expensive for genotyping large numbers of samples (McCord and Migneault, 2015). Later, two probes, *Bru*1A1 and *Bru*1A2, were developed to detect the *Bru*1 locus using an unlabelled probe melting technique, which showed a 94.7% successful rate after genotyping 344 sugarcane clones, and 100% consistency between *Bru*1A and unlabelled probe genotyping (McCord and Migneault, 2015). The development of precise markers to diagnose the *Bru*1 locus has built a solid foundation for MAS in sugarcane breeding programmes.

Raboin et al. (2006) constructed two genetic maps by using a sugarcane segregation population (198 progeny) between R570 and MQ76-53. An AFLP marker was found associated with red stalk colour at 6.5 cM and was mapped on the MQ76-53 map using interval marker analysis. Meanwhile, the other major gene controlling sugarcane brown rust resistance, *Bru*2, was also localized on MQ76-53 (Raboin et al., 2006). These two successful examples of identifying traits controlled by a single gene or a major gene suggest that identification of new major genes is achievable in species with a complex genome structure, such as sugarcane.

4.2 QTL mapping of complex traits

Complex traits are also known as polygenic traits or quantitative traits, which are controlled by multiple genes as well as being influenced by the environment (Tanksley, 1988). Usually, these traits are continuously distributed in a segregating population. For example, crop yield (Huang et al., 2016; Pastina et al., 2012; Rami et al., 1998; Singh et al., 2013) and some pathogen resistances (Costet et al., 2012b; Klosowski et al., 2013; Tabien et al., 2002) are complex traits. Genetic variations in complex traits result from the segregation of numerous QTL, in which each explains a portion of the phenotypic variation (Mackay, 2001). QTL mapping using biparental population is the process of constructing linkage maps and identifying genomic regions associated with traits of interests (Collard et al., 2005).

QTL mapping in sugarcane is a challenge. Due to its high polyploid genome and heterozygosity, multiple alleles (up to 12 alleles) coexist for any one gene in sugarcane. Only when an allele exceeds the average effect of all other segregating alleles in the genetic background can it be perceived (Raboin et al., 2006). QTL mapping has proven to be an effective method for discovering the loci underlying important traits in sugarcane, such as disease resistance (Aitken, 2016; Aljanabi et al., 2007; Costet et al., 2012b; Nibouche et al., 2012) and yield-related traits (Aitken et al., 2008; Alwala et al., 2009; Pastina et al., 2012; Singh et al., 2013) (Table 3). The performance of QTL detection is usually influenced by interactions among QTL, harvest year and location. For example, Pastina et al. (2012) evaluated cane yield (CY), sugar yield, fibre percentage and sucrose content in progeny from a biparental cross at two locations for three consecutive harvest years. A total of 30 QTL showed interactions of QTL by harvest year, QTL by location and QTL by harvest year and by location.

In 2007, a genetic linkage map of the sugarcane cultivar M134/75 was constructed by evaluating 227 segregating clones for yellow spot resistance. These were generated by a biparental cross between M134/75 (resistant) and R570 (susceptible). Using single marker analysis, a single major QTL was shown to be linked to an AFLP marker. This QTL could explain 23.8% of the phenotypic variation of sugarcane yellow spot resistance (Aljanabi et al., 2007). In another study, focusing on sugarcane biomass and sucrose content-related traits, 46 QTL have been identified using the interval mapping approach, with eight QTL for sucrose content, 11 for fibre percentage, 13 for CY and 14 for sugar yield (Pastina et al., 2012).

By using 207 segregating progeny derived from a cross between two elite sugarcane cultivars, Co86011 and CoH70, Singh et al. (2013) identified 24 regions harbouring 31 QTL for sugar yield-related traits using composite interval mapping analysis. This explained 9–19% of phenotypic variation. With the enrichment of genetic and genomic information, additional markers will be developed and more QTL for important sugarcane traits will be identified; it will then be possible to apply these in sugarcane breeding programmes for MAS.

Association mapping is an alternative to traditional linkage analysis for QTL detection (Zhu et al., 2008). By using a natural population with historical recombination, association mapping offers higher mapping resolution and greater allele numbers, and requires less research time compared with cross populations (Yu and Buckler, 2006). Association between DNA markers and disease resistance was investigated in sugarcane, and 26–59% of the phenotypic variation could be explained by several significant markers (Wei et al., 2006). Fartek et al. (2011) also utilized association mapping to study sugarcane yellow leaf virus disease on 344 sugarcane cultivars with AFLP and DArT markers, and identified

Table 3 Mapping of complex traits in sugarcane

	Traits	Mapping population	Progeny size	Marker type	Effect (explained phenotypic variation)	Linked markers	References
Disease resistance	Smut resistance	BC1 *S. spontaneum* clone and a sugarcane cultivar		AFLP, SSR, DArT	36–45%	NA	Aitken, 2016
	Yellow spot resistance	M134/75 (r) × R570 (s)	227	AFLP, SSR	23.8%	1 QTL linked with an AFLP and a SSR	Aljanabi et al., 2007
	Spotted stem borer (SSB) resistance	Selfing of R570	147	AFLP, SSR, RFLP	6–10%	9 QTAs	Nibouche et al., 2012
	Yellow leaf virus resistance	R570 × MQ76-53	196	AFLP, SSR, RFLP	32%	1 major QTA	Costet et al., 2012b
Yield related	Sugar content	Green German × IND 81-146;PIN84-1 × Muntok Java	264;239	RFLP	3.6–20.6%	36 regions	Ming et al., 2001
	Plant height, stalk number, stalk diameter and brix	R570 × R570	295	AFLP	30–55%	45 additive S-QTAs, 14 M-QTAs, and 41 independent di-genic interactions	Hoarau et al., 2002
	Sugar yield, pol, stalk weight, stalk number, fibre content, ash content	Green German × IND 81-146;PIN84-1 × Muntok Java	264;239	RFLP	3.8–16.2%	102 regions (61 linked, 41 unlinked)	Ming et al., 2002

(Continued)

Table 3 (*Continued*)

Traits	Mapping population	Progeny size	Marker type	Effect (explained phenotypic variation)	Linked markers	References
Brix and pol	Q165 × IJ76-514	230	AFLP, SSR	3–9%	37 QTL	Aitken et al., 2006
Stalk weight, stalk diameter, stalk number, stalk length, total biomass	Q165 × IJ76-514	227	AFLP, SSR	4–10%	27 regions	Aitken et al., 2008
Brix, pol	Louisiana Striped × 'SES 147B'	100	AFLP, SRAP, TRAP	22.1–48.4%	11 QTL	Alwala et al., 2009
Cane yield, sugar yield, fibre per cent, and sucrose content	SP80-180 × SP80-4966	100	RFLP, SSR	NA	46 QTL	Pastina et al., 2012
Seedling brix, brix, sucrose per cent in juice, stalk number, stalk length, stalk diameter, internodes, number of green leaves	Co86011 × CoH70	207	SSR	2.4–19%	31 QTL	Singh et al., 2013

16 significant markers and 24 significant markers, respectively, from two locations. Later, Débibakas et al. (2014) performed a genome-wide association study on sugarcane yellow leaf disease in 189 samples of sugarcane breeding germplasm by using 3949 DArT and AFLP markers. Of these, six markers were found to be associated with resistance to the disease, explaining 9–14% of the disease variation, respectively. Apart from disease resistance traits, association mapping studies also produced results on yield-related traits (Gouy et al., 2015), such as stalk height, stalk number, leaf chlorophyll content, leaf chlorophyll florescence, brix, fresh weight (Todd et al., 2014), cane weight and tillers per plant (Bilal et al., 2015). Racedo et al. performed an association analysis on CY and SC, which revealed 43 markers significantly associated with CY and 38 markers significantly associated with SC (Racedo et al., 2016). Recently, 18 061 of restriction-associated DNA sequencing markers were developed for genotype–phenotype association for 23 traits, out of which, 15 markers were significantly associated (Riascos et al., 2016). However, the available algorithms used for these association analyses in sugarcane are mostly designed for diploid with no complicated dosage effects. Therefore, a sophisticated validation should be carried out on these identified associations. After further verification, these significantly associated markers can be utilized to introduce robust traits into sugarcane breeding programmes.

5 Applications of the *Bru*1 marker

Sugarcane brown rust, caused by the fungus *Puccinia melanocephala* H. & P. Sydow., has spread over almost all sugarcane planting areas in the world since 1950 (Comstock et al., 1992; Patel et al., 1950). This disease can cause substantial yield loss for sugarcane (Comstock et al., 1992). Since the discovery of *Bru*1 (Daugrois et al., 1996), research on this major resistance gene has made significant progress in the application of *Bru*1 markers in sugarcane MAS breeding. After the identification of R12H16 and 9O20-F4 co-segregating completely with *Bru*1 (Raboin et al., 2006), these two brown rust-resistant markers have been widely used in sugarcane breeding. A total of 1072 clones, representing the genetic backgrounds of commercial cultivars in Florida, USA, were evaluated with R12H16 and 9O20-F4. Among these clones, 27% (287) were positive for *Bru*1. The frequency of detection of *Bru*1 markers increased from 15% (1975–85) to 47% (2002–12). Moreover, in the second clonal stage, 44% of the clones (667 out of 1512) contained *Bru*1, and 87% of them (580 out of 667) were disease-asymptomatic (Glynn et al., 2013).

The *Bru*1 locus has been used as a major gene source of brown rust resistance, and *Bru*1 markers have been widely used to diagnose this trait globally in breeding programmes. In Estación Experimental Agroindustrial Obispo Colombres (EEAOC, Argentina), a total of 129 sugarcane accessions were phenotypically evaluated and subsequently screened using *Bru*1 diagnostic markers (R12H16 and 9O20-F4). A total of 49 clones were observed with brown rust resistance based on phenotypic evaluation, of which eight contained the *Bru*1 locus. This indicates that additional brown rust-resistant genes exist. In addition, to assess the overall frequency of *Bru*1 in a local sugarcane collection, 190 other accessions were evaluated using *Bru*1 diagnostic markers, of which 13 (7%) contained the *Bru*1 locus. Thus, of the 319 accessions in these two studies, 21 (6.6%) clones were identified as containing the *Bru*1 locus (Racedo et al., 2013). In China, detection of the *Bru*1 marker was applied in a breeding programme, from which 101 clones were evaluated. Of the 77 clones resistant to brown rust based on phenotype, 48 (62.3%) genotypes were found to contain the *Bru*1 locus, while *Bru*1 was absent from the remaining 24 susceptible parental clones, further indicating that other brown rust-resistant genes were present in the breeding programme (Li et al., 2016). These studies confirmed the contribution of the *Bru*1 locus to brown rust resistance in the breeding programme and also helped the researchers to explore new sources of resistance.

Simply relying on *Bru*1 may increase the potential for *Bru*1 resistance to be overcome by adapting brown rust pathogen populations that are under the continual selection pressure of a single source of resistance. In addition, it was reported that one *Bru*1-positive clone, L 10-146, showed susceptibility to brown rust. This could be caused by the suppression of *Bru*1 activity by other genetic factors (Parco et al., 2014). Knowledge of the various *Bru*1 frequencies and its performance in different sugarcane genetic backgrounds in the world will encourage breeders to make more efficient usage of *Bru*1. More importantly, alternative sources of resistance genes and additional molecular markers linked to brown rust resistance are needed in order to broaden brown rust resistance (Li et al., 2016; Parco et al., 2014).

6 Conclusion and future trends

In the past 20 years, significant breakthroughs have been achieved in genetic mapping and the dissection of simply inherited and complex traits in sugarcane. These have improved understanding of sugarcane's genome structure and the inheritance patterns of important traits. With the development of genetic maps, several genes and QTL controlling agronomic traits have been identified. Molecular markers linked to traits of interest have been used to expedite the selection of superior cultivars in sugarcane breeding programmes. However, marker density is still extremely low and genetic maps of sugarcane are far from being saturated. Current genetic mapping in sugarcane largely relies on SD markers, and very few statistical models can directly incorporate high-dosage makers into such analysis. SD markers can account for only a portion of the polymorphisms in the mapping populations, and focusing on SD markers may mean that a biased sampling of the genome is covered (Garcia et al., 2013). Together with SD markers, exploration of high-dosage markers is helpful for filling the gap in the linkage maps constructed in previous studies. More importantly, there is no reason to assume that the genes or QTL of interest are SD, or that they are coupled with the markers identified for mapping important agronomic traits. Without doubt, a critical improvement will be to increase the power of genetic analysis software and develop comprehensive statistical models tailored for polyploidy sugarcane, which will detect high-dosage markers and apply them in linkage analysis.

In spite of the importance of the reference genome of sugarcane, it is currently not available. NGS has brought about a paradigm shift in our understanding of genetics and genomics on account of its high-throughput and low-cost features. Unfortunately, this revolutionary technology cannot fulfil the need to successfully draft a sugarcane genome, on account of the additional challenge of assembling a large genome with a high degree of heterozygosity, which is rich in repetitive sequences. The TGS takes advantage of sequencing ultra-long reads (average read lengths ranging from 6–8 Kbp and reaching up to 30–150 Kbp) (Bleidorn, 2016), and could critically improve assembly of complex genomes of sugarcane. Until the sugarcane reference genome becomes available, the sorghum genome can still be utilized as a reference for analysing short read sequences of a representative part of the genome through exome-seq, GBS or methylation filtration, in order to detect genetic variations and interrogate the genome structure. Except for high-throughput genotyping, NGS data can call markers with all possible dosages, which means that high-dosage markers could be included in linkage analysis and QTL mapping of sugarcane. Aitken et al. developed 345k SNP chips from NGS data on 16 sugarcane genotypes selected for their contribution to the Australian breeding programme (Aitken et al., 2016). Although still in the development stage, this high-throughput genotyping technology will be a powerful tool for rapid sugarcane genotyping. Therefore, it will speed up research into dissecting important agronomic traits and genomic selection in sugarcane breeding programmes. Additionally, transcriptomics has been used to delineate whole-genome expression profiling in sugarcane. Long reads generated from TGS can be used to differentiate different transcript isoforms, which can be further associated with sugarcane haplotype alleles. The results will help us understand the evolution and interactions among these homo(eo)logs, inherited from their ancestor genomes.

7 Where to look for further information

More details on sequencing technologies and strategies including Sanger sequencing technology, next-generation sequencing, BAC-by-BAC strategy and whole-genome shotgun (WGS) strategy can be found in Feuillet et al. (2011). The major institutions focusing on the genome sequencing and genetic mapping include but not limited to Dr. Ming's groups at the University of Illinois, the United States; Fujian Agriculture and Forest University, China; Dr. Souza's group at the University of São Paulo, Brazil; Dr. Aitken's group at CSIRO Agriculture, Australia; Dr. D'Hont's group at CIRAD, France and Dr. Henry's group at The University of Queensland, Australia.

8 References

Aitken, K. S. 2016, 'Introgression of a large effect QTL for smut resistance inherited from *S. spontaneum*', In: *Plant and Animal Genome XXIV Conference*.

Aitken, K. S., Jackson, P. A. and McIntyre, C. L. 2005, 'A combination of AFLP and SSR markers provides extensive map coverage and identification of homo(eo)logous linkage groups in a sugarcane cultivar', *Theoretical and Applied Genetics*, vol. 110, no. 5, pp. 789–801.

Aitken, K. S., Jackson, P. A. and McIntyre, C. L. 2007, 'Construction of a genetic linkage map for *Saccharum officinarum* incorporating both simplex and duplex markers to increase genome coverage', *Genome*, vol. 50, no. 8, pp. 742–56.

Aitken, K. S., Hermann, S., Karno, K., Bonnett, G. D., McIntyre, L. C. and Jackson, P. A. 2008, 'Genetic control of yield related stalk traits in sugarcane', *Theoretical and Applied Genetics*, vol. 117, no. 7, pp. 1191–203.

Aitken, K. S., McNeil, M. D., Hermann, S., Bundock, P. C., Kilian, A., Heller-Uszynska, K., et al. 2014, 'A comprehensive genetic map of sugarcane that provides enhanced map coverage and integrates high-throughput Diversity Array Technology (DArT) markers', *BMC Genomics*, vol. 15, no. 1, pp. 1.

Aitken, K. S., Berkman, P. J. and Rae, A.L. 2015, 'The Australian contribution to the R570 sugarcane genome', In: *Proceedings of the International Society of Sugar Cane Technologists*, p. 43.

Aitken, K., Berkman, P. and Rae, A. 2016, 'The first sugarcane genome assembly: How can we use it?', In: *Proceedings of the Australian Society of Sugar Cane Technologists*, p. 38.

Aljanabi, S. M., Honeycutt, R. J., McClelland, M. and Sobral, B. W. 1993, 'A genetic linkage map of *Saccharum spontaneum* L. 'SES 208'', *Genetics*, vol. 134, no. 4, pp. 1249–60.

Aljanabi, S. M., Honeycutt, R. J. and Sobral, B. W.S. 1994, 'Chromosome assortment in *Saccharum*', *Theoretical and Applied Genetics*, vol. 89, no. 7, pp. 959–63

Aljanabi, S. M., Parmessur, Y., Kross, H., Dhayan, S., Saumtally, S., Ramdoyal, K., et al. 2007, 'Identification of a major quantitative trait locus (QTL) for yellow spot (*Mycovellosiella koepkei*) disease resistance in sugarcane', *Molecular Breeding*, vol. 19, no. 1, pp. 1–14.

Alkan, C., Sajjadian, S. and Eichler, E. E. 2011, 'Limitations of next-generation genome sequence assembly', *Nature Methods*, vol. 8, no. 1, pp. 61–5.

Alwala, S. and Kimbeng, C. 2010, 'Molecular genetic linkage mapping in *Saccharum*: Strategies, resources and achievements', In: R. Henry and C. Kole (Eds), *Genetics, Genomics and Breeding of Sugarcane*, Science Publishers, New Hampshire, pp. 69–96.

Alwala, S., Kimbeng, C. A., Veremis, J. C. and Gravois, K. A. 2008, 'Linkage mapping and genome analysis in a *Saccharum* interspecific cross using AFLP, SRAP and TRAP markers', *Euphytica*, vol. 164, no. 1, pp. 37–51.

Alwala, S., Kimbeng, C. A., Veremis, J. C. and Gravois, K. A. 2009, 'Identification of molecular markers associated with sugar-related traits in a *Saccharum* interspecific cross', *Euphytica*, vol. 167, no. 1, pp. 127–42.

Andru, S., Pan, Y., Thongthawee, S., Burner, D. M. and Kimbeng, C. A. 2011, 'Genetic analysis of the sugarcane (*Saccharum* spp.) cultivar 'LCP 85-384'. I. Linkage mapping using AFLP, SSR, and TRAP markers', *Theoretical and Applied Genetics*, vol. 123, no. 1, pp. 77–93.

Arabidopsis Genome Initiative 2000, 'Analysis of the genome sequence of the flowering plant *Arabidopsis thaliana*', *Nature*, vol. 408, no. 6814, pp. 796.

Arruda, P. 2001, 'Sugarcane transcriptome: A landmark in plant genomics in the tropics', *Genetics and Molecular Biology*, vol. 24, no. 1–4, pp. 1–296.

Asano, T., Tsudzuki, T., Takahashi, S., Shimada, H. and Kadowaki, K. 2004, 'Complete nucleotide sequence of the sugarcane (*Saccharum officinarum*) chloroplast genome: A comparative analysis of four monocot chloroplast genomes', *DNA Research*, vol. 11, no. 2, pp. 93–9.

Asnaghi, C., Paulet, F., Kaye, C., Grivet, L., Deu, M., Glaszmann, J. C., et al. 2000, 'Application of synteny across *Poaceae* to determine the map location of a sugarcane rust resistance gene', *Theoretical and Applied Genetics*, vol. 101, no. 5–6, pp. 962–9.

Asnaghi, C., Roques, D., Ruffel, S., Kaye, C., Hoarau, J., Telismart, H., Girard, J. C., et al. 2004, 'Targeted mapping of a sugarcane rust resistance gene (*Bru*1) using bulked segregant analysis and AFLP markers', *Theoretical and Applied Genetics*, vol. 108, no. 4, pp. 759–64.

Balsalobre, T. W.A., Pereira, G. D.S., Margarido, G. R.A., Gazaffi, R., Barreto, F. Z., Anoni, C. O., et al. 2017, 'GBS-based single dosage markers for linkage and QTL mapping allow gene mining for yield-related traits in sugarcane', *BMC Genomics*, 18, 72.

Bartholomé, J., Mandrou, E., Mabiala, A., Jenkins, J., Nabihoudine, I., Klopp, C., Schmutz, J., et al. 2015, 'High-resolution genetic maps of Eucalyptus improve *Eucalyptus grandis* genome assembly', *New Phytologist*, vol. 206, no. 4, pp. 1283–96.

Berkman, P. J., Bundock, P. C., Casu, R. E., Henry, R. J., Rae, A. L. and Aitken, K. S. 2014, 'A survey sequence comparison of *Saccharum* genotypes reveals allelic diversity differences', *Tropical Plant Biology*, vol. 7, no. 2, pp. 71–83.

Bertioli, D. J., Cannon, S. B., Froenicke, L., Huang, G., Farmer, A. D., Cannon, E. K., et al. 2016, 'The genome sequences of *Arachis duranensis* and *Arachis ipaensis*, the diploid ancestors of cultivated peanut', *Nature Genetics*, 48(4), 438–46.

Bilal, M., Saeed, M., Nasir, I. A., Tabassum, B., Zameer, M., Khan, A., et al. 2015, 'Association mapping of cane weight and tillers per plant in sugarcane', *Biotechnology & Biotechnological Equipment*, vol. 29, no. 4, pp. 617–23.

Blair, M. W., Muoz, C., Garza, R. and Cardona, C. 2006, 'Molecular mapping of genes for resistance to the bean pod weevil (*Apion godmani* Wagner) in common bean', *Theoretical and Applied Genetics*, vol. 112, no. 5, pp. 913–23.

Bleidorn, C. 2016, 'Third generation sequencing: Technology and its potential impact on evolutionary biodiversity research', *Systematics and Biodiversity*, vol. 14, no. 1, pp. 1–8.

Bower, N. I., Casu, R. E., Maclean, D. J., Reverter, A., Chapman, S. C. and Manners, J. M. 2005, 'Transcriptional response of sugarcane roots to methyl jasmonate', *Plant Science*, vol. 168, no. 3, pp. 761–72.

Bundock, P. C., Casu, R. E. and Henry, R. J. 2012, 'Enrichment of genomic DNA for polymorphism detection in a non-model highly polyploid crop plant', *Plant Biotechnology Journal*, vol. 10, no. 6, pp. 657–67.

Carson, D. L. and Botha, F. C. 2000, 'Preliminary analysis of expressed sequence tags for sugarcane', *Crop Science*, vol. 40, no. 6, pp. 1769–79.

Casu, R. E., Grof, C. P., Rae, A. L., McIntyre, C. L., Dimmock, C. M. and Manners, J. M. 2003, 'Identification of a novel sugar transporter homologue strongly expressed in maturing stem vascular tissues of sugarcane by expressed sequence tag and microarray analysis', *Plant Molecular Biology*, vol. 52, no. 2, pp. 371–86.

Casu, R. E., Dimmock, C. M., Chapman, S. C., Grof, C. P., McIntyre, C. L., Bonnett, G. D., et al. 2004, 'Identification of differentially expressed transcripts from maturing stem of sugarcane by *in silico* analysis of stem expressed sequence tags and gene expression profiling', *Plant Molecular Biology*, vol. 54, no. 4, pp. 503–17.

Casu, R. E., Jarmey, J. M., Bonnett, G. D. and Manners, J. M. 2007, 'Identification of transcripts associated with cell wall metabolism and development in the stem of sugarcane by Affymetrix GeneChip Sugarcane Genome Array expression profiling', *Functional & Integrative Genomics*, vol. 7, no. 2, pp. 153–67.

Chen, Z. J., Scheffler, B. E., Dennis, E., Triplett, B. A., Zhang, T., Guo, W., et al. 2007, 'Toward sequencing cotton (*Gossypium*) genomes', *Plant Physiology*, vol. 145, no. 4, pp. 1303–10.

Chen, J., Lao, F., Chen, X., Deng, H., Liu, R., He, H., et al. 2015, 'DNA marker transmission and linkage analysis in populations derived from a sugarcane (*Saccharum* spp.) × *Erianthus arundinaceus* hybrid', *PloS One*, vol. 10, no. 6, pp. e0128865.

Collard, B., Jahufer, M., Brouwer, J. B. and Pang, E. 2005, 'An introduction to markers, quantitative trait loci (QTL) mapping and marker-assisted selection for crop improvement: The basic concepts', *Euphytica*, vol. 142, no. 1–2, pp. 169–96.

Comstock, J. C., Shine Jr., J. M. and Raid, R. N. 1992, 'Effect of early rust infection on subsequent sugarcane growth', *Sugar Cane*, no. 4, pp. 7–9.

Cordeiro, G. M., Eliott, F., McIntyre, C. L., Casu, R. E. and Henry, R. J. 2006, 'Characterisation of single nucleotide polymorphisms in sugarcane ESTs', *Theoretical and Applied Genetics*, vol. 113, no. 2, pp. 331–43.

Costet, L., Le Cunff, L., Royaert, S., Raboin, L., Hervouet, C., Toubi, L., et al. 2012a, 'Haplotype structure around *Bru*1 reveals a narrow genetic basis for brown rust resistance in modern sugarcane cultivars', *Theoretical and Applied Genetics*, vol. 125, no. 5, pp. 825–36.

Costet, L., Raboin, L., Payet, M., D'Hont, A. and Nibouche, S. 2012b, 'A major quantitative trait allele for resistance to the *Sugarcane yellow leaf virus* (*Luteoviridae*)', *Plant Breeding*, vol. 131, no. 5, pp. 637–40.

Cui, Y., Zhang, F., Xu, J., Li, Z. and Xu, S. 2015, 'Mapping quantitative trait loci in selected breeding populations: A segregation distortion approach', *Heredity*, vol. 115, no. 6, pp. 538–46.

da Silva, J., Honeycutt, R. J., Burnquist, W., Al-Janabi, S. M., Sorrells, M. E., Tanksley, S. D., et al. 1995, '*Saccharum spontaneum* L.'SES 208'genetic linkage map combining RFLP-and PCR-based markers', *Molecular Breeding*, vol. 1, no. 2, pp. 165–79.

Dahlquist, E. 2013, *Biomass as Energy Source: Resources, Systems and Applications*, CRC Press, Taylor & Francis, London.

Daugrois, J. H., Grivet, L., Roques, D., Hoarau, J. Y., Lombard, H., Glaszmann, J. C., et al. 1996, 'A putative major gene for rust resistance linked with a RFLP marker in sugarcane cultivar 'R570'', *Theoretical and Applied Genetics*, vol. 92, no. 8, pp. 1059–64.

Davey, J. W., Hohenlohe, P. A., Etter, P. D., Boone, J. Q., Catchen, J. M. and Blaxter, M. L. 2011, 'Genome-wide genetic marker discovery and genotyping using next-generation sequencing', *Nature Reviews Genetics*, vol. 12, no. 7, pp. 499–510.

Débibakas, S., Rocher, S., Garsmeur, O., Toubi, L., Roques, D., D'Hont, A., et al. 2014, 'Prospecting sugarcane resistance to sugarcane yellow leaf virus by genome-wide association', *Theoretical and Applied Genetics*, vol. 127, no. 8, pp. 1719–32.

D'Hont, A. 2005, 'Unraveling the genome structure of polyploids using FISH and GISH; examples of sugarcane and banana', *Cytogenetic and Genome Research*, vol. 109, no. 1–3, pp. 27–33.

D'Hont, A. and Glaszmann, J. C. 2001, 'Sugarcane genome analysis with molecular markers, a first decade of research', In: *Proceedings of the International Society of Sugarcane Technologists*, volume 24, pp. 556–9.

D'Hont, A., Grivet, L., Feldmann, P., Glaszmann, J. C., Rao, S. and Berding, N. 1996, 'Characterisation of the double genome structure of modern sugarcane cultivars (*Saccharum* spp.) by molecular cytogenetics', *Molecular and General Genetics MGG*, vol. 250, no. 4, pp. 405–13.

D'Hont, A., Ison, D., Alix, K., Roux, C. and Glaszmann, J. C. 1998, 'Determination of basic chromosome numbers in the genus *Saccharum* by physical mapping of ribosomal RNA genes', *Genome*, vol. 41, no. 2, pp. 221–5.

Edmé, S. J., Miller, J. D., Glaz, B., Tai, P. Y. and Comstock, J. C. 2005, 'Genetic contribution to yield gains in the Florida sugarcane industry across 33 years', *Crop Science*, vol. 45, no. 1, pp. 92–7.

Edmé, S. J., Glynn, N. G. and Comstock, J. C. 2006, 'Genetic segregation of microsatellite markers in *Saccharum officinarum* and *S. spontaneum*', *Heredity*, vol. 97, no. 5, pp. 366–75.

Fartek, B., Rocher, S., Dbibakas, S., Hervouet, C., Royeart, S., Toubi, L., et al. 2011, 'Association mapping of sugarcane resistance to the sugarcane yellow leaf disease', In: *Proceedings of the International Society of Sugar Cane Technologists*, p. 58.

FAO. 2013, '*FAOSTAT*', FAO Statistical Databases. Available at: http://faostat3.fao.org/ (Last accessed 18 December 2015).

FAO. 2014, '*FAOSTAT*', FAO Statistical Databases. Available at: http://faostat3.fao.org/ (Last accessed 07 June 2016).

Feuillet, C., Leach, J. E., Rogers, J., Schnable, P. S. and Eversole, K. 2011, 'Crop genome sequencing: Lessons and rationales', *Trends in Plant Science*, vol. 16, no. 2, pp. 77–88.

Fierst, J. L. 2015, 'Using linkage maps to correct and scaffold de novo genome assemblies: Methods, challenges, and computational tools', *Frontiers in Genetics*, vol. 6, pp. 220.

Fisher, R. A. 1949, 'The linkage problem in a tetrasomic wild plant, *Lythrum salicaria*', *Hereditas*, vol. 35, no. S1, pp. 225–33.

Fitch, M. M. and Moore, P. H. 1996, 'Haploids of sugarcane', In: S. M. Jain, S. K. Sopory and R. E. Veilleux (Eds), *In Vitro Haploid Production in Higher Plants. Volume 3: Important Selected Plants*, Springer Netherlands, the Netherlands, pp. 1–16.

Garcia, A., Kido, E. A., Meza, A. N., Souza, H., Pinto, L. R., Pastina, M. M., et al. 2006, 'Development of an integrated genetic map of a sugarcane (*Saccharum* spp.) commercial cross, based on a maximum-likelihood approach for estimation of linkage and linkage phases', *Theoretical and Applied Genetics*, vol. 112, no. 2, pp. 298–314.

Garcia, A. A., Mollinari, M., Marconi, T. G., Serang, O. R., Silva, R. R., Vieira, M. L., et al. 2013, 'SNP genotyping allows an in-depth characterisation of the genome of sugarcane and other complex autopolyploids', *Scientific Reports*, vol. 3, no. 3399, pp. 1–10.

Garsmeur, O., Charron, C., Bocs, S., Jouffe, V., Samain, S., Couloux, A., Droc, G., et al. 2011, 'High homologous gene conservation despite extreme autopolyploid redundancy in sugarcane', *New Phytologist*, vol. 189, no. 2, pp. 629–42.

Garsmeur, O., Potier, B., Aitken, K. S., Berkman, P. J., Droc, G., Charron, C., et al. 2015, 'Toward a reference sequence of the gene-rich part of the highly polyploid sugarcane genome', In: *Plant and Animal Genome XXIV Conference*.

Glynn, N. C., Laborde, C., Davidson, R. W., Irey, M. S., Glaz, B., D'Hont, A., et al. 2013, 'Utilization of a major brown rust resistance gene in sugarcane breeding', *Molecular Breeding*, vol. 31, no. 2, pp. 323–31.

Goff, S. A., Ricke, D., Lan, T., Presting, G., Wang, R., Dunn, M., Glazebrook, J., et al. 2002, 'A draft sequence of the rice genome (*Oryza sativa* L. ssp. *japonica*)', *Science*, vol. 296, no. 5565, pp. 92–100.

Goldemberg, J. 2008, 'The Brazilian biofuels industry', *Biotechnology for Biofuels*, vol. 1, no. 1, p. 1.

Gouy, M., Rousselle, Y., Chane, A. T., Anglade, A., Royaert, S., Nibouche, S., et al. 2015, 'Genome wide association mapping of agro-morphological and disease resistance traits in sugarcane', *Euphytica*, vol. 202, no. 2, pp. 269–84.

Grativol, C., Regulski, M., Bertalan, M., McCombie, W. R., Silva, F. R., Zerlotini Neto, A., et al. 2014, 'Sugarcane genome sequencing by methylation filtration provides tools for genomic research in the genus *Saccharum*', *The Plant Journal*, vol. 79, no. 1, pp. 162–72.

Grivet, L. and Arruda, P. 2002, 'Sugarcane genomics: Depicting the complex genome of an important tropical crop', *Current Opinion in Plant Biology*, vol. 5, no. 2, pp. 122–7.

Grivet, L., D'Hont, A., Roques, D., Feldmann, P., Lanaud, C. and Glaszmann, J. C. 1996, 'RFLP mapping in cultivated sugarcane (*Saccharum* spp.): Genome organization in a highly polyploid and aneuploid interspecific hybrid', *Genetics*, vol. 142, no. 3, pp. 987–1000.

Grivet, L., Glaszmann, J., Vincentz, M., Da Silva, F. and Arruda, P. 2003, 'ESTs as a source for sequence polymorphism discovery in sugarcane: Example of the *Adh* genes', *Theoretical and Applied Genetics*, vol. 106, no. 2, pp. 190–7.

Gualberto, J. M., Mileshina, D., Wallet, C., Niazi, A. K., Weber-Lotfi, F. and Dietrich, A. 2014, 'The plant mitochondrial genome: Dynamics and maintenance', *Biochimie*, vol. 100, pp. 107–20.

Guimares, C. T., Honeycutt, R. J., Sills, G. R. and Sobral, B. W. 1999, 'Genetic maps of *Saccharum officinarum* L. and *Saccharum robustum* Brandes & Jew. ex grassl', *Genetics and Molecular Biology*, vol. 22, no. 1, pp. 125–32.

Guo, S., Zhang, J., Sun, H., Salse, J., Lucas, W. J., Zhang, H., et al. 2013, 'The draft genome of watermelon (*Citrullus lanatus*) and resequencing of 20 diverse accessions', *Nature Genetics*, vol. 45, no. 1, pp. 51–8.

Hoarau, J., Offmann, B., D'hont, A., Risterucci, A., Roques, D., Glaszmann, J., et al. 2001, 'Genetic dissection of a modern sugarcane cultivar (*Saccharum* spp.). I. Genome mapping with AFLP markers', *Theoretical and Applied Genetics*, vol. 103, no. 1, pp. 84–97.

Huang, X., Wei, X., Sang, T., Zhao, Q., Feng, Q., Zhao, Y., et al. 2010, 'Genome-wide association studies of 14 agronomic traits in rice landraces', *Nature Genetics*, vol. 42, no. 11, pp. 961–7.

Huang, X., Yang, S., Gong, J., Zhao, Q., Feng, Q., Zhan, Q., et al. 2016, 'Genomic architecture of heterosis for yield traits in rice', *Nature*, vol. 537, no. 7622, pp. 629–33.

Jaillon, O., Aury, J., Noel, B., Policriti, A., Clepet, C., Casagrande, A., et al. 2007, 'The grapevine genome sequence suggests ancestral hexaploidization in major angiosperm phyla', *Nature*, vol. 449, no. 7161, pp. 463–7.

Jia, J., Zhao, S., Kong, X., Li, Y., Zhao, G., He, W., et al. 2013, '*Aegilops tauschii* draft genome sequence reveals a gene repertoire for wheat adaptation', *Nature*, vol. 496, no. 7443, pp. 91–5.

Kai, W., Nomura, K., Fujiwara, A., Nakamura, Y., Yasuike, M., Ojima, N., et al. 2014, 'A ddRAD-based genetic map and its integration with the genome assembly of Japanese eel (*Anguilla japonica*) provides insights into genome evolution after the teleost-specific genome duplication', *BMC Genomics*, vol. 15, no. 1, pp. 233.

Kido, E. A., Ferreira Neto, J. R. C., Silva, R. L. de O., Pandolfi, V., Guimaraes, A. C.R., Veiga, D. T., et al. 2012, 'New insights in the sugarcane transcriptome responding to drought stress as revealed by SuperSAGE', *The Scientific World Journal*, vol. 2012, 821062.

Kim, C., Lee, T., Compton, R. O., Robertson, J. S., Pierce, G. J. and Paterson, A. H. 2013, 'A genome-wide BAC end-sequence survey of sugarcane elucidates genome composition, and identifies BACs covering much of the euchromatin', *Plant Molecular Biology*, vol. 81, no. 1–2, pp. 139–47.

Klosowski, A. C., Bespalhok Filho, J. C., Ruaro, L. and de Mio, L. M. 2013, 'Inheritance of resistance to orange rust (*Puccinia kuehnii*) in sugarcane families from crosses between parents with different orange rust reactions', *Sugar Technology*, vol. 15, no. 4, pp. 379–83.

Kuhl, J. C., Bradeen, J. M. and Kole, C. 2011, 'Mapping and tagging of simply inherited traits', In: J.M. Bradeen and K. Chittaranjan (Eds), *Genetics, Genomics and Breeding of Potato*, Science Publishers, New Hampshire, pp. 90–112.

Le Cunff, L., Garsmeur, O., Raboin, L. M., Pauquet, J., Telismart, H., Selvi, A., et al. 2008, 'Diploid/polyploid syntenic shuttle mapping and haplotype-specific chromosome walking toward a rust resistance gene (*Bru*1) in highly polyploid sugarcane (2n~ 12x~ 115)', *Genetics*, vol. 180, no. 1, pp. 649–60.

Levin, I. and Schaffer, A. A. 2013, 'Mapping and tagging of simply inherited traits', In: B. E. Liedl, J. A. Labate J. R. Stommel, et al. (Eds), *Genetics, Genomics and Breeding of Tomato*, CRC Press, Boca Raton, FL, USA, pp. 109–49.

Lewin, H. A., Larkin, D. M., Pontius, J. and O'Brien, S. J. 2009, 'Every genome sequence needs a good map', *Genome Research*, vol. 19, no. 11, pp. 1925–8.

Li, R., Li, Y., Fang, X., Yang, H., Wang, J., Kristiansen, K., et al. 2009, 'SNP detection for massively parallel whole-genome resequencing', *Genome Research*, vol. 19, no. 6, pp. 1124–32.

Li, W., Wang, X., Huang, Y., Zhang, R., Shan, H., Yin, J., et al. 2016, 'Molecular detection of *Bru*1 gene and identification of brown rust resistance in Chinese sugarcane germplasm', *Sugar Tech*, vol. 19, no. 2, pp. 183–90.

Liu, P., Chandra, A., Que, Y., Chen, P., Grisham, M. P., White, W. H., et al. 2016, 'Identification of quantitative trait loci controlling sucrose content based on an enriched genetic linkage map of sugarcane (*Saccharum* spp. hybrids) cultivar 'LCP 85-384'', *Euphytica*, vol. 207, no. 3, pp. 527–49.

Ma, H., Schulze, S., Lee, S., Yang, M., Mirkov, E., Irvine, J., Moore, P., et al. 2004, 'An EST survey of the sugarcane transcriptome', *Theoretical and Applied Genetics*, vol. 108, no. 5, pp. 851–63.

Mackay, T. F. 2001, 'The genetic architecture of quantitative traits', *Annual review of genetics*, vol. 35, pp. 303–39.

Marcussen, T., Sandve, S. R., Heier, L., Wulff, B. B., Steuernagel, B., Mayer, K. F., et al. 2014, 'A chromosome-based draft sequence of the hexaploid bread wheat (*Triticum aestivum*) genome', *Science*, vol. 345, pp. 1250092.

Mascher, M. and Stein, N. 2014, 'Genetic anchoring of whole-genome shotgun assemblies', *Frontiers in Genetics*, vol. 5, pp. 208.

Mascher, M., Richmond, T. A., Gerhardt, D. J., Himmelbach, A., Clissold, L., Sampath, D., et al. 2013, 'Barley whole exome capture: A tool for genomic research in the genus *Hordeum* and beyond', *The Plant Journal*, vol. 76, no. 3, pp. 494–505.

McCord, P. H. and Migneault, A. J. 2015, 'Genotyping sugarcane for the brown rust resistance locus *Bru*1 using unlabeled probe melting', *Sugar Tech*, vol. 18, no. 4, pp. 401–6.

Mendel, G. 1865, 'Experiments in plant hybridization', available at: http://old.esp.org/foundations/genetics/classical/gm-65-a.pdf (Last accessed October 2017).

Miller, J. R., Dilley, K. A., Harkins, D. M., Torralba, M. G., Moncera, K. J., Beeri, K., et al. 2017, 'Initial genome sequencing of the sugarcane CP 96-1252 complex hybrid', *F1000Research*, vol. 6, 688.

Ming, R., Liu, S., Lin, Y., da Silva, J., Wilson, W., Braga, D., et al. 1998, 'Detailed alignment of *Saccharum* and sorghum chromosomes: Comparative organization of closely related diploid and polyploid genomes', *Genetics*, vol. 150, no. 4, pp. 1663–82.

Ming, R., Liu, S., Moore, P. H., Irvine, J. E. and Paterson, A. H. 2001, 'QTL analysis in a complex autopolyploid: Genetic control of sugar content in sugarcane', *Genome Research*, vol. 11, no. 12, pp. 2075–84.

Ming, R., Wang, Y., Draye, X., Moore, P., Irvine, J. and Paterson, A. 2002, 'Molecular dissection of complex traits in autopolyploids: Mapping QTLs affecting sugar yield and related traits in sugarcane', *Theoretical and Applied Genetics*, vol. 105, no. 2–3, pp. 332–45.

Mudge, J., Andersen, W. R., Kehrer, R. L. and Fairbanks, D. J. 1996, 'A RAPD genetic map of *Saccharum officinarum*', *Crop Science*, vol. 36, no. 5, pp. 1362–6.

Nibouche, S., Raboin, L. M., Hoarau, J., D'Hont, A. and Costet, L. 2012, 'Quantitative trait loci for sugarcane resistance to the spotted stem borer *Chilo sacchariphagus*', *Molecular Breeding*, vol. 29, no. 1, pp. 129–35.

Nogueira, F. T., de Rosa Jr., V. E., Menossi, M., Ulian, E. C. and Arruda, P. 2003, 'RNA expression profiles and data mining of sugarcane response to low temperature', *Plant Physiology*, vol. 132, no. 4, pp. 1811–24.

Oliveira, K. M., Pinto, L. R., Marconi, T. G., Margarido, G. R., Pastina, M. M., Teixeira, L. H.M., et al. 2007, 'Functional integrated genetic linkage map based on EST-markers for a sugarcane (*Saccharum* spp.) commercial cross', *Molecular Breeding*, vol. 20, no. 3, pp. 189–208.

Page, J. T., Huynh, M. D., Liechty, Z. S., Grupp, K., Stelly, D., Hulse, A. M., et al. 2013, 'Insights into the evolution of cotton diploids and polyploids from whole-genome re-sequencing', *G3: Genes| Genomes| Genetics*, vol. 3, no. 10, pp. 1809–18.

Palhares, A. C., Rodrigues-Morais, T. B., Van Sluys, M., Domingues, D. S., Maccheroni, W., Jordo, H., et al. 2012, 'A novel linkage map of sugarcane with evidence for clustering of retrotransposon-based markers', *BMC Genetics*, vol. 13, no. 1, pp. 1.

Parco, A. S., Avellaneda, M. C., Hale, A. H., Hoy, J. W., Kimbeng, C. A., Pontif, M. J., et al. 2014, 'Frequency and distribution of the brown rust resistance gene *Bru*1 and implications for the Louisiana sugarcane breeding programme', *Plant Breeding*, vol. 133, no. 5, pp. 654–9.

Park, J., Benatti, T. R., Marconi, T., Yu, Q., Solis-Gracia, N., Mora, V., et al. 2015, 'Cold responsive gene expression profiling of sugarcane and *Saccharum spontaneum* with functional analysis of a cold inducible *Saccharum* homolog of NOD26-like intrinsic protein to salt and water stress', *PloS One*, vol. 10, no. 5, pp. e0125810.

Pastina, M. M., Malosetti, M., Gazaffi, R., Mollinari, M., Margarido, G., Oliveira, K. M., et al. 2012, 'A mixed model QTL analysis for sugarcane multiple-harvest-location trial data', *Theoretical and Applied Genetics*, vol. 124, no. 5, pp. 835–49.

Patel, M. K., Kamat, M. N. and Padhye, Y. A. 1950, 'A new record of *Puccinia* on sugarcane in Bombay', *Current Science*, vol. 19, no. 4, p. 122.

Paterson, A. H., Bowers, J. E., Bruggmann, R., Dubchak, I., Grimwood, J., Gundlach, H., et al. 2009, 'The *Sorghum bicolor* genome and the diversification of grasses', *Nature*, vol. 457, no. 7229, pp. 551–6.

Paterson, A. H., Souza, G., Van Sluys, M., Ming, R. and D'Hont, A. 2010, 'Structural genomics and genome sequencing', In: R. Henry and C. Kole (Eds), *Genetics, Genomics and Breeding of Sugarcane*, Science Publishers, New Hampshire, pp. 149–65.

Pinto, L. R., Oliveira, K. M., Ulian, E. C., Garcia, A. A.F. and de Souza, A. P. 2004, 'Survey in the sugarcane expressed sequence tag database (SUCEST) for simple sequence repeats', *Genome*, vol. 47, no. 5, pp. 795–804.

Piperidis, G., D'hont, A. and Hogarth, D. M. 2001, 'Chromosome composition analysis of various *Saccharum* interspecific hybrids by genomic in situ hybridisation (GISH)', In: *Proceedings of the International Society of Sugar Cane Technologists*, volume 24, pp. 556-559.

Potato Genome Sequencing Consortium 2011, 'Genome sequence and analysis of the tuber crop potato', *Nature*, vol. 475, no. 7355, pp. 189–95.

Qu, L. and Hancock, J. F. 2001, 'Detecting and mapping repulsion-phase linkage in polyploids with polysomic inheritance', *Theoretical and Applied Genetics*, vol. 103, no. 1, pp. 136–43.

Raboin, L., Oliveira, K. M., Lecunff, L., Telismart, H., Roques, D., Butterfield, M., et al. 2006, 'Genetic mapping in sugarcane, a high polyploid, using bi-parental progeny: Identification of a gene controlling stalk colour and a new rust resistance gene', *Theoretical and Applied Genetics*, vol. 112, no. 7, pp. 1382–91.

Racedo, J., Perera, M. F., Bertani, R., Funes, C., González, V., Cuenya, M. I., et al. 2013, '*Bru*1 gene and potential alternative sources of resistance to sugarcane brown rust disease', *Euphytica*, vol. 191, no. 3, pp. 429–36.

Racedo, J., Gutiérrez, L., Perera, M. F., Ostengo, S., Pardo, E. M., Cuenya, M. I., et al. 2016, 'Genome-wide association mapping of quantitative traits in a breeding population of sugarcane', *BMC Plant Biology*, vol. 16, no. 1, pp. 142.

Rami, J., Dufour, P., Trouche, G., Fliedel, G., Mestres, C., Davrieux, F., et al. 1998, 'Quantitative trait loci for grain quality, productivity, morphological and agronomical traits in sorghum (*Sorghum bicolor* L. Moench)', *Theoretical and Applied Genetics*, vol. 97, no. 4, pp. 605–16.

Reffay, N., Jackson, P. A., Aitken, K. S., Hoarau, J., D'Hont, A., Besse, P., et al. 2005, 'Characterisation of genome regions incorporated from an important wild relative into Australian sugarcane', *Molecular Breeding*, vol. 15, no. 4, pp. 367–81.

Riaño-Pachón, D. M. and Mattiello, L. 2017, 'Draft genome sequencing of the sugarcane hybrid SP80-3280', *F1000Research*, vol. 6, p. 861.

Riaño-Pachón, D. M., Mattiello, L. and da Cruz, L. P. 2016. 'Surveying the complex polyploid sugarcane genome sequence using synthetic long reads', Technical Report. Laboratório Nacional de Ciência e Tecnologia do Bioetanol, Sao Paulo.

Riascos, J. J., Salazar, F. A., Loaiza, C. D., Lpez, J., Espitia, H. F., Franco, C. M., et al. 2016, 'Advances in the utilization of Next Generation Sequencing data in the Colombian sugarcane breeding program', In: *Proceedings of the International Society of Sugar Cane Technologists*, p. 1143.

Roach, B. T. 1989, 'Origin and improvement of the genetic base of sugarcane', In: *Proceedings of Australian Society of Sugar Cane Technologists*, p. 34.

Rossi, M., Araujo, P. G., Paulet, F., Garsmeur, O., Dias, V. M., Chen, H., et al. 2003, 'Genomic distribution and characterization of EST-derived resistance gene analogs (RGAs) in sugarcane', *Molecular Genetics and Genomics*, vol. 269, no. 3, pp. 406–19.

Schadt, E. E., Turner, S. and Kasarskis, A. 2010, 'A window into third-generation sequencing', *Human Molecular Genetics*, vol. 19, no. R2, pp. R240.

Schmutz, J., Cannon, S. B., Schlueter, J., Ma, J., Mitros, T., Nelson, W., et al. 2010, 'Genome sequence of the palaeopolyploid soybean', *Nature*, vol. 463, no. 7278, pp. 178–83.

Schnable, P. S., Ware, D., Fulton, R. S., Stein, J. C., Wei, F., Pasternak, S., et al. 2009, 'The B73 maize genome: Complexity, diversity, and dynamics', *Science*, vol. 326, no. 5956, pp. 1112–15.

Shearman, J. R., Sonthirod, C., Naktang, C., Pootakham, W., Yoocha, T., Sangsrakru, D., et al. 2016, 'The two chromosomes of the mitochondrial genome of a sugarcane cultivar: Assembly and recombination analysis using long PacBio reads', *Scientific Reports*, vol. 6, no. 31533, pp. 1–7.

Silva, J. A. d. and Bressiani, J. A. 2005, 'Sucrose synthase molecular marker associated with sugar content in elite sugarcane progeny', *Genetics and Molecular Biology*, vol. 28, no. 2, pp. 294–8.

Silva, J. A. d., Sorrells, M. E., Burnquist, W. L. and Tanksley, S. D. 1993, 'RFLP linkage map and genome analysis of *Saccharum spontaneum*', *Genome*, vol. 36, no. 4, pp. 782–91.

Simpson, A. J. and Perez, J. F. 1998, 'ONSA, the São Paulo virtual genomics institute', *Nature Biotechnology*, vol. 16, pp. 795–6.

Singh, R. K., Singh, S. P., Tiwari, D. K., Srivastava, S., Singh, S. B., Sharma, M. L., et al. 2013, 'Genetic mapping and QTL analysis for sugar yield-related traits in sugarcane', *Euphytica*, vol. 191, no. 3, pp. 333–53.

Smith, D. R., Quinlan, A. R., Peckham, H. E., Makowsky, K., Tao, W., Woolf, B., et al. 2008, 'Rapid whole-genome mutational profiling using next-generation sequencing technologies', *Genome Research*, vol. 18, no. 10, pp. 1638–42.

Song, J., Yang, X., Resende Jr., M. F., Neves, L. G., Todd, J., Zhang, J., et al. 2016, 'Natural allelic variations in highly polyploid *Saccharum* complex', *Frontiers in Plant Science*, vol. 7, pp. 804.

Souza, G. M., Simoes, A. C.Q., Oliveira, K. C., Garay, H. M., Fiorini, L. C., Gomes, F. D. S., et al. 2001, 'The sugarcane signal transduction (SUCAST) catalogue: Prospecting signal transduction in sugarcane', *Genetics and Molecular Biology*, vol. 24, no. 1–4, pp. 25–34.

Souza, G. M., Berges, H., Bocs, S., Casu, R., D'Hont, A., Ferreira, J. E., et al. 2011, 'The sugarcane genome challenge: Strategies for sequencing a highly complex genome', *Tropical Plant Biology*, vol. 4, no. 3–4, pp. 145–56.

Tabien, R., Li, Z., Paterson, A., Marchetti, M., Stansel, J. and Pinson, S. 2002, 'Mapping QTLs for field resistance to the rice blast pathogen and evaluating their individual and combined utility in improved varieties', *Theoretical and Applied Genetics*, vol. 105, no. 2–3, pp. 313–24.

Tanksley, S. D. 1988, 'Resolution of quantitative traits into Mendelian factors by using a complete linkage map of restriction fragment length polymorphisms', *Nature*, vol. 335, no. 6192, pp. 721–6.

Tanksley, S. D. and Nelson, J. C. 1996, 'Advanced backcross QTL analysis: A method for the simultaneous discovery and transfer of valuable QTLs from unadapted germplasm into elite breeding lines', *Theoretical and Applied Genetics*, vol. 92, no. 2, pp. 191–203.

Thomas, C. M., Vos, P., Zabeau, M., Jones, D. A., Norcott, K. A., Chadwick, B. P., et al. 1995, 'Identification of amplified restriction fragment polymorphism (AFLP) markers tightly linked to the tomato *Cf-9* gene for resistance to *Cladosporium fulvum*', *The Plant Journal*, vol. 8, no. 5, pp. 785–94.

Todd, J., Glaz, B., Wang, J., Sood, S. and Comstock, J. 2014, 'Phenotyping of association mapping panel of sugarcane and related grasses', In: *Plant and Animal Genome Conference Proceedings*, p. 184.

Tomato Genome Consortium 2012, 'The tomato genome sequence provides insights into fleshy fruit evolution', *Nature*, vol. 485, no. 7400, pp. 635–41.

Tomkins, J. P., Yu, Y., Miller-Smith, H., Frisch, D. A., Woo, S. S. and Wing, R. A. 1999, 'A bacterial artificial chromosome library for sugarcane', *Theoretical and Applied Genetics*, vol. 99, no. 3–4, pp. 419–24.

Varshney, R. K., Nayak, S. N., May, G. D. and Jackson, S. A. 2009, 'Next-generation sequencing technologies and their implications for crop genetics and breeding', *Trends in Biotechnology*, vol. 27, no. 9, pp. 522–30.

Vettore, A. L., Silva, F. R. d., Kemper, E. L. and Arruda, P. 2001, 'The libraries that made SUCEST', *Genetics and Molecular Biology*, vol. 24, no. 1–4, pp. 1–7.

Vettore, A. L., da Silva, F. R., Kemper, E. L., Souza, G. M., da Silva, A. M., Ferro, M. I., et al. 2003, 'Analysis and functional annotation of an expressed sequence tag collection for tropical crop sugarcane', *Genome Research*, vol. 13, no. 12, pp. 2725–35.

Vicentini, R., Bottcher, A., dos Santos Brito, M., dos Santos, A.B., Creste, S., de Andrade Landell, M. G., et al.2015, 'Large-scale transcriptome analysis of two sugarcane genotypes contrasting for lignin content', *PLoS One*, vol. 10, no. 8, pp. e0134909.

Vincentz, M., Cara, F.A., Okura, V.K., da Silva, F. R., Pedrosa, G. L., Hemerly, A. S., et al. 2004, 'Evaluation of monocot and eudicot divergence using the sugarcane transcriptome', *Plant Physiology*, vol. 134, no. 3, pp. 951–9.

Waclawovsky, A. J., Sato, P. M., Lembke, C. G., Moore, P. H. and Souza, G. M. 2010, 'Sugarcane for bioenergy production: An assessment of yield and regulation of sucrose content', *Plant Biotechnology Journal*, vol. 8, no. 3, pp. 263–76.

Wang, J., Roe, B., Macmil, S., Yu, Q., Murray, J. E., Tang, H., et al. 2010, 'Microcollinearity between autopolyploid sugarcane and diploid sorghum genomes', *BMC Genomics*, vol. 11, pp. 261.

Wang, X., Wang, H., Wang, J., Sun, R., Wu, J., Liu, S., Bai, Y., et al. 2011, 'The genome of the mesopolyploid crop species *Brassica rapa*', *Nature Genetics*, vol. 43, no. 10, pp. 1035–9.

Watt, D. A., McCormick, A. J., Govender, C., Carson, D. L., Cramer, M. D., Huckett, B. I., et al. 2005, 'Increasing the utility of genomics in unravelling sucrose accumulation', *Field Crops Research*, vol. 92, no. 2, pp. 149–58.

Wei, X., Jackson, P. A., McIntyre, C. L., Aitken, K. S. and Croft, B. 2006, 'Associations between DNA markers and resistance to diseases in sugarcane and effects of population substructure', *Theoretical and Applied Genetics*, vol. 114, no. 1, pp. 155–64.

Wu, K. K., Burnquist, W., Sorrells, M. E., Tew, T. L., Moore, P. H. and Tanksley, S. D. 1992, 'The detection and estimation of linkage in polyploids using single-dose restriction fragments', *Theoretical and Applied Genetics*, vol. 83, no. 3, pp. 294–300.

Wu, P., Zhou, C., Cheng, S., Wu, Z., Lu, W., Han, J., et al. 2015, 'Integrated genome sequence and linkage map of physic nut (*Jatropha curcas* L.), a biodiesel plant', *The Plant Journal*, vol. 81, no. 5, pp. 810–21.

Yang, D. E., Zhang, C. L., Zhang, D. S., Jin, D. M., Weng, M. L., Chen, S. J., et al. 2004, 'Genetic analysis and molecular mapping of maize (*Zea mays* L.) stalk rot resistant gene *Rfg1*', *Theoretical and Applied Genetics*, vol. 108, no. 4, pp. 706–11.

Yang, X., Sood, S., Glynn, N., Islam, M. S., Comstock, J., and Wang, J. 2017, 'Constructing high-density genetic maps for polyploid sugarcane (*Saccharum* spp.) and identifying quantitative trait loci controlling brown rust resistance', *Molecular Breeding*, vol. 37, p. 116. DOI: 10.1007/s11032-017-0716-7

Young, N. D. and Tanksley, S. D. 1989, 'RFLP analysis of the size of chromosomal segments retained around the *Tm*-2 locus of tomato during backcross breeding', *Theoretical and Applied Genetics*, vol. 77, no. 3, pp. 353–9.

Yu, J. and Buckler, E. S. 2006, 'Genetic association mapping and genome organization of maize', *Current Opinion in Biotechnology*, vol. 17, no. 2, pp. 155–60.

Yu, J., Hu, S., Wang, J., Wong, G. K., Li, S., Liu, B., et al. 2002, 'A draft sequence of the rice genome (*Oryza sativa* L. ssp. *indica*)', *Science*, vol. 296, no. 5565, pp. 79–92.

Zhang, J., Nagai, C., Yu, Q., Pan, Y., Ayala-Silva, T., Schnell, R. J., Comstock, J. C., et al. 2012, 'Genome size variation in three *Saccharum* species', *Euphytica*, vol. 185, no. 3, pp. 511–19.

Zhao, L., Yuanda, L., Caiping, C., Xiangchao, T., Xiangdong, C., Wei, Z., et al. 2012, 'Toward allotetraploid cotton genome assembly: Integration of a high-density molecular genetic linkage map with DNA sequence information', *BMC Genomics*, vol. 13, no. 1, pp. 539.

Zhu, C., Gore, M., Buckler, E. S. and Yu, J. 2008, 'Status and prospects of association mapping in plants', *The Plant Genome*, vol. 1, no. 1, pp. 5–20.

Chapter 2

Advances in understanding of sugarcane plant growth and physiology

Frederik C. Botha, Sugar Research Australia and the University of Queensland, Australia

1 Introduction

Sugarcane produces the highest biomass yield among the world's four most produced crops (rice, wheat, maize and sugarcane), and is ranked third in quantity of plant calories in the human diet (Moore and Botha, 2013). As a result of its very high biomass production, well-established farming, harvesting and processing technologies, sugarcane is a leading candidate for bioenergy production and as a feedstock for bio-refineries.

However, productivity improvements in sugarcane have been negligible in the past three decades, and production costs and production statistics are reflecting decreased yields globally (FAO, 2014). In all cases, increased production is linked to expansion of production area rather than to increases in yield (Jackson, 2005).

Despite sugarcane's growing importance, the progress in understanding crop physiology to the level in other crops has not been achieved in sugarcane. There are two main reasons for this. First, most sugarcane-specific research worldwide stems from work done by private- and industry-owned research institutions, some of which have closed their doors or are in danger of doing so. Second, sugarcane is genetically and metabolically famous for its complexity.

Commercial sugarcane (*Saccharum* L. spp.) has the capacity to convert up to 2% of incident solar energy into biomass (Robertson et al., 1996; Muchow et al., 1997; Moore et al., 1997). The current maximum yields attained in commercial production is only 50% of the experimental maximum yield and 30% of the theoretical maxima (Moore, 2005). The highest reported sucrose content of whole sugarcane culms is also more than 10% higher than current commercial realised yields (Berding, 1997).

http://dx.doi.org/10.19103/AS.2017.0035.03

In its most basic form crop yield can be summarised as ensuring that at each stage of growth and development the rate of supply of assimilates from the 'source' leaves to the 'sink' of growing or filling organ must meet the potential rate of growth or filling of the organ (Lawlor and Paul, 2014; Reynolds et al., 2011). Similarly, supply must meet demand if the potential yield of the sink is to be attained (Watt et al., 2013).

In contrast to the major staple and fruit crops, source–sink relationships and metabolic control of these processes in sugarcane is still poorly understood. However, some recent advances in the past two decades, and their potential impacts are discussed in this chapter.

Continued investment into these complex metabolic interactions hold much promise to overcome the current impasse on progress in crop yield. The progress in understanding and manipulation of sugarcane flowering has also set the scene for significant improvements in sugarcane breeding by controlled and planned crossing programmes.

2 The importance of plant physiology

Plant physiology is the study of the chemical and physical processes associated with living plant tissue. In such a living system, materials and energy are in constant flux and regulated by complex internal signal cascades operated by internal cues and external biotic and abiotic factors.

Plant and crop physiology bridge the gap between biochemistry and agricultural productivity. Biochemistry operates on short time and small-scale experimentation and analyses while crop production is focused on abiotic and biotic conditions effecting yield over large areas. The full integration of these small-scale highly detailed processes into large-scale complex systems is difficult. The success or failure to achieve this integration is directly linked to the level of understanding of the underpinning metabolic mechanisms. These processes are usually imperfectly understood because of the significant variation in data between small-scale measurements under controlled conditions and field-based production conditions. This is due to the complex interaction between processes under the more variable field conditions, and is also due to our limited understanding on how to scale observations at the cellular level to canopy and crop performance.

With the large amounts of data that are generated in the 'omics' era the challenge is increasingly to put this disparate information together as an integrated system at the molecular level. Such integration requires a systems biology approach (for review Rohwer and Uys, 2013). Much of these analyses heavily rely on computational modelling to understanding the dynamic aspects of plant metabolism. As shown later in this chapter mathematical modelling of these complex metabolic processes and crop performance offers opportunities of examining and explaining such systems. Metabolic modelling is highly dependent on detailed knowledge of the biochemical processes that underpin the particular crop physiological responses (Foulkes et al., 2011).

Crop production and advances in agriculture are tightly linked to our understanding of plant physiology. Without an understanding of plant physiology it becomes virtually impossible to make sound predictions of how the crop will respond to different environments, or how certain genetic modifications through conventional breeding or gene manipulation will alter crop performance.

A good understanding of the underpinning biochemical mechanisms supports and encourages links to breeding and crop selection using quantitative trait loci (QTL) and molecular marker techniques.

3 Carbon fixation, partitioning and biomass of sugarcane

Sugarcane's high biomass yield and accumulation of extraordinary quantities of sucrose have attracted much attention and investigation (for review Moore and Maretzki, 1996).

Translocation of reduced carbon occurs through the phloem and is driven by a turgor pressure difference between the sink and source. Such a pressure gradient is dependent on the creation of a high sugar concentration in the phloem in the source tissue and an unloading or utilisation of the sugar in the phloem at the sink. The flow rate of an assimilate through the phloem is related to the hydrostatic pressure difference between the source and the sink. The hydraulic conductivity is negatively linearly dependent on the path, and hence in sugarcane, the leaves closest to a particular internode of the culm are most important in supplying photoassimilates (Hartt et al., 1963; McCormick et al., 2006). The sugarcane plant could therefore be viewed as a series of source–sink compartments at different stages of development.

However, it is well established that photoassimilates from source leaves move basipetally for several internodes before a portion is redirected acropetally to shoot apices (Hartt et al., 1963). Unlike storage sinks in other commodities the internodes of sugarcane have a limited capacity to store sucrose and are dependent on the volume of the internode (Rae et al., 2006).

More than 95% of the carbon exported from the sugarcane leaf is in the form of sucrose (Hartt et al., 1963), and is translocated via phloem to the different sink tissues. These include both the culm immature stem internodes (meristematic growth sink) and fully extended internodes (storage sink). In many plants, meristematic sinks are source-limited and storage sinks are sink-limited (Smith and Stitt, 2007). This would imply that for sugarcane, the immature stalks would be constrained by the extent of available photosynthate, while the mature sugarcane stalks would be limited by their capacity to import sucrose from leaves (Wang et al., 2013).

It is therefore not surprising that much of the emphasis in past studies were focused on the metabolism in the sink (culm) tissues. As later discussed, much of this work had a strong focus on sink strength.

3.1 Supply and demand

The control of flux through a metabolic system is regulated not only by demand but also by the sensitivity of supply to demand (Cornish-Bowden et al., 1995; Hofmeyr and Cornish-Bowden, 2000). Supply–demand models can be linear or cyclic (Hofmeyr and Rohwer, 2011). Control of sink and source activities in sugarcane can be described in an analogous manner (Watt et al., 2013). In this description it is argued that sucrose accumulation in the sugarcane culm is only possible because the supply exceeds the demand for growth and respiration. The important points to consider from this work include:

- The demand function contains three major components, growth, respiration and storage. Flux in the photosynthetic supply would therefore be controlled by these three factors.
- The slope of the supply and demand functions could have different sensitivities to change in a particular intermediate (elasticity coefficients). In sugarcane it is likely that the elasticity (sensitivity) of demand in the mature internode is low.

- Metabolic flux is shared between many enzymes and it is unlikely that the control of flux is exercised by any one enzyme. A change in the activity of any single enzyme in a pathway is more likely to change the concentration of associated metabolites than to alter the flux through the pathway.
- The widely held view that 'rate limiting' steps in metabolism, and conclusion that these are important in the control of flux is incorrect. In fact, in a metabolic pathway, the control of flux through a metabolic pathway exerted by a single enzyme is variable and never high, that is, the enzyme control coefficient of a single enzyme is not high.
- To alter flux through a pathway, it is likely that the activities of many or all of the enzymes in a pathway must be altered. Two possible ways for this to occur could be through parallel posttranslational modification of many enzymes or 'regulation by demand' in which products exert feedback regulation through a signalling mechanism(s).

This innovative description brought some important insight and generally forces a reassessment in interpretation and even future planning of sugarcane research. It demonstrates that the interaction through feedforward and feedback regulation between the supply (source) and demand (sink) needs to receive much more attention. Many studies have shown the dominant influence of sink activity on source photosynthesis through sugar-induced feedback inhibition (for review Watt et al., 2013).

3.2 Source metabolism

Photosynthesis plays a key role in the supply function and therefore on biomass production and crop yield. Sugarcane can achieve peak dry matter yields up to 100 tonne dry matter (ha year)$^{-1}$ (Inman-Bamber, 2013). This is at least in part due to the fact that sugarcane utilise C4 photosynthesis, which is known to confer superior radiation-, nitrogen-, and water-use efficiencies in the warm tropical and subtropical climatic conditions (Sage et al., 2013).

In C4 plants, CO_2 is fixed by phosphoenolpyruvate carboxylase (PEPC) in the mesophyll cells to produce C4 acids which are translocated to the bundle sheath cells where RuBisCO is located. The form of C4 acid translocated to the bundle sheath cells varies between species, and based on this C4 plants are described in three types. The Nicotinamide adenine dinucleotide phosphate-malic enzyme (NADPME), Nicotinamide adenine dinucleotide-malic enzyme (NADME) and phosphoenolpyruvate carboxykinase (PEPCK). Sugarcane is classified as an NADP-ME type plant.

To understand canopy photosynthesis, it is important to note that new leaves continue to develop on the culm until the plant is harvested, or dies especially after damage or flowering when the apical meristem switches from in-determinant to determinant growth. Up to 60 leaves can be produced in an annual cropping cycle. The rate of leaf appearance is cultivar dependent and is driven primarily by temperature, but it can also be modified by water stress (Bonnett, 2013; Inman-Bamber, 2013). The base temperature (Tbase = daily mean temperature below which the process under consideration ceases) ranges from 8°C to 10°C.

The time and temperature conditions occurring between the appearance of two successive leaves is termed the phyllochron and is measured in heat units or thermal time (°Cd). Thermal time is the sum of daily effective temperatures (daily mean air temperature – Tbase). On average sugarcane requires 144°Cd for the emergence of each leaf (Inman-Bamber, 2013).

Leaf senescence in usually triggered after emergence of seven mature leaves and the rate of senescence is approximately 10% lower than the rate of leaf production and hence between 9 and 12 leaves can be present on the culm. Competition for light, water availability and leaf sugar levels are important factors controlling the rate of leaf senescence.

Photosynthesis is highest in the youngest leaves of a sugarcane plant and decreases with increasing leaf age, regardless of plant age or environmental conditions. A similar decrease is evident in the entire canopy where the photosynthetic rates of 4-month-old plants exceed those of 8-month-old plants (Allison et al., 1997).

A detailed description of sugarcane photosynthesis is beyond the scope of this chapter but can be found in good review papers (Furbank, 2011; Sage et al., 2013) and is summarised in Fig. 1.

Historically sugarcane was described as using the NADP-malic enzyme type of C4 photosynthesis. However, increasingly molecular and biochemical data suggest that the three decarboxylation types may not be rigidly determined and that flexibility between the pathways could be regulated by developmental and environmental cues (Furbank, 2011; Pick et al., 2011).

There are two possible mechanisms through which aspartate can contribute to the bundle sheath CO_2 pool. Until recently a model was favoured where OAA generated from aspartate is re-reduced to malate in the bundle sheath and then decarboxylated by either the chloroplast by NADP-malic enzyme or the mitochondrial NAD-MDH. However,

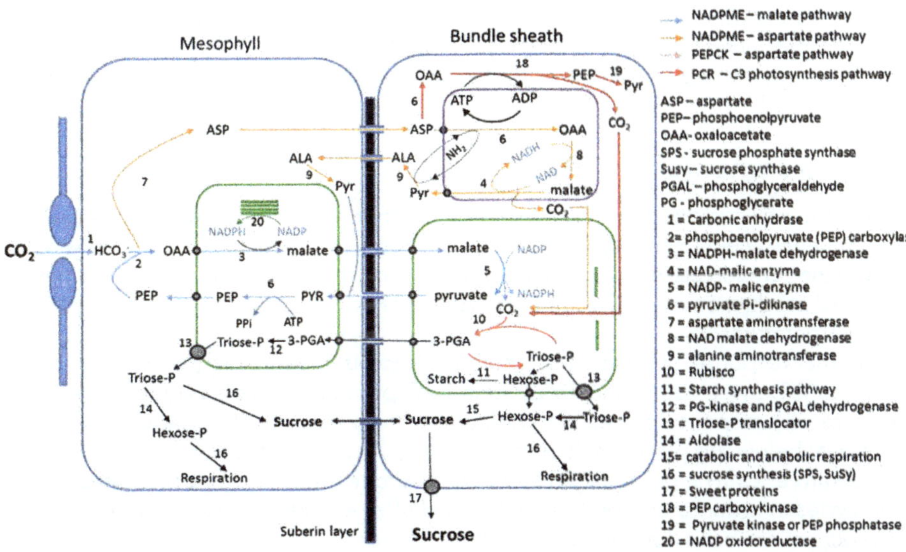

Figure 1 Photosynthetic mechanism in sugarcane. There are two pathways for production of the translocated C4-acid to the bundle sheath, and three decarboxylation reactions in the bundle sheath. The dominant translocation pathway is through malate with aspartate as the minor pathway. Significant flexibility exists between the carboxylation and decarboxylation reactions. Chloroplasts and thylakoids are coloured green and the mitochondria purple. Note the lack of stacked grana in the bundle sheet chloroplasts.

molecular analyses demonstrated that PEP-carboxykinase was one of the most highly expressed bundle sheath transcripts in sugarcane leaves (Calsa and Figueira, 2007). If PEP-carboxykinase is involved, the OAA produced by aspartate aminotransferase could be directly decarboxylated in the cytosol. The ATP that is required in this system could come from the chloroplast or from oxidative phosphorylation (Furbank, 2011).

In comparison with other C4 species our information on the process in sugarcane is more limited. However, the fact that there is a linear relationship between intercepted photosynthetic active radiation (PAR) and total dry matter production highlights the importance of photosynthesis in sugarcane yield.

The physiological and biochemical causes of photosynthesis variation among cultivars and leaf types are not well characterised. The leaves of grasses represent a complex system because there is a cellular developmental gradient along the leaf blade, with the basal cells being undifferentiated and immature and the cells towards the tip becoming more mature and specialised (Mattiello et al., 2015).

Early work on the photosynthesis rates in sugarcane varieties reported values (20 to 40 mol m^{-2} s^{-1}) that were not typical of C4-like; in fact it was only marginally better than C_3-species. An evaluation of all literature suggest that the range within commercial varieties actually vary between 16 to 61 mol m^{-2} s^{-1} with an average around 40 mol m^{-2} s^{-1} (Sage et al., 2013). Maximum photosynthesis rates in the wild *Saccharum* species are higher than in cultivated commercial sugarcane varieties (Irvine, 1975). This difference might be linked to leaf nitrogen status as discussed later.

Like with many other crops, early studies on leaf photosynthesis failed to indicate a significant correlation between photosynthesis and yield (Irvine, 1983; Irvine, 1975). However, recent studies indicate a positive correlation among photosynthesis, crop yield and biomass production (Long et al., 2015), suggesting that increasing photosynthesis is a potential way to enhance sugarcane productivity (Marchiori et al., 2014; De Souza et al., 2008).

3.2.1 Environmental factors

The environmental factors that influence photosynthetic activity (Sage et al., 2013) will directly influence yield. Because of the lack of progress to significantly increase yield, many industries are forced into expansion of production areas. Often these expansions are into marginal production areas. Two of the most important constraints are water availability and low temperatures.

The thermal response of sugarcane photosynthesis is a parabolic pattern with broad thermal optima between 30°C and 35°C. Sugarcane photosynthesis is very sensitive to low temperature with response curves extrapolating to zero around 10°C (Sage et al., 2013). Temperatures below this point result in chilling injury. As a consequence of its sensitivity to low temperature, cultivation of sugarcane in frost-prone areas often necessitates shorter growing cycles in order to escape the detrimental after-effects of frost damage.

The biochemical basis for the low temperature sensitivity is not clear but probably related to less efficient energy transfer to reaction centres and slower reduction of the primary electron acceptors of photosystem II (PSII) (Ebrahim et al., 1998) and the activity of the enzymes NADP-malate dehydrogenase (NADP-MDH), pyruvate, orthophosphate dikinase (PDKK) and PEP carboxylase (Du et al., 1999). Because of the sensitivity of these enzymes to low temperature it was suggested that disruption of CO2-supply to the bundle sheath is a primary limitation caused by chilling injury in sugarcane.

There are significant varietal differences in chilling tolerance, and at least some of this tolerance relates to better temperature stability of NADP-MDH and PDKK (Du et al., 1999) and protection of the electron transport system of the chloroplast (van Heerden, 2014). With more supportive evidence these parameters might be useful in selecting for cold-tolerant sugarcane genotypes.

3.2.2 Nitrogen use efficiency

Most productive crop species have high leaf nitrogen (N) contents which allows high rates of photosynthesis. Photosynthesis varies linearly with leaf N content, up to a saturation point (Sage et al., 2013). In sugarcane, photosynthesis varies linearly with leaf N content between 30 and 120 mmol N m^{-2}.

The nitrogen response curve of maize is much steeper than for commercial sugarcane varieties. Noteworthy is that some of the sugarcane progenitors have nitrogen response curves close to that of maize. Also maximum photosynthetic rates in sugarcane requires twice as much nitrogen as in maize. The photosynthetic nitrogen use efficiency (PNUE) of sugarcane varieties varies significantly and is 20–60% lower than that of maize. The relative low variability (<10%) in PNUE in maize and high variability (up to 40%) in sugarcane probably suggest much better selection for this trait in maize than sugarcane.

For sugarcane it is important to note that *Saccharum spontaneum* tends to maintain high leaf N and hence operate near saturation of photosynthesis while this is not the case for *Saccharum officinarum* and the commercial sugarcane varieties. Recurrent selection back to the *S. officinarum* genetic background could therefore have significant consequences in PNUE.

In many sugarcane farming systems fertiliser is applied at the beginning of the growth cycle. In these cases, the leaf nitrogen concentration begins to decrease after the first few months of growth. In addition, leaf nitrogen concentration decreases with crop age, irrespective of time of year, and thus cannot be ascribed to seasonal variation in weather. The decrease in leaf N, although smaller, is evident even in crops where double nitrogen levels were applied later as a top dressing.

The pronounced depletion of leaf N in the plants as they mature results in a parallel decrease in photosynthesis (Bull, 1969; Allison et al., 1997). If efforts are made to ensure leaf N is high, either by measuring young plants or by following determined efforts to super-load leaves with N through heavy fertilisation, then the peak photosynthetic rates can be high and approach those reported for the best performing C4 crops.

Collectively these data suggest that high photosynthetic performance requires either maintenance of high leaf nitrogen or an increase in PNUE. In a sugarcane production system focused on high sugar content, maintaining high leaf N status is a problem as this results in growth at the expense of sugar accumulation.

Two different types of data suggest that there is 'spare' photosynthetic capacity if some of the sucrose can be utilised for synthesis of other components. Firstly, the significant gains in total biomass achieved in breeding attempts where sucrose is not a main selection criterion (Giamalva et al., 1984; Viator and Richard Jr., 2012). Secondly, the fact that production of other sugars in the sugarcane culm can increase total carbon yield (for reviews Patrick et al., 2013; Wang et al., 2013) points to feedback of sucrose on leaf metabolism. In addition, genetic manipulation that results in increases in sucrose (early ripening) in the internodes (for review Wang et al., 2013) leads to biomass penalties consistent with sucrose feedback control on photosynthesis.

3.2.3 Feedback inhibition

Sucrose is not only the major long-distance transport form of C in most plants; it is also part of an important signalling pathway involving sucrose and its hydrolytic products glucose and fructose. The levels of these sugars are sensed by both intracellular and extracellular mechanisms resulting in the control of a range of physiological and molecular processes (Smeekens et al., 2010). Generally, carbon depletion is associated with enhanced expression of photosynthesis while accumulation of these sugars tend to suppress photosynthesis and stimulate storage processes (Smeekens et al., 2010; Koch, 1996).

Feedback regulation of leaf sucrose metabolism and photosynthetic activity has been described for C3 and C4 species including sugarcane, indicating that it is probably a universal control mechanism in plants (Watt et al., 2013).

Indications are that sugarcane leaves have a large 'spare' capacity for adapting to increased demand, indicating that significant feedback regulation of photosynthesis occurs (McCormick et al., 2008c). Progenitors of the sugarcane commercial hybrids with low sucrose accumulation capacity, such as *S. spontaneum* and *Saccharum sinense*, have lower photosynthetic rates than higher sucrose-accumulating commercial *Saccharum* spp. hybrids (Irvine, 1975; Nose et al., 1994). However, this is also related to the fact that *S. spontaneum* tends to maintain high leaf N (for review Sage et al., 2013).

In the sugarcane leaf a doubling in end of the day sucrose levels results in both a decline in photosynthetic electron transport and CO_2 fixation, and a decrease in stomatal aperture (McCormick et al., 2008a). The expression of 74 genes is altered as a consequence of the elevated sugar levels. Decreased expression of genes related to photosynthesis and increased expression of genes involved in assimilate partitioning, cell wall synthesis, phosphate metabolism and stress were observed. Genes in both the C3 and C4 pathways were downregulated, indicating that the elevated sugar levels affect metabolism in both mesophyll and bundle sheath cells.

The genes involved in the light reactions of photosynthesis, especially those linked to chlorophyll binding and the photosystems (PSI and PSII) are very sensitive to increased sugar levels. In fact, leaf physiological ageing in sugarcane was promoted as a result of sucrose accumulation during culm maturation as a consequence of decreased sink-demand for photosynthate (McCormick et al., 2008b).

3.2.4 Leaf senescence and sucrose

A factor, till recently largely ignored, is the role leaf sugar levels play in controlling leaf senescence. Although there are feedback mechanism to control leaf photosynthesis, it is evident that sucrose will build up in the leaf over time, with highest concentrations in the older leaves. Leaf senescence can be significantly delayed if sugar build up in the leaf is reduced (McCormick et al., 2008b). Sugar accumulation in the leaf is linked to the lack of sink strength, that is, sucrose in the phloem and in the internodes. There appears to be a threshold value of sugar in the internodes attached to and above the leaf that triggers senescence.

It is interesting to note that sucrose accumulation in the leaf has been associated with leaf yellowing induced by pathogens (Yan et al., 2009; Lehrer et al., 2007), or other factors (Marquardt et al., 2016).

3.3 Sink metabolism

High 'demand' and efficiency to load, transport of sugars, off-loading and sucrose metabolism at the sink are key drivers of crop yield. This chapter does not consider the loading and transport of sugars in any detail as it is well covered in recent reviews (Grof et al., 2013; Bihmidine et al., 2013).

Export of photoassimilates from the leaf and unloading of the phloem commences when the leaf is fully expanded and elongation of the internode commences (Moore and Maretzki, 1996). Sugarcane largely follows a symplastic unloading pathway and therefore efficiency of transport is highly dependent on favourable gradients in sucrose concentration and hydrostatic pressure in the phloem. In maize and barley, the estimated sucrose concentrations in the sieve elements of the phloem range between 0.8 and 1 M and is likely to be similar in sugarcane (Grof et al., 2013). The prevailing sucrose concentration in elongating internodes is almost an order of magnitude lower and hence a substantial concentration difference exists to drive diffusion through a symplastic route.

3.3.1 Sink strength

What constitutes sink strength of the magnitude of the 'demand function' has been the source of longstanding arguments. However, it is widely accepted that it is the competitive ability of an organ to import photoassimilates and that this is the product of the sink size and sink activity (Bihmidine et al., 2013; Herbers and Sonnewald, 1998).

As an internode develops, the size of individual cells, but not the number increases (Rae et al., 2006). There is a limited window for increasing the length of each internode (Rae et al., 2006; Lingle, 1999) and it would seem that there is little genetic variability for this trait in sugarcane. There are two clear phases of sink size increase in the internodes. Using a base temperature of 18°C internode elongation stops after 150°Cd while internode elongation is very sensitive to water stress (Inman-Bamber, 2013). This is entirely consistent with internode expansion being driven by dry matter accumulation and cell expansion.

By contrast, dry matter accumulation continues for up to 800°Cd (Lingle et al., 2009; Lingle, 1997). By the time internode length stopped increasing, the dry weight of the internodes was still less than half its final weight. After cessation of internode elongation the internodes continue to expand radially, resulting in volume increases up to internode 15 (Sinclair et al., 2005).

Sucrose and total sugar levels start to increase before internode elongation stops but the rate rapidly increases after 200°Cd. Sucrose accumulation reaches only a plateau after 1000°Cd. Water content of internodes decreased in an inverse manner to sucrose content. In the young elongating internodes, the water content is around 920 g kg^{-1} (92%) fresh weight and decrease to around 700 g kg^{-1} (70%) fresh weight in the mature internodes.

This demonstrates that internodes without any attached leaves are still active sinks and that a strong demand function stays in place after leaf senescence of the attached leaf. As the mature internodes dry weight is maintained it also shows that sucrose imports in these internodes at least balances that used for maintenance respiration and remobilisation to the rest of the plant.

When leaf 6 is pulsed with labelled CO_2 only 18% of the exported label is present in internode 6. Most of the exported label is present in internode 8 (Fig. 2). The data suggest that more than 70% of the exported sucrose is present in internodes 6, 7 and 8. Only 11% of the sucrose moves into the internodes above the labelling site. It is important to note

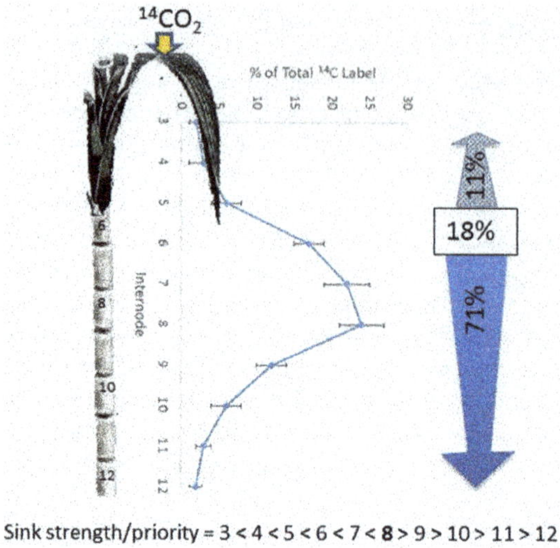

Figure 2 Distribution of radio-label in the sugarcane culm 24 h after a pulse of leaf 6 with $^{14}CO_2$. The blue arrows indicate the flow of label above (younger internodes), and below (older internodes) of internode 6. The percentage label in the internodes was used to ascribe the relative sink priority to individual internodes.

that within 24 hours after pulse-feeding of sugarcane leaves the roots produce labelled CO_2, indicating that the sucrose produced in the leaf serves as a carbon source for the entire stalk and root system (Hartt et al., 1963).

3.3.2 Carbon partitioning in the internode

Sink strength and hence sucrose import into the internodes is dependent on maintaining a low sucrose concentration in the cytosol (for review Wang et al., 2013). An important point to note is that sucrose accumulation initially occurs in the symplastic compartment and probably by loading of the sucrose into the vacuole. However, as sucrose accumulation proceeds, especially after internode length expansion ceases, sugars are moved into the vacuole and also leaked into the apoplastic space that can occupy as much as 20% of the total internode volume (for review Moore, 1995). The fact that the sucrose concentration in the apoplast is the same as in the symplast supports the suggestion that it is primarily from the storage parenchyma cells.

There are three important components of the 'demand' function in the internode, namely, use of sucrose for biosynthesis (cell wall and other cellular constituents), respiration and storage of sucrose in the vacuole.

The dry matter composition of the internodes changes significantly during the elongation and biomass accumulation phases (Fig. 3a). The water-insoluble component (primarily fibre), and non-sucrose-water-soluble fraction (reducing sugars, amino acids and organic acids), represents approximately 90% of the total dry matter during the internode elongation phase (up to internode 5). Sucrose accumulation starts already in the young internodes but accelerates sharply when internode elongation stops. In internode 10, 50% of the dry mass is sucrose and this elevated sucrose came at the expense of the other water solubles and fibre.

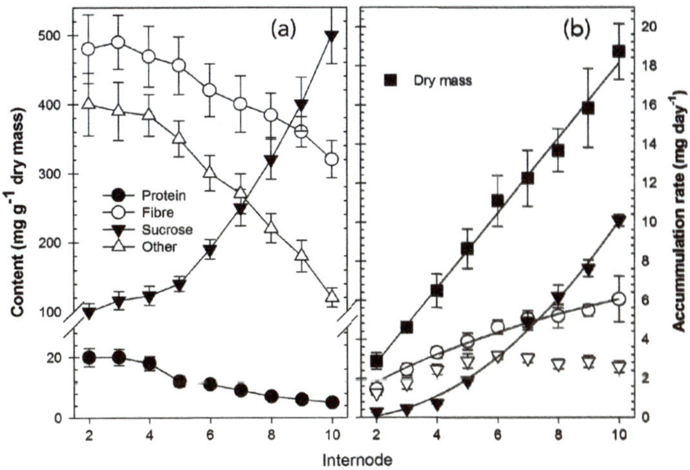

Figure 3 Changes in biomass composition (a) and accumulation rate of individual biomass components (b) for internodes 2 (youngest) to 10.

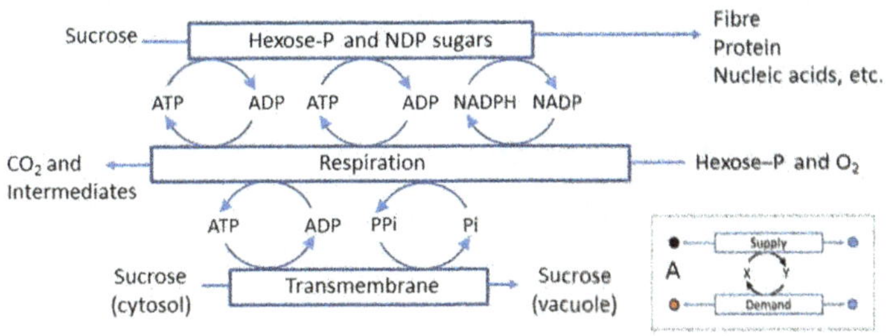

Figure 4 Metabolic supply and demand reaction blocks can be linked in a cyclic manner as illustrated in (A) (for details see Hofmeyr and Rohwer 2011). This model is likely the way cellular metabolism is driven in the sugarcane internodes. Even in the simplest model of direct loading of sucrose into the vacuole without prior cleavage, respiration will play a pivotal role in ensuring effective carbon movement into the sink.

The rate of total dry mass accumulation is linear. This suggests that sink strength is maintained and the ratio between carbon input and the rate of respiratory CO_2 loss is constant (Fig. 3b). None of the other components change in a linear fashion, demonstrating the change in carbon partitioning between the different biomass components as the internode develop and matures.

In the top ten internodes the incoming sucrose is rapidly broken down for respiration and then re-synthesised which allows for a dynamic balance between storage and respiration or for other uses (Wendler et al., 1990). During the sucrose degradation and resynthesis cycle, carbon is partitioned into other competing metabolic sinks including respiratory pathways, cycling through hexose pools, organic acids and amino acids, proteins and cell walls (Botha et al., 1996; Whittaker and Botha, 1997).

The supply–demand function in the internodes probably better fits a cyclic model rather than a linear control model (Fig. 4). The storage and mobilisation of sucrose for synthesis of other cell constituents, including polymers, will be driven by the availability of ATP and reducing equivalents in the form of NADPH and NADH. In this model respiration is core to the metabolism of sucrose and maintenance of a low sucrose concentration in the parenchyma cytosol to ensure continued movement from the source leaves into the internodes. Respiration is therefore a key component of maintaining sink strength.

Partitioning into these competing pools are not occurring at the same rate during development of the internodes (Whittaker and Botha, 1997; Botha and McDonald, 2010). The pattern of loss of radio-label from the internodes is consistent with a model where the incoming sucrose is cleaved, or hydrolysed, ensuring that a portion of the label remains in cellular constituents that can participate in cellular metabolism and be lost as CO_2.

An important point in the partitioning of carbon and sucrose storage is the competition for UDP-glucose for cell wall synthesis, sucrose synthesis and respiration. In a supply–demand model, it can be argued that sucrose cleavage to UDP-glucose serve as the supply function for the three demand functions, anabolic and catabolic respiration, cell-wall synthesis and sucrose resynthesis.

The sucrose metabolism in the sugarcane internode is characterised by substrate cycles (futile cycles) involving the continuous synthesis and degradation of sucrose and the recycling of metabolic intermediates between the pools of hexose phosphates and triose phosphates in the cytosol (Bindon and Botha, 2002). Between 10 and 30% of carbon entering sucrose and the hexose phosphate pools is subjected to such substrate cycling (Bindon and Botha, 2002; van der Merwe et al., 2010). These cycles are thought to increase the sensitivity of the system to respond more quickly to a multitude of stimuli that may affect the respiratory flux and, in this way, play an important role in regulation of sucrose utilisation and accumulation.

3.3.3 Importance of respiration

In the young internodes still in the phase where cell elongation occurs little of the incoming sucrose remains as sucrose and is largely distributed to the cellular respiration, anabolic and catabolic respiration. In older tissue the breakdown of sucrose is reduced and the majority of the carbon is stored as sucrose (Bindon and Botha, 2002; Whittaker and Botha, 1997; Ebrahim et al., 1998; Komor, 2000).

The respiratory rate in those internodes still in the elongation phase (up to internode 6) remains fairly constant but then rapidly decrease by approximately 50% over the next three internodes (Whittaker and Botha, 1997; Bindon and Botha, 2002).

Genetic manipulation by down regulation of soluble neutral invertase (SNI) or pyrophosphate dependent phosphofructokinase (PFP), both key for sucrose hydrolysis and glycolytic carbon flow, led to enhanced sucrose accumulation in the sugarcane culms (Rossouw et al., 2010; Groenewald and Botha, 2008; van der Merwe et al., 2010). However, in both cases the net result was a significant reduction in biomass accumulation. This phenotype could be the result of reducing the sucrose concentration gradient between the leaf and sink tissues by slowing down the production of ATP, reducing equivalents and carbon skeletons to enable maximum extension. It probably could also reduce the ability to import sucrose into the vacuole.

3.3.4 Maximising sucrose accumulation

A key question that plant breeders and physiologist have been wrestling with is what the maximum sucrose level is that the cells in mature sugarcane culm can accumulate. Important in this is to understand the relationships between total sugars, moisture content and fibre content in the sugarcane culm. As pointed out previously water content in the mature culm of sugarcane is around 70%. This would then imply that the maximum amount of dry mass is 30%, with more than 90% being allocated largely to fibre and sucrose. Not surprising therefore is that there is a strong negative correlation between fibre and sucrose across the *Saccharum* complex (Bull and Glasziou, 1963). Modern sugarcane varieties have fibre contents around 13% which would then suggest that the maximum sucrose content would be approximately 15% (assuming that the two components make up 95% of the total dry mass). Assuming that 5% would be a minimum fibre content to ensure adequate mechanical strength it was suggested that there is a physiological limit to sugar accumulation in the genus *Saccharum* around 27% of the fresh weight.

There is a remarkable similarity in sucrose content on a dry mass basis in published data on some Australian, South African, Mauritian and Hawaiian cultivars. It would appear that the maximum sucrose content is around 48–50% of total dry weight (Muchow et al., 1997; Robertson et al., 1996). This would suggest that sucrose content above 15% is not achieved in current sugarcane varieties.

Modern sugarcane varieties are all interspecific hybrids and represent generation seven back crosses. It is interesting to highlight some of the most noticeable differences in sucrose accumulation between the modern varieties and the first generation cultivars that came out of the interspecific crosses between *S. officinarum* and *S. spontaneum* (Lingle et al., 2009). .

Modern cultivars accumulated sugar faster, and the final sucrose concentration was higher, in the internodes than the older varieties. The water content in the modern varieties were lower than in older ones. However, the elevated sucrose levels were not only due to the lower water content but also due to increased total sugar content.

Until recently it was believed that the amount of sucrose that can be accumulated is strongly linked to an osmotic sensing and that the apparent ceiling in commercial sucrose levels is linked to a physical constraint (Jackson, 2005).

A number of transgenic approaches have shown that the total sucrose and carbohydrate pools can be significantly increased through the addition of an additional molecular sink (for review Patrick et al., 2013). Introduction of a transgene to drive fructan production resulted in an 84% increase in total sugar content without any loss in the amount of sucrose. Production of isomaltulose in transgenic sugarcane resulted in a doubling of sucrose content (Wu and Birch, 2007). This is made possible by a reduction in the water content of the internodes by 10–12%.

These results challenge the long-standing assumption that high-sugar genotypes require at least 70% moisture content in mature internodes. This shifts the focus for work towards enhanced sugar content from increased sugar:fibre partitioning to increased sugar:water ratio, with substantial scope for gains.

3.4 The impact of source–sink relationships on yield

3.4.1 Ripening

Sucrose accumulation in the sugarcane culm can be divided into two phases. During the first phase of the season the sucrose content of the basal internodes is increasing and in the second phase the basal internodes have reached a maximum and the only gains then come from redirecting and altering carbon partitioning to the top internodes.

Ripening of sugarcane refers to a practice or treatment that results in an increase in sucrose content (%cane) and an improvement in sucrose yield (tonne sucrose ha^{-1}). This is very different from a practice where the culms become dehydrated due to water limitation with an associated increase in sucrose concentration. Often such extreme conditions are associated with a penalty on total biomass and hence a reduction in the sucrose yield per area land (for review van Heerden et al., 2013).

So for true ripening of sugarcane conditions are sought that will reduce carbon flow into non-sucrose in the internodes while still maintaining a net supply of carbon to the internodes. This could be achieved by factors such as water and nutrient stress, temperature, and chemicals that reduce biosynthetic activities and respiration in the internodes more than that of leaf photosynthesis and phloem transport to the internodes. In many production areas the environmental drivers for ripening are suboptimal during the harvest season and hence management strategies have been in the development to enhance ripening.

In practice, sugarcane ripening can be achieved through either manipulation of water supply or chemical treatment. The crop response to these ripening treatments depend on the physiological status of the crop, the environmental conditions at the time of treatment and the nature of the chemicals being used.

3.4.2 Reduced growth phenomenon

The radiation use efficiency (RUE) of a crop can be defined as the ratio of biomass accumulated to intercepted radiation. Crop species differ in RUE, but despite these differences, many species sustain very consistent RUE values throughout the cropping cycle and therefore a strong linear regression exists between biomass accumulation and radiation. Crop models, including that for sugarcane, therefore often use constant values of potential RUE when simulating plant growth during the season.

However, there are many examples where constant RUE in sugarcane crops is not achieved throughout the full crop cycle. In affected crops there is a distinct uncoupling of the linear relationship between cumulative intercepted radiation and biomass accumulation, with the latter declining well before the harvest date which is referred to as the reduced growth phenomenon (RGP) (for review van Heerden et al., 2010). Depending on the time of planting the yields can be reduced between 12 and 62% by the RGP. The severity of RGP differs between varieties.

The physiological basis for the RGP is still poorly understood. However, recent advances in three potential causes are worthwhile exploring. Firstly, when water is not a limiting factor, specific leaf nitrogen content is a major determinant of leaf photosynthetic rate and, therefore, of RUE. There is typically a progressive decrease in leaf nitrogen as the sugarcane crop age (Hartt and Burr, 1965; Allison et al., 1997). Although it is commonly believed that low plant N close to harvest is necessary to increase sucrose content, it is possible that SLN below a certain threshold may limit CO2 assimilation capacity and therefore biomass accumulation. The restriction of photosynthesis by inadequate leaf nitrogen could reach sizeable proportions later during the crop life cycle. A decline of specific leaf nitrogen is strongly associated with the occurrence of the RGP. Special care needs to be taken to ensure that leaf nitrogen content late in the cropping cycle does not induce the RGP.

Secondly, high foliar sugar content in sugarcane as a result of the high sugar content of the culm later in the cropping cycle could be involved in the feedback inhibition of photosynthesis (Hartt and Burr, 1965; McCormick et al., 2008c). Recent work on sugarcane suggested that it is not foliar sucrose content *per se* that exerts feedback regulatory control over photosynthesis, but rather foliar hexose content. It is important to note that high culm sucrose as found in the second season of a two-year crop and at the end of the winter period is often linked to the induction of the RGP.

Thirdly, a substantial portion of photoassimilates is consumed by respiration. Maintenance respiration (R_m) will be much higher in a high biomass than low biomass crop. As biomass increases with crop age, the proportion of respiring to photosynthesising tissue increases, causing a gradual decline in net productivity. High respiratory demand at the end of the winter period could exacerbate the low carbon availability as a result of possible feedback inhibition on photosynthesis.

All these mechanisms will be exacerbated in high early sucrose varieties and they might be more vulnerable to the RGP.

3.4.3 Lodging

Lodging of sugarcane occurs when the culm bends, or even snaps, or the stool tips out of the ground when the root anchorage fails. Lodging typically increases with increasing stalk height and hence it is often suggested that crop erectness could be improved by selecting clones with shorter stalks (Berding et al., 2005, and references therein). However, as pointed out in earlier sections this will have a negative impact on crop yield.

Lodging reduces yield through lower biomass production and a reduction in cane quality. This is caused by a reduction in radiation interception, radiation use efficiency, stalk smothering and stalk death (Berding et al., 2005; Park et al., 2005).

Lodging results in a 11–15% reduction in biomass, and 15–35% loss in sugar yield respectively (Singh et al., 2002; van Heerden et al., 2015).

Although it suggested that at least a part of the RGP is linked to lodging, installing of scaffolding to prevent lodging, the typical RGP was not evident. In addition, younger crops grew no faster than the scaffolded older crop showing that the reduced growth at time of lodging is not just due to the age of the crop. This also suggests that the losses as a result of lodging is more than just the RGP.

4 Flowering of sugarcane

A better understanding of the mechanisms that control flowering in sugarcane was one of the major advances in sugarcane biology. In nature, flowering is a critical event in the life cycle of plants, marking the transition from vegetative to reproductive growth.

Various crops are grown solely for the fruit and seed that they produce. Hence the transition from the vegetative growth stage to flowering has been studied in great detail by plant physiologists. However, flowering is also of crucial importance for plant breeding programmes as this leads to the production of seed, which carry all the genetic variation from one generation to the next.

During the development of the shoot the fate of the apical meristem will change at some point to start the production of new structures that culminate in the production of flowers. These steps that drive the phase change in the apical meristem are referred to as floral evocation. In most plant species flowering can only occur after the tissue has developed through a juvenile into an adult stage (for a review, see Moore and Berding, 2013).

Propensity to flower is a genetically determined trait of sugarcane species and clones. Under conditions normally conducive to flowering the proportion of flowering culms among clones will range from 0 to 100%.

The initiation of flowering is achieved by mature plants receiving the correct combination of day length (photoperiod), temperature, water and nutrients. Photoperiod is the most critical of these. The switch from vegetative to flowering requires a series of photoperiod cycles that occur in autumn. In addition, the subsequent development of the sugarcane panicle and florets requires a series of progressively shorter day lengths. Sugarcane is classified as an intermediate day length plant (IDP) or a quantitative short-day plant (SDP) with multiple inductive stages.

Because sugarcane flowers during a period of shortening days, the differentiation of the flower bud occurs from February to April, and the emergence of panicles occurs from April to July in the Southern Hemisphere. In the Northern Hemisphere, these conditions occur between July and August and from September to November, respectively.

4.1 Negative impact on yield

Because sugarcane inflorescence is a determinant panicle, induction of flowering stops the reproductive culms from producing internodes or leaves. Sugarcane flowering therefore is regarded as harmful to sucrose accumulation, as it forms a new strong sink which competes for sucrose, resulting in a reduction of harvestable biomass. Depending on the genotype, and proportion of culms that flower, losses in sugarcane productivity and sucrose content up to 10% have been attributed to flowering.

In some varieties, flowering culms develop central columns of collapsed parenchyma tissue (pithiness) which dry out from the top of the culm. This results in a reduction of extractable juice and a proportional increase in culm fibre. In addition, this pith column allows the rapid spread of certain stem-rotting organisms and even diseases such as yellow spot.

The preferred method for limiting flowering in commercial fields is to grow nonflowering varieties. Sugarcane breeders have been fairly successful in breeding and selection for the nonflowering trait (Moore and Berding, 2013). This is especially important in the more temperate regions where the cropping cycle tends to be longer than 18 months so that the crop might be exposed to two flowering cycles.

4.2 Importance in plant breeding

Significant variation in flowering intensity over cropping cycles, sensitivity of the process to micro-environmental variation and physiological age suggest that there is a very strong genotype-environment interaction (GxE) in flowering control. Variable flowering is a major constraint on variety improvement programmes. Unpredictability in when a clone will flower, and the extent of fertile pollen production makes selection of parents for targeted crosses very difficult.

Use of a controlled environment to improve pollen fertility of flowers initiated under ambient subtropic autumn conditions in heated glasshouses maintained at 21°C, drastically improved pollen production and fertility (for a review, see Moore and Berding, 2013).

Minimal photoperiod regimes to achieve flowering and pollen production incorporated the following:

- Avoid days where the maximum temperature is higher than 32°C.
- Use artificial sunrises and sunsets in photoperiodic facilities which have superior photoperiodic precision and accuracy.

- A photoperiodic regime with a commencing night length of 11 h 05 min, increasing by 30 or 45 seconds per night.

5 Crop and metabolic modelling of sugarcane

Crop models have many current and potential uses for answering questions in research, crop management and policy. These models are useful for analysing and predicting genotype, environmental and management effects on crop performance, and resource dynamics. As research tools, models integrate knowledge and data across disciplines, thereby assisting the synthesis of new knowledge. Models also enable scientists to examine scientific hypotheses and investigate the effect of unprecedented agricultural and ecological conditions.

As a management tool, modelling can support strategic decisions such as analysing the effects of land use and climate change on resource sustainability and agricultural productivity. Tactical decisions, such as choice of cultivar, plant density and fertiliser application could also benefit from modelling and many examples exist of operational decision support. Currently there are two main sugarcane crop models namely the APSIM-Sugarcane model and the CANEGRO model. Modules for these models are continuously being developed and improved as more knowledge about sugarcane physiology is produced. An example would be the development of a module that simulate lodging (van Heerden et al., 2015).

One of the exciting new developments in sugarcane physiology was the introduction of kinetic modelling of metabolic pathways (Rohwer and Botha, 2001). These models are a powerful tool for better understanding the control and regulation of sugarcane metabolism and for the development of strategies to manipulate it. These models are developed around a network stoichiometry that could be based on similar crop-specific species. Without any additional information, such a model would be referred to as structural. The expansion in our knowledge regarding kinetic information about the enzymes for specific pathways enables the development of dynamic models that allows time-course simulation, steady-state analysis, and metabolic control analysis (for a review, see Rohwer and Uys, 2013).

The initial steady-state sugarcane growth model was expanded to simulate metabolism in each internode separately (Uys et al., 2007). In the latest version this model was extended to a continuous growth model that includes phloem flow (Rohwer and Uys, 2013). These models allow identification of control points within metabolism and could pinpoint potential targets for biotechnological intervention.

Some of the interesting information that was derived from the metabolic models include:

- The most important factors affecting sucrose accumulation are the synthesis and breakdown of sucrose in futile cycles and the rate of cross-membrane transport of sucrose.
- Sucrose moves down a concentration gradient from the leaves in the symplast, where it is transported against a concentration gradient into the vacuole.
- There is a net gain in carbohydrate accumulation in the realistic model, despite an increase in futile cycling with internode maturity.
- Sucrose concentrations are most sensitive to the transport and breakdown steps.
- Hexokinase and neutral invertases have a major impact on 'futile cycling' in the cytosol.

6 Conclusion and future trends

The future of a successful sugarcane industry will be the ability to more efficiently grow sugarcane, and fully utilise all of the biomass for extraction of valuable products and bioenergy production. In such a system, future germplasm improvement, and production, will focus on maximising energy yield (Joule) per unit area in the shortest possible cropping cycle. The success will be hinged not only on the production cost, but also on the quality of the feedstock. However, regardless of the specific application or focus, the two most important limiting factors for sugarcane industries today, and in the future, is inadequate cane supply and the quality of the feedstock (for review Botha and Moore, 2013).

There are five major challenges for future technology development in sugarcane:

- The sugar–fibre dilemma. In current commercial varieties these two fractions together occupy more than 95% of the total dry matter in the culm and this leaves little room for further gains in any one of the two components per unit mass. Sucrose accumulation occurs late during culm maturation, and much of the current breeding strategies is to select 'early ripening' varieties. An early switch to sucrose storage could have a negative impact on biomass yield and is probably associated with a loss in ratoonability (Botha and Moore, 2013; Alvarez et al., 2009; Carvalho-Netto et al., 2014). The high turnover of sucrose in the young internodes drives cell wall biosynthesis, fuel respiration and ensure maximum internode growth which eventually have a huge impact on internode volume and how sucrose can be stored. Not only is this breakdown required to provide the carbon skeletons to support growth; it also ensures that a steep sucrose gradient is maintained between the leaves and the internodes which drives phloem transport and is crucial for biomass accumulation.
- The apparent lack of variability in modern varieties for water content in the mature culm. All modern varieties contain approximately 70% water in the mature culm, yet there is significant variation for this trait in the *Saccharum* germplasm. Genetic manipulation of sugarcane illustrated the significant gains in sucrose content that can be made in sugarcane by reducing the water content. The physiological basis for the variation in these traits are not well understood. The significance of this and what the sensing mechanisms are needs to be established.
- The feedback inhibition of sugars on leaf photosynthesis, and the enhancement of leaf senescence by high sucrose in the internodes. This illustrates that increased biomass production can only occur if sucrose accumulation is delayed as long as possible, or alternatively if the feedback regulation mechanism can be altered.
- The inefficient nitrogen photosynthetic use efficiency of sugarcane. Ways should be sought to improve this trait by improving nitrogen uptake and distribution in the crop and by ensuring a more efficient provision of nitrogen to the crop.
- Control of carbon partitioning especially the utilisation of UDP-glucose and the Hexose-P pools in the sugarcane culm. Despite the growth in understanding of carbon partitioning our knowledge about the genetic complexity and gene expression control is rudimentary. However, it is evident that the genetic control is very complex, probably more so in this highly polyploid crop, than in other commodities.

Significant technological advances in the past decade will contribute largely to the pace at which a better understanding of sugarcane physiology is realised. The main gains in our understanding of sugarcane physiology and crop performance will come from full integration of the power of '-omics' and nanotechnologies. We are witnessing an exponential expansion in knowledge from the use of high-throughput genetic analyses, protein and metabolome data, and increasingly from high-throughput phenomics generated through satellite, unmanned aerial systems (UAS) and unmanned ground vehicles (UGV) equipped with high-resolution sensors. These technologies are generating bigdata and allow for a complete revolution in trial design and determination of crop performance, often in realtime.

Sugarcane is increasingly falling further behind other commodities due to the failing to fully embrace these modern technologies. It is therefore encouraging to see the rapidly growing international database on the sugarcane transcriptome data, and attempts to sequence the sugarcane genome and analyses of the metabolome. The full integration of all this information will have a significant impact on future variety and farming system development.

As the data of mRNA transcripts, protein expression levels and metabolites levels increase, it will allow significant improvement of the current metabolic models. The more we populate these models, the closer they become to a replica of the real system and the more accurate their predictive power.

All this information will have an impact on sugarcane plant breeding and the sugarcane farming system. Increasingly we can see breeding strategies for sugarcane focused on two different applications. First, varieties that are being selected and optimised for both sugar and fibre (current modern varieties) and second, varieties bred, selected and cultivated for fibre content and optimisation of the GJ/ha/cropping cycle. Composition and amount of lignocellulose would be important in the latter type varieties.

7 Abbreviations

ADP	adenosine-5′-diphosphate
ALA	alanine
ASP	aspartate
ATP	adenosine-5′-triphosphate
IDP	intermediate day length plant
MDH	malate dehydrogenase (EC 1.1.1.37)
ME	malic enzyme
NAD-ME	nicotinamide adenine dinucleotide malic enzyme (EC 1.1.1.39)
NADP-ME	nicotinamide adenine dinucleotide phosphate malic enzyme (EC 1.1.1.40)
OAA	oxaloacetate
oCd	heat units or thermal time
PAR	photosynthetic active radiation
PDKK	pyruvate orthophosphate dikinase (EC 2.7.9.1)
PEP	phosphoenolpyruvate
PEPCK	phosphoenolpyruvate carboxykinase (EC 4.1.1.49)
PFP	pyrophosphate-dependent phosphofructokinase (EC 2.7.1.90)
PG	phosphoglycerate
PGA	3-phosphoglycerate
PGAL	phosphoglyceraldehyde
Pi	inorganic phosphate
PNUE	photosynthetic nitrogen use efficiency

PPi	inorganic pyrophosphate
PSI	photosystem I
PSII	photosystem II
Pyr	pyruvate
RGP	reduced growth phenomenon
Rm	maintenance respiration
RuBisCo	Ribulose-1,5-bisphosphate carboxylase oxygenase (EC 4.1.1.39)
SDP	short-day plant
SNI	soluble neutral invertase (EC 3.2.1.26)
SPS	sucrose phosphate synthase (EC 2.4.1.14)
SuSy	sucrose synthase (EC 2.4.1.13)
Tbase	daily mean temperature below which the process under consideration ceases
UAS	unmanned aerial systems
UDP	Uridine diphosphate
UGV	unmanned ground vehicles

8 Where to look for further information

To fully understand the impact of these new developments, and their potential impact on future sugarcane improvement and production, a good understanding of current sugarcane production systems is required. These differ significantly between agro-climatic zones and countries. For a general overview the reader is referred to Alexander (1973). The farming system deployed can have a major impact on source–sink relationships and biomass yield (Alexander, 1985).

Sugarcane physiology is complex and differ in many ways from that of other crops. The most thorough overview covering all aspects of sugarcane physiology requires careful integration of the old and modern expansions into our knowledge base (Moore, 2014).

It is impossible to fully understand crop physiology without knowledge regarding the structure and function relationship. This is well covered in Bonnett (2013) and Rae et al. (2013).

There are two rapidly emerging areas of knowledge that will have a huge impact on how we view genetic inheritance, and how we apply agrochemicals to the crop. The first area is the new insights into the important role of epigenetics in crop performance (Alvarez-Venegas et al., 2014; Chen et al., 2010). The second emerging area is that of nanotechnologies (Fraceto et al., 2016). Nanotechnolgy has the potential to fundamentally change the way in which we manage plant nutrition, control pest and disease, and regulate plant growth.

Lastly, we are poised for probably the most exciting time of knowledge expansion regarding sugarcane physiology and production. This will flow from full integration of the power of '-omics' and nanotechnologies and high-throughput genetic, metabolic and phenotyping (https://rirdc.infoservices.com.au/items/16-046).

9 References

Alexander, A. G. (1973), *Sugarcane Physiology: A Study of the Saccharum Source-to-Sink System*, Amsterdam: Elsevier Scientific.

Alexander, A. G. (1985) *The Energy cane Alternative*, New York: Elsevier.
Allison, J. C. S., Williams, H. T. and Pammenter, N. W. (1997), Effect of specific leaf nitrogen content on photosynthesis of sugarcane. *Annals of Applied Biology* 131: 339–50.
Alvarez-Venegas, R. I., De la Peña, C. and Casas-Mollano, J. A. (2014), *Epigenetics in Plants of Agronomic Importance Fundamentals and Applications:Transcriptional Regulation and Chromatin Remodelling in Plants*, Cham: Cham: Springer International Publishing.
Alvarez, J., Deren, C. W. and Glaz, B. (2009), Sugarcane Selection for Sucrose and Tonnage Using Economic Criteria.
Berding, N. (1997), Clonal improvement of sugarcane based on selection for moisture content: fact or fiction. *Proceedings Australian Society for Sugarcane Technologists* 19: 245–53.
Berding, N., Hurney, A. P., Salter, B., et al. (2005), Agronomic impact of suckerdevelopment in sugarcane under different environmental conditions. *Field Crops Research* 92: 203–17.
Bihmidine, S., Hunter, C. T., 3rd, Johns, C. E., et al. (2013), Regulation of assimilate import into sink organs: update on molecular drivers of sink strength. *Front Plant Science* 4: 177.
Bindon, K. A. and Botha, F. C. (2002), Carbon allocation to the insoluble fraction, respiration and triose-phosphate cycling in the sugarcane culm. *Physiologia Plantarum* 116: 12–19.
Bonnett, G. D. (2013), Developmental stages (Phenology). In Moore, P. H. and Botha, F. C. (eds), *Sugarcane: Physiology, Biochemistry, and Functional Biology*, New York: John Wiley & Sons Ltd, 35–53.
Botha, F. C. and McDonald, Z. A. (2010), Carbon partitioning in the sugarcane stalk. *Proceedings Australian Society of Sugarcane Technologists* 32: 486–96.
Botha, F. C. and Moore, P. H. (2013), Biomass and bioenergy. *Sugarcane: Physiology, Biochemistry, and Functional Biology*, Chichester, UK: John Wiley & Sons Ltd, 521–40.
Bull, T. A. (1969), Photosynthetic efficiencies and photorespiration in Calvin cycle and C4-dicarboxylic acid plants. *Crop Science* 9: 726–9.
Bull, T. A. and Glasziou, K. (1963), The evolutionary significance of sugar accumulation in Saccharum. *Australian Journal of Biological Sciences* 16(4): 737.
Calsa, T., Jr. and Figueira, A. (2007), Serial analysis of gene expression in sugarcane (Saccharum spp.) leaves revealed alternative C4 metabolism and putative antisense transcripts. *Plant Molecular Biology* 63(6): 745–62.
Carvalho-Netto, O. V., Bressiani, J. A., Soriano, H. L., et al. (2014), The potential of the energy cane as the main biomass crop for the cellulosic industry. *Chemical and Biological Technologies in Agriculture* 1(1): 20.
Chen, M., Lv, S. and Meng, Y. (2010), Epigenetic performers in plants‡: Review of plant epigenetics. *Development, Growth & Differentiation* 52(6): 555–66.
Cornish-Bowden, A., Hofmeyr, J. H. S. and Cardenas, M. L. (1995), Strategies for Manipulating Metabolic Fluxes in Biotechnology. *Bioorganic Chemistry* 23(4): 439–49.
De Souza, A. P., Gaspar, M., Da Silva, E. A., et al. (2008), Elevated CO_2 increases photosynthesis, biomass and productivity, and modifies gene expression in sugarcane. *Plant, Cell & Environment* 31(8): 1116–27.
Du, Y. C., Nose, A. and Wasano, K. (1999), Effects of chilling temperature on photosynthetic rates, photosynthetic enzyme activities and metabolite levels in leaves of three sugarcane species. *Plant Cell and Environment* 22(3): 317–24.
Ebrahim, M. K., Zingsheim, O., El-Shourbagy, M. N., et al. (1998), Growth and sugar storage in sugarcane grown at temperatures below and above optimum. *Journal of Plant Physiology* 153(5–6): 593–602.
FAO. (2014), Global sugarcane production statistics. http://faostat.fao.org/site/567/DesktopDefault.aspx?PageID=567#ancor.
Foulkes, M. J., Slafer, G. A., Davies, W. J., et al. (2011), Raising yield potential of wheat. III. Optimizing partitioning to grain while maintaining lodging resistance. *Journal of Experimental Botany* 62: 469–86.
Fraceto, L. F., Grillo, R., de Medeiros, G. A., et al. (2016), Nanotechnology in agriculture: which innovation potential does it have? *Frontiers in Environmental Science* 4(20).

Furbank, R. T. (2011), Evolution of the C4 photosynthetic mechanism: are there really three C4 acid decarboxylation types? *Journal of Experimental Botany* 62(9): 3103–8.

Giamalva, M. J., Clarke, S. J. and Stein, J. M. (1984), Sugarcane hybrids of biomass. *Biomass* 6(1–2): 61–8.

Groenewald, J-H. and Botha, F. C. (2008), Down-regulation of pyrophosphate: fructose 6-phosphate 1-phosphotransferase (PFP) activity in sugarcane enhances sucrose accumulation in immature internodes. *Transgenic Research* 17(1): 85–92.

Grof, C. P. L., Byrt, C. S. and Patrick, J. W. (2013), Phloem transport of resources. In Moore, P. and Botha, F. (eds), *Sugarcane: Physiology, Biochemistry, and Functional Biology*, John Wiley & Sons Ltd, New York, 267–305.

Hartt, C. E. and Burr, G. O. (1965), Factors affecting photosynthesis in sugarcane. *Proceedings of the International Society of Sugarcane Technologists* 12: 590–609.

Hartt, C. E., Kotschak, H. P., Forbes, A., et al. (1963), Translocation of 14 C in sugarcane. *Plant Physiology* 38: 305–18.

Herbers, K. and Sonnewald, U. (1998), Molecular determinants of sink strength. *Current Opinion in Plant Biology* 1(3): 207–16.

Hofmeyr, J-H. S. and Rohwer, J. M. (2011), Chapter twenty-five – supply–demand analysis: a framework for exploring the regulatory design of metabolism. In Daniel Jameson MVaHVW (ed.), *Methods in Enzymology*, Academic Press, Cambridge, 533–54.

Hofmeyr, J. H. and Cornish-Bowden, A. (2000), Regulating the cellular economy of supply and demand. *FEBS Letters* 476: 47–51.

Inman-Bamber, G. (2013), Sugarcane yields and yield-limiting processes. *Sugarcane: Physiology, Biochemistry, and Functional Biology*, John Wiley & Sons Ltd, New York, 579–600.

Irvine, J. E. (1975), Relations of photosynthetic rates and leaf and canopy characters to yield. *Crop Science* 15: 671–6.

Irvine, J. E. (1983), Sugarcane. *Potential Productivity of Field Crops Under Different Environments*, International Rice Research Institute, Los Banos, Philippines, 361–81.

Jackson, P. A. (2005), Breeding for improved sugar content in sugarcane. *Field Crops Research* 92(2–3): 277–90.

Koch, K. E. (1996), Carbohydrate-modulated gene expression in plants. *Annual Review of Plant Physiology and Plant Molecular Biology* 47: 509–40.

Komor, E. (2000), The physiology of sucrose storage in sugarcane. In Gupta, A. K. and Kaur, N. (eds), *Carbohydrate Reserves in Plants: Synthesis and Regulation*, Elsevier, Amsterdam, 35–53.

Lawlor, D. W. and Paul, M. J. (2014), Source/sink interactions underpin crop yield: the case for trehalose 6-phosphate/SnRK1 in improvement of wheat. *Frontiers in Plant Science* 5: 418.

Lehrer, A. T., Moore, P. H. and Komor, E. (2007), Impact of sugarcane yellow leaf virus (ScYLV) on the carbohydrate status of sugarcane: comparison of virus-free plants with symptomatic and asymptomatic virus-infected plants. *Physiological and Molecular Plant Pathology* 70(4–6): 180–8.

Lingle, S. E. (1997), Seasonal internode development and sugar metabolism in sugarcane. *Crop Science* 37: 1222–7.

Lingle, S. E. (1999), Sugar metabolism during growth and development in sugarcane internodes. *Crop Science* 39(2): 480–6.

Lingle, S. E., Viator, R. P., Johnson, R. M., et al. (2009), Recurrent selection for sucrose content has altered growth and sugar accumulation in sugarcane. *Field Crops Research* 113(3): 306–11.

Long Stephen, P., Marshall-Colon, A. and Zhu, X-G. (2015), Meeting the global food demand of the future by engineering crop photosynthesis and yield potential. *Cell* 161(1): 56–66.

Marchiori, P. E. R., Machado, E. C. and Ribeiro, R. V. (2014), Photosynthetic limitations imposed by self-shading in field-grown sugarcane varieties. *Field Crops Research* 155(0): 30–7.

Marquardt, A., Scalia, G., Joyce, P., et al. (2016), Changes in photosynthesis and carbohydrate metabolism in sugarcane during the development of Yellow Canopy Syndrome (YCS). *Functional Plant Biology* (submitted).

Mattiello, L., Riano-Pachon, D., Bassi, D., et al. (2015), Physiological and transcriptional analyses of developmental stages along sugarcane leaf. *BMC Plant Biology* 15.

McCormick, A., Cramer, M. and Watt, D. (2008a), Differential expression of genes in the leaves of sugarcane in response to sugar accumulation. *An International Journal Devoted to Original Research in Tropical Plants* 1(2): 142–58.

McCormick, A. J., Cramer, M. D. and Watt, D. A. (2006), Sink strength regulates photosynthesis in sugarcane. *New Phytologist* 171(4): 759–70.

McCormick, A. J., Cramer, M. D. and Watt, D. A. (2008b), Culm sucrose accumulation promotes physiological decline of mature leaves in ripening sugarcane. *Field Crops Research* 108(3): 250–8.

McCormick, A. J., Cramer, M. D. and Watt, D. A. (2008c), Regulation of photosynthesis by sugars in sugarcane leaves. *Journal of Plant Physiology* 165(17): 1817–29.

Moore, P. H. (1995), Temporal and spatial regulation of sucrose metabolism in the sugarcane stem. *Australian Journal of Plant Physiology* 22: 661–79.

Moore, P. H. (2005), Integration of sucrose accumulation processes across hierarchical scales: towards developing an understanding of the gene-to-crop continuum. *Field Crops Research* 92(2–3): 119–35.

Moore, P. H. and Berding, N. (2013), Flowering. *Sugarcane: Physiology, Biochemistry, and Functional Biology*, John Wiley & Sons Ltd, New York, 379–410.

Moore, P. H. and Botha, F. C. (2013), *Sugarcane: Physiology, Biochemistry, and Functional Biology*, John Wiley & Sons, Inc, Chichester, UK.

Moore, P. H., Botha, F. C., Furbank, R. T., et al. (1997), Potential for overcoming physio-chemical limits to sucrose accumulation. *Intensive sugarcane production: meeting the challenges beyond 2000. Proceedings of the Sugar 2000 Symposium, Brisbane, Australia, 20–23 August 1996*: 141–55.

Moore, P. H. and Maretzki, A. (1996), Sugarcane. In Zamski, E. and Schaffer, A. A. (eds), *Photoassimilate Distribution in Plants and Crops: Source-Sink Relationships*, Marcel Dekker Inc., New York, 643–69.

Moore, P. H. B. F. C. (2014), *Sugarcane: Physiology, Biochemistry, and Functional Biology*. Available at: http://search.ebscohost.com/login.aspx?direct=true&scope=site&db=nlebk&db=nlabk&AN=647252.

Muchow, R. C., Evensen, C. I., Osgood, R. V., et al. (1997), Yield accummulation in irrigated sugarcane: Utilisation of intercepted radiation. *Agronomy Journal* 89: 646–52.

Nose, A., Uehara, M., Kawamitsu, Y., et al. (1994), Variations in leaf gas exchange traits of Saccharum including feral sugarcane, Saccharum spontaneum L. *Japanese Journal of Crop Science* 63(3): 489–95.

Park, S. E., Robertson, M. and Inman-Bamber, N. G. (2005), Decline in the growth of a sugarcane crop with age under high input conditions. *Field Crops Research* 92(2–3): 305–20.

Patrick, J. W., Botha, F. C. and Birch, R. G. (2013), Metabolic engineering of sugars and simple sugar derivatives in plants. *Plant Biotechnology Journal* 11(2): 142–56.

Pick, T. R., Brautigam, A., Schluter, U., et al. (2011), Systems analysis of a maize leaf developmental gradient redefines the current C-4 model and provides candidates for regulation. *Plant Cell* 23.

Rae, A. L., Bonnett, G. D. and Karno (2006), Understanding stem development and sucrose accumulation to increase CCS. *Proceedings Australian Society for Sugarcane Technologists* 28: 2–5.

Rae, A. L., Martinelli, A. P. and Dornelas, M. C. (2013), Anatomy and morphology. *Sugarcane: Physiology, Biochemistry, and Functional Biology*, John Wiley & Sons Ltd, New York, 19–34.

Reynolds, M., Bonnett, D., Chapman, S. C., et al. (2011), Raising yield potential of wheat. I. Overview of a consortium approach and breeding strategies. *Journal of Experimental Botany* 62: 439–52.

Robertson, M. J., Wood, A. W. and Muchow, R. C. (1996), Growth of sugarcane under high input conditions in tropical Australia. I. Radiation use, biomass accumulation and partitioning. *Field Crops Research* 48(1): 11–25.

Rohwer, J. M. and Botha, F. (2001), Analysis of sucrose accumulation in the sugar cane culm on the basis of in vitro kinetic data. *Biochemical Journal* 358: 437–45.

Rohwer, J. M. and Uys, L. (2013), Systems biology and metabolic modeling. *Sugarcane: Physiology, Biochemistry, and Functional Biology*, John Wiley & Sons Ltd, New York, 601–22.

Rossouw, D., Kossmann, J., Botha, F. C., et al. (2010), Reduced neutral invertase activity in the culm tissues of transgenic sugarcane plants results in a decrease in respiration and sucrose cycling and an increase in the sucrose to hexose ratio. *Functional Plant Biology* 37(1): 22–31.

Sage RF, Peixoto MM and Sage TL. (2013) Photosynthesis in Sugarcane. *Sugarcane: Physiology, Biochemistry, and Functional Biology*. John Wiley & Sons Ltd, 121–54.

Sinclair TR, Gilbert RA, Perdomo RE, et al. (2005) Volume of individual internodes of sugarcane stalks. *Field Crops Research* 91(2): 207–15.

Singh, G., Chapman, S. C., Jackson, P. A., et al. (2002), Lodging reduces sucrose accumulation of sugarcane in the wet and dry tropics. *Australian Journal of Agricultural Research* 53(11): 1183.

Smeekens, S., Ma, J., Hanson, J., et al. (2010), Sugar signals and molecular networks controlling plant growth. *Current Opinion in Plant Biology* 13(3): 273–8.

Uys, L., Botha, F. C., Hofmeyr, J.-H. S., et al. (2007), Kinetic model of sucrose accumulation in maturing sugarcane culm tissue. *Phytochemistry* 68(16–18): 2375–92.

van der Merwe, M. J., Groenewald, J. H., Stitt, M., et al. (2010), Downregulation of pyrophosphate: D-fructose-6-phosphate 1-phosphotransferase activity in sugarcane culms enhances sucrose accumulation due to elevated hexose-phosphate levels. *Planta* 231(3): 595–608.

van Heerden, P. D. R. (2014), Differential acclimation capacity to frost in sugarcane varieties grown under field conditions. *Plant Growth Regulation* 72(2): 181–7.

van Heerden, P. D. R., Donaldson, R. A., Watt, D. A., et al. (2010), Biomass accumulation in sugarcane: unravelling the factors underpinning reduced growth phenomena. *Journal of Experimental Botany* 61(11): 2877–87.

van Heerden, P. D. R., Eggleston, G. and Donaldson, R. A. (2013), Ripening and Postharvest Deterioration. *Sugarcane: Physiology, Biochemistry, and Functional Biology*, John Wiley & Sons Ltd, New York, 55–84.

van Heerden, P. D. R., Singels, A., Paraskevopoulos, A., et al. (2015), Negative effects of lodging on irrigated sugarcane productivity – an experimental and crop modelling assessment. *Field Crops Research* 180(0): 135–42.

Viator, R. P. and Richard, Jr E. P. (2012), Sugar and energy cane date of planting effects on cane, sucrose, and fiber yields. *Biomass and Bioenergy* 40(0): 82–5.

Wang, J., Nayak, S., Koch, K., et al. (2013), Carbon partitioning in sugarcane (Saccharum species). *Frontiers in Plant Science* 4.

Watt, D. A., McCormick, A. J. and Cramer, M. D. (2013), Source and sink physiology. In Moore, P. H. and Botha, F. C. (eds), *Sugarcane: Physiology, Biochemistry, and Functional Biology*, John Wiley & Sons Ltd, Chichester, UK, 483–520.

Whittaker, A. and Botha, F. (1997), Carbon partitioning during sucrose accumulation in sugarcane internodal tissue. *Plant Physiology* 115(4): 1651–9.

Wu, L. and Birch, R. G. (2007), Doubled sugar content in sugarcane plants modified to produce a sucrose isomer. *Plant Biotechnology Journal* 5(1): 109–17.

Yan, S. L., Lehrer, A. T., Hajirezaei, M. R., et al. (2009), Modulation of carbohydrate metabolism and chloroplast structure in sugarcane leaves which were infected by Sugarcane Yellow Leaf Virus (SCYLV). *Physiological and Molecular Plant Pathology* 73: 78–87.

Chapter 3

Ensuring and exploiting the genetic diversity of sugarcane

Anna L. Hale, USDA-ARS, USA; Phillip Jackson, CSIRO, Australia; and James R. Todd, USDA-ARS, USA

1 Introduction

Modern sugarcane (*Saccharum* spp.; 2n = 100 = 130) cultivars are derived from interspecific hybridization, primarily involving *Saccharum officinarum* L. and *Saccharum spontaneum*, L. (Ming et al. 2006). In the late 1800s, breeders in Java, Indonesia, recognized the value of interspecific hybrids and began to intentionally hybridize the two species. The interspecific hybrids were considerably more vigorous than the noble canes (*Saccharum officinarum*) that were being grown commercially at the time. Furthermore, these early hybrids were resistant to the sereh disease (a putative virus of uncertain etiology), which was severely affecting the local industry. Interspecific cultivars from early hybridization efforts in Java and India were imported into every major sugarcane-producing country and almost completely replaced noble cane cultivars worldwide. The majority of cultivars in current commercial sugarcane breeding programmes trace back to a few key interspecific hybrids that were developed during the early 1900s.

Concerns about the narrow sampling of ancestral clones in modern sugarcane breeding programmes are frequently expressed by sugarcane breeders (Roach 1989). This has prompted periodic attempts by breeders, since the 1960s and continuing to the present day, at so-called 'base-broadening' programmes (Berding and Roach 1987; Jackson et al. 2014; Roach 1986; Sundaram et al. 2010). The terms 'base-broadening breeding', 'basic breeding' and 'introgression breeding' have been used interchangeably in the sugarcane literature, although introgression breeding tends to refer to efforts targeting the incorporation of

http://dx.doi.org/10.19103/AS.2017.0035.04

particular traits into commercial breeding programmes (e.g. adaptation to stress from use of *Saccharum spontaneum*). Such programmes have been initiated in many countries and continue in a range of countries including Barbados, Brazil, Australia, Fiji, Thailand, India, China, Japan and the USA.

Introgression programmes were modelled after the early breeding efforts in Java and India. The first introgression attempts targeted wide hybridization between *Saccharum officinarum* and *Saccharum spontaneum*, but as stepwise gains were obtained, breeders began to cross cultivated sugarcane with wild accessions. Traditional introgression programmes begin by crossing a commercial sugarcane genotype with a wild accession (typically *Saccharum spontaneum*) and generating what sugarcane breeders commonly refer to as an 'F_1 population'. This, typically low-sucrose population, is then selected for traits of interest. Selected genotypes are then crossed back to a commercial sugarcane parent (generally with high sucrose content) to generate the first backcross population (BC_1). The cycle is repeated and subsequent backcrosses (BC_2, BC_3, BC_4, etc.) are made until commercially acceptable levels of sucrose content are recovered in the selected progeny (typically BC_2 generations and higher). Early generations of progeny following initial crosses between sugarcane and its wild relatives are characterized by sucrose levels too low and fibre levels too high for commercial production in sugar industries. However, there is interest in such early generation material as a feedstock in future bioenergy production systems. Sugar levels progressively increase and fibre levels decrease in subsequent crossing generations (Hsu and Shih 1989; Mullins 1989). This is a long-term process that requires many years, and even decades, to complete (Berding and Roach 1987; Miller and Tai 1992).

Despite early successes, and continued introgression breeding, an examination of pedigrees of cultivars in the most prominent sugarcane breeding programmes reveals what appears to be a narrow genetic base (Tew 1987). For example, pedigree data indicated that in 1995, 90% or more cultivars in Louisiana had two ancestors in common and 90% of cultivars in Florida had ten ancestors in common (Deren et al. 1995). However, even though a limited number of parents were used to initiate breeding programmes, the coefficient of parentage between US mainland sugarcane clones and their ancestral clones is relatively low, and the genetic base is not as narrow as previously thought. Factors contributing to genetic diversity in modern cultivars probably include the polyploid and heterozygous nature of the crop, its interspecific origin, the unequal contribution of parents when crossed and use of wide hybridization and introgression work in modern breeding programmes (Deren et al. 1995; Lima et al. 2002). Molecular markers confirm that moderate genetic diversity exists in the crop (Lima et al. 2002; Pan et al. 2003a,b). These markers indicate that a majority of alleles contributing to variability in the germplasm come from the *Saccharum spontaneum* even though a majority of the genetic makeup of sugarcane is of *Saccharum officinarum* origin (Janoo et al. 1999).

Although at least moderate genetic variability exists in current sugarcane cultivars, most of the available genetic diversity in germplasm collections remains untested and underutilized. Brown et al. (2006), using microsatellite markers, indicated that the clones of *Saccharum spontaneum* formed a group that distinctly separated from commercial and *Saccharum officinarum* clones, indicating that there is still much variability to be gained from this germplasm. This apparent distinction among species and commercial clones indicates that crossing between groups would likely produce progeny with extreme variability (Brown et al. 2006). The genera *Erianthus, Miscanthus, Sclerostachya and Narenga* represent other sources of potential genetic variability for sugarcane. These genera along with *Saccharum* constitute a closely related interbreeding group referred to as the '*Saccharum complex*' and were assumed to be involved in the origin of sugarcane (Daniels and Roach 1987).

By contrast, at least two studies suggest that a high proportion of genetic diversity in collections of *Saccharum officinarum* have already been captured within commercial sugarcane breeding programmes (Janoo et al. 1999; Aitken et al. 2006). However, some significant untapped variability remains within this species, particularly from New Caledonia and New Guinea.

Some important cultivars have emerged from introgression breeding since the 1960s. One example is the use of the *Saccharum spontaneum* clone 'US56-15-8' (collected in northern Thailand), which led to the cultivar LCP85-384 in Louisiana in the fourth backcross generation. This variety contributed greatly to the sugarcane industry and became an ancestor of many other cultivars produced from that programme (Arro et al. 2006; Gravois et al. 2008, 2010, 2011; Hale et al. 2012, 2013; Milligan et al. 1994; Tew et al. 2005; White et al. 2014). Another example is the use of the wild *Saccharum spontaneum* clone 'Mandalay' in Australia. This clone was collected in Myanmar and was first crossed with POJ2878. Selected progeny were then backcrossed in two further cycles to produce the parent clone QN66-2008. The latter clone became a parent of a series of very important cultivars for the Australian sugarcane industry (Berding et al. 2000).

This chapter first provides an overview of sugarcane germplasm collections and then describes introgression-related research and breeding efforts in the past focused on the use of *Saccharum spontaneum*, *Erianthus* and other species. Difficulties and challenges that need to be overcome for successful outcomes from introgression breeding are emphasized. Finally, a few possible future directions are considered, given experiences to date and emerging opportunities.

2 Germplasm collections

A large germplasm collection, designated as 'The World Collection' by the International Society of Sugar Cane Technologists, (ISSCT), is located at the USDA-ARS Subtropical Horticultural Research Station in Miami, Florida, and has provided clones upon request to sugarcane breeders worldwide. This collection is part of the U.S. National Plant Germplasm System, information base of which is available online (https://npgsweb.ars-grin.gov/gringlobal/search.aspx). Other collections, typically available to breeders within a given country, also exist. Notable examples include collections in India (maintained by the Sugarcane Breeding Institute) and Indonesia (maintained by the Indonesian Sugarcane Research Institute), which include duplicated clones from the World Collection, and the National Nursery of Sugarcane Germplasm Resources in China (maintained by Yunnan Academy of Agricultural Sciences). These germplasm repositories maintain a vast store of genetic diversity potentially available to sugarcane breeders. The maintenance of a large number of diverse clones is similar to having an insurance policy for sugar industries worldwide against new diseases or pests which may appear and broadly affect modern sugarcane cultivars. Wild germplasm has been used to increase disease resistance of sugarcane in the past (Jeswiet 1930). The eminent discovery of new diseases and the resurgence of old ones (Rott 2000; Glynn et al. 2010; Viswanathan and Rao 2011) will likely motivate further utilization of this unselected material in the future in search of resistance genes. Wild germplasm maintained in collections has also been used to increase adaptability (Nair 2008), cold tolerance (Tai 1981), ratooning ability (Matsuoka and Stolf 2012) and cane yield (Breaux and Legendre 1978). These collections contain many, to date, untapped genes and traits which collectively increase the available genetic variability, increasing the genetic base of the *Saccharum* complex, and potentially underpinning genetic gains in breeding programmes well into the future (Gilbert et al. 1999; Barrett and Kidwell 1998; Bered et al. 2002). Because of the complex and interspecific origins of sugarcane, the collections are quite large and contain not only *Saccharum* species, but also related genera used by sugarcane breeders such as *Erianthus, Miscanthus* and *Narenga* (Daniels and Roach 1987; Dillion et al. 2007). The wide genetic variation contained in these repositories also has the potential to serve as a future source of genes related to abiotic stress resistance.

While these collections have been in existence for many years, their genetic potential has clearly not been fully evaluated. As methods for evaluation have been developed, scientists have made numerous attempts to assess the phenotypic and genotypic diversity contained in these genetic goldmines (Cordeiro et al. 2003; Edme et al. 2005; Alwala et al. 2006; Dillon et al. 2007; Nayak et al. 2014; Todd et al. 2014). Some diversity studies have focused on creating a smaller, core collection, which contains a majority of the diversity found in the collection as a whole, but is more logistically manageable (Holbrook et al. 1993). The smaller size of the core collection allows for multiple replications, more robust datasets, rationalized characterization studies and a better understanding of the relationship between germplasm (Upadhyaya and Ortiz 2001). In addition, core collections can be used as diversity panels for identifying marker trait associations for important traits such as yield, disease resistance and fibre content (Perera et al. 2012). Several core collections have been established, including a collection of *Saccharum officinarum* clones maintained at the Sugarcane Breeding Institute Research Centre in Cannanore, India (Balakrishnan et al. 2000), a collection of *Saccharum spontaneum* clones maintained at the world collection in Miami, Florida (Tai and Miller 2001), as well as a later core collection which utilized a majority of the species maintained in Miami, Florida (Nayak et al. 2014; Todd et al. 2014).

Assessing the diversity within the collections is important, but a critical and perhaps more urgent need is maintaining the integrity and accessibility of these collections. A difficulty arises because maintenance of a bulky and vegetative crop like sugarcane is expensive, while at the same time the potential benefits are long term and uncertain. In addition, benefits from any collection which is available to industries worldwide are not claimed by any single country, and therefore it is sometimes difficult to make a case for sustained financial support from any individual country. There is a case perhaps for a consortium comprising a number of major sugarcane industries worldwide to cooperate to maintain and coordinate a true 'world' collection available to all industries for mutual long-term benefit. This may not necessarily be in one location, but could be coordinated with a common database, management protocols and access agreements.

3 Introgression of *Saccharum spontaneum*

Introgression breeding efforts in recent decades continue to be heavily focused on the use of *Saccharum spontaneum*. Advantages with the use of this species are that it readily produces fertile seed in crosses with sugarcane, and some progeny clones have shown relatively good vigour, ratooning performance and tolerance to stresses such as cold temperature or water stress. The major disadvantage is the extremely low sucrose content in progeny, requiring multiple backcrosses to high-sucrose parents as explained in Section 1. Traits of interest from *Saccharum spontaneum* include dense plant cane stands, profuse tillering, good ratooning ability, disease resistance, stem borer resistance and stress tolerance (e.g. flood, salt, drought, cold) (Berding and Roach 1987; Dunckelman and Breaux 1972; Hale et al. 2014, 2016; Matsuoka and Garcia 2011; Roach 1978; Sugimoto et al. 2012; Viator et al. 2012; Walker 1972; White et al. 2011). The approach taken by breeding programmes to identify parental clones of *Saccharum spontaneum* has varied. Some programmes (e.g. the USDA-ARS-Houma, LA) chose to screen parental material before crossing, while others chose to start with a large set of basic clones focusing on maximizing the introduction of new variability. An example of the latter was an approach referred to as incorporation (Simmonds 1993) conducted in Barbados which involved a starting population of many *Saccharum officinarum* and *Saccharum spontaneum* clones.

The small number of parents upon which current sugarcane breeding programmes worldwide are based, and the exclusive use of commercial cultivars derived from these genotypes as the recurrent parent in base-broadening programmes has contributed to modern cultivars with little cytoplasmic diversity. Deren et al. (1995) reported that the cytoplasmic origins of 85 Louisiana and Florida cultivars contained only three sources of cytoplasm, all of which were of *Saccharum officinarum* origin. These results are similar to those of Mangelsdorf (1983) and Coto et al. (2002) who reported the cytoplasmic genetic base of commercial cultivars were derived from only three to five *Saccharum spontaneum* clones. The declaration of *Saccharum spontaneum* as a potentially problematic weed in some countries has led breeders to avoid its use as a female parent because of its propensity to sometimes self-pollinate and produce pure *Saccharum spontaneum* seedlings which are difficult to remove in the field (Bonnett et al. 2014). By using the species as a male, breeders are able to ensure that resulting seedlings are not a result of self-pollination of the *Saccharum spontaneum* parent. Furthermore, an unusual phenomenon, known as female restitution, is exhibited in hybrids derived from crosses where *Saccharum officinarum* is used as a female parent, and *Saccharum spontaneum* is used as the male. Hybrids derived from such a cross exhibit 2n + n chromosome numbers, where the entire genome of the *Saccharum officinarum* parent is retained (Bremer 1961). This unusual type of inheritance means that commercial-type high-sucrose cane is recovered more quickly during introgression when making the cross in this direction. The reciprocal cross (*Saccharum spontaneum* × *Saccharum officinarum*) generally results in n + n transmission, requiring more generations of backcrossing to recover commercial sucrose and agronomic types. Molecular markers now allow for breeders to distinguish, at the seedling stage, true hybrids from selfs when a *Saccharum spontaneum* is used as a parent in a cross. This has allowed the use of this species as a female parent in some crosses and allowed for the possibility of introducing new sources of cytoplasm into the breeding programme. A cultivar with *Saccharum spontaneum* cytoplasm was first released in the United States in 2013 (Hale et al. 2013).

A range of reports on use of *Saccharum spontaneum* in recent decades have been published with some general conclusions as follows:

- Some early generation progenies derived from *Saccharum spontaneum* have provided good biomass yield, particularly in ratoon crops (Anderson et al. 2016;

Hale et al. 2013; Leon et al. 2015; Richard and Anderson 2014; Terajima et al. 2007; Wang et al. 2008), although it should be noted that some studies have involved small plots in which competition effects may be important in affecting cane yield in some clones.

- As noted by Roach (1984) much breeding work involving *Saccharum spontaneum* has been conducted under the vague objective of 'base broadening'. This objective in itself has little value and may not provide a focused basis for achieving practical commercial results. However, against this viewpoint, it is noted that the base-broadening programme in Barbados and programmes in Louisiana and Australia, all have produced commercially important cultivars even though the motivation was primarily to simply incorporate new sources of genetic diversity rather than targeting specific traits or genes.
- Some *Saccharum spontaneum* clones provide good sources of resistance to diseases such as sugarcane mosaic (Koike 1980), red rot (Hale et al. 2010; Natarajan et al. 2001) and Sugarcane yellow leaf virus (Costet et al. 2012) and to pests (Jackson and Dunckelman 1974; White et al. 2001; White et al. 2011).
- Some *Saccharum spontaneum* clones provide sources of tolerance to environmental stress such as cold (Hale et al. 2014, 2016, 2017; Miller et al. 2005) and waterlogging tolerance (Sooksathan et al. 1992; Viator et al. 2012). Work has also been conducted to identify molecular markers associated with cold tolerance (Khan et al. 2013).
- Variable results have been found for the heritability of traits from *Saccharum spontaneum* clones in introgression breeding programmes. Roach (1978) reported that sucrose level in *Saccharum spontaneum* clones provided a good prediction of progeny performance when crossed with sugarcane (*Saccharum officinarum* or commercial-type clones), while cane yield was a moderate predictor. Wang et al. (2008) found the performance of *Saccharum spontaneum* parent clones to be a poor predictor of progeny performance for both sucrose content and yield, although stalk weight was well predicted from the *Saccharum spontaneum* parents. Heritability estimates have also been reported for cold tolerance (Hale et al. 2016, 2017) and indicate that the parental clone used can be a good predictor of late-season cold tolerance and a moderate predictor of tolerance of eyes to prolonged freezing conditions.
- Although very preliminary, there are some indications that *Saccharum spontaneum* clones sourced from the northern Thailand–Burma region could provide superior breeding material. Commercial successes in Australia and the United States have arisen from clones selected from this region (as noted in references above). The impressive phenotypic characteristics of this material were highlighted by Sooksathan et al. (1992). Heinz (1980) noted the high yields of *Saccharum spontaneum* clones collected from the northern tip of Thailand (19–20° latitude) and the apparent propensity to pass these favourable traits onto progeny.

4 Introgression of *Erianthus*

There has been considerable interest and activity in the past 30 years in trying to use clones from the genus *Erianthus* in sugarcane breeding programmes. However, to our knowledge no commercial cultivars derived from *Erianthus* have yet been released.

Characteristics such as high vigour, drought and waterlogging resistance, good ratooning ability and disease resistance have been attributed to *Erianthus* and to *E. arundinaceus* in particular (Jackson and Henry 2011). However, there is little supporting data available apart from anecdotal observations of the obviously vigorous growth of *Erianthus* spp. plants relative to other species in germplasm collections, and in the wild. It should also be noted that while *Erianthus* plants may often look large, a high level of pithiness in the stalks may lead to low stalk density and low overall biomass yields. Published reports showing data from well-controlled field experiments are rare and we are only aware of one. Matsou et al. (2001) reported on a study showing superior performance of *Erianthus* (species not indicated) compared with Napier grass and maize under both dry and wet field conditions. These authors also reported results from pot experiments with adjusted water tables, suggesting favourable growth of *Erianthus* under waterlogged conditions and ability of roots to grow well-developed aerenchyma (which could assist in root oxygenation). The ability of *Erianthus* to extend its root system during the rainy season, and to take up water from deeper soil layers during the dry season, was suggested as the mechanism for greater drought tolerance in *Erianthus*. The same study also suggested superior adaptation of *Erianthus* spp. under low N conditions and low soil pH.

There are several reports of high levels of resistance in *E. arundinaceus* to insects that are commercially destructive in sugarcane, including major borer pests (Sardana 2002; White et al. 2013) and white grub (Allsopp and Cox 2002; Mukunthan and Nirmala 2002). These sources of insect resistance could be important in an introgression breeding programme targeting sugarcane improvement. It has also been suggested that *Erianthus* could be a valuable source of red rot resistance. A screening project conducted by Hale et al. (2010) of a wide range of *Saccharum* and *Erianthus* accessions showed that inoculated stalks of all *Erianthus* accessions possessed a high level of resistance to the disease. Furthermore, sugarcane clones produced by crossing an *E. arundinaceus* clone with a *Saccharum officinarum* clone were reported by Ram et al. (2001) as having superior low-temperature tolerance and red rot resistance.

There are several published reports of the production of hybrids between *Saccharum* and *E. arundinaceus* (D'Hont et al. 1995; Piperidis et al. 2000, 2010; Ram et al. 2001; Cai et al. 2005; Nair et al. 2006; Lalitha and Premachandran 2007), *E. rockii* (Aitken et al. 2007), *E. ravennae* (Janaki Ammal 1941) and *E. cilaris* (Kandasami 1961). A range of efforts which have not been published in scientific literature have also occurred. Interestingly all successful hybrids reported between *Saccharum* and *Erianthus* in recent decades that have been validated using DNA markers have been between *Saccharum officinarum* and *Erianthus*, rather than between modern sugarcane cultivars (containing both *Saccharum officinarum* and *Saccharum spontaneum* chromosomes) and *Erianthus*.

The use of *Erianthus* in sugarcane improvement programmes has been complicated by the difficulty in obtaining fertile progeny when crossed with *Saccharum*, combined with difficulty in identifying true hybrids among populations of seedlings arising from these crosses. DNA markers have shown that many putative hybrids were actually selfs or arose from pollen contamination (D'Hont et al. 1995; personal communication with numerous breeders). This demonstrates the value of verifying hybrids with markers at the first opportunity to avoid wasted efforts and possible loss of rare true hybrids. While there has been limited successful hybridizations, fertile crosses between these hybrids and *Saccharum* are difficult to obtain (Piperidis et al. 2000; personal communication with numerous breeders). These difficulties may be attributed to the large genetic dissimilarity between *Saccharum* and *Erianthus*, which appears even larger than for other genera such as *Miscanthus* (Sobral et al. 1994; Alix et al. 1998; Cai et al. 2005).

Other factors suggested as potentially contributing to the lack of success of introgression of *Erianthus* into sugarcane breeding programmes include chromosome erosion during crossing and backcrossing, and lack of recombination between the chromosomes of the two genera (D'Hont et al. 1995; Piperidis et al. 2010). Reduced transmission of chromosomes in crosses between *Saccharum* and *Erianthus* or the hybrids and *Saccharum* were reported by D'Hont et al. (1995), Piperidis et al. (2000) and Piperidis et al. (2010).

Despite these difficulties there is at least one report of fertile hybrids between *Saccharum* and *Erianthus*, and progeny produced from these hybrids and sugarcane, opening up opportunity for introgression of *Erianthus* genome components into sugarcane breeding programmes (Cai et al. 2005). To date there have been few reports backed with statistically supported data demonstrating the performance of *Saccharum* × *Erianthus* hybrids or their derivatives.

5 Introgression of other genera and species

Miscanthus has been of considerable interest to sugarcane breeders. This genus is distributed across Tahiti, Indonesia, China, Siberia, Japan, India, Southern Africa and Nepal (Adati and Shioani 1962; Hodkinson et al. 2002) and is found growing in diverse environments (Lo and Sun 1968). Traits of interest to sugarcane breeders from this genus include cold tolerance, drought resistance and downy mildew and smut resistance (Irvine 1977; Lo et al. 1978).

Breeders in Taiwan have had an interest in breeding with *Miscanthus* since the 1950s (Chen 1953; Chen et al. 1983), but with the closure of this programme many years ago it is now unclear whether validated hybrids were produced or whether any true hybrids contributed to development of commercial varieties. Tai et al. (1991) evaluated juice quality traits of hybrid and backcross progeny of *Saccharum* and *Miscanthus* and found that the mean sucrose content of F_2 and BC_1 progenies was higher than that of the F_1 hybrids, but stalk diameter remained small. Burner (1997) described n + 2n transmission when sugarcane was crossed to *M. sinensis*, and n + n transmission when crossed with *M. japonicas*.

The recent emphasis on bioenergy production has corresponded with a renewed interest with intercrossing *Saccharum* and *Miscanthus*. The USDA-ARS in Houma, Louisiana, has made crosses with *Miscanthus* in the past and has a few verified hybrids (unpublished). Recently, Texas A&M University reported that hybrids between the two genera, named 'Miscanes', showed promise as donors for drought and cold tolerance. This group verified the hybrid nature of the clones and continues to work with breeding with these towards cultivars for bioenergy feedstock (Park et al. 2011). Burner et al. (2015) compared yields of giant *Miscanthus*, giant reed and 'miscane' grown under both irrigated and rainfed conditions and concluded that in the cold region where the test was conducted, the miscane did not have enough cold tolerance to produce vigorous ratoon crops.

Hybridization between *Saccharum* and *Sorghum* was reported in 1999, but like previous attempts, recovery of viable seedlings was low, with only five seedlings recovered from 3670 pollinated florets (Nair 1999). Another attempt at hybridization between *Saccharum* and *Sorghum* was reported by Hodnett et al. (2010). While past attempts at hybridization were met with poor seed set, a new approach has been used with an inbred line of *Sorghum bicolor* that had a mutation which removed inhibition of alien pollen allowing for easy production of interspecific hybrids. Embryo rescue was used on the pollinated seed, resulting in seedling recovery. However, attempts to backcross the hybrids to *Sorghum* apparently have not been successful.

6 Conclusions and future trends

Sugarcane is a genetically complex crop, and its complexity makes genetics research and breeding challenging. However, the ability to intercross within the *Saccharum* complex makes it possible to utilize a wide range of germplasm to genetically improve the crop. Sugarcane industries today only remain profitable because of the ability to exploit the diversity within the related wild species.

Various introgression breeding efforts in sugarcane in the past 50 years, albeit with highly variable success rates, have proven that introgression breeding, potentially, can have an enormous and long-lasting impact. However, despite some important successes, anecdotal evidence suggests that introgression breeding in the past 40–50 years in many programmes has not been as successful as was hoped or expected by its proponents, with the major successes being considered 'hit and miss' and many efforts not leading to commercial results. Several reasons for this have been suggested (Roach 1984, 1989), primarily relating to inferior traits in the wild donor clones and difficulties in selecting and combining the appropriate desirable portions of both the wild type and the recurrent parents during subsequent selection cycles. The futility of the general goal of 'introducing genetic diversity' has also been emphasized.

Other problems noted include a lack of genetic information to confirm the hybrid nature of clones from interspecific hybridization and challenges of obtaining synchronous flowering (Berding and Roach 1987; Roach 1984, 1989). Some of these problems have largely been overcome in recent decades with the use of DNA markers and photoperiod treatments. In particular, the use of DNA markers in validating parentage has been of particular value as illustrated in examples reported by D'Hont et al. (1995), D'Hont and Glaszmann (2001), Pan et al. (2004), Cai et al. (2005) and Wang et al. (2009).

The overall major challenge associated with the introgression of wild germplasm into highly selected and commercially adapted cultivars in sugarcane breeding remains the same as in other crops: the wild germplasm brings with it many undesirable traits (especially low sugar content and high fibre) which need to be selected against using cycles of backcrossing to the highly bred and commercially superior parental material, while at the same time desirable traits and genes from the wild donor may be diluted or lost with successive generations. In the case of sugarcane, the major undesirable trait introduced with use of wild canes is clearly low sucrose content. Following initial crosses made with wild clones (e.g. *Saccharum spontaneum*), usually two or more backcrosses to elite commercial-type parents are required, with lengthy intervening field evaluation and selection programmes between each crossing cycle, before commercial-type progeny are developed. Therefore, the process of introgression in sugarcane using conventional breeding procedures is relatively long and risky, which deters investment.

Personal communication with many sugarcane breeders indicates divided opinion about the priority to place on introgression of new wild germplasm compared with focusing only on highly bred clones, although most acknowledge the potential longer-term value. The development of more effective approaches than those used in recent decades has the potential to hasten the process and lead to a higher frequency of successful introgression-based varieties. The use of DNA markers in selection may offer some pathways, and while such approaches are being used in other crops (Reyna and Sneller 2001; Steele et al. 2006) and suggested in sugarcane (Alwala et al. 2009), there do not yet appear to be successful examples of sugarcane cultivar development using this method. The optimal approach in sugarcane breeding may be to initially identify markers linked to both favourable and unfavourable QTLs in promising backcross populations derived from wild donor clones. These markers could then be used in selecting among progeny populations during regeneration of such populations. In addition to the underlying challenge of marker development on a genetically complex crop such as sugarcane, another issue is the relatively slow throughput of the DNA analysis on large populations and the logistical challenge of having this information available in a timely manner for breeders to effectively utilize the data. Recent developments in DNA marker technology, particularly genotyping by sequencing approaches, are expected to provide better opportunities for sugarcane than earlier methods.

The nature of introgression breeding requires a long-term commitment by funding agencies and project scientists. Many programmes and funding agencies invest in short-term introgression projects and therefore often stop before any measurable commercial impact is achieved. While short-term investments are important for germplasm characterization, and stepwise gains can be achieved within a few years, the potential and future commercial impact of introgression breeding cannot be achieved within a short timeframe. In order for introgression breeding programmes to succeed, long-term plans and commitment of experienced staff need to be sustained and supported.

7 Where to look for further information

Notable examples of institutes maintaining sugarcane introgression breeding programmes include: USDA ARS Sugarcane Research Unit, Houma, Louisiana, USA; West Indies Central Sugarcane Breeding Station, Barbados; Sugarcane Breeding Institute (within the Indian Council of Agricultural Research), India; Guangzhou Sugarcane Industry Research Institute, Guangzhou, China; and Sugar Research Australia, Brisbane, Australia. Information on progress in sugarcane introgression breeding from many countries is usually provided in publications reporting discussions held at the International Society of Sugar Cane Technologists congress and Germplasm and Breeding Section workshops (www.issct.org).

8 References

Adati, S. and Shiotani, I. 1962. The cytotaxonomy o the genus *Miscanthus* and its phylogenetic status. *Mie Univ. Ac. Agric. Bull.* 25, 1–24.

Aitken, K., Li, J.-C., Jackson, P., Piperidis, G. and McIntyre, C. 2006. AFLP analysis of genetic diversity within *Saccharum officinarum* and comparison with sugarcane cultivars. *Crop Pasture Sci.* 57, 1167–84.

Aitken, K., Li, J., Wang, L., Qing, C., Fan, Y. H. and Jackson P. 2007. Characterization of intergeneric hybrids of *Erianthus rockii* and *Saccharum* using molecular markers. *Genet. Resour. Crop. Evol.* 54, 1395–405.

Alix, K., Baurens, F. C., Paulet, F., Glaszmann, J. C. and D'Hont A. 1998. Isolation and characterization of a satellite DNA family in the *Saccharum* complex. *Genome* 41(6), 854–64.

Allsopp, P. G. and Cox, M. C. 2002. Sugarcane clones vary in their resistance to sugarcane whitegrubs. *Aust. J. Agric. Res.* 53, 1111–36.

Alwala, S., Suman, A., Arro, J. A., Veremis, J. C. and Kimbeng C. A. 2006. Target region amplification polymorphism (TRAP) for assessing genetic diversity in sugarcane germplasm collections. *Crop Sci.* 46(1), 448–55.

Alwala, S., Kimbeng, C. A., Veremis, J. C. and Gravois, K. A. 2009. Identification of molecular markers associated with sugar-related traits in a *Saccharum* interspecific cross. *Euphytica* 167, 127–42.

Anderson, W. F., Sarath, G., Edme, S., Casler, M. D., Mitchell, R. B., Tobias, C. M., Hale, A. L., Sattler, S. E. and Knoll, J. E. 2016. Dedicated herbaceous biomass feedstock genetics and development. *Bioenergy Res.* 9, 399–411.

Arro, J., Veremis, J., Kimbeng, C. and Botanga, C. 2006. Genetic diversity and relationships revealed by AFLP markers among *Saccharum spontaneum* and related species and genera. *J. Am. Soc. Sugar Cane Technol.* 26, 101–15.

Balakrishnan, R., Nair, N. V. and Sreenivasan, T. V. 2000. A method for establishing a core collection of *Saccharum officinarum* L. germplasm based on quantitative-morphological data. *Genet. Resour. Crop Evol.* 47(1), 1–9.

Barrett, B. and Kidwell, K. 1998. AFLP-based genetic diversity assessment among wheat cultivars from the Pacific Northwest. *Crop Sci.* 38, 1261–71.

Berding, N. and Roach, B. T. 1987. Germplasm collection, maintenance, and use. In: Heinz, D. (Ed.), *Sugarcane Improvement Through Breeding*. Elsevier, Amsterdam, pp. 143–210.

Berding, N., Moore, P. H. and Smith, G. R. 2000. Advances in breeding technology for sugarcane. In: Keating B. A. and Wilson J. (Eds), *Intensive Sugarcane Production: Meeting the Challenges Beyond 2000*. CAB International, UK, pp. 141–56.

Bered, F., Barbosa-Neto, J. F. and de Carvalho F. I. 2002. Genetic variability in common wheat germplasm based on coefficients of parentage. *Genet. Mol. Biol.* 25(2), 211–15.

Bonnett, G., Kushner, J. and Saltonstall K. 2014. The reproductive biology of *Saccharum spontaneum* L.: Implications for management of this invasive weed in Panama. *NeoBiota* 20, 61.

Breaux, R. and Legendre, B. 1978. Biomass and sucrose production in sugarcane: Divergent selection strategies. In: *Agronomy Abstracts*. American Society of Agronomy, p. 48.

Brown, J. S., Schnell, R. J., Power, E. J., Douglas, S. L. and Kuhn, D. N. 2006. Analysis of clonal germplasm from five *Saccharum* species: *S. barberi, S. robostum, S. officinarum, S. sinense, and S. spontaneum.* A study of inter- and intra species relationships using microsatellite markers. *Genet. Resour. Crop Evol.* 54, 627–48.

Burner, D. M. 1997. Chromosome transmission and meiotic behaviour in various sugarcane crosses. *J. Am. Soc. Sugar Cane Tech.* 17, 38–40.

Burner, D. M., Hale, A. L., Carver, P., Pote, D. H. and Fritschi, F. B. 2015. Biomass yield comparisons of giant miscanthus, giant reed, and miscane grown under irrigated and rainfed conditions. *Ind. Crops Prod.* 76, 1025–32.

Bremer G. 1961. Problems in breeding and cytology of sugarcane. *Euphytica* 10, 59–78.

Cai, Q., Aitken, K., Deng, H., Chen, X., Fu, C., Jackson, P. and McIntyre, C. 2005. Verification of the introgression of *Erianthus arundinaceus* germplasm into sugarcane using molecular markers. *Plant Breed.* 124, 322–8.

Chen, C. F. 1953. Genetical analysis of morphological characters of the OMM-type hybrid obtained from the POJ 2725 × *Miscanthus japonicas* cross. *Proc. Int. Soc. Sugarcane Technol.* 8, 533–46.

Chen, W. H., Huang, Y. J., Shen, I. S. and Shih, S. C. 1983. Utilization of *Miscanthus* germplasm in sugarcane breeding in Taiwan. *Proc. Int. Soc. Sugarcane Technol.* 18(2), 641–8.

Cordeiro, G. M., Pan, Y. B. and Henry, R. J. 2003. Sugarcane microsatellites for the assessment of genetic diversity in sugarcane germplasm. *Plant Sci.* 165(1), 181–9.

Costet, L., Raboin, L. M., Payet, M., D'Hont, A. and Nibouche, S. 2012. A major quantitative trait allele for resistance to the Sugarcane yellow leaf virus (Luteoviridae). *Plant Breed.* 131(5), 637–40.

Coto, O., Cornide, M. T., Calvo, D., Canales, E., D'Hont, A. and de Prada, F. 2002. Genetic diversity among wild sugarcane germplasm from Laos revealed with markers. *Euphytica* 123, 121–30.

Daniels, J. and Roach, B. T. 1987. Taxonomy and evolution. In: Heinz, D. (Ed.), *Sugarcane Improvement Through Breeding, Volume* 7. Elsevier, Amsterdam, pp. 7–84.

Deren, C., Alvarez, J. and Glaz, B. 1995. Use of economic criteria for selecting clones in a sugarcane breeding program. *Proc. Int. Soc. Sugarcane Technol.* 21, 437–47.

D'Hont, A. and Glaszmann, J. C. 2001. Sugarcane genome analysis with molecular markers, a first decade of research. *Proc. Int. Soc. Sugarcane Technol.* 24, 556–9.

D'Hont, A., Rao, P., Feldmann, P., Grivet, L., Islam-Faridi, N., Taylor, P. and Glaszmann, J. 1995. Identification and characterisation of sugarcane intergeneric hybrids, *Saccharum officinarum* × *Erianthus arundinaceus*, with molecular markers and DNA in situ hybridisation. *Theor. Appl. Genet.* 91, 320–6.

Dillon, S. L., Shapter, F. M., Henry, R. J., Cordeiro, G., Izquierdo, L. and Lee, L. S. 2007. Domestication to crop improvement: Genetic resources for *Sorghum* and *Saccharum* (Andropogoneae). *Ann. Bot.* 100(5), 975–89.

Dunckelman, P. and Breaux, R. 1972. Breeding sugarcane varieties for Louisiana with new germplasm. *Proc. Int. Soc. Sugarcane Technol.* 14, 233–9.

Edmé, S. J., Miller, J. D., Glaz, B., Tai, P. Y. and Comstock, J. C. 2005. Genetic contribution to yield gains in the Florida sugarcane industry across 33 years. *Crop Sci.* 45(1), 92–7.

Gilbert, J. E., Lewis, R. V., Wilkinson, M. J. and Caligari, P. D. S. 1999. Developing an appropriate strategy to assess genetic variability in plant germplasm collections. *Theor. Appl. Genet.* 98, 24–8.

Glynn, N. C., Dixon, L. J., Castlebury, L. A., Szabo, L. J. and Comstock, J. C. 2010. PCR assays for the sugarcane rust pathogens *Puccinia kuehnii* and *P. melanocephala* and detection of a SNP associated with geographical distribution in *P. kuehnii. Plant Pathol.* 59(4), 703–11.

Gravois, K. A., Bischoff, K. P., Milligan, S. B., Martin, F. A., Hoy, J. W., Reagan, T. E., Kimbeng, C. A., LaBorde, C. M. and Hawkins, G. L. 2008. Registration of 'L 97-128' Sugarcane. *J. Plant Reg.* 2(1), 24–8.

Gravois, K. A., Bischoff, K. P., LaBorde, C. M., Hoy, J. W., Reagan, T. E., Pontif, M. J., Kimbeng, C. A., Hawkins, G. L., Sexton, D. R. and Fontenot, D. P. 2010. Registration of 'L 01-283' Sugarcane. *J. Plant Reg.* 4(3), 183–8.

Gravois, K. A., Bischoff, K. P., Pontif, M. J., LaBorde, C. M., Hoy, J. W., Reagan, T. E., Kimbeng, C. A., Legendre, B. L., Hawkins, G. L., Sexton, D. R. and Fontenot, D. P. 2011. Registration of 'L 01-299' Sugarcane. *J. Plant Reg.* 5(2), 191–5.

Hale, A. L., Hoy, J. W. and Veremis, J. C. 2010. Identification of sources of resistance to sugarcane red rot *Proc. Int. Soc. Sugarcane Technol.* 27, 1–8.

Hale, A., White, W., Dufrene, E., Tew, T., Grisham, M., Pan, Y.-B. and Miller, J. 2012. HoCP 04-838–A new sugarcane variety for Louisiana. *J. Am. Soc. Sugar Cane Technol.* 32, 84–5.

Hale, A. L., Dufrene Jr., E. O., Tew, T. L., Pan, Y.-B., Viator, R. P., White Jr., P. M., Veremis, J. C., White, W. H., Cobill, R., Richard, E. P., Rukavina, H. and Grisham, M. P. 2013. Registration of 'Ho 02-113' sugarcane. *J. Plant Reg.* 7(1), 51–7.

Hale, A. L., Viator, R. P. and Veremis, J. C. 2014. Identification of freeze tolerant *Saccharum spontaneum* accessions through a pot-based study for use in sugarcane germplasm enhancement for adaptation to temperate climates. *Biomass Bioenergy* 61, 53–7.

Hale, A. L., Viator, R. P., Eggleston, G., Hodnett, G., Stelly, D. M., Boykin, D. and Miller, D. K. 2016. Estimating broad-sense heritability and investigating the mechanism of genetic transmission of cold tolerance using mannitol as a measure of post-freeze juice degradation in sugarcane and energycane (*Saccharum* spp.). *J. Ag. Food Chem.* 64, 1657–63.

Hale, A. L., Viator, R. P., Kimbeng, C. and Veremis, J. C. 2017. Use of artificially-induced freezing temperatures to identify freeze tolerance in above-ground buds of *Saccharum* and *Erianthus* accessions. *Euphytica* 213, 46.

Heinz, D. J. 1980. Thailand S. spontaneum hybrid progeny as a new germplasm source in Hawaii. *Proc. Int. Soc. Sugarcane Technol.* 17, 1347–56.

Hodkinson, T. R., Chase, M. W., Lledo, M. D., Salamin, N. and Renvoize, S. A. 2002. Phylogenetics of *Miscanthus, Saccharum*, and related genera (*Sacchainae, Andropogoneae, Poacea*) based on DNA sequences from ITS nuclear ribosomal DNA and plastid trnL intron and trnL-F intergenic spacers. *J. Plant Res.* 115, 381–92.

Hodnett, G. L., Hale, A. L., Packer, D. J., Stelly, D. M., Da Silva, J. and Rooney, W. L. 2010. Elimination of a reproductive barrier facilitates intergeneric hybridization of a *Sorghum bicolor* and *Saccharum*. *Crop Sci.* 50(4), 1188–95.

Holbrook, C. C., Anderson, W. F. and Pittman, R. N.. 1993. Selection of a core collection from the U.S. germplasm collection of peanut. *Crop Sci.* 33(4), 859–61.

Hsu, S. and Shih, S. 1989. Development of prominent breeding stocks through nobilization of *Saccharum spontaneum. Proc. Int. Soc. Sugarcane Technol.* 20, 911–16.

Irvine, J. E. 1977. Identification of cold tolerance in *Saccharum* and related genera through refrigerated freezing screening. *Proc. Int. Soc. Sugarcane Technol.* 16, 147–56.

Jackson, R. and Dunckelman, P. 1974. Relative resistance of *Saccharum spontaneum* clones to the sugarcane borer. *Proc. Int. Soc. Sugarcane Technol.* 14(1), 513–15.

Jackson, P. and Henry, R. J. 2011. Erianthus. In: Kole, C. (Ed.), *Wild Crop Relatives: Genomic and Breeding Resources, Industrial Crops*. Springer-Verlag Berlin Heidelberg, pp. 97–107.

Jackson, P., Hale, A., Bonnett, G. and Lakshmanan, P. 2014. Sugarcane. In: Pratap, A., Kumar, J. (Eds), *Alien Gene Transfer in Crop Plants*, Volume 2. Springer, New York, NY, , pp. 317–45.

Janaki Ammal, E. K. 1941. Intergeneric hybrids of *Saccharum. J. Genet.* 41, 217–53.

Jannoo, N., Grivet, L., Seguin, M., Paulet, F., Domaingue, R., Rao, P., Dookun, A., d'Hont, A. and Glaszmann, J. 1999. Molecular investigation of the genetic base of sugarcane cultivars. *Theor. Appl. Genet.* 99, 171–84.

Jeswiet, J. 1930. The development of selection and breeding of the sugar cane in Java. *Proc. Int. Soc. Sugarcane Technol.* 3, 44–57.

Kandasami, P. A. 1961. Interspecific and intergeneric hybrids of *Saccharum spontaneum* L. 1. Functioning of gametes. *Cytologia (Tokyo)* 26, 117–23.

Khan, N. A., Bedre, R., Parco, A., Bernaola, L., Hale, A., Kimbeng, C., Pontif, M. and Baisakh, N. 2013. Identification of cold-responsive genes in energycane for their use in genetic diversity analysis and future functional marker development. *Plant Sci.* 211, 122–31.

Koike, H. 1980. Evidence of resistance to *Saccharum spontaneum* and *Saccharum*-related genera to sugarcane mosaic virus strains H and I. *Proc. Int. Soc. Sugarcane Technol.* 17, 1523–7.

Lalitha, R. and Premachandran, M. N. 2007. Meiotic Abnormalities in Intergeneric Hybrids between *Saccharum spontaneum* and *Erianthus arundinaceus* (Gramineae). *Cytologia*. 72(3), 337–43.

Leon, R. G., Gilbert, R. A. and Comstock, J. C. 2015. Energycane *(Saccharum spp × S. spontaneum* L.) biomass production, replication, and weed risk assessment scoring in the humid tropics and sub-tropics. *Agron. J.* 107, 323–9.

Lima, M. L. A., Garcia, A. A. F., Oliveira, K. M., Matsuoka, S., Arizono, H., de Souza Jr., C. L. and de Souza A. P. 2002. Analysis of genetic similarity detected by AFLP and coefficient of parentage among genotypes of sugar cane (*Saccharum* spp.). *Theor. Appl. Genet.* 104, 30–8.

Lo, C. C. and Sun, S. 1968. Collecting wild cane in Taiwan. *Proc. Int. Soc. Sugarcane Technol.* 8, 1047–55.

Lo, C. C., Chia, Y. H., Chen, W. H., Shang, K. C., Shen, I. S. and Shih, S. C. 1978. Collecting Miscanthus germplasm in Taiwan. *Proc. Int. Soc. Sugarcane Technol.* 16, 59–69.

Mangelsdorf, A. J. 1983. Cytoplasmic diversity in relation to pests and pathogens. *Int. Soc. Sugar Cane Technol. Sugarcane Breeders' Newsl.* 45, 45–9.

Matsuo, K., Chuenpreecha, T., Ponragdee, W. and Wonwiwachai, C. 2001. Eco-physiological characteristics and yielding ability of *Erianthus* spp. *JIRCAS Research Highlights* 2001, 12–13.

Matsuoka, S. and Garcia, A. A. F. 2011. Sugarcane underground organs: Going deep for sustainable production. *Trop. Plant Biol.* 4, 22–30.

Matsuoka, S. and Stolf R. 2012. Sugarcane tillering and ratooning: Key factors for a profitable cropping. In: Goncalves, J. F. and Correia, K. D. (Eds), *Sugarcane: Production, Cultivation and Uses*. Nova Science Publishers, Inc., Hauppauge, New York, USA, pp. 137–57.

Miller, J. and Tai, P. 1992. Use of plant introductions in sugarcane cultivar development. In: Shands, H. L. and Weisner, L. E. (Eds), *Use of Plant Introductions in Cultivar Development Part 2*, CSSA Spl Pub no. 20. CSSA, Madison, WI, USA, pp. 137–49.

Miller, J., Tai, P., Edme, S., Comstock, J., Glaz, B. and Gilbert, R. 2005. Basic germplasm utilization in the sugarcane development program at Canal Point; FL; USA. *Proc. Int. Soc. Sugarcane Technol.* 2, 532–6.

Milligan, S., Martin, F., Bischoff, K., Quebedeaux, E., Dufrene, E., Quebedeaux, K. and Hoy, J., Reagan, T., Legendre, B. and Miller, J. 1994. Registration of 'LCP 85-384'sugarcane. *Crop Sci.* 34, 819–20.

Ming, R., Moore, P. H., Wu, K. K., D'Hont, A., Glaszmann, J. C., Tew, T. L., Mirkov, T. E., da Silva, J., Jifon, J., Rai, M., Schnell, R. J., Brumbley, S. M., Lakshmanan, P., Comstock, J. and Paterson, A. H. 2006. Sugarcane Improvement through breeding and biotechnology. *Plant Breed. Rev.* 27, 17–118.

Mukunthan, N. and Nirmala, R. 2002. Tolerance to white grub, *Holotrichia serrata* Fabr. In *Saccharum* and *Erianthus* spp. *Sugar Cane Int.* 22, 26–8.

Mullins, R. T. 1989. Studies of Genotype by Environment (GE) Interactions in Sugar Cane. Ph.D. Thesis, University of Queensland, Brisbane, Australia.

Nair, N. 2008. Sugarcane breeding Institute, Coimbatore: A perspective. *Sugar Tech.* 10(4): 285–92.

Nair, N. V., Nair, S., Sreenivasan, T. and Mohan, M. 1999. Analysis of genetic diversity and phylogeny in *Saccharum* and related genera using RAPD markers. *Genet. Resour. Crop Evol.* 46, 73–9.

Nair, N. V., Selvi, A., Sreenivasan, T. V., Pushpalatha, K. N. and Mary, S. 2006. Characterization of intergeneric hybrids of *Saccharum* using molecular markers. *Genet. Resour. Crop Evol.* 53(1), 163–9.

Natarajan, U. S., Balasundaram, N., Ramana Rao, T. C., Padmanaban, P., Mohanraj, D., Karthigeyan, S. and Dhamodaran, S. 2001. Role of *Saccharum spontaneum* in imparting stable resistance against sugar cane red rot. *Sugar Cane Int.* 2001, 17–20.

Nayak, S. N., Song, J., Villa, A., Pathak, B., Ayala-Silva, T., Yang, X., Todd, J., Glynn, N. C., Kuhn, D. N., Glaz, B. and Gilbert, R. A. 2014. Promoting utilization of Saccharum spp. genetic resources through genetic diversity analysis and core collection construction. *PloS One* 9(10), e110856.

Pan, Y. B., Cordeiro, G. M., Richard Jr., E. P. and Henry, R. J. 2003a. Molecular genotyping of sugarcane clones with microsatellite DNA markers. *Maydica* 48, 319–29.

Pan, Y. B., Miller, J. D., Schnell, R. J., Richard, E. P. and Wei, Q. 2003b. Application of microsatellite and RAPD fingerprints in the Florida sugarcane variety program. *Sugar Cane Int.* 2003, 19–28.

Pan, Y.-B., Burner, D. M., Wei, Q., Cordeiro, G. M., Legendre B. L. and Henry, R. J. 2004. New *Saccharum* hybrids in *S. spontaneum* cytoplasm developed through a combination of conventional and molecular breeding approaches. *Plant Genet. Resour.* 2, 131–9.

Park, J. W., Yu, Q, Gracia, N. S., Acuna, G. M. and da Silva J. A. 2011. Development of new intergeneric cane hybrids, Miscanes, as a source o biomass feedstock or biofuel production. In: *Plant and Animal Genomes XIX Conference*. Abstract.

Perera, M. F., Arias, M. E., Costilla, D., Luque, A. C., García, M. B., Romero, C. D., Racedo, J., Ostengo, S., Filippone, M. P., Cuenya, M. I. and Castagnaro, A. P. 2012. Genetic diversity assessment and genotype identification in sugarcane based on DNA markers and morphological traits. *Euphytica* 185(3), 491–510.

Piperidis, G., Christopher, M. J., Carroll, B. J., Berding, N. and D'hont A. 2000. Molecular contribution to selection of intergeneric hybrids between sugarcane and the wild species *Erianthus arundinaceus*. *Genome* 43(6), 1033–7.

Piperidis, N., Chen, J. W., Deng, H. H., Wang, L. P., Jackson, P. and Piperidis, G. 2010. GISH characterization of *Erianthus arundinaceus* chromosomes in three generations of sugarcane intergeneric hybrids. *Genome* 53(5), 331–6.

Ram, B., Sreenivasan, T. V., Sahi, B. K. and Singh, N. 2001. Introgression of low temperature tolerance and red rot resistance from *Erianthus* in sugarcane. *Euphytica*. 122(1), 145–53.

Reyna, N. and Sneller, C. 2001. Evaluation of marker-assisted introgression of yield QTL alleles into adapted soybean. *Crop Sci.* 41, 1317–21.

Richard, E. P. and Anderson, W. F. 2014. Sugarcane, energy cane and napier grass. In: Karlen D. L. (Ed.), *Cellulosic Energy Cropping Systems*. Wiley, New Delhi, India, pp. 91–108

Roach B. 1978. Utilization of *Saccharum spontaneum* in sugarcane breeding. *Proc. Int. Soc. Sugarcane Technol.* 16, 43–58.

Roach, B. 1984. Conservation and use of the genetic resources of sugarcane. *Sugar Cane* 2, 7–11.

Roach, B. 1986. Evaluation and breeding use of sugarcane germplasm. *Proc. Int. Soc. Sugarcane Technol.* 19, 492–502.

Roach, B. 1989. Origin and improvement of the genetic base of sugarcane. *Proc. Aust. Soc. Sugar Cane Technol.* 11, 34–47.

Rott, P, RA Bailey, JC Comstock, BJ Croft, AS Saumtally. 2000. *A Guide to Sugarcane Diseases*. CIRAD and ISSCT, Montpellier, France.

Sardana, H. R. 2002. Evaluation of *Saccharum barberi, S. sinense* and *Erianthus* spp. Clones of Sugarcane for susceptibility to major borer pests. *Ann. Plant Prot. Sci. (India)* 10, 42–4.

Simmonds, N. 1993. Introgression and incorporation. Strategies for the use of crop genetic resources. *Biol. Rev.* 68, 539–62.

Sobral, B. W., Braga, D. P., LaHood, E. S. and Keim, P. 1994. Phylogenetic analysis of chloroplast restriction enzyme site mutations in the *Saccharinae* Griseb. Subtribel of the *Andropogoneae* Dumort. tribe. *Theor. Appl. Genet.* 87(7), 843–53.

Sooksathan, K., Nagatomi, S. and Naritoom, K. 1992. *Saccharum* germplasm collected in Thailand and its significance. *Proc. Int. Soc. Sugarcane Technol.* 21, 324–37

Steele, K., Price, A., Shashidhar, H. and Witcombe, J. 2006. Marker-assisted selection to introgress rice QTLs controlling root traits into an Indian upland rice variety. *Theor. Appl. Genet.* 112, 208–21.

Sugimoto, A., Terajima, Y., Terauchi, T., Ponragdee, W., Ohara, S., Tagane, S., Sansayawichai, T., Ishida, T., Ando, S. and Matsuoka M. 2012. Developing new types of sugarcane by hybridization between commercial sugarcane cultivars and wild relatives. *Plant Genetic Resources for Food and Agriculture in Asia and the Pacific: Impacts and Future Directions*, 11.

Sundaram, S., Krishnamurthi, M., Rajeshwari, S., Sekar, S. and Shanmuganathan, M. 2010. Genetic base broadening of sugarcane (*Saccharum* spp.) by introgression of genes through intergeneric hybridization. *Proc. Int. Soc. Sugarcane Technol.* 27, 1–9.

Tai, P. 1981. Freezing cold-tolerant parental clones of sugarcane. *Agron. J.* 73(3), 423–6.

Tai, P. Y. and Miller, J. D. 2001. A core collection for *Saccharum spontaneum* L. from the World Collection of Sugarcane. *Crop Sci.* 41(3), 879–85.

Tai, P. Y. P., Gan, H., He, H. and Miller, J. D. 1991. Phenotypic characteristics of F2 and BC1 progenies from sugarcane intergeneric crosses. *J. Am. Soc. Sugar Cane Tech.* 11, 38–47.

Terajima, Y., Matsuoka, M., Irei, S., Sakaigaichi, T., Fukuhara, S., Ujihara, K., Ohara, S. and Sugimoto, A. 2007. Breeding for high-biomass sugarcane and its utilization in Japan. *Proc. Int. Soc. Sugarcane Technol.* 26, 759–62.

Tew, T. L. 1987. New varieties. In: Heinz D. J. (Ed.), *Sugarcane Improvement Through Breeding*. Elsevier, Amsterdam, pp. 559–94.

Tew, T. L., White, W. H., Legendre, B. L., Grisham, M. P., Dufrene, E. O., Garrison, D. D., Veremis, J. C., Pan Y. B., Richard Jr., E. P. and Miller, J. D. 2005. Registration of 'HoCP 96-540' sugarcane. *Crop Sci.* 45, 785–6.

Todd, J., Wang, J., Glaz, B., Sood, S., Ayala-Silva, T., Nayak, S. N., Glynn, N. C., Gutierrez, O. A., Kuhn, D. N., Tahir, M. and Comstock, J. C. 2014. Phenotypic characterization of the Miami World Collection of sugarcane (*Saccharum* spp.) and related grasses for selecting a representative core. *Genet. Resour. Crop Evol.* 61(8), 1581–96.

Upadhyaya, H. D. and Ortiz, R. 2001. A mini core subset for capturing diversity and promoting utilization of chickpea genetic resources in crop improvement. *Theor. Appl. Genet.* 102(8), 1292–8.

Viator, R. P., White Jr., P. M., Hale, A. J. and Waguespack, H. L. 2012. Screening for tolerance to periodic flooding for cane grown for sucrose and bioenergy. *Biomass Bioenergy* 44, 56–63.

Viswanathan, R. and Rao, G. 2011. Disease scenario and management of major sugarcane diseases in India. *Sugar Tech* 13(4): 336–53.

Walker, D. I. T.. 1972. Utilization of noble and *Saccharum spontaneum* germplasm in the West Indies. *Proc. Int. Soc. Sugarcane Technol.* 14, 224–32.

Wang, L. P., Jackson, P. A., Lu, X., Fan, Y. H., Foreman, J. W., Chen, X. K., Deng, H. H., Fu, C., Ma, L. and Aitken, K. S. 2008. Evaluation of sugarcane × *Saccharum spontaneum* progeny for biomass composition and yield components. *Crop Sci.* 48, 951–61.

Wang, X.-H., Yang, Q.-H., Li, F.-S., He, L.-L. and He, S.-C. 2009. Molecular identification of spp.× hybrids using sequence-characterized amplified region markers. *Crop Sci.* 49, 864–70.

White, W., Miller, J., Milligan, S., Burner, D. and Legendre, B. 2001. Inheritance of sugarcane borer resistance in sugarcane derived from two measures of insect damage. *Crop Sci.* 41, 1706–10.

White, W. H., Dufrene Jr., E. O., Verdun, D. L., Duet Jr., M. J., Hale, A. L., Landry, C. L., Grisham, M. P., Warnke, K. Z. and Pan, Y. B. 2014. Ho 07-613 – A potential new sugarcane variety for Louisiana. *J. Am. Soc. Sugar Cane Technol.* 34, 61.

White, W. H., Hale, A. L., Veremis, J. C., Tew, T. L. and Richard, E. P. Jr. 2011. Registration of two sugarcane germplasm clones with antibiosis to the sugarcane borer (Lepidoptera: Crambidae). *J. Plant Reg.* 5(2), 248–53.

White, W. H., Richard, R. T. and Hale, A. L. 2013. Erianthus: A sugarcane relative with potential as a source of resistance to the stem borer *Diatraea saccharalis* (F.) (Lepidoptera: Crambidae). *Proc. Int. Soc. Sugarcane Technol.* 28, 825–32.

Chapter 4

Advances in conventional sugarcane breeding programmes

Phillip Jackson, CSIRO, Australia

1 Introduction

The breeding and selection of disease-resistant sugarcane varieties with high cane yield and high commercially extractable sucrose content has underpinned the development of profitable sugar industries worldwide. In most countries, development and identification of better new varieties remain among the highest priorities for research investment by industry stakeholders. Generally, farmers in most countries eagerly obtain new varieties to test on their farms and rapidly adopt those which are more profitable to them.

The term 'conventional' means different things to different people, and perhaps some would suggest the title for this chapter 'advances in conventional breeding' is an oxymoron. Usually in the past, the terms 'conventional' and 'traditional breeding' were used to refer to methods involving the generation of new genotypes from cross-pollination and selection based on measurements of traits in field trials. The use of DNA-based technologies such as DNA markers was normally excluded from a definition of 'conventional' breeding. However, for about the last 10–20 years, in many countries, DNA markers have been used routinely in many breeding programmes for identifying clones, verifying or determining parentage of clones and for better understanding genetic relationships in germ plasm. As the importance of DNA-based technologies for adding value to selection and breeding programmes has increased, the distinction between 'conventional' breeding and biotechnology has become more blurred. For this chapter, 'conventional breeding' excludes research and development involving transgenic clones. Since another chapter in this book deals specifically with DNA markers, details on DNA technology development in relation to sugarcane are not presented here. However, some aspects of DNA markers in relation to the interface with more traditional breeding methods are discussed.

http://dx.doi.org/10.19103/AS.2017.0035.05

It is difficult, if not impossible, to be aware of, let alone discuss, all significant innovations in sugarcane breeding programmes around the world. This is at least in part because in many programmes, numerous innovations are tested and adopted but not published in the scientific literature or communicated, or at least take some time to be widely communicated. It is also recognised that useful advances, even well-published advances, have occurred in some programmes but not adopted in other programmes. This chapter therefore includes some commentary based on anecdotal observations of a range of sugarcane breeding programmes in different countries where 'new' advances are being made, or could be made, using 'old' methodologies.

The genetic origin of sugarcane and past progress in sugarcane breeding are important in considering the current status and future directions in breeding programmes. This chapter first provides an overview of the history of sugarcane breeding for background context, but mainly focuses on reports published in the past 30 years. Introgression of wild species into adapted sugarcane has been an ongoing focus for sugarcane breeders using conventional breeding approaches. A diverse audience is assumed, including those less experienced with sugarcane breeding, as well as more experienced practitioners. It therefore includes some discussion on basic concepts to help understand reporting on recent specialist research.

Despite advances in many methods and ideas in particular aspects of sugarcane breeding, worldwide, sugarcane cultivars continue to be developed using basic underlying breeding programme structures and some areas of methodology that have been broadly similar for about 100 years, as discussed in the following section. These systems have undoubtedly provided an effective and vital contribution to the development of cultivars for all sugar industries. However, there are concerns among industry stakeholders in some (probably most) countries with established breeding programmes that genetic gains for yield and sugar content have slowed in the most recent 3–4 decades. This concern appears justified. There would seem to be a serious need for sugarcane breeders to critically reassess the breeding and selection systems they have been using. This chapter concludes with some hypotheses of the limitations of traditional approaches in sugarcane breeding programmes and how these may be addressed.

2 Historical context

2.1 Taxonomy and origin of sugarcane

The taxonomy of sugarcane and its close relatives is controversial with the delineation of related species being unclear and debated (Hodkinson et al., 2002; Irvine, 1999; Kellogg, 2013). Sugarcane cultivars belong to the genus *Saccharum*, in the Andropogoneae tribe. Members of this group are usually polyploid, and classification into species is unclear at least partly because of interspecific and intergeneric hybridisation. The term 'Saccharum complex' was coined by Mukherjee (1957) and has since been regularly used to describe the group of taxa possibly involved in the origin of sugarcane. Daniels and Roach (1987) provided a comprehensive review of the members of this complex, which included those classified into the genera of *Saccharum*, *Erianthus*, *Miscanthus*, *Narenga* and *Sclerostachya*.

Traditionally six species have been recognised within the genus *Saccharum* (Daniels and Roach, 1987). Two of these species, (*S. spontaneum* and *S. robustum*), are found growing in the wild, while the other four (*S. officinarum*, *S. barberi*, *S. sinense* and *S. edule*) exist mostly just in cultivation. This division of species was questioned by Irvine (1999) who argued that there is little basis for the separation and that these six species should be represented by only two: *S. spontaneum*, and the all others combined to be called *S. officinarum*. However, most sugarcane breeders continue to commonly refer to the six species as traditionally identified.

S. officinarum, S. barberi and *S sinense* clones were propagated in gardens in New Guinea, India and China, respectively, for many centuries (Daniels and Roach, 1987). The development of factory-based sugar industries from the 1800s used selected clones of *S. officinarum*. This species originated in New Guinea and/or nearby Melanesian or Polynesian islands (Brandes, 1958; Mukherjee, 1957) and is hypothesised as evolving from the wild *S. robustum* (Grivet et al., 2006). *S. officinarum* has low fibre content and low levels of impurities in juice (non-sucrose-dissolved solids). These attributes contribute superior processing characteristics (see the following sections).

S. spontaneum ($2n = 36–128$) is distributed widely in the tropics and subtropics of Asia and Africa and is found in a wide diversity of habitats (Amalraj et al., 2006; Daniels and Roach, 1987; Roach, 1978). It is a highly variable species that is found in the wild, and varies in appearance from short bushy type to large stemmed clone over 5 m in height. Most *S. spontaneum* clones have thin stalks and leaves, low sugar content and high fibre content. *S. barberi and S. sinense* appear to have developed from interspecific hybrids involving *S. officinarum* and *S. spontaneum* (D'Hont et al., 2002), with possible introgression from other genera (Brown et al., 2007).

The ability of sugarcane to cross-pollinate and produce viable seeds was apparent by the late 1800s, and following this, breeding programmes operated by professional breeders were quickly developed (Daniels and Roach, 1987). The highest priority objectives of the first professionally directed sugarcane breeding programmes were maintaining or improving disease and pest resistance, and improving commercially extractable sugar content. In raw sugar production, high levels of soluble impurities and fibre in cane cause loss of sucrose in processing and increase juice extraction costs, respectively. Because of this, the first commercially produced sugarcane varieties, of the species *S. officinarum*, had high sucrose content and low fibre and impurity levels. As described in Section 2.2, a major progression in sugarcane improvement arose with introgression of components of the *S. spontaneum* genome and other species into *S. officinarum* cultivars. *S. spontaneum* provided sources of disease resistance, vigour, ratooning ability and better yields under abiotic stresses.

2.2 Brief history of sugarcane breeding

Historically, three main phases have been identified in professionally directed sugarcane breeding programmes conducted since the late nineteenth century (Roach, 1989). The first phase was intraspecific hybridisation of selected *S. officinarum* clones. Generally the goal was more disease-resistant forms of *S. officinarum* with good factory qualities, and improved yield. Apart from the favourable processing attributes of *S. officinarum* described earlier, clones from this species are also usually characterised by poor vigour under abiotic stresses, poor ratooning ability and susceptibility to a range of diseases.

The second phase, which occurred between 1900 and around the 1930s, involved interspecific hybridisation between *S. officinarum* and *S. spontaneum* clones. Hybrids of these two species were generally more vigorous and more tolerant to a range of environmental stresses and had stronger ratooning, compared with the prior *S. officinarum* cultivars. However, breeders found it necessary to backcross the hybrids to *S. officinarum* to 'dilute' the undesirable characters in the wild canes, particularly low sugar content (Roach, 1989). This backcrossing process is termed 'nobilisation' by sugarcane breeders.

The introgression of genome components from *S. spontaneum* into *S. officinarum* in sugarcane breeding programmes represented one of the most important and successful examples of introgression of wild relatives in any major crop species. Introgression of *S. spontaneum* into *S. officinarum* using deliberate hybridisation and selection within sugarcane breeding programmes occurred in both Indonesia and India independently. The most famous clone produced from the early introgression programme in Java was POJ2878, also known as the 'Java wonder cane'. Development of this pivotal clone (selected in 1921) was described by Mangelsdorf (1960). It is reported that within eight years of being selected, this variety occupied over 400 000 acres, or 90%, of the production area in Indonesia and produced a 35% yield advantage over the varieties it replaced (Mangelsdorf, 1960). This clone is exceptional. Its use in subsequent breeding efforts has resulted in it being in the ancestry of most commercial cultivars around the world today.

In India the first recorded deliberate cross between *S. officinarum* and *S. spontaneum* was made at the Sugarcane Breeding Institute, Coimbatore, in 1912 (Nair, 2008). This cross between Vellai (*S. officinarum*) and a wild *S. spontaneum* produced the interspecific hybrid Co205. This clone was released as a commercial cultivar in Punjab, India, in 1918 and reportedly yielded 50% more than *the S. barberi* cultivars it replaced in that region (Thuljaram Rao, 1987; cited in Selvi et al., 2005). Other interspecific hybrids were also produced in India at around the same time and were crossed with *S. officinarum* clones and other hybrids, including POJ clones introduced from Java. The work in Coimbatore led to the production of other important cultivars suited to environments in India and worldwide. Many of these cultivars feature in the ancestries of important sugarcane cultivars worldwide today.

The third phase of sugarcane improvement, beginning around the 1930–40s and continuing today, generally involved exploiting the material produced in the second phase (Roach, 1989). This has involved inter-crossing among the original hybrids, and recurrent crossing and selection among progeny with increasingly larger populations. Most parents in breeding programmes throughout the world today are based on a relatively small number of ancestors derived from the early interspecific hybrids produced in the second phase (Arceneaux, 1967). Only a few additional introgressions of *S. spontaneum* clones have occurred in the past 50 or more years and only in some programmes (see Section 2.3). Modern cultivated sugarcane varieties therefore are complex interspecific hybrids primarily involving *S. officinarum* and *S. spontaneum*, and in most cases involving some contributions from *S. robustum, S. sinense* and *S. barberi* (Daniels and Roach, 1987; Irvine, 1999).

2.3 Current sugarcane breeding programmes

There are currently around 50 significant sugarcane breeding programmes around the world, mostly specifically targeting sugarcane industries within individual countries (Machado (2001), http://www.sugarcanevariety.org). The three traits of highest overall importance in all breeding programmes are resistance to local diseases and pests, high commercially extractable sucrose content and high cane yield in plant and ratoon crops. Many programmes also pursue traits associated with ease of manual harvesting (e.g. propensity for dead leaves to easily fall off, large stalks) and crop management (e.g. fast canopy cover to control weeds).

A detailed account of a range of many practical aspects of sugarcane breeding, which have remained largely unchanged for many decades, was provided in chapters within Heinz (1987). While individual breeding programmes in different countries have their own particular features, most, if not all currently, follow the general scheme outlined in Fig. 1. While the clonal selection systems (i.e. the sequence of selection trials between stage 1 and release of cultivars) are often well described in documentations published from breeding programmes, the arrangements for managing parents are usually less clear.

Usually elite clones identified as performing well for key traits in advanced stages of selection are used as parents in further generation cycles (Fig. 1). These are crossed in some format and seedlings evaluated for key traits in the first stage of selection. Based on progeny performance, the breeding value of the parents can be estimated, and decisions made regarding the retention or otherwise of each parent. Different breeding programmes use different methods to select parents but there are very few (if any) reports on optimising the above breeding system, and it is questionable even whether the overall approach is even suited to sugarcane (Wei and Jackson, 2016), and this is discussed further in Section 4.

An important feature of most effective breeding programmes is the introduction of germ plasm from outside of the 'closed loop' depicted in Fig. 1. This can occur by introducing elite clones either from other, usually foreign, breeding programmes or from introgression breeding programmes using wild germ plasm. Sugarcane breeding programmes around the world have historically maintained good relations, and this has contributed much to the exchange of improved parental germ plasm (Simmonds, 1993). The value of introducing foreign clones as parental material is clear from examining the parent clones of major cultivars worldwide (http://www.sugarcanevariety.org). Piperidis et al. (2008) described the active import of foreign varieties into the Australian sugarcane breeding, and noted that while most foreign varieties fail to perform directly as cultivars, many prove to be valuable parents. The large contribution of clones from the USDA breeding programme

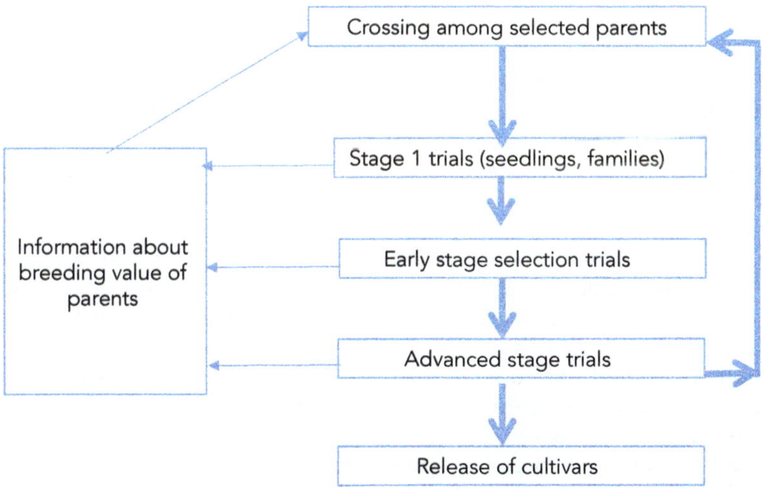

Figure 1 Schematic flow chart of structure of a typical traditional sugarcane breeding programme. Heavy lines indicate flow of clones and light lines indicate key information flow (copied substantially from Wei and Jackson, 2016).

in Florida ('CP' varieties) as parental material worldwide is apparent from parentage of important cultivars (http://www.sugarcanevariety.org) and is generally recognised (D'Hont et al., 2016).

Although the value of using elite foreign clones as parents is well recognised, anecdotal evidence suggests that many programmes currently do not devote significant resources to this activity. Part of the explanation in some countries may be due to concerns about importing new diseases combined with the unavailability or expense of appropriate quarantine facilities. Since sugarcane is propagated by vegetative cuttings, diseases can easily be transmitted with germ plasm movement. In the past 20 years or so, a wide range of molecular techniques have been developed for effective detection of diseases which will assist in sugarcane quarantine (e.g. Comstock et al., 2007; Joomun et al., 2013; Ricaud et al., 2012; Thompson et al., 2012). However, sugarcane quarantine operations require specialist high-level skills and are costly, and therefore usually only large organisations have their own quarantine service. In the past decade, several high-quality professional sugarcane quarantine services available to potential international clients have been established, and these should help facilitate safe and efficient international variety exchange between breeding programmes. These include services offered by CIRAD (http://visacane.cirad.fr/en; Guinet-Brial et al., 2015), the Mauritius Sugarcane Industry Research Institute (http://www.msiri.mu; Salem Saumtally, pers. comm.2017) and the South African Sugarcane Research Institute (http://www.sasa.org.za; Stuart Rutherford, pers. comm.2017).

The major effort, in terms of resources, in sugarcane breeding is the growing and evaluating of clones generated from crossing in field trials (Skinner et al., 1987). Following crossing, seed is germinated and seedlings usually grown in glasshouses before being transplanted into the field following well-established methods as described by Breaux and Miller (1987). Pierre et al. (2014) found the optimum temperature for germination to be between 27 and 36°C. Elite clones are generally rare and large populations are considered necessary to find superior individuals. Most breeding programmes plant thousands of seedlings each year (typically 20 K to over 1 M) (Loureiro et al., 2011), from less than one hundred to over a thousand different crosses. A multi-stage selection process follows, whereby the aim is to select a much smaller number of clones with superior attributes to move to the next stage. The number of selection stages usually varies between 4 and 6 and takes between 8 and 15 years from crossing to cultivar release. In general, in the first stage, selection is conducted on the basis of visual assessment of individual clones, and in some programmes, using family selection as an indirect selection criterion for selecting superior clones (see details in Section 3). During each stage the aim is to select a much smaller number of clones that appear to have the highest economic value. In subsequent selection stages, more accurate evaluation (through larger plots, more replicates, more measurements and more sites or years) is conducted on the smaller set of superior clones.

While rarely published, an operational feature of nearly all well-run breeding programmes which has developed in recent decades is the use of modern relational database software systems which are used to concurrently support a range of tasks associated with conducting breeding programmes, such as decisions on parental selection and crosses, preparation of field plans for selection trials, preparation of labels for all operations, storage of all types of data and statistical analysis and multiple data visualisation and reporting needs. A good example of such software specifically developed for sugarcane breeding programmes was reported by Lethbridge and

Cox (2010). The value of such software should not be underestimated – it facilitates efficient teamwork and operational conduct of breeding programmes by staff including technical staff, breeders and managers; facilitates optimal use of data for decision-making by breeders and others and provides secure long-term storage of data. The best software utilises barcode labelling, is integrated with machinery such as harvesting and weighing units and planters, and incorporates geographical positioning information (i.e. from satellite data) for recording locations of plots during planting and for checking if samples or flowers are being cut from correct locations, among many other applications. Ongoing development and adoption of database software including use with smartphones is expected to progressively improve operational procedures in many breeding programmes in the coming years.

3 Advances in breeding and selection systems

3.1 Objectives and traits of economic importance

Usually the ultimate objective of sugarcane breeding programmes is to release varieties which improve the profitability of the sugar industry being targeted. To optimise efforts towards this objective, breeders need to determine the optimal weightings that should be applied to each trait being selected for. A first step towards this involves identifying all traits influencing industry stakeholders and determining the relative economic value of variation in each trait, preferably in quantitative terms (Wei et al., 2006). As industries change, the economic value of traits may change. In recent decades weightings of some traits have changed in response to developments such as the introduction of mechanical harvesting, increased use (or consideration) of sugarcane for energy production and change in agronomic practices.

In all sugarcane breeding programmes worldwide the key target traits are resistance to important local diseases and pests, commercially extractable sucrose content, cane yield, acceptable fibre content and ratooning performance. In some programmes other traits affecting costs of harvesting or crop management are of importance. These are considered in the following section, although consideration of diseases and pests is covered separately in Section 3.4.

To better understand the potential impact of different traits on industry profitability, some basic background information on sugarcane composition and processing is given here. Sugarcane stalks can be partitioned into water (typically 70–75% of stalk fresh weight), fibre (non-soluble solids, 10–15% of fresh weight) and brix (soluble solids, 10–20% of fresh weight) (Chen and Chou, 1993). Brix in mature cane consists of sucrose, usually estimated using polarimetry and termed pol (70–95% of brix) and impurities (the reducing sugars glucose and fructose, other organic constituents and inorganic matter called ash). Impurities are important because they affect sugar partitioned to relatively lower-value molasses during processing (Chen and Chou, 1993). In a sugar production system, molasses is an end product containing residual sucrose which cannot profitably be separated as raw sugar crystals. Unless the molasses is used for ethanol production, increased loss of sucrose via molasses has a negative impact on industry profitability because molasses usually has a much lower value than raw sugar. The method used to calculate commercial cane sugar content (CCS, %) in sugarcane in Australia (Bureau of

Sugar Experiment Stations, 1984) illustrates the approximate impact of changing levels of impurities on sugar production:

CCS (%) = pol in cane (%) – impurities in cane (%)/2

In some reports it is inferred that brix (% total soluble solids) is closely related to sucrose levels. However, this may not necessarily be the case, particularly for early generation interspecific hybrids which have low and variable purity levels (e.g. Wang et al, 2008). In the production of ethanol, impurities present as fermentable sugars, most likely as glucose and fructose, will contribute to ethanol yield.

The recovery of sucrose is affected by the nature of the impurities. Increasing the ratio of reducing sugars to ash lowers the purity of molasses and thus decreases the loss of sucrose via molasses (Miller et al., 1998) but these impacts are minor. Genetic variation in impurities affecting sugar quality, particularly ash (Hogarth and Kingston, 1983; Jackson et al., 2008a; Mullins and Roach, 1985) and colour (Jackson et al., 2004, 2006), has been reported. However, the economic impact of variation in impurities appears to be small in relation to the impact of variation in the major productivity traits of cane yield and sucrose content (unpublished studies). Therefore inclusion of impurity-related traits does not materially affect selection particularly in early stages of selection.

The optimal weighting to apply to fibre in sugarcane breeding programmes has been the subject of much discussion in the last 10–15 years. High fibre content in cane has an adverse impact on sugar production systems through a combination of either greater loss of sucrose in bagasse or slower rate of crushing (Kent, 1997; Thaval and Kent, 2012). Therefore fibre has a slightly negative weighting in the selection of varieties for sugar production. However, recently there has been a growing interest by sugar mills in co-production of energy products (electricity or ethanol) from bagasse or trash, and in such situations fibre may have a higher value (e.g. Cardona et al., 2010; Dias et al., 2012). Several authors have suggested that in future, sugarcane could be selected specifically for high total biomass or fibre yields (Alexander, 1985; Tew and Cobill, 2008) or energy yield (Botha, 2009; Botha and Moore, 2014). While this may occur in greenfield developments (i.e. in areas not growing sugarcane before) it is unlikely to be adopted in existing sugarcane production regions in the foreseeable future for two reasons (Jackson, 2013). First, there exists a large capital investment in factory infrastructure already producing raw sugar. This situation, combined with expected approximate parity between sugar and liquid biofuel prices, means that sugar production will remain dominant in these regions. A parity (on average over a long-term period) between sugar and liquid biofuel price could be expected because of flexibility of some industries (including the world's largest producer Brazil) to swing a proportion of cane juice into either raw sugar or ethanol depending on market price. Secondly, the cost of producing liquid biofuel from easily extracted fermentable sugars from cane is much less, and is likely to remain so, than lignocellulosic production of biofuel from fibre.

However, if products from fibre such as electricity or biofuels have a positive value in future, a positive weighting on fibre in selection indices may be appropriate. Some different example scenarios have been presented (Jackson, 2013). A positive weighting for fibre may have important implications for sugarcane introgression breeding programmes in that early generation progeny will have a higher relative economic value than in production systems targeting only sucrose production, and therefore introgression of new traits from wild canes may be easier than in the past.

Specific fibre characteristics have been considered a potential target in breeding programmes. Fibre is composed of lignin, hemicellulose and cellulose as the main components of cell walls and their production is mediated by complex pathways. In possible future industries producing biofuel from fibre (lignocellulosic biomass), the efficiency of conversion and yields may be affected by fibre composition (Chong and O'Shea, 2012; Himmel et al., 2007; Vermerris and Abril, 2015). Some investigations of genetic variation in the composition of fibre in sugarcane have been conducted (Chong and O'Shea, 2013; Lingle and Thomson, 2012). However, identification of targets for breeding is complicated by uncertainty about which processes will be widely deployed in commercial practice.

Research to better define the relative weighting to apply to cane yield and sugar content in selection has been conducted (Deren et al., 1995; Jackson, 2005; Ong'ala et al., 2013; Wei et al., 2008). While some breeders may continue to set 'sugar yield' (i.e. the product of cane yield and sugar content) as a selection criterion, it has been clearly shown that a higher weighting on sugar content compared with the same proportionate change (e.g. 1%) in cane yield is appropriate for two reasons. First, any improvement in sugar content is economically more valuable than the same proportional improvement in cane yield (Cock et al., 2000; Jackson et al., 2000; Wei et al., 2006). This is because increases in cane yield are associated with higher marginal costs (due to harvesting, transport and milling costs) compared with increases in sugar content. Secondly, in early stages of selection involving small plots, sugar content usually has a higher broad-sense heritability and is less prone to competition effects and may therefore be measured with greater accuracy than cane yield (Jackson and McRae, 2001; Skinner et al., 1987).

It has been generally acknowledged that sugarcane breeding programmes have historically made relatively limited improvement in sugar content (Section 4.1) and possible reasons for this have been discussed (Jackson, 2005). A number of experimental-scale programmes have been conducted specifically targeting intense selection for sucrose or brix content, without concurrent intense selection for yield as is normally done in commercial breeding programmes, in different countries including Barbados (Kennedy, 2004), the United States (Louisiana) (Breaux, 1987; Legendre, 1995) and Australia (Cox et al., 1994). These have usually shown that some gains in sucrose content can be made at least in initial generations. However, the cane yield and the final commercial impacts of products from some of these programmes have not been clearly documented in publications. Anecdotal information suggests that the final commercial impact of most of these efforts may be limited because of deficiencies in cane yield of the materials generated.

Flowering has been selected against in many breeding programmes because of possible negative impacts on yield and commercial value particularly late in the harvesting period (Moore and Berding, 2014), although its value in selection is controversial. Berding and Hurney (2005) reported on a lighting study in the tropics which contained clones that did not flower freely and some that did, and attributed a 6.8% gain in cane yield and 3% gain in CCS to the prevention of flowering. However, the question of whether flowering should be weighted in selection has rarely been objectively examined. In a study in the Australian breeding programme, genetic correlations between flowering and cane yield and CCS measured across many sites (including in the tropics) and different times of the year were mostly zero or very low in cases, and it was argued therefore that zero weighting should be given to flowering propensity in selection (Jackson et al., 2012).

Harvesting costs are a high proportion of costs in sugarcane production systems, and a range of traits in sugarcane affect these costs (Fig. 2). Cultivars with a high propensity for dead leaves to fall off, large stalks and low level of stalk pithiness are particularly

Figure 2 Some traits such as stalk diameter and easily removable dead leaves are weighed more highly in breeding programmes targeting industries where cane is cut by hand (top), compared with industries which have adopted mechanical harvesting.

desirable for minimising harvesting costs, although these traits are more important for manual harvesting compared with mechanical harvesting. Yang et al. (2017) determined economic weightings for a range of traits for production systems with either manual or mechanical harvesting and showed that different clones should be selected depending on the production system being targeted. Manual harvesting will be phased out in coming decades in many countries because of unsustainable increases in costs of labour for cane harvesting. Therefore, because of the long time between early stage selection and wide cultivar adoption (over 10–15 years), breeders need to consider industry conditions well into the future rather than only at the time selection is done.

Colourants and ash content in cane vary between varieties and have been considered as potential selection criteria because of economic impacts on raw sugar quality (both colour and ash) and on impacts on sugar losses in molasses (ash) (Jackson et al., 2006, 2008b; Mullins and Roach, 1985). There are less (if any) publications in the international scientific literature on the economic impact of genetic variation in colour and ash, particularly on raw sugar quality, at least in part because of industry sensitivity to disclosing relevant data which may affect marketing. However, in general it can be stated that the economic impacts of variation among clones are very small compared with variation in other traits and therefore sugarcane breeders rarely devote significant effort to measuring and selecting these traits.

3.2 Advances in parental selection and crossing

As indicated in Section 2.3, most parents in commercial sugarcane breeding programmes are initially selected based on good performance for key traits in selection trials. A range of studies have been reported in sugarcane suggesting that narrow-sense heritability for cane yield is low, that is, additive genetic variance represents a low proportion of total genetic variance in populations generated in commercial sugarcane breeding programmes (Barbosa et al., 2005; Hogarth et al., 1981; Pisaroglo de Carvalho et al., 2014; Tyagi and Lal, 2005). Variable but generally higher proportions of genetic variance present as additive variance for brix or sucrose levels (Cox et al., 1994; Hogarth, 1987; Jackson, 2005; Pisaroglo de Carvalho et al., 2014). These results, particularly for cane yield, indicate that the selection of clones based on performance in selection trials is unlikely to be effective in improving breeding value. This raises concerns with the general breeding scheme, as shown in Fig. 1, wherein the selection of clones at least initially is restricted to those performing the best in selection trials. Low narrow-sense heritability also implies that progeny evaluation to estimate general combining ability should provide a superior indicator of breeding value. However, because a high proportion of variation among crosses is attributable to specific combining ability effects in most sugarcane breeding programmes (Hogarth, 1987; Hogarth et al., 1981; Miller, 1977; Rattey et al., 2004), several crosses (perhaps 5 or more) with different parents are required to provide a reliable estimate of general combining ability. Therefore estimating the breeding value of parents via progeny performance is expensive and time consuming, and its assessment is therefore usually restricted to a relatively small number of clones (100s rather than 1000s) per year in most commercial breeding programmes.

Estimates of family performance for cane yield and sugar content in initial stages of selection are used to estimate breeding values (Atkin et al., 2009; Simmonds, 1996). Restricted maximum likelihood and best linear unbiased prediction (BLUP) have become the methods of choice for estimating breeding values and are now widely used in animal and plant genetic evaluation (Bernardo, 2002; Henderson, 1984; Mrode and Thompson, 2005). A key advantage of BLUP is that it can combine all sources of information and automatically weigh the information accordingly, and can deal with highly unbalanced data. These methods are now being used in sugarcane breeding programmes to estimate breeding values using family performance data, with additional pedigree data contributing to precision of estimates (Atkin et al., 2009).

Unfortunately there appears to be very few, if any, robust investigations assessing overall breeding strategies to optimise sugarcane breeding programmes. Questions related to optimising criteria for selecting and discarding parents annually do not appear to have been seriously addressed. This issue is considered further in Section 5.

Methodology to make crosses, store seeds and raise seedlings varies among different countries with some approaches described in Heinz (1987) and some methods used in the Australian sugarcane breeding programme illustrated in Fig. 3. Reliable flowering of selected parents has long been a widely recognised problem for sugarcane breeders particularly in subtropical regions. Clones often vary widely in flowering propensity (Moore and Nuss, 1987), and inducement of flowering in shy flowering clones is obviously required for making crosses. However, in most breeding programmes, breeders have been adept at improving flowering in clones and synchronising flowering, enabling a wider range of crosses than would be possible under natural flowering conditions. Various methods to enhance flowering and manipulate time of flowering have been developed (reviewed by

Figure 3 Bi-parental crosses setup (top) and resulting seedlings before and during transplanting (middle), prior to being planted to the field (bottom).

Moore and Nuss (1987) and more recently by Berding and Hurney (2005), Berding (2005), Berding et al. (2010) and Moore and Berding (2014)). Temperature has been found to be an important factor given the likely inhibitory effect of high temperatures on flowering. It has been suggested that in hot tropical regions, flowering facilities should be located in cooler mountain areas or photoperiod treatments should be managed so that initiation and flowering occur in cooler times of the year. While there is large genetic variation in flowering propensity and response to key environmental factors, studies indicate that a combination of environmental conditions are collectively required for a high level of flowering, including appropriate photoperiod regime, temperature range (possibly between approximately 18° and 32°), adequate moisture (particularly at initiation) and low N levels.

3.3 Clonal selection systems

3.3.1 Advances in optimising selection system designs

As indicated in Section 2, the process of selecting superior clones from original seedling populations through a multistage system of selection field trials is a critical but expensive component of all sugarcane breeding programmes. Selection systems aim to choose superior clones from large populations as efficiently as possible. However, the selection is complicated by the fact that genotypes are ranked on the basis of phenotype which is often dominated by non-genetic factors. Various studies have investigated sources of variation in selection trials to optimise the selection process. In sugarcane, experimental error variance and competition variance are greater than the genetic variance in small-plot trials (Jackson and McRae, 2001; Skinner et al., 1987). This is particularly relevant at the first stage of selection because of the extremely small 'plot' size (i.e. a single plant usually spaced between 0.2 and 0.5 m). Any measurements or ratings made in seedlings tend to be imprecise because of likely large experimental error effects and interplant completion effects (Ram and Chaudhary, 2000; Ram et al., 1997; Skinner et al., 1987). Generally such studies have measured low broad-sense heritability (<0.2) for key traits like cane yield. This indicates a low proportion of observed phenotypic variation is attributable to repeatable genetic effects. However, most such studies also indicate that good genetic gains from selection of seedlings can be made.

The use of family selection in the earliest stages of selection has been evaluated and advocated from the 1980s onwards (Hogarth, 1987; Hogarth et al., 1990; Hogarth and Mullins, 1989; Sukarso, 1986; Wu and Tew, 1989) and subsequently supported in further studies in a range of countries (Kimbeng and Cox, 2003; Leite et al., 2009; Pedrozo et al., 2011; Shanthi et al., 2008). In most cases, a family comprises random clones (progeny) from a bi-parental cross.

Family selection involves making measurements of traits (e.g. cane yield or sugar content) in 'family' plots containing multiple progeny clones from the same family planted end-to-end. The aim is to obtain an estimate of the whole (family) plot value for trait being measured rather than for the individual clones within the plot. Family plots are usually replicated to improve precision of estimates of family means and to allow for estimates of error variance. Families can also be evaluated across different locations to sample genotype × environment (GE) interaction effects which may be advantageous where genotype × site interactions are known to be large (Bissessur et al., 2000; Jackson et al., 1995b).

The potential value of using family means (compared with data from individual clones) lies in lower error variance and higher broad-sense heritability for families than for individual clones in the early stages of selection. Measurement of family means also provides a way to objectively and rapidly estimate breeding values of parents providing possible benefits for parental improvement schemes (Atkin et al., 2009; Stringer et al., 2011). Anecdotal observation suggests that the use of family selection in commercial sugarcane breeding programmes is tending to increase with time.

However, family selection may not provide superior gains compared with the selection of individual clones in all situations (Jackson et al., 1995a). The use of family means has been questioned on the basis that the best families are those which can provide the best individual clones, rather than those with the best mean performance of members. However, a high correlation between family means and proportion of elite clones has been observed (Chang and Milligan, 1992; Kimbeng and Cox, 2003), providing support to the use of family means in selection. However, it should also be emphasised that in selecting clones in early stages, the choice for a breeder is not whether to use family *or* individual clonal selection but rather the relative weighting to put on data relating to family mean and individual clones when selecting clones at any stage of selection. In this context, optimal selection index methods (White and Hodge, 2013) can be used to optimise weightings of different sources of data. A similar, but more subjective and arbitrary approach involves using family means to weigh the numbers of clones selected from families (Kimbeng and Cox, 2003).

Several methods of data analysis to partition experimental error variance in order to obtain more precise estimates of genetic effects in selection trials for a range of crops have been developed in recent decades (e.g. Cullis et al., 1998; Smith et al., 2005). Stringer and Cullis (2002) developed a modelling approach using sugarcane selection trial data which involved partitioning of variability into that due to spatial trends (e.g. associated with variation in soil fertility or moisture availability) and inter-plot competition, which is now routinely applied in the Australian sugarcane breeding programme.

GE interaction is an important source of variation in sugarcane yield and can strongly influence selection in breeding programmes (Hogarth, 1971; Kang and Miller, 1984; Milligan et al., 1990; Jackson and Hogarth, 1992; Ramburan et al., 2011; Jackson et al., 2007). Most studies have shown that for cane yield and sugar content genotype × location (GL) interactions are more important than genotype × crop-year (GY) and genotype × location × crop-year (GLY) interactions (Bull et al., 1992; Milligan et al., 1990; Jackson and Hogarth, 1992, Chen et al., 2012), although there are exceptions (Kimbeng et al., 2002). Partitioning of GE interaction variances has been used to help optimise resource allocation in breeding programmes such as optimising the trade-offs between numbers of sites, crop-year cycles and replicates (e.g. Brown and Glaz, 2001; Kimbeng et al., 2009). The use of bi-plots to visually depict relationships among genotypes and test environments has become more popular in many crops and used in examining sugarcane selection trial data (Glaz and Kang, 2008; Queme et al., 2010).

Arguably one of the most important end goals of GE studies should be to understand the underlying biological causes. However, the few studies in sugarcane that have attempted to relate patterns of GE interactions to other factors (Chen et al., 2012; Jackson and Galvez, 1996; Ramburan et al., 2011) have not provided strong resolution of the underlying causes. The general importance of GL interactions as opposed to GY interactions and the study of Jackson and Galvez (1996) suggested a possible stronger role for soil-related factors than weather-related factors. Research conducted by Basnayake et al. (2012) also indicated that variation due to GE interaction between well-watered and water stress environments was very small, except where the water stress was extremely severe.

For selection there are numerous trade-offs to be considered in terms of optimising the design of individual selection stages or of the entire system for a given available budget. For example, for a given stage of selection, bigger plots or more replicates will increase precision of measuring each clone while reducing the numbers of clones that can be evaluated. Similarly, more clones evaluated in early stages of selection can increase intensity of selection and therefore gains possible from these stages, but reduce size of effort in later stages of selection. Optimising the entire selection system is therefore a difficult and complex problem. Many decisions are still made intuitively by breeders. However, some attempts to optimise systems in particular programmes have been made using stochastic simulation modelling based on assumptions on key statistics (e.g. size of genetic variance, error variance, competition variance, etc.) (Calija et al., 2001; Hu et al., 2009).

3.3.2 Advances in phenotyping

A basic requirement in the selection of sugarcane breeding programmes is methodology for accurate and low-cost measurement of large numbers of clones in field trials (Fig. 4). Much of the focus is on measurements of cane yield, sugar content, disease resistance and

Figure 4 A critical and major component of sugarcane breeding programmes are field trials for assessment of important traits, particularly cane yield and commercially extractable sucrose content.

fibre content, and in most breeding programmes the methods used have remained largely unchanged for many decades. However, recent technology developments have arisen in some areas and are discussed in the following sections.

High-throughput methodology to measure commercially recoverable sucrose content in sugarcane is clearly important given the economic importance of this trait (Section 3) and the fact it is not prone to competition in small plots. In the earliest stages of selection, most breeding programmes continue to use brix, measured in a small sample of juice obtained from stabbing the stalk, as a surrogate for estimating the commercial sucrose content of the whole stalk. This method has traditionally been considered to provide a reliable indicator of relative whole stalk brix and has a large labour-saving advantage over alternative methods involving cutting and carrying stalks to a laboratory for processing (Hebert, 1959). However, there are concerns regarding lack of precision of measuring brix using field sampling methods and consequently some breeding programmes exclusively use whole-stalk sampling and processing

In programmes using family selection, whole stalks are usually taken from a random sample of about ten stalks from each family plot in the first stage of selection. Nearly all breeding programmes sample whole stalks from the field for laboratory processing from at least middle stages of selection onwards because of a desire to obtain precise estimates of commercial sucrose content. Typically when stalks are sampled and taken from the field for processing in a laboratory, brix and pol are measured from juice extracted from crushed cane. While the juice measurements are rapid, measurement of fibre is generally slower and more labour intensive. In early stages of selection, variation in fibre content does not make large differences in overall ranking of clones for overall economic value. Therefore impact of selection on obtaining fibre measurements (at relatively high cost) is usually small and the additional expense and effort is not worthwhile.

At-line near-infrared (NIR) spectroscopic techniques and equipment for measuring sucrose and fibre content and other constituents in samples of shredded cane from selection trial plots have been developed (Berding et al., 1989; Berding and Marston, 2010). However, a disadvantage of NIR technology is that the initial capital investment and ongoing maintenance costs (on a total cost per unit measurement basis) may be greater

Figure 5 A mechanical harvester and tractor-drawn weigh bin (with electronic load cells for accurate weighing) being used for fast and efficient measurement of cane yield in 10-m long plots in selection trials.

than traditional methods estimating brix and pol in the laboratory using hired labour, even in countries where labour costs are high. NIR does currently offer lower costs when fibre is to be measured, but the additional gains in most selection stages from adding measured fibre in selection indices are usually minimal (as indicated in the paragraph above). A possible advantage of NIR is that predictions of many other composition attributes may be made simultaneously, if these are of interest in the breeding programme, and at no significant extra cost using already collected spectra. NIR methods have been developed to analyse lignocellulosic composition of sugarcane (Chong and O'Shea, 2013; Oxley et al., 2012; Rodríguez-Zúñiga et al., 2014) which may be important for potential future production systems producing biofuel from bagasse. NIR measurements/techniques have also been developed for predicting resistance of clones to various pests and diseases including Fiji leaf gall disease (Purcell et al., 2009), smut (Churchill et al., 2006; Purcell et al., 2010) and *Eldana* borers (Rutherford, 1998). However, it seems that in some of these cases, lack of confidence in the extrapolation of the predictions has limited routine application in breeding programmes.

Measurement of cane yield in most breeding programmes continues to be via traditional labour-intensive methods involving either estimates obtained from stalk counts and weighing samples of stalks or hand-harvesting and weighing entire plots. However, where mechanised commercial harvesting occurs, machine weighing of cane billets that are collected from mechanical harvesting of small plots (Fig. 5) has been successfully introduced (Hogarth and Mullins, 1989). The prospect for mechanised sampling of cane billets has been long advocated and investigated (Berding and Bull, 1997; McRae et al., 1998). However, there is no evidence of routine adoption. The combination of automated sampling of billets and in-field NIR analysis of samples would seem to offer large potential benefits, and Nawi et al. (2014) provided a review of emerging technologies and approaches which may be taken into account in future.

3.4 Disease and pest resistance

Breeding and selecting of varieties that are resistant to important diseases has long been a primary goal of sugarcane breeding programmes. In fact susceptibility of existing cultivars to particular diseases has been the major motivation in the initiation of many sugarcane breeding efforts historically (Walker, 1987). The use of resistant cultivars and the regulation of which varieties are grown by farmers in a region (i.e. avoiding susceptible varieties which may build inoculum pressures) have over time proven to be powerful strategies to prevent economic losses in sugarcane industries (Croft et al., 2000; Rutherford et al., 2003). Breeding for disease resistance has often been highly effective because resistance to many diseases has been observed to have moderate-to-high (>0.5) broad- and narrow-sense heritability (Hogarth, 1987; Hogarth et al., 1993a,b; Tai et al., 1981), although in some cases such as for red rot resistance (Yin et al., 1996) low heritability (<0.3) has also been reported. Modern sugarcane breeding programmes maintain procedures for routine screening of clones for important diseases, and choice of parents and cross combinations is also usually strongly affected by consideration of disease resistance (e.g. as one criteria in selection of parents, and avoiding crosses between two parents which are both susceptible to the same disease).

Comprehensive lists and information on diseases of sugarcane in different parts of the world have been provided in texts such as Rott (2000) and Ricaud et al. (2012). In most countries there are generally only three or four diseases considered to be of major importance. While these vary in different countries, diseases which are arguably the most important and which affect many sugarcane breeding programmes worldwide currently include ratoon stunting disease (RSD), smut, mosaic, red rot and brown rust.

The justification for any programme (breeding or management) to control a disease is based initially on an assessment of economic losses, either realised or anticipated (Walker, 1987). Many studies to determine the level of yield loss due to different diseases, including losses associated with different levels of disease susceptibility in different varieties, have been carried out in recent decades, and these have helped breeding programmes assign relative priority to place on resistance to disease in selection (Hoy and Hollier, 2009; Magarey and Bull, 2003; Magarey et al., 2002, 2004; Rassaby et al., 2003; Viswanathan and Balamuralikrishnan, 2005). In considering the importance of diseases as an ongoing target in sugarcane breeding programmes, it is vital to consider the risk of changes to yield impact if susceptible varieties were released due to increased inoculum and pathogen populations (Rutherford et al., 2003; Walker, 1987). This includes considering potential impact on varieties which are intermediate in resistance and susceptibility which may be unaffected by low inoculum pressures.

RSD, caused by a bacterium within xylem vessels, is possibly the most economically important disease of sugarcane worldwide (Comstock, 2002; Davis and Bailey, 2000). As discussed in the following section, the effect of this disease on the performance of many sugarcane breeding programmes remains unclear but may be major in many cases, and control of RSD through breeding programmes remains controversial. In some countries anecdotal communications with local sugarcane industry personnel and researchers can indicate that the importance of RSD is not recognised or even denied, a view which can be subsequently reversed after proper investigations using appropriate disease detection methods. This situation arises because no obvious external visible symptoms are associated with RSD, except poor or uneven yield, which only becomes evident if diseased cane is grown alongside disease-free cane. Internal visible symptoms may also be unreliable or non-existent in some varieties. Yield losses are greatest under water stress and can be up to 40% (Comstock, 2002). Genetic variation in resistance to RSD has been reported with resistant varieties identified (Comstock et al., 1997; Comstock, 2002; Roach and Jackson, 1992). Different opinions have been expressed in recent decades among sugarcane breeders regarding the role of varietal resistance in managing RSD. One view, held by some breeders (e.g. in Australia), is that because the disease can be effectively controlled using good farm management (i.e. use of disease-free seed and avoiding contamination from infected blades), it is inappropriate to actively select for resistance (Croft et al., 2000). This view is held because selecting for RSD resistance would impose a large drag on selection gains possible for the majority of farmers who follow good management practices and use disease-free planting material (because varieties which are elite apart from susceptibility to RSD would be discarded). It is also considered that publicising resistance of some varieties may act as a disincentive to some farmers to keep these varieties disease-free, which may lead to increased inoculum presence (e.g. from infected planting and harvesting equipment), increasing probability of infection of neighbouring blocks and farms which may contain varieties susceptible or intermediate in resistance to RSD. Conversely, the arguments for breeding for resistance may arise in industries where the adoption of clean seed and hygiene practices by farmers is not widespread for different reasons, and the fact that variety resistance can play a practical and important role in reducing the impact of the disease (Comstock, 2002).

It would seem important for breeding programmes (or their industry funders) to take a decision on a clear strategy on addressing RSD, but anecdotal observation suggests that this is often not the case. On the one hand if management of RSD through farm management practices is pursued instead of variety resistance, it is important to maintain 'clean' (disease free) cane throughout the entire selection trial system to avoid ranking of clones for yield being affected by the disease. This strategy has been successfully adopted and maintained in the Australian sugarcane breeding programme for many years. Conversely, if it is decided to manage RSD at least partly through variety resistance, it is desirable to ensure uniform infection of selection trials, preferably at an earliest stage possible. In this way clones which are resistant or which can tolerate the disease will be favoured from selection. This approach has been taken by the RIDESA breeding programme in Brazil. An unsatisfactory option, which occurs in some programmes, is to not control infection of selection trials by RSD. In this case clones may become inoculated with RSD in an *ad hoc* or accidental manner (e.g. by a worker using a cane knife which may be infected with the bacterium from prior cutting of cane in an infected block of cane) resulting in an uneven impact of the disease in selection trials and leading to unreliable yield responses and selection trial data. A related problem may arise in breeding programmes targeting regions where RSD exists in commercial production but when an otherwise elite but RSD susceptible clone is not infected by RSD during the selection process. In this case the clone may be released for commercial production but ultimately reduces commercial performance because of gradual infection by RSD.

Breeding programmes differ in how much reliance is placed on natural infection within selection trials to select for disease resistance or whether specially managed trials are used in which clones are artificially inoculated with the disease. The latter approach appears to be increasingly used in modern breeding programmes in advanced stages of selection and has an advantage that high and uniform disease pressure can be ensured, resulting in more reliable results and effective selection (Berding et al., 1997). Such trials are vital prior to the release of cultivars for diseases which may be well managed with existing cultivars but where release of susceptible varieties would be expected to increase inoculum and disease incidence increase. However, relying on natural infection may clearly be effective when there is strong naturally occurring inoculum pressure for a disease. This may exist particularly for some fungal diseases such as smut or rust if susceptible cultivars are grown on large areas near to selection trials, and the planting of susceptible cultivars around selection trials may be done for this purpose deliberately. Because of the high cost and labour involved, special trials where clones are inoculated are usually limited to clones in advanced stage trials.

Soil pathogens (mostly unknown organisms) and nematodes which may build up over time in any continuous monoculture cropping systems may be responsible for widespread losses in sugarcane yield worldwide (Garside et al., 2001b; Pankhurst et al., 2005; Stirling et al., 2001). In Australia the soil fungal pathogen *Pachymetra* is of importance and targeted screening of clones is now routinely conducted to improve the effectiveness of selection and breeding for resistance (Magarey and Bull, 2003). However, resistance to other soil pathogens is generally only selected indirectly in sugarcane breeding programmes through the impact of these organisms on results (on yield or other measurements) obtained from selection trials on affected soil.

One concerning issue in some breeding programmes may be the use of methods to rate relative resistance of varieties which do not necessarily provide comparable results across different environments. Arguably the best approach for obtaining disease resistance ratings is to derive resistance ratings based on comparison with a set of 'standard' clones. Standard clones are chosen based on considerable experience of clone reaction to the disease, and covering a range of resistance ratings (a 1–9 scale was recommended, where 1 is very highly resistant, and 9 is very highly susceptible) (following the ISSCT rating scale as proposed by Hutchinson (1969)). A linear regression or other linear relationship between the pre-known ratings of the standards (*x*-axis), and their level of infection within the resistance rating trial (*y*-axis) is then determined. Resistance ratings for the experimental varieties are then estimated from the intersection of a horizontal line at a height corresponding to the observed level of infection in the same trial. This approach is used to correct for large differences in the overall level of disease infection or symptoms commonly occurring in environments used for screening trials. A much less satisfactory approach (although unfortunately still used in some programmes) is allocation of resistance ratings based on assessments of absolute level of infection, which may be affected strongly by environmental effects rather than genetic effects.

There have been a number of studies identifying DNA markers linked to disease resistance, but so far widespread application of markers routinely in breeding programmes for screening for resistance is not apparent. One constraint is cost but this may change in future if markers for disease resistance are incorporated with marker-assisted breeding approaches for a range of traits. Molecular genetic approaches have led to the identification of Bru1, a major gene conferring broad spectrum resistance to brown rust, and diagnostic markers for its detection (Asnaghi et al., 2004; Costet et al., 2012; Daugrois et al., 1996). These markers have been used in breeding programmes to help breed more durable rust resistance by combining Bru1 with other sources of resistance (Glynn et al., 2013). This approach may be used in future for other diseases with major gene effects.

In general there has been less emphasis on breeding and selection for pest resistance than disease resistance in sugarcane, and fewer studies reported on understanding the inheritance of resistance. However, variety resistance has played a role in integrated management of insect pests in some industries, and this may tend to increase in the future with a generally growing move away from pesticides due to economic, environmental and resistance pressures. A range of studies have been reported on the important pest sugarcane borer (*Diatraea saccharalis*), with a high narrow-sense heritability being reported (Kimbeng et al., 2006; White et al., 2001). One factor apparently contributing to resistance to this pest is high fibre content (White et al., 2006) which may be undesirable for overall economic value for sugar production, although White et al. (2011) suggested that an alternative mechanism of antibiosis was found in some clones. Variety resistance has also been important for the management of the major insect pest endemic to Africa, *Eldana saccharina* (Keeping, 1999), and routine screening trials have been a feature of the South African breeding programme (Keeping, 2006; Nuss, 1991). White grubs (or 'cane grubs', Coleoptera and Scarabaeidae) are a serious pest in some sugarcane industries, and although variety resistance has been reported (Allsopp and Cox, 2002), to date there does not appear to be sustained efforts in sugarcane breeding programmes to select for this trait.

Plant-parasitic nematodes are a major pest to sugarcane worldwide (Cadet et al., 2005), although their importance may not be appreciated in many places for similar reasons as for RSD described earlier. There appears to be limited genetic variation for resistance in most sugarcane cultivars, but some resistance has been found in *Erianthus* and *S. spontaneum* clones and progeny derived from some of these, indicating a potential breeding pathway towards resistance (Bhuiyan et al., 2016; Croft et al., 2015).

4 Current concerns

4.1 Long-term genetic gain

There is clear evidence from a range of countries and sources of data that historical genetic gains through sugarcane breeding have been large with major industry impacts (Acreche et al., 2015; Burnquist et al., 2010; Cock, 2003; Cox and Stringer, 2007; Edmé et al., 2005; Jackson, 2005). It is also extremely obvious in many industries that addressing important disease problems by the release of resistant varieties has played a vital role in industry survival and profitability. From several studies on historical gains, as well as anecdotal information, it is apparent that improvements in cane yield are significant, while any improvements in sugar content appear relatively minor (Jackson, 2005).

Despite the clear historical contributions by sugarcane breeding programmes, a question perhaps increasingly asked by sugar industry stakeholders worldwide is whether rates of genetic gain in sugarcane are slowing or even plateauing. These concerns reflect low or no apparent industry productivity gains in recent decades. They are fuelled also by comparisons with productivity gains in some other crops, including sugar beet, which show much larger gains, with examples shown in Fig. 6. Figure 6 shows cane yield data but a similar trend is observed with sugar content which has also plateaued (data not shown), and has been noted in the past (Jackson, 2005). Interpreting industry productivity data trends is complicated by multiple potentially confounding factors. In sugarcane, factors such as adoption of mechanical harvesting in some countries, expansion to marginal land, more ratoon crop cycles and long-term degradation of soil fertility from a cropping monoculture have affected productivity trends (Garside et al., 1997, 2001; Wilson and Leslie, 1997). However, despite complexities in interpretation, large and very obvious contributions by genetic improvement have occurred in most major crops such as wheat, corn, cotton and

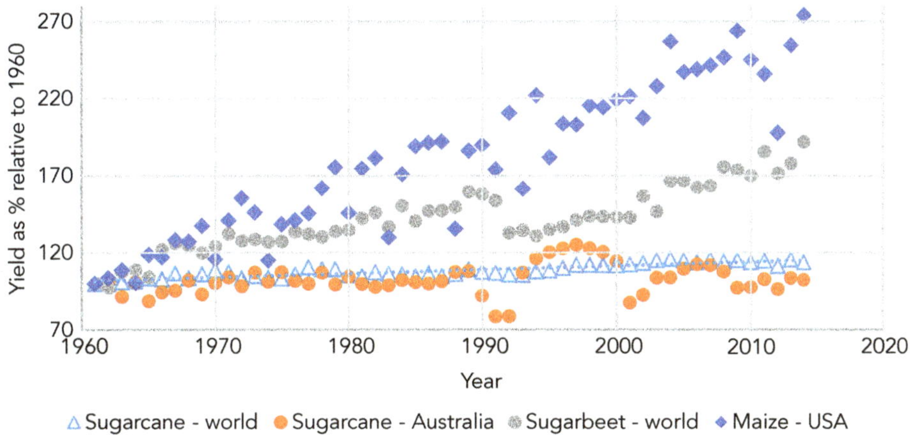

Figure 6 Realised yield gains in several cropping industries. Source: FAOSTAT http://www.fao.org/faostat.

sugar beet (Fischer et al., 2012). Meanwhile, in many sugarcane industries, genetic gains in yield or sugar content are not obvious to growers and certainly not as large as some other important crops (Fig. 6). A low rate of improvement in sugar yield in the past decade in Brazil, was also noted by Dal-Bianco et al. (2012).

In Section 2, the basic structure of modern sugarcane breeding programmes was illustrated (Fig. 1). Longer-term genetic gains over multiple breeding and selection cycles within this scheme are dependent on both an effective clonal selection system and effective selection of parents. In relation to parental selection, it is apparent in many breeding programmes that many major current cultivars arise from quite old parents, with some widely grown cultivars in some countries arising from parents that were grown as seedlings in the 1950s and 1960s (http://sugarcanevariety.org/default.aspx). From Fig. 1, it is apparent that *if* good gains were made in each 10-year breeding cycle over the last 50 years or so, parents which are old (e.g. 40–50 years from being first planted as a seedling) should be incapable of producing the best cultivars, in contrast to what is being observed. Most breeders certainly do adequately use and test new high performing clones (identified based on the performance in selection trials) as parents. The problem appears to be more that the best new high-performing clones are not on average substantially better in breeding value to the best older (even >50 years) parents. This issue has been highlighted for at least ten years (e.g. Rattey et al., 2004).

On reviewing studies investigating practices in sugarcane breeding programmes, it is apparent that relatively little effort or objective analysis seems to have been devoted to optimising parental improvement schemes. A likely limitation to faster progress in sugarcane breeding programmes is the high proportion of nonadditive genetic variance affecting yield and sugar content. The general breeding scheme shown in Fig. 1 does not deal well with nonadditive variance in relation to parental selection. High nonadditive genetic variance means that phenotypic performance of clones poorly predicts breeding value, and therefore selection of elite clones based on the performance in trials during each clonal selection cycle does not necessarily provide clones with highest breeding values. Two possible strategies may help address this issue. First, the use of DNA markers and genomic selection may provide faster and more accurate predictions of breeding values, as discussed in the following sections. Second, a breeding scheme similar to reciprocal recurrent selection which can take advantage of nonadditive effects (Hallauer and Darrah, 1985; Pandey and Gardner, 1992) may be appropriate. In sugarcane this could involve establishing two small groups of parental clones with those in each group recognised (from historical data) as having high specific combining ability effect in combination with most of the clones in the opposite group. Clones from crosses made *within* each of these populations could be used in crosses with clones selected from the opposite population. Breeding values of these clones based only on progeny performance from crosses between the two groups can then be determined, and those with the highest breeding value retained. Thus parental evaluation and improvement would occur within each group based on combining ability with the opposite group. While seedlings from both within-group crosses and between-group crosses would be evaluated for the performance in multistage field trials as currently (with best ones potentially released to industry as currently), crosses between the groups would be expected to produce the best crosses and most commercial releases.

Whatever the best breeding approach for sugarcane is, there would seem to be currently a need and opportunity in sugarcane breeding programmes to explore and critically and scientifically assess different breeding strategies than those used in the last 50 years.

4.2 Use of DNA markers

The future of all animal and plant breeding programmes is expected to increasingly involve the combined application of 'conventional' breeding approaches supplemented with use of DNA markers, and sugarcane breeding is no exception. However some progress and prospects for the integration of conventional breeding and DNA markers are briefly discussed here.

DNA markers have already been widely utilised in sugarcane breeding programmes for the purpose of identifying clones (Pan, 2010; Piperidis et al., 2004; Selvi et al., 2003), verifying parentage (Besse et al., 1996; Cai et al., 2005; D'Hont et al., 1995; McIntyre and Jackson, 2001) and in genetic diversity studies that can help in targeted use of germ plasm collections (Aitken et al., 2006; Jannoo et al., 1999; Nair et al., 1999; Selvi et al., 2006). These impacts have been useful and important. However, the widespread expectation in the 1990s that DNA markers would be used for improved selection among seedling populations for key traits such as yield, sugar content and disease resistance is yet to emerge in commercial sugarcane breeding programmes.

Progress in genetics and genomics in sugarcane has tended to lag behind other crops of comparable economic value at least partly because of the genome complexity. However, a large effort has been maintained in this area and progress has been reviewed elsewhere (Grivet and Arruda, 2002; Henry and Kole, 2010). Genetic linkage maps are difficult to construct because of the high level of polyploidy and random pairing of homologous chromosomes. The usage of single-dose markers (i.e. markers present in a single copy, and therefore segregate 1:1 in gametes) provided an avenue for the construction of linkage maps based on this subset of markers (Wu et al., 1992). A range of QTL mapping studies have been conducted on sugarcane (Pastina et al., 2010). Major advances in DNA marker technology have occurred in the last few decades and continue to rapidly develop. This is opening new opportunities for data analysis and marker applications.

In recent years, genomic selection has emerged as a promising approach for using DNA markers in animal and plant breeding programmes, and major advances are being made in some industries such as dairy (Hayes et al., 2009; Heffner et al., 2009; Jannink et al.,

2010). Genomic selection uses all marker information simultaneously to predict each trait of interest, rather than relying on precise estimates of effects of a small number of markers. In a study on genomic selection in sugarcane (Gouy et al., 2013), correlations between predictions using marker data and observed values for commercial traits ranged from 0.11 to 0.62. These levels of accuracy are interpreted as being promising for this approach to contribute to sugarcane breeding.

It has been suggested that the use of DNA markers in sugarcane breeding, particularly genomic selection, may be of high value in predicting breeding values (Wei and Jackson, 2016). This is because markers are more likely to be of value in predicting additive genetic effects which are the major genetic effects related to the breeding value of parents. It was shown that use of DNA markers could more accurately predict the breeding value of clones compared with traditional approaches of using phenotype, presumably because phenotype may be strongly affected by nonadditive genetic variance. Therefore marker predictions may provide a faster and more accurate option for improving breeding value compared with selection of clones based on phenotype as done traditionally (Fig. 1) (Wei and Jackson, 2016).

5 Future trends and conclusion

Sugarcane breeding programmes based on the so-called 'conventional' cross-pollination and selection have clearly made vital and massive contributions to sugar industries in all countries in the past and will undoubtedly continue to do so. Investment in sugarcane breeding programmes will remain a high priority for investment by all major sugarcane industries. A major reason for this investment will be as an insurance against disease and pest outbreaks. Incremental improvements in cane yield, sugar content, ratooning ability and other traits affecting production and harvesting costs are also expected, but concerns are arising regarding possible slowing rates of improvement in sugar content and cane yield in comparison with other crops, as discussed earlier.

Over the past 40 years or so, there has been a considerable investment in research aimed at improving the rates of gain in sugarcane breeding programmes. Most of these investments have contributed to an ongoing continual and incremental improvement in knowledge and execution of sugarcane breeding. However, despite many studies, the overall methods employed in commercial sugarcane breeding programmes have remained largely unchanged without major innovations and 'disruptive' changes.

Arguably one limitation to major innovation in sugarcane breeding programmes to date has been lack of involvement by private companies including multinational companies specialising in genetic improvement of crops. Sugarcane breeding programmes are largely conducted by nationally focused government or sugar industry–funded organisations, many of which enjoy monopolies within their particular industries. This contrasts with some crops such as corn and sugar beet (among many others) which are serviced by several strongly competing and often highly innovative privately funded programmes. A lack of private investment, and therefore lack of competition, in sugarcane breeding historically arises at least in part because (unlike with some crops such as those using F1 hybrids) farmers easily replant sets of new varieties without necessarily compensating the breeding company producing the variety. While sugarcane farmers worldwide enjoy low or no costs for the intellectual property rights associated with the varieties they use, this situation and associated lack of private investment and competition in supply of new varieties is unfortunately probably very detrimental to longer-term rates of genetic gain in sugarcane. It is well documented by economists how the need to innovate, including in the breeding of new varieties, is often driven by market competition (Carlin et al., 2004). By contrast, it is well documented that monopoly providers, and particularly government organisations, often have little incentive to take risks and fail to innovate strongly (Potts, 2009). An international system of Plant Breeder's Rights has been developed to address these problems. Enabling commercial rights of new varieties to be owned by breeding companies provides incentives for private investment (https://en.wikipedia.org/wiki/Plant breeders'_rights). Plant Breeder's Rights are now being utilised by many sugarcane breeding programmes (Cox et al., 2001), but it still remains to be seen whether this alone will lead to a greater private investment, competition, innovation and genetic gains.

Concurrently with ongoing investment in conventional sugarcane breeding programmes, major investments in molecular genetics research and applications to sugarcane improvement have emerged and grown over the last 30 or more years. Large advances have been made in DNA marker technology, development of transgenic plants and basic understanding of molecular genetics, following pioneering research such as that reported in sugarcane by Bower and Birch (1992), Wu et al. (1992) and D'Hont et al. (1996), amongst many others. However, so far all sugarcane cultivars continue to be developed through 'conventional' breeding methods, and it is still unclear as to when a major contribution from molecular biology technologies or better physiological understanding in cultivar development is going to arise. For investors interested in maximising rates of genetic improvement, balancing between investment in traditional (non-molecular genetics-orientated) breeding programmes and molecular genetics research and application is an important but difficult issue. Of concern to many conventional sugarcane breeders is (i) the susceptibility of governments or industry investors to 'transformational' molecular biology–based project proposals, in which the levels of technical risks of not delivering commercial outcomes are high but are either misjudged or not communicated, and (ii) the apparently low proportion of students pursuing a career in plant breeding (compared with molecular biology). Related to the latter point, it is clear that in many institutional settings, careers of young scientists may often benefit more by working (and publishing) on 'cutting-edge' technologies, compared with traditional breeding which may deliver fewer publications but higher industry impact. This creates concerns regarding attracting, retaining and training very high-quality scientists to lead breeding programmes in the future.

Based on their past and current achievements, traditional breeding programmes will continue to provide incremental improvements in productivity and solutions for important disease outbreaks. Progress in molecular genetics technologies, including those in sugarcane, is occurring at a rapid rate, and it is inevitable that practical impacts will arise within the next decade or two. However, highly competent breeding programmes based on conventional approaches are likely to remain important for a range of reasons, not least of which is that any new biotechnology approaches will almost certainly only deliver commercial outcomes by enhancing already well-functioning 'conventional' breeding programmes.

6 References

Acreche, M. M., Saez, J. V. and Vera, J. C. 2015. Physiological bases of genetic gains in sugarcane yield in Argentina. *Field Crops Research* 175, 80–86.

Aitken, K., Li, J.-C., Jackson, P., Piperidis, G. and McIntyre, C. 2006. AFLP analysis of genetic diversity within Saccharum officinarum and comparison with sugarcane cultivars. *Crop and Pasture Science* 57, 1167–84.

Alexander, A. G. 1985. *The Energy Cane Alternative*. Elsevier Science Publishers BV, Amsterdam, Netherlands.

Allsopp, P. G. and Cox, M. C. 2002. Sugarcane clones vary in their resistance to sugarcane whitegrubs. *Crop and Pasture Science* 53, 1111–36.

Amalraj, V. A., Balakrishnan, R., Jebadhas, A. W. and Balasundaram, N. 2006. Constituting a core collection of Saccharum spontaneum L. and comparison of three stratified random sampling procedures. *Genetic Resources and Crop Evolution* 53, 1563–72.

Arceneaux, G. 1967. Cultivated sugarcanes of the world and their botanical derivation. *Proceedings of the International Society of Sugar Cane Technologists* 12, 844–54.

Asnaghi, C., Roques, D., Ruffel, S., Kaye, C., Hoarau, J.-Y., Telismart, H., Girard, J., Raboin, L., Risterucci, A. and Grivet, L. 2004. Targeted mapping of a sugarcane rust resistance gene (Bru1) using bulked segregant analysis and AFLP markers. *Theoretical and Applied Genetics* 108, 759–64.

Atkin, F. C., Dieters, M. J. and Stringer, J. K. 2009. Impact of depth of pedigree and inclusion of historical data on the estimation of additive variance and breeding values in a sugarcane breeding program. *Theoretical and Applied Genetics* 119, 555–65.

Barbosa, M. H. P., de Resende, M. D. V., Bressiani, J. A., da Silveira, L. C. I. and Peternelli, L. A. 2005. Selection of sugarcane families and parents by Reml/Blup. *Crop Breeding and Applied Technology* 5, 443.

Basnayake, J., Jackson, P., Inman-Bamber, N. and Lakshmanan, P. 2012. Sugarcane for water-limited environments. Genetic variation in cane yield and sugar content in response to water stress. *Journal of Experimental Botany* 63, 6023–33.

Berding, N. 2005. Poor and variable flowering in tropical sugarcane improvement programmes. Diagnosis and resolution of a major breeding impediment. In: *XXV Jubilee Congress of International Society of Sugarcane Technologists*. Guatemala.

Berding, N., Brotherton, G., LeBrocq, D. and Skinner, J. 1989. Application of near infrared reflectance (NIR) spectroscopy to the analysis of sugar cane in clonal evaluation trial. *Proceedings of the Australian Society Sugar Cane Technologists* 11, 8–15.

Berding, N. and Bull, J. K. 1997. Australian sugarcane improvement: Current concerns. In: *Proceedings-Australian Society of Sugar Cane Technologists*. Editor: BT Egan. Watson Ferguson and Company, Brisbane, Australia. pp. 93–102.

Berding, N. and Hurney, A. P. 2005. Flowering and lodging, physiological-based traits affecting cane and sugar yield: What do we know of their control mechanisms and how do we manage them? *Field Crops Research* 92, 261–75.

Berding, N. and Marston, D. H. 2010. Operational validation of the efficacy of Spectracane™, a high-speed analytical system for sugarcane quality components. *Proceedings of the Australian Society Sugar Cane Technologists* 32, 445–59.

Berding, N., Moore, P. and Smith, G. 1997. Advances in breeding technology for sugarcane. In: Intensive Sugarcane Production: Meeting the Challenge Beyond 2000. Editors: B. Keating and J. Wilson. CAB International.

Berding, N., Pendrigh, R. S. and Dunne, V. 2010. Pursuing higher efficacy for managed photoperiodic initiation of sugarcane flowering in the tropics. *Proceedings of the Australian Society Sugar Cane Technologists* 32, 234–50.

Bernardo, R. 2002. *Breeding for Quantitative Traits in Plants*. Woodbury, Minnesota: Stemma Press.

Besse, P., McIntyre, C. and Berding, N. 1996. Ribosomal DNA variations in Erianthus, a wild sugarcane relative (Andropogoneae-Saccharinae). *Theoretical and Applied Genetics* 92, 733–43.

Bhuiyan, S., Croft, B., Stirling, G., Wong, E., Jackson, P. and Cox, M. 2016. Assessment of resistance to root-lesion and root-knot nematodes in Australian hybrid clones of sugarcane and its wild relatives. *Australasian Plant Pathology* 45, 165–73.

Bissessur, D., Tilney-Bassett, R., Chong, L. L. S., Domaingue, R. and Julien, M. 2000. Family × environment and genotype× environment interactions for sugarcane across two contrasting marginal environments in Mauritius. *Experimental Agriculture* 36, 101–14.

Botha, F. 2009. Energy yield and cost in a sugarcane biomass system. *Proceedings of the Australian Society of Sugar Cane Technologists* 31, 1–10.

Botha, F. C. and Moore, P. H. 2014. Biomass and bioenergy. In: *Sugarcane: Physiology, Biochemistry, and Functional Biology*, Editors: PH Moore and FC Botha. Wiley Blackwell, Ames, Iowa, USA. pp. 521–40.

Bower, R. and Birch, R. G. 1992. Transgenic sugarcane plants via microprojectile bombardment. *The Plant Journal* 2, 409–16.

Brandes, E. 1958. Origin, classification and characteristics. In: *Sugarcane*, Editors: E Artschumer and E Brandes. US Dept. of Agriculture. pp. 1–35.

Bureau of Sugar Experiment Stations 1984. *The Standard Laboratory Manual for Australian Sugar Mills. Volume 1. Principles and Practices.* Bureau of Sugar Experiment Stations, Indooroopilly, Australia.

Breaux, R. 1987. Breeding for enhance sucrose contente of sugarcane in Louisiania. *Field Crops Research* 9, 59–67.

Breaux, R. D. and Miller, J. D. 1987. Seed handling, germination and seedling propagation. In: *Sugarcane Improvement Through Breeding*, Editor: DJ Heinz. Elsevier, Amsterdam. pp. 385–407.

Brown, J. S. and Glaz, B. 2001. Analysis of resource allocation in final stage sugarcane clonal selection. *Crop Science* 41, 57–62.

Brown, J. S., Schnell, R., Power, E., Douglas, S. L. and Kuhn, D. N. 2007. Analysis of clonal germplasm from five Saccharum species: S. barberi, S. robustum, S. officinarum, S. sinense and S. spontaneum. A study of inter-and intra species relationships using microsatellite markers. *Genetic Resources and Crop Evolution* 54, 627–48.

Bull, J., Hogarth, D. and Basford, K. 1992. Impact of genotype multiply environment interaction on response to selection in sugarcane. *Australian Journal of Experimental Agriculture* 32, 731–7.

Burnquist, W., Redshaw, K. and Gilmour, R. 2010. Evaluating sugarcane R&D performance: evaluation of three breeding programs. *Proceedings of the International Society of Sugar Cane Technologists* 27, 1–15.

Cadet, P., Spaull, V., Luc, M., Sikora, R. and Bridge, J. 2005. Nematode parasites of sugarcane. In: *Plant Parasitic Nematodes in Subtropical and Tropical Agriculture*, Editors: M Luc, RA Sikora, J Bridge. CAB International, Wallington, UK. pp. 645–74.

Cai, Q., Aitken, K., Deng, H., Chen, X., Fu, C., Jackson, P. and McIntyre, C. 2005. Verification of the introgression of Erianthus arundinaceus germplasm into sugarcane using molecular markers. *Plant Breeding* 124, 322–8.

Calija, V., Higgins, A. J., Jackson, P. A., Bielig, L. M. and Coomans, D. 2001. An operations research approach to the problem of the sugarcane selection. *Annals of Operations Research* 108, 123–42.

Cardona, C., Quintero, J., Paz, I. 2010. Production of bioethanol from sugarcane bagasse: Status and perspectives. *Bioresource Technology* 101, 4754–66.

Carlin, W., Schaffer, M. and Seabright, P. 2004. A minimum of rivalry: Evidence from transition economies on the importance of competition for innovation and growth. Contributions in Economic Analysis & Policy 3.

Chang, Y. and Milligan, S. 1992. Estimating the potential of sugarcane families to produce elite genotypes using univariate cross prediction methods. *Theoretical and Applied Genetics* 84, 662–71.

Chen, J. C. and Chou, C. C. 1993. *Cane Sugar Handbook: A Manual for Cane Sugar Manufacturers and their Chemists.* John Wiley and Sons, Inc., New York.

Chen, X., Jackson, P., Shen, W., Deng, H., Fan, Y., Li, Q., Hu, F., Wei, X. and Liu, J. 2012. Genotype× environment interactions in sugarcane between China and Australia. *Crop and Pasture Science* 63, 459–66.

Chong, B. F. and O'Shea, M. G. 2013. Advancing energy cane cell wall digestibility screening by near-infrared spectroscopy. *Applied Spectroscopy* 67, 1160–4.

Chong, B. F. and O'Shea M. G. 2012. Developing sugarcane lignocellulosic biorefineries: Opportunities and challenges. *Biofuels* 3, 307–19.

Churchill, J., Croft, B., Kokot, S. and O'Shea, M. 2006. Determining plant ratings without field trials II-Evaluation and prediction of clonal smut resistance status by morphological and spectroscopic examinations of sugarcane bud scale surfaces. In: *Proceedings-Australian Society Of Sugar Cane Technologists*, Volume 2006. Editor: DM Hogarth. PK Editorial Services, Brisbane, Australia, p. 232.

Cock, J. H. 2003. Sugarcane growth and development. *International Sugar Journal* 105, 540–52.

Cock, J. H., Luna, C. A. and Palma, A. 2000. The trade-off between total harvestable production and concentration of the economically useful yield component: Cane tonnage and sugar content. *Field Crops Research* 67, 257–62.

Comstock, J., Perdomo, R., Powell, G. and Wang, Z. 1997. Ratoon stunting disease in Florida sugarcane fields: relationship between disease incidence and cultivar resistance. *Journal American Society Sugarcane Technologists, Baton Rouge* 17, 95–101.

Comstock, J. C. 2002. Ratoon stunting disease. *Sugar Tech* 4, 1–6.

Comstock, J. C., Croft, B. J., Rao, G., Saumtally, S. and Victoria, J. I. 2007. A review of the 2006 international society of sugar cane technologists-pathology workshop. *Proceedings of International Society of Sugar Cane Technologists* 26, 2020–5.

Costet, L., Le Cunff, L., Royaert, S., Raboin, L.-M., Hervouet, C., Toubi, L., Telismart, H., Garsmeur, O., Rousselle, Y., Pauquet, J. 2012. Haplotype structure around Bru1 reveals a narrow genetic basis for brown rust resistance in modern sugarcane cultivars. *Theoretical and Applied Genetics* 125, 825–36.

Cox, M., Hogarth, D. and Hansen P. 1994. Breeding and selection for high early season sugar content in a sugarcane (Saccharum spp. hybrids) improvement program. *Crop and Pasture Science* 45, 1569–75.

Cox, M., Piperidis, G., Kimbeng, C. and Hansen, R. 2001. Investment in varietal development. *Proceedings of the Australian Society Sugar Cane Technologists* 23, 142–7.

Cox, M. and Stringer, J. 2007. Benchmarking genetic gains from new cultivars in Queensland using productivity data. *Proceedings of the International Society of Sugar Cane Technologists* 26, 624–31.

Croft, B., Bhuiyan, S., Magarey, R., Piperidis, G., Wong, E., Wickramasinghe, P., Bull, J., Cox, M., Stirling, G. and Foreman, J. 2015. New sources of resistance to major diseases from wild relatives of sugarcane. *Proceedings of the Australian Society Sugar of Cane Technologists* 37, 218–26.

Croft, B., Magarey, R. and Whittle, P. 2000. *Disease management. Manual of canegrowing*. Brisbane: Bureau of Sugar Experiment Stations, pp. 263–89.

Cullis, B., Gogel, B., Verbyla, A. and Thompson, R. 1998. Spatial analysis of multi-environment early generation variety trials. *Biometrics* 54, 1–18.

D'Hont, A., Grivet, L., Feldmann, P., Glaszmann, J., Rao, S. and Berding N. 1996. Characterisation of the double genome structure of modern sugarcane cultivars (Saccharum spp.) by molecular cytogenetics. *Molecular and General Genetics MGG* 250, 405–13.

D'Hont, A., Jackson P. A., Thong-Chane, A., Barau, L. and Hoarau, J.-Y. 2016. Pushing the frontiers of sugarcane improvement: A summary of the joint ISSCT 11th germplasm and breeding and 8th molecular biology workshops. In: *Proceedings of the 29th Congress of the International Society of Sugar Cane Technologists*. Chiang Mai: ISSCT, pp. 1937–44.

D'Hont, A., Paulet, F., Glaszmann, J. C. 2002. Oligoclonal interspecific origin of 'North Indian'and 'Chinese' sugarcanes. *Chromosome Research* 10, 253–62.

D'Hont, A., Rao P, Feldmann P, Grivet L, Islam-Faridi N, Taylor P, Glaszmann J. 1995. Identification and characterisation of sugarcane intergeneric hybrids, Saccharum officinarum × Erianthus arundinaceus, with molecular markers and DNA in situ hybridisation. *Theoretical and Applied Genetics* 91, 320–6.

Dal-Bianco, M., Carneiro, M. S., Hotta, C. T., Chapola, R. G., Hoffmann, H. P., Garcia, A. A. F., Souza, G. M. 2012. Sugarcane improvement: How far can we go? *Current Opinion in Biotechnology* 23, 265–70.

Daniels, J. and Roach, B. T. 1987. Taxonomy and evolution. In: *Sugarcane Improvement Through Breeding*, Editor: D. J. Heinz. Volume 7. Amsterdam: Elsevier, pp. 7–84.

Daugrois, J., Grivet, L., Roques, D., Hoarau, J., Lombard, H., Glaszmann, J. and D'Hont, A. 1996. A putative major gene for rust resistance linked with a RFLP marker in sugarcane cultivar 'R570'. *Theoretical and Applied Genetics* 92, 1059–64.

Davis, M. J. and Bailey R. A. 2000. Ratoon stunting. A guide to sugarcane diseases, p. 340.

Deren, C., Alvarez, J. and Glaz, B. 1995. Use of economic criteria for selecting clones in a sugarcane breeding program. *Proceedings of the International Society of Sugar Cane Technologists* 21, 437–47.

Dias, M. O., Junqueira, T. L., Cavalett, O, Cunha, M. P., Jesus, C. D., Rossell, C. E., Maciel Filho, R. and Bonomi, A. 2012. Integrated versus stand-alone second generation ethanol production from sugarcane bagasse and trash. *Bioresource Technology* 103, 152–61.

Edmé, S. J., Miller, J. D., Glaz, B., Tai, P. Y. and Comstock, J. C. 2005. Genetic contribution to yield gains in the Florida sugarcane industry across 33 years. *Crop Science* 45, 92–7.

Fischer, R., Byerlee, D. and Edmeades, G. 2012. *Crop Yields and Global Food Security*. Canberra: Australian Center for International Agricultural Research.

Garside, A., Bell, M. and Magarey, R. 2001. Monoculture yield decline – fact not fiction. *Proceedings of the International Society of Sugar Cane Technologists* 24, 16–20.

Garside, A., Smith, M., Chapman, L., Hurney, A. and Magarey, R. 1997. The yield plateau in the Australian sugar industry: 1970–1990. In: Intensive Sugarcane Production: Meeting the Challenge Beyond 2000. Editors: B. Keating and J. Wilson. CAB International.

Glaz, B. and Kang, M. S. 2008. Location contributions determined via GGE biplot analysis of multienvironment sugarcane genotype-performance trials. *Crop Science* 48, 941–50.

Glynn, N. C., Laborde, C., Davidson, R. W., Irey, M. S., Glaz, B., D'Hont, A. and Comstock J. C. 2013. Utilization of a major brown rust resistance gene in sugarcane breeding. *Molecular Breeding* 31, 323–31.

Gouy, M., Rousselle, Y., Bastianelli, D., Lecomte, P., Bonnal, L., Roques, D., Efile, J.-C., Rocher, S., Daugrois, J. and Toubi, L. 2013. Experimental assessment of the accuracy of genomic selection in sugarcane. *Theoretical and Applied Genetics* 126, 2575–86.

Grivet, L. and Arruda, P. 2002. Sugarcane genomics: Depicting the complex genome of an important tropical crop. *Current Opinion in Plant Biology* 5, 122–7.

Grivet, L., Glaszmann, J. and D'Hont, A. 2006. Molecular evidence of sugarcane evolution and domestication. In: T. J. Motley, N. Zerega and H. Cross (Eds), *Darwin's Harvest: New Approaches to the Origins, Evolution and Conservation of Crops*. Columbia University Press, New York, pp. 49–66.

Guinet-Brial, I., Girard, J., Roumagnac, P., Daugrois, J., Fernandez, E. and Rott, P. 2015. Visacane: An innovative quarantine tool for the exchange of pest and disease free sugarcane germplasm. *International Sugar Journal* 117, 890–6.

Hallauer, A. R. and Darrah, L. L. 1985. Compendium of recurrent selection methods and their application. *Critical Reviews in Plant Sciences* 3, 1–33.

Hayes, B., Bowman, P., Chamberlain, A. and Goddard, M. 2009. Invited review: Genomic selection in dairy cattle: Progress and challenges. *Journal of Dairy Science* 92, 433–43.

Hebert, L. 1959. Method of determining Brix with the hand refractometer in selecting sugar cane seedlings. *Proceedings of the International Society of Sugar Cane Technologists* 10, 768–75.

Heffner, E. L., Sorrells, M. E. and Jannink, J.-L. 2009. Genomic selection for crop improvement. *Crop Science* 49, 1–12.

Heinz, D. J. 1987. *Sugarcane Improvement Through Breeding*. Elsevier, Amsterdam, Netherlands.

Henderson, C. R. 1984. *Applications of Linear Models in Animal Breeding*. University of Guelph, Ontario, Canada.

Henry, R. J. and Kole, C. 2010. *Genetics, Genomics and Breeding of Sugarcane*. CRC Press, Boca Raton, FL.

Himmel, M. E., Ding, S.-Y., Johnson, D. K., Adney, W. S., Nimlos, M. R., Brady, J. W. and Foust, T. D. 2007. Biomass recalcitrance: engineering plants and enzymes for biofuels production. *Science* 315, 804–7.

Hodkinson, T. R., Chase, M. W., Lledó, D. M., Salamin, N. and Renvoize S. A. 2002. Phylogenetics of Miscanthus, Saccharum and related genera (Saccharinae, Andropogoneae, Poaceae) based on DNA sequences from ITS nuclear ribosomal DNA and plastid trnL intron and trnL-F intergenic spacers. *Journal of Plant Research* 115, 381–92.

Hogarth, D. 1971. Quantitative inheritance studies in sugar-cane. I. Estimation of variance components. *Crop and Pasture Science* 22, 93–102.

Hogarth, D. 1987. Genetics of sugarcane. In: D. J. Heinz (Ed.), *Sugarcane Improvement Through Breeding*. Elsevier, Amsterdam, Netherlands, pp. 255–71.

Hogarth, D., Braithwaite, M. and Skinner, J. 1990. Selection of sugarcane families in the Burdekin district. *Proceedings of the Australia Society of Sugar Cane Technologists* 12, 99–104.

Hogarth, D. and Kingston, G. 1983. The inheritance of ash in juice from sugar cane. *Proceedings of the Australia Society of Sugar Cane Technologists* 5, 21–7.

Hogarth, D. and Mullins, R. 1989. Changes in the BSES plant improvement programme. *Proceedings of the International Society of Sugar Cane Technologists* 20, 956–61.

Hogarth, D., Reimers, J., Ryan, C. and Taylor, P. 1993a. Quantitative inheritance of Fiji disease resistance in sugarcane. *Field Crops Research* 34, 175–86.

Hogarth, D., Ryan, C. and Taylor, P. 1993b. Quantitative inheritance of rust resistance in sugarcane. *Field Crops Research* 34, 187–93.

Hogarth, D., Wu, K. and Heinz, D. 1981. Estimating genetic variance in sugarcane using a factorial cross design. *Crop Science* 21, 21–5.

Hoy, J. and Hollier, C. 2009. Effect of brown rust on yield of sugarcane in Louisiana. *Plant Disease* 93, 1171–4.

Hu, F., Jackson, P. and Basford, K. 2009. Developing optimal selection systems in sugarcane breeding programs. *Sugarcane International* 27, 118–30.

Hutchinson, P. 1969. Note on disease resistance ratings for sugarcane varieties. *Proceedings of the International Society of Sugar Cane Technologists* 13, 1087–9.

Irvine, J. 1999. Saccharum species as horticultural classes. *Theoretical and Applied Genetics* 98, 186–94.

Jackson, P. 2013. Energy cane. In: M. C. Saha, H. S. Bhandari and J. H. Bouton (Eds), *Bioenergy Feedstocks: Breeding and Genetics*. Wiley-Blackwell, Ames, Iowa, USA. pp. 117–49.

Jackson, P., Bonnett, G., Chudleigh, P., Hogarth, M., Wood, A. and Hogarth, D. 2000. The relative importance of cane yield and traits affecting CCS in sugarcane varieties. In: D. M. Hogarth (Ed.), *Proceedings of the 2000 Conference of the Australian Society of Sugar Cane Technologists*. PK Editorial Services, Brisbane, pp. 23–9.

Jackson, P., Bull, J. and McRae, T. 1995a. The role of family selection in sugarcane breeding programs and the effect of genotype × environment interactions. *Proceedings of the International Society of Sugar Cane Technologists* 22, 261–70.

Jackson, P., Chapman, S., Rattey, A., Wei, X. M. and Cox. M. 2007. Genotype × region interactions, and implications for sugarcane breeding programs. *Proceedings of the Australian Society of Sugarcane Technologists* 29, 122–30.

Jackson, P. and Galvez, G. 1996. Role of variable soil nutrient levels in causing genotype × site interactions in sugarcane. In: *Sugarcane*, Research towards efficient and sustainable production. Editors: JR Wilson, DM Hogarth, JA Campbell, AL Garside. CSIRO Division of Tropical Crops and Pastures, Brisbane, Australia, pp. 52–4.

Jackson, P. and Hogarth, D. 1992. Genotype × environment interactions in sugarcane. I. Patterns of response across sites and crop-years in North Queensland. *Australian Journal of Agricultural Research* 43, 1447–59.

Jackson, P. and McRae, T. 2001. Selection of sugarcane clones in small plots. *Crop Science* 41, 315–22.

Jackson, P., McRae, T. and Hogarth, M. 1995b. Selection of sugarcane families across variable environments I. Sources of variation and an optimal selection index. *Field Crops Research* 43, 109–18.

Jackson, P., Wei, X. and Atkin, F. 2012. Genetic correlations between flowering propensity and cane yield and CCS and implications for sugarcane breeding programs.

Jackson, P. A. 2005. Breeding for improved sugar content in sugarcane. *Field Crops Research* 92, 277–90.

Jackson, P. A., O'Shea M. G., Rattey A. R., Bonnett G. D., Lindeman P. F., Cox M. C., Vickers J. E. and Morgan T. 2006. Effect of genotype and genotype× nitrogen rate interactions on color in juice and raw sugar from sugarcane. *Crop Science* 46, 886–92.

Jackson, P. A., O'Shea, M. G., Lindeman, P. F., Rattey, A., Vickers, J. and Bonnett, G. 2004. Effects of varieties and nitrogen fertiliser rate on colour and conductivity levels in sugarcane. In: Hogarth, D. M. (Ed.), *Proceedings-Australian Society of Sugar Cane Technologists*. PK Editorial Services, Brisbane, p. 47.

Jackson, P. A., Schroeder, B. L., Rattey, A. R., Wood, A. and O'Shea, M. G. 2008a. Management of ash/impurity ratio in sugarcane: Relative effects of genotypes, and N and K fertiliser rates. *Crop and Pasture Science* 59, 795–801.

Jackson, P. A., Schroeder, B. L., Rattey, A. R., Wood, A. and O'Shea, M. G. 2008b. Management of ash/impurity ratio in sugarcane: Relative effects of genotypes, and N and K fertiliser rates. *Crop and Pasture Science* 59, 795–801.

Jannink, J.-L., Lorenz, A. J. and Iwata, H. 2010. Genomic selection in plant breeding: from theory to practice. *Briefings in Functional Genomics* 9, 166–77.

Jannoo, N., Grivet, L., Seguin, M., Paulet, F., Domaingue, R., Rao, P., Dookun, A., D'Hont, A. and Glaszmann, J. 1999. Molecular investigation of the genetic base of sugarcane cultivars. *Theoretical and Applied Genetics* 99, 171–84.

Joomun, N., Parmessur, Y., Antoine, M. and Dookun-Saumtally, A. 2013. Screening for Sugarcane yellow leaf virus in quarantine in Mauritius. *Proceedings of the International Society of Sugar Cane Technologists* 28, 12.

Kang, M. and Miller, J. 1984. Genotype × environment interactions for cane and sugar yield and their implications in sugarcane breeding. *Crop Science* 24, 435–40.

Keeping, M. 1999. Field screening of sugarcane varieties for resistance to the stalk borer, Eldana saccharina (Lepidoptera: Pyralidae). *Proceedings of the South African Sugarcane Technologists Association* 73, 102–3.

Keeping, M. G. 2006. Screening of South African sugarcane cultivars for resistance to the stalk borer, Eldana saccharina Walker (Lepidoptera: Pyralidae). *African Entomology* 14, 277.

Kellogg, E. A. 2013. Phylogenetic relationships of Saccharinae and Sorghinae. In: A. H. Patterson (Ed.), *Genomics of the Saccharinae*. Springer, Dordrecht, Netherlands. pp. 3–21.

Kennedy, A. 2004. Prospects for breeding varieties for the Caribbean with very high sucrose content. In: *Proceedings of the 28th West Indies Sugar Technologists Conference*, Barbados, pp. 6–11.

Kent, G. 1997. Modelling the extraction processes of milling trains. In: B. T. Egan (Ed.), *Proceedings-Australian Society of Sugar Cane Technologists*. Watson Ferguson and Company, Brisbane, pp. 315–21.

Kimbeng, C., Rattey, A. and Hetherington, M. 2002. Interpretation and implications of genotype by environment interactions in advanced stage sugarcane selection trials in central Queensland. *Australian Journal of Agricultural Research* 53, 1035–45.

Kimbeng, C., White, W., Miller, J. and Legendre, B. 2006. Sugarcane resistance to the sugarcane borer: Response to infestation among progeny derived from resistant and susceptible parents. *Sugar Cane International* 24, 14–21.

Kimbeng, C. A. and Cox, M. C. 2003. Early generation selection of sugarcane families and clones in Australia: A review. *Journal American Society of Sugarcane Technologists* 23: 2039.

Kimbeng, C. A., Zhou, M. M. and Da Silva, J. A. 2009. Genotype × environment interactions and resource allocation in sugarcane yield trials in the Rio Grande valley region of Texas. *Journal of the American Society of Sugar Cane Technologists* 29, 11–24.

Legendre, B. 1995. Potential for increasing sucrose content of sugarcane varieties in Louisiana through breeding. In: *Proceedings XXI Congress of ISSCT*, Bangkok (Thailand), 5–14 March 1992.

Leite, M. S. d. O., Peternelli, L. A., Barbosa, M. H. P., Cecon, P. R. and Cruz, C. D. 2009. Sample size for full-sib family evaluation in sugarcane. *Pesquisa Agropecuária Brasileira* 44, 1562–74.

Lethbridge, P. and Cox, M. 2010. Managing variety improvement program operations using an integrated database and handheld computers to improve effi ciency and reduce errors. In:

R. C. Bruce (Ed.), *Proceedings of the Australian Society of Sugar Cane Technologists*. Watson Ferguson and Company, Brisbane, pp. 251–61.

Lingle, S. E. and Thomson, J. L. 2012. Sugarcane internode composition during crop development. *BioEnergy Research* 5, 168–78.

Loureiro, M. E., Barbosa, M. H., Lopes, F. J. and Silvério F. O. 2011. Sugarcane breeding and selection for more efficient biomass conversion in cellulosic ethanol. In: M. S. Buckeridge and G. H. Goldman (Eds), *Routes to Cellulosic Ethanol*. Springer, Dordrecht, Netherlands, pp. 199–239.

Machado, G. 2001. *Sugarcane Variety Notes – An International Directory*, seventh revision, Piracicaba, Brazil: Guilherme Rossi Machado, Februray.

Magarey, R. and Bull, J. 2003. Relating cultivar Pachymetra root rot resistance to sugarcane yield using breeding selection trial analyses. *Animal Production Science* 43, 617–15.

Magarey, R., Bull, J., Neilsen, W. and Magnanini, A. 2002. The use of breeding trials to estimate disease-induced yield losses and to refi ne selection strategies. In: D. M. Hogarth (Ed.), *Proceedings-Australian Society of Sugar Cane Technologists*. P. K. Editorial Services, Brisbane, pp. 113–323.

Magarey, R., Neilsen, W. and Bull, J. 2004. The effect of orange rust on sugarcane yield in breeding selection trials in central queensland: 1999–2001. *Proceedings of the Australian Society of Sugar Cane Technologists* 26 (CD ROM).

Mangelsdorf, A. 1960. Sugar-cane breeding methods. In: *Proceedings of the International Society of Sugarcane Technologists, 10th Congress*, Hawaii, 1959, pp. 694–701.

McIntyre, C. and Jackson, P. 2001. Low level of selfing found in a sample of crosses in Australian sugarcane breeding programs. *Euphytica* 117, 245–9.

McRae, T., Bull, J. and Robotham, B. 1998. Billet samples for measuring sugar content in variety trials. In: *Proceedings-Australian Society Of Sugar Cane Technologists*. Editor: DM Hogarth. Watson Ferguson and Company, pp. 211–17.

Miller, J. 1977. Combining ability and yield component analyses in a five-parent diallel cross in sugarcane. *Crop Science* 17, 545–7.

Miller, K., Ingram, G. and Murry, J. 1998. Exhaustion characteristics of Australian molasses. In: D. M. Hogarth (Ed.), *Proceedings-Australian Society of Sugar Cane Technologists*. Watson Ferguson and Company Company, Brisbane, pp. 506–13.

Milligan, S., Gravois, K., Bischoff, K. and Martin, F. 1990. Crop effects on genetic relationships among sugarcane traits. *Crop Science* 30, 927–31.

Moore, P. H. and Berding, N. 2014. Flowering. In: *Sugarcane: Physiology, Biochemistry, and Functional Biology*, Editors: PH Moore and FC Botha. Wiley-Blackwell, Ames, Iowa,, USA. pp. 379–410.

Moore, P. H. and Nuss, K. 1987. Flowering and flower synchronization. In: D. J. Heinz (Ed), *Sugarcane Improvement Through Breeding*, Volume 11. Elsevier, Amsterdam, Netherlands, pp. 273–311.

Mrode, R. A. and Thompson, R. 2005. *Linear Models for the Prediction of Animal Breeding Values*. Wallingford, UK: CAB International.

Mukherjee, S. K. 1957. Origin and distribution of Saccharum. *Botanical Gazette* 119, 55–61.

Mullins, R. and Roach, B. 1985. Genetic origins of ash in sugarcane juice. *Proceedings of the Australian Society of Sugar Cane Technologists* 6, 43–51.

Nair, N. 2008. Sugarcane breeding institute, Coimbatore: A perspective. *Sugar Tech* 10, 285–92.

Nair, N. V., Nair, S., Sreenivasan, T. and Mohan, M. 1999. Analysis of genetic diversity and phylogeny in Saccharum and related genera using RAPD markers. *Genetic Resources and Crop Evolution* 46, 73–9.

Nawi, N. M., Chen, G. and Jensen, T. 2014. In-field measurement and sampling technologies for monitoring quality in the sugarcane industry: A review. *Precision Agriculture* 15, 684–703.

Nuss, K. 1991. Screening sugarcane varieties for resistance to eldana borer. *Proceedings of the South African Sugar Technologists' Association* 65, 92–5.

Ong'ala, J., Wawire, N., Jamoza, J., Mainna, P., Ong'injo, E. and Otieno, V. 2013. An economic selection index that combines cane yield and sugar content in identifying superior sugarcane clones in Kenya. In: *African Crop Science Conference Proceedings*, Volume 11. African Crop Science Society, Uganda, pp. 739–43.

Oxley, J., Fong, C., Sant, G. G. and O'Shea, M. G. 2012. Accelerating the characterisation of sugarcane biomass using near-infrared (NIR) spectroscopic techniques. In: *Proceedings of the 34th Conference of the Australian Society of Sugar Cane Technologists held at Cairns, Queensland, Australia*, pp. 1–4.

Pan, Y.-B. 2010. Databasing molecular identities of sugarcane (Saccharum spp.) clones constructed with microsatellite (SSR) DNA markers. *American Journal of Plant Sciences* 1, 87.

Pandey, S. and Gardner, C. 1992. Recurrent selection for population, variety, and hybrid improvement in tropical maize. *Advances in Agronomy* 48, 1–87.

Pankhurst, C., Blair, B., Magarey, R., Stirling, G., Bell, M., Garside, A. and Venture, S. Y. D. J. 2005. Effect of rotation breaks and organic matter amendments on the capacity of soils to develop biological suppression towards soil organisms associated with yield decline of sugarcane. *Applied Soil Ecology* 28, 271–82.

Pastina, M. M., Pinto, L. R., Oliveira, K. M., Souza, A. and Garcia, A. A. F. 2010. Molecular mapping of complex traits. In: R. J. Henry and C. Kole (Eds), *Genetics, Genomics and Breeding of Sugarcane*. CRC Press, Boca Raton, FL, pp. 117–48.

Pedrozo, C. Â., Barbosa, M. H. P., da Silva, F. L., de Resende, M. D. V. and Peternelli, L. A. 2011. Repeatability of full-sib sugarcane families across harvests and the efficiency of early selection. *Euphytica* 182, 423–30.

Pierre, J., Rae, A. and Bonnett, G. 2014. Abiotic limits for germination of sugarcane seed in relation to environmental spread. *Tropical Plant Biology* 7, 100–10.

Piperidis, G., Cox, M., Thompson, N. and Croft, B. 2008. The role of foreign varieties in the Australian sugar industry. *Proceedings of the Australian Society of Sugar Cane Technologists*. Editor: R. Bruce. 30, 155–61.

Piperidis, G., Rattey, A., Taylor, G. and Cox, M. 2004. DNA markers: A tool for identifying sugarcane varieties. In: D. M. Hogarth (Ed.), *Proceedings-Australian Society of Sugar Cane Technologists*. PK Editorial Services, Brisbane, p. 36.

Pisaroglo de Carvalho, M., Gezan, S. A., Peternelli, L. A. and Pereira Barbosa, M. H. 2014. Estimation of additive and nonadditive genetic components of sugarcane families using multitrait analysis. *Agronomy Journal* 106, 800–8.

Potts, J. 2009. The innovation deficit in public services: The curious problem of too much efficiency and not enough waste and failure. *Innovation* 11, 34–43.

Purcell, D., Oxley, J., Cox, M., Croft, B. and O'Shea, M. 2010. On-site rapid screening for sugarcane smut resistance using near-infrared (NIR) spectroscopy. *Proceedings of the Australian Society of Sugar Cane Technologists* 32, 366–74.

Purcell, D. E., O'Shea, M. G., Johnson, R. A. and Kokot, S. 2009. Near-infrared spectroscopy for the prediction of disease ratings for Fiji leaf gall in sugarcane clones. *Applied Spectroscopy* 63, 450–7.

Queme, J., Orozco, H. and Melgar, M. 2010. GGE biplot analysis used to evaluate cane yield of sugarcane (Saccharum spp.) cultivars across sites and crop cycles. *Proceedings of the International Society of Sugar Cane Technologists* 27, 584–90.

Ram, B. and Chaudhary, B. 2000. Individual and simultaneous selection for brix yield in seedling populations of sugarcane. *Sugar Cane International*, 12–17.

Ram, B., Chaudhary, B. and Singh, S. 1997. Response to indirect selection in ratoon of sugarcane seedlings. *Australian Journal of Agricultural Research* 48(2), 207–13.

Ramburan, S., Zhou, M. and Labuschagne, M. 2011. Interpretation of genotype× environment interactions of sugarcane: Identifying significant environmental factors. *Field Crops Research* 124, 392–9.

Rassaby, L., Girard, J.-C., Letourmy, P., Chaume, J., Irey, M., Lockhart, B., Kodja, H. and Rott, P. 2003. Impact of Sugarcane yellow leaf virus on sugarcane yield and juice quality in Réunion Island. *European Journal of Plant Pathology* 109, 459–66.

Rattey, A., Jackson, P., Wei, X., Cox, M. and Stringer, J. 2004. Opportunities to increase rates of parent improvement in Australian sugarcane breeding programs. In: D. M. Hogarth (Ed.), *Proceedings-Australian Society of Sugar Cane Technologists*. PK Editorial Services, p. 42.

Ricaud, C., Egan, B., Gillaspie, A. and Hughes, C. 2012. *Diseases of Sugarcane: Major Diseases*. Elsevier, Amsterdam, Netherlands.

Roach, B. 1978. Utilization of Saccharum spontaneum in sugarcane breeding. *Proceedings of the International Society of Sugar Cane Technologists* 16, 43–58.

Roach, B. 1989. Origin and improvement of the genetic base of sugarcane. *Proceedings of the Australian Society of Sugar Cane Technologists* 11, 34–47.

Roach, B. and Jackson, P. 1992. Screening sugar cane clones for resistance to ratoon stunting disease. *Sugar Cane*, 2: 2–12.

Rodríguez-Zúñiga, U. F., Farinas, C. S., Carneiro, R. L., Da Silva, G. M., Cruz, A. J. G., Giordano. R. d. L. C., de Campos Giordano, R. and de Arruda Ribeiro, M. P. 2014. Fast determination of the composition of pretreated sugarcane bagasse using near-infrared spectroscopy. *BioEnergy Research* 7, 1441–53.

Rott, P. 2000. *A Guide to Sugarcane Diseases.* CIRAD, Montpellier, France.

Rutherford, R. 1998. Prediction of resistance in sugarcane to stalk borer Eldana saccharina by near-infrared spectroscopy on crude budscale extracts: Involvement of chlorogenates and flavonoids. *Journal of Chemical Ecology* 24, 1447–63.

Rutherford, R., McFarlane, S., Van Antwerpen, T. and McFarlane, K. 2003. Use of varieties to minimise losses from sugarcane diseases in South Africa. *Proceedings of the South African Sugar Technologists' Association* 77, 180–8.

Selvi, A., Nair, N., Balasundaram, N. and Mohapatra, T. 2003. Evaluation of maize microsatellite markers for genetic diversity analysis and fingerprinting in sugarcane. *Genome* 46, 394–403.

Selvi, A., Nair, N., Noyer, J., Singh, N., Balasundaram, N., Bansal, K., Koundal, K. and Mohapatra, T. 2005. Genomic constitution and genetic relationship among the tropical and subtropical Indian sugarcane cultivars revealed by AFLP. *Crop Science* 45, 1750–7.

Selvi, A., Nair, N., Noyer, J., Singh, N., Balasundaram, N., Bansal, K., Koundal, K. and Mohapatra, T. 2006. AFLP analysis of the phenetic organization and genetic diversity in the sugarcane complex, Saccharum and Erianthus. *Genetic Resources and Crop Evolution* 53, 831–42.

Shanthi, R., Bhagyalakshmi, K., Hemaprabha, G., Alarmelu, S. and Nagarajan, R. 2008. Relative performance of the sugarcane families in early selection stages. *Sugar Tech* 10, 114–18.

Simmonds, N. 1993. Introgression and incorporation. Strategies for the use of crop genetic resources. *Biological Reviews* 68, 539–62.

Simmonds, N. 1996. Family selection in plant breeding. *Euphytica* 90, 201–8.

Skinner, J., Hogarth, D. and Wu, K. 1987. Selection methods, criteria and indices In: D. Heinz (Ed.) *Sugarcane Improvement Through Breeding.* Amsterdam: Elsevier. Editor: DJ Heinz. pp. 409–54.

Smith, A., Cullis, B. R. and Thompson, R. 2005. The analysis of crop cultivar breeding and evaluation trials: an overview of current mixed model approaches. *The Journal of Agricultural Science* 143, 449–62.

Stirling, G., Blair, B., Pattemore, J., Garside, A. and Bell, M. 2001. Changes in nematode populations on sugarcane following fallow, fumigation and crop rotation, and implications for the role of nematodes in yield decline. *Australasian Plant Pathology* 30, 323–35.

Stringer, J., Cox, M., Atkin, F., Wei, X. and Hogarth, D. 2011. Family selection improves the efficiency and effectiveness of selecting original seedlings and parents. *Sugar Tech* 13, 36–41.

Stringer, J. and Cullis, B. 2002. Application of spatial analysis techniques to adjust for fertility trends and identify interplot competition in early stage sugarcane selection trials. *Crop and Pasture Science* 53, 911–18.

Sukarso, G. 1986. Assessment of family selection on original seedlings of sugarcane at Pasuruan. *Proceedings of the International Society of Sugar Cane Technologists* 19, 440–6.

Tai, P., Miller, J. and Dean, J. 1981. Inheritance of resistance to rust in sugarcane. *Field Crops Research* 4, 261–8.

Tew, T. L. and Cobill, R. M. 2008. Genetic improvement of sugarcane (*Saccharum* spp.) as an energy crop. In: W. Vermerris (Ed.), *Genetic Improvement of Bioenergy Crops.* Springer, Dordrecht, Netherlands, pp. 273–94.

Thaval, O. P. and Kent, G. A. 2012. An enhanced mill extraction model. In: R. C. Bruce (Ed.), *Proceedings of the 34th Conference of the Australian Society of Sugar Cane Technologists.* Watson Ferguson and Company. Volume 34 (CD).

Thompson, N., Wilson, E. and Croft, B. 2012. Sugarcane quarantine disease screening in Australia. *International Sugar Journal* 114, 577–83.

Thuljaram Rao, J. 1987. Sugarcane origin, taxonomy, breeding and varieties. In: K. M. Naidu, T. V. Srinivasan and M. N. Premchandran (Eds), *Sugarcane Varietal Improvement*. Sugarcane Breeding Institute, Coimbatore, India, pp. 83–113.

Tyagi, A. P. and Lal, P. 2005. Line × tester analysis in sugarcane (Saccharum officinarum). *The South Pacific Journal of Natural and Applied Sciences* 23, 30–6.

Vermerris, W. and Abril, A. 2015. Enhancing cellulose utilization for fuels and chemicals by genetic modification of plant cell wall architecture. *Current Opinion in Biotechnology* 32, 104–12.

Viswanathan, R. and Balamuralikrishnan, M. 2005. Impact of mosaic infection on growth and yield of sugarcane. *Sugar Tech* 7, 61–5.

Walker, D. 1987. Breeding for disease resistance. In: D. J. Heinz (Ed.), *Sugarcane Improvement Through Breeding*. Elsevier, Amsterdam, Netherlands, pp. 455–502.

Wang, L. P., Jackson, P. A., Lu, X., Fan, Y. H., Foreman, J. W., Chen, X. K., Deng, H. H., Fu, C., Ma, L. and Aitken, K. S. 2008. Evaluation of sugarcane × progeny for biomass composition and yield components. *Crop Science* 48, 951–61.

Wei, X. and Jackson, P. 2016. Addressing slow rates of long term genetic gain in sugarcane. *Proceedings of the International Society of Sugarcane Technologists* 29. 1923–30.

Wei, X., Jackson, P., Cox, M. and Stringer, J. 2006. Maximising economic benefi ts to the whole sugarcane industry from the BSES-CSIRO sugarcane improvement program. In: *Proceedings Australian Society of Sugar Cane Technologists*, Volume 2006. PK Editorial Services, Brisbane. p. 181.

Wei, X., Jackson, P., Stringer, J. and Cox, M. 2008. Relative economic genetic value (rEGV)–an improved selection index to replace net merit grade (NMG) in the Australian sugarcane variety improvement program. *Proceedings of the Australian Society of Sugar Cane Technologists* 30, 174–81.

White, T. L. and Hodge, G. R. 2013. *Predicting Breeding Values with Applications in Forest Tree Improvement*. Springer Science & Business Media, Dordrecht, Netherlands.

White, W., Miller, J., Milligan, S., Burner, D. and Legendre, B. 2001. Inheritance of sugarcane borer resistance in sugarcane derived from two measures of insect damage. *Crop Science* 41, 1706–10.

White, W., Tew, T. and Richard Jr., E. 2006. Association of sugarcane pith, rind hardness, and fiber with resistance to the sugarcane borer. *Journal of American Society Sugar Cane Technologies* 26, 87–100.

White, W. H., Hale, A. L., Veremis, J. C., Tew, T. L. and Richard E. P. 2011. Registration of two sugarcane germplasm clones with antibiosis to the sugarcane borer (Lepidoptera: Crambidae). *Journal of Plant Registrations* 5, 248–53.

Wilson, G. and Leslie, J. 1997. Productivity trends in sugarcane in the wet tropics. In: B. T. Egan (Ed.), *Proceedings-Australian Society of Sugar Cane Technologists*. Watson Ferguson and Company, Brisbane, pp. 21–9.

Wu, K., Burnquist, W., Sorrells, M., Tew, T., Moore, P. and Tanksley, S. 1992. The detection and estimation of linkage in polyploids using single-dose restriction fragments. *Theoretical and Applied Genetics* 83, 294–300.

Wu, K. and Tew, T. 1989. Evaluation of sugarcane crosses by family yields. *Proceedings of Congress of the International Society of Sugarcane Technologists* 20, 926–31.

Yang, K., Jackson, P., Wei, X., Chen, X., Liu, J., Zhao, J., Wu, C. and Fan, Y. 2017. Selecting for multiple traits in complex production systems: A case study of sugarcane in China. *Agricultural Systems* 152, 110–20.

Yin, Z., Hoy, J. and Milligan, S. 1996. Evaluation and heritability of resistance to sugarcane red rot. *Phytopathology* 86, 662–7.

Chapter 5

Advances in marker-assisted breeding of sugarcane

Per McCord, USDA-ARS, USA

1 Introduction

Although sugarcane has been undergoing the process of selection for thousands of years, modern controlled breeding of sugarcane dates back to the 1880s (reviewed in Roach 1989). The ability of sugarcane to produce sexual offspring was initially observed in Barbados in 1858, but was not then exploited for breeding. Later, nearly simultaneous observations in Java in 1885 and in Barbados in 1888 initiated an interest in modern sugarcane breeding. The first crosses were made amongst *Saccharum officinarum* genotypes. Controlled interspecific crossing and nobilization (backcrossing to *S. officinarum*) began in Java soon thereafter as a way to introgress disease resistance from wild species (such as *S. spontaneum*) into an *S. officinarum* background and was performed in response to an outbreak of sereh disease, now thought to be caused by a virus. The breeding took place at Proofstation Oest Java, and the varieties were given the prefix POJ. POJ hybrids displayed a marked increase in vigour (biomass), ratooning ability and disease resistance compared to the traditional 'noble' (*S. officinarum*) canes used for sugar production. These hybrids were exported to other sugar-growing regions beginning in the 1920s and had a major impact on cane sugar production (Dodds 1938; Roach 1989; Gravois and Bischoff 2008). In addition, breeding programmes were established around the world. Today, active sugarcane crossing and/or breeding programmes are present in over 25 countries.

http://dx.doi.org/10.19103/AS.2017.0035.06

Sugarcane breeding until very recently has been traditional, that is, based solely on phenotype. Even now, marker-assisted breeding of sugarcane is in its infancy compared with that of row crops such as maize and soybean. A major reason for this is the complex genetics of sugarcane. The primary donor species of modern sugarcane are high-level polyploids. *S. officinarum* is apparently an octoploid (2n = 80), with a basic chromosome number of 10 (D'Hont et al. 1995). *S. spontaneum*, on the other hand, has a basic chromosome number of 8 (D'Hont et al. 1998) and is a highly variable species, with chromosome numbers ranging from 2n = 40 to 2n = 128 (Sreenivasan et al. 1987). When *S. officinarum* is used as the female parent in a hybrid between the two species (as was done by the early breeders), 2n + n transmission of chromosomes is generally observed. This is apparently followed by varying levels of aneuploidy in succeeding generations when the hybrids are crossed amongst themselves (as in a breeding programme), with the result that modern sugarcane hybrids have high chromosome numbers (100–130) that vary even within clones (D'Hont et al. 1996).

In addition to the genetic challenges, relatively high intrinsic yield and slow breeding cycle of sugarcane and the limited research budgets of many sugar-producing areas are all likely factors for the limited use of DNA technology in sugarcane. However, sugarcane geneticists, molecular biologists and breeders have been exploring and utilizing marker-assisted breeding techniques for more than two decades. Furthermore, the increased power and reduced cost per data point of the latest technologies are expected to result in increased use of marker-assisted breeding in sugarcane. In this chapter, I will first review the uses of DNA marker technology in fingerprinting and diversity analysis of sugarcane. This will be followed by a review of the development of linkage maps and initial trait/quantitative trait locus (QTL) mapping, including the *Bru*1 locus for resistance to brown rust (*Puccinia melanocephala*). Finally, I will review the use of newer next-generation sequencing (NGS)–based technologies in sugarcane, including genome-wide association analysis (GWAS) and genomic selection (GS).

2 Sugarcane fingerprinting and variety identification

Because sugarcane is vegetatively propagated, cultivars exist as stable clones (except for mutations) essentially indefinitely. It is often important to be able to positively identify a particular clone or genotype. Morphological markers were traditionally used, but these require practice and skill to interpret, do not have a high throughput and can be affected by environmental conditions. Molecular markers are more stable, straightforward and adaptable to high-throughput assays. The earliest use of molecular markers in sugarcane research, reported by Heinz (1969), was an attempt in Hawaii to distinguish two morphologically identical clones originally from New Guinea. Isozyme fingerprints using esterase and peroxidase were identical between the two clones, helping to eliminate a duplication error that had existed for many years. Piperidis et al. (2004) summarized the development and use of a sugarcane DNA fingerprint database in Australia based on simple sequence repeat (SSR) fragment analysis. The SSR fingerprints (from five SSR primer pairs) of 35 clones were compared with data from 28 morphological traits. Results showed complete agreement between the two methods, and the SSR database has been used to identify unlabelled/mislabelled varieties and provide information on uniqueness needed for obtaining intellectual property protection. Similar databases have been or are in use in sugarcane breeding programmes around the world, for example, in Brazil (Maccheroni et al. 2007), Louisiana (Pan 2010) and South Africa (Joshi and Albertse 2013).

3 Sugarcane diversity analysis

One of the major uses of molecular markers in sugarcane industry has been to estimate genetic diversity. Although it is not marker-assisted breeding *per se*, diversity analysis is important to ensure that sufficient genetic variability is present in a breeding population to provide a base for genetic gain. As for variety identification, the first markers used for diversity analysis were isozymes, not DNA markers. Glaszmann et al. (1989) examined the isozyme patterns of nine enzymes across 39 individuals representing *S. officinarum*, *S. robustum*, *S. spontaneum* and *Erianthus*. The isozyme patterns were able to distinguish the *Erianthus* clone and the *S. spontaneum* accessions from each other and *S. officinarum* and *S. robustum*, but were not able to separate clones from the latter two species from each other. This evidence supported the suggestion made earlier by Artschwager and Brandes (1958) that *S. robustum* is the wild ancestor of domesticated *S. officinarum*. DNA markers soon superseded isozymes for diversity analysis. Glaszmann et al. (1990) utilized restriction fragment length polymorphisms (RFLPs) within nuclear ribosomal DNA to examine the variability in a diverse pool of *Saccharum*, *Erianthus* and *Miscanthus* accessions. A key result of this study was the identification of great diversity present in *S. spontaneum* accessions. D'Hont et al. (1993) utilized RFLP markers from the mitochondrial and plastid genomes to analyse cytoplasmic diversity within the *Saccharum* complex. The plastid-derived probes yielded no diversity amongst any of the *Saccharum* accessions; only the *Erianthus* and *Miscanthus* accessions could be distinguished. The mitochondrial probes revealed a large amount of diversity amongst *S. spontaneum* accessions; very little diversity (2 patterns) amongst *S. robustum* accessions and no diversity within the *S. officinarum*, *S. sinense* or *S. edule* accessions. Interestingly, only one of the *S. robustum* mitochondrial fingerprints was present within the *S. officinarum* accessions, suggesting that the *S. robustum* progenitors of *S. officinarum* were restricted to one of the groups. Lu et al. (1994) analysed relationships amongst wild *Saccharum* and early cultivated sugarcanes by RFLP using heterologous probes from maize and were able to distinguish *S. officinarum* and *S. robustum* accessions, unlike the earlier isozyme-based efforts of Glaszmann et al. (1989). In addition, they demonstrated greater diversity within *S. spontaneum* accessions, in agreement with Glaszmann et al. (1990) and provided evidence that *S. barberi* and *S. sinense* are derived from interspecific hybridization between *S. officinarum* and *S. spontaneum*. Jannoo et al. (1999) conducted a large diversity analysis of 109 modern hybrids (mostly from Barbados and Mauritius) and 53 *S. officinarum* accessions (51 of these were found to be unique). Twelve RFLP probes utilized in conjunction with 1–2 restriction enzymes per probe resulted in over 386 polymorphic markers. The greatest diversity was observed amongst the *S. officinarum* accessions from New Guinea, the proposed centre of origin for *S. officinarum*. Despite the fact that relatively few *S. officinarum* accessions have contributed to modern sugarcane varieties, nearly 90% of the polymorphic *S. officinarum* markers were also detected in the modern hybrids. Cluster analysis separated the accessions into two groups, one comprising most of the *S. officinarum* accessions and only one hybrid, and the other containing the remaining hybrids and 14 *S. officinarum* accessions. The cluster analysis also separated the hybrids from Barbados and Mauritius into two groups, most likely due to the different *S. spontaneum* accessions used in the early ancestral hybrids that contributed to the two programmes. A multivariate dimensional reduction technique known as factor analysis of correspondences was also used to analyse the marker data, which revealed similar patterns *vis-à-vis* the cluster analysis.

Amplified fragment-length polymorphism or AFLP markers (Vos et al. 1995) and SSR markers are more prevalent and easy to generate than RFLPs and were preferred for later diversity analyses. AFLP markers require no prior sequence information, and a single primer combination can produce dozens of markers. On the other hand, SSR markers require prior sequence information, making them more expensive to identify, but their sequence specificity makes them much more easily transferable between populations than AFLPs.

Aitken et al. (2006a) conducted a large diversity study of 421 accessions using AFLP markers. Most of the accessions (270) were of *S. officinarum*, while the remaining 151 accessions were prior/current commercial cultivars from Australia, clones used as parents at some point in Australian sugarcane breeding programmes or ancestors of those parents. A total of 657 usable markers were generated from five AFLP primer combinations. Cluster analysis of the data set identified ten main clusters, with a single cluster containing 224 (83%) of *S. officinarum* accessions. Another cluster contained 149 (99%) of the cultivars/parents. Subgroups tended to group clones related by pedigree. Principal component analysis (PCA) of the data set also divided the accessions into two major groups, one containing most of the cultivars and the other containing most of the *S. officinarum* clones. In addition, plotting the first and second principal components (PCs) revealed no strong population structure amongst true *S. officinarum* clones (those with 80 chromosomes), while plotting the first and third PCs separated *S. officinarum* clones from Hawaii and other Pacific islands from those originating from New Guinea and Indonesia. Within the cultivars, plotting the first and second PCs again revealed no obvious structure, while plotting the first and third PCs identified loose clusters associated with pedigree relationships. In general more diversity was present within the cultivars than within the *S. officinarum* accessions, with this diversity being contributed largely by the *S. spontaneum* chromosomes in the hybrids. With respect to the *S. officinarum* clones, which contribute the majority of the genome present in modern sugarcane cultivars, little support for population structure was revealed by the study, but the authors suggested that the large number of polymorphic markers found within the group suggested a larger amount of diversity within the species than previously thought. In addition, the number of polymorphic markers from New Guinea accessions supported earlier hypotheses that the island is the centre of origin of the species. Ten per cent of the markers found in *S. officinarum* accessions were not found in the cultivar group, suggesting that most of the diversity within *S. officinarum* is represented within hybrid cultivars, at least those contributing to Australian germplasm.

In terms of the number of accessions analysed, one of the largest sugarcane diversity studies undertaken was that of Nayak et al. (2014). From the World Collection of Sugarcane and Related Grasses (WCSRG) in Miami, Florida, 1002 accessions were screened with 36 SSR markers, resulting in 209 scoreable alleles. A phylogenetic tree (created by neighbour joining analysis) and principal coordinate analysis (PCoA) were used to characterize the genetic diversity. Population structure was further explored via the STRUCTURE V2.2 algorithm (Pritchard et al. 2000), and diversity within subpopulations was analysed by analysis of molecular variance (AMOVA). The phylogenetic tree and the plot of the first

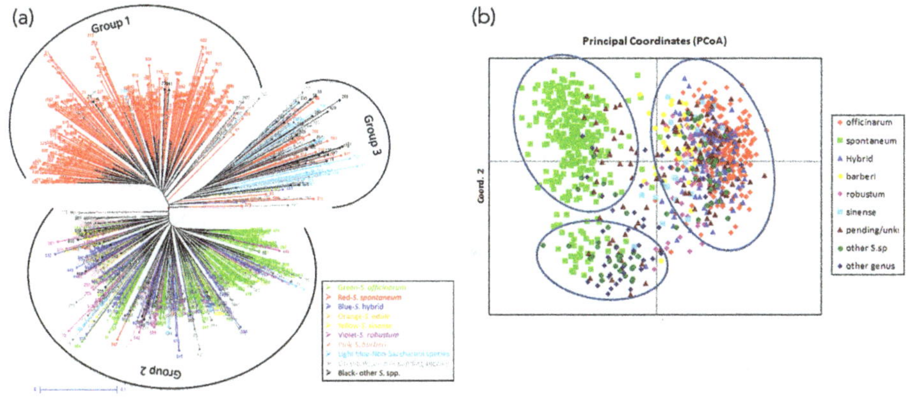

Figure 1 (a) SSR-based phylogenetic tree of 1002 *Saccharum* and related species accessions from the World Collection of Sugarcane and Related Grasses (WCSRG) at Miami, Florida. Adapted from Nayak et al. (2014). (b) Plot of first two principal coordinates of SSR-based diversity analysis of WCSRG accessions. Adapted from Nayak et al. (2014).

two principal coordinates divided the accessions into three main groups (Fig. 1a and b). The first group (from the phylogenetic analysis) contained the majority of *S. spontaneum* accessions. The second group contained almost all *Saccharum* hybrids, in addition to *S. officinarum*, *S. robustum*, *S. sinense*, *S. barberi* and *S. edule* accessions. The third group was comprised mainly of unidentified and non-*Saccharum* species (*Erianthus*, *Miscanthus*, sorghum, etc.). Population structure analysis (using STRUCTURE V2.2) also identified three subpopulations within the collection, although a significant number of accessions (27%) were not clearly assigned to a single subpopulation. Results of AMOVA suggested that there was significant genetic variance, and that this variance was higher within subgroups than amongst subgroups. Based on the cluster analysis used for the phylogenetic tree, a core collection containing 300 accessions was also identified. The core collection represents only about 30% of the total number of accessions, while retaining most of the genetic diversity present in the entire collection.

4 Linkage mapping and QTL analysis

Genetic linkage mapping based on analysis of biparental progeny has long been considered a tool for geneticists, helping them to understand the structure of a genome, localize markers (morphological, isozyme or DNA based) to particular chromosomes or linkage groups and identify markers linked to both qualitative and quantitative traits. If the trait variability accounted for by the markers (or QTL linked to the markers) is sufficiently large, such markers can be used in marker-assisted breeding (reviewed by Collard et al. 2005; Collard and Mackill 2008; Buerstmayr et al. 2009). In sugarcane, mapping efforts (including linkage and physical mapping, and QTL identification) have been aided by the genetic resources available in sorghum, a closely related diploid species.

4.1 *Bru*1

The discovery, characterization and utilization of markers for *Bru*1, which provides resistance to sugarcane brown rust (*Puccinia melanocephala* H. & P. Sydow), is the primary example of trait mapping and subsequent marker-assisted breeding in sugarcane. The first linkage map in sugarcane was developed by Grivet et al. (1996) using RFLP and isozyme markers. Selfed progeny (n = 77) of the sugarcane cultivar 'R570' were used to develop the linkage map. Segregation analysis of 408 markers identified 96 linkage groups, based on markers linked in coupling (i.e. on the same homologous chromosome). By grouping linkage groups containing markers from the same probe, a composite map was generated that combined most of the groups (85/96) into composite groups, putatively made up of homologous/homeologous chromosomes.

This map, along with additional markers identified but not used in the mapping process, was used by Daugrois et al. (1996) in combination with field and greenhouse evaluations of brown rust response to identify an RFLP linked to a major locus for brown rust resistance. The segregation of resistant and susceptible progeny in the mapping population was not significantly different from that expected of a simplex (single copy), dominant gene. A second group of selfed progeny of 'R570' was also analysed for both brown rust response and the genotype of the linked marker. These data confirmed both the nature of the resistance factor (simplex dominant) and the linkage of the RFLP marker to the locus. By pooling the two data sets, the authors concluded that the RFLP marker (*CDSR29–H5*) was located 9.5 cM from the resistance locus. Follow-on research by Asnaghi et al. (2001) confirmed that the resistance factor in 'R570' made it resistant to brown rust isolates from a broad range of geographic locations. Asnaghi et al. (2000) attempted to locate more markers in the vicinity of the resistance factor using syntenic markers from sorghum, rice and maize. The linkage map of 'R570' was improved as a result, with the locus now assigned to a linkage group, but no markers were found with a tighter linkage than *CDSR29-H5*. A map-based effort to clone the resistance gene (now named *Bru*1) was initiated by Asnaghi et al. (2004). A much larger population of selfed 'R570' progeny (n = 695) was screened for rust resistance. In addition, DNA markers (AFLPs in this case) were utilized to screen pooled DNA of resistant and susceptible individuals from the new population, a technique known as bulked segregant analysis (Michelmore et al. 1991). Eight new markers were identified within a 10 cM interval, and two of the markers flanked *Bru*1 at a distance of approximately 2 cM on either side. The cloning effort in a multipronged approach by Le Cunff et al. (2008) was continued. This approach leveraged the synteny between sugarcane, rice and sorghum to gain positional information for *Bru*1 and identify additional markers. In addition, Le Cunff et al. increased even further the size of the selfed 'R570' population (n = 1600) in an effort to increase map resolution around *Bru*1. The result was a partial physical map (based on sugarcane BAC clones) of the *Bru*1 region, markers that flanked *Bru*1 at 0.28 cM and 0.14 cM on each side of the locus and 13 markers that cosegregated with the trait. In 2012, Costet et al. identified markers that were directly applicable in breeding for brown rust resistance. They directly utilized RFLP and AFLP markers from the *Bru*1-focused studies of Asnaghi et al. (2004) and Le Cunff et al. (2008) and the general 'R570' mapping study of Hoarau et al. (2001) and derived SSR, PCR and CAPS (cleaved amplified polymorphic sequence) markers from data generated by Le Cunff et al. (2008) and Garsmeur et al. (2011). These markers were used to genotype an internationally derived panel of 380 sugarcane cultivars and breeding clones that was also screened for brown rust response under natural infection conditions. Two markers, one CAPS and the other a dominant PCR marker, cosegregated with *Bru*1 on the original 'R570' map, were present in most resistant accessions in the international panel (85.6%) and were absent from all susceptible accessions. The results of this investigation (and the many leading up to it) were significant in several ways. First, *Bru*1 was shown to contribute to brown rust resistance in a wide range of cultivars; it is not limited to a small number of breeding programmes. Second, the locus confers resistance that is broad spectrum, and, at this point, durable. Finally, markers were now available that could be used directly in marker-assisted breeding of sugarcane to obtain cultivars with brown rust resistance.

Glynn et al. (2013) utilized the *Bru*1 markers to characterize the germplasm at the breeding station in Canal Point, Florida. Out of 1072 clones, 27% tested positive for the markers cosegregating with *Bru*1. The distribution was skewed between Florida ('CP') and Louisiana ('Ho') material, in that 42% of the Florida clones tested positive, versus only 6% of Louisiana clones. Parco et al. (2014) also screened Louisiana germplasm for *Bru*1 and obtained similar results, finding only five *Bru*1-positive clones out of 117 cultivars and elite clones (4.3%). Glynn et al. (2013) also screened 485 Florida genotypes for brown rust response. Although the percentage was not as high as that reported by Costet et al. (2012), most resistant clones were positive for *Bru*1, indicating that the locus is the major source of brown rust resistance in Florida germplasm. In contrast, Racedo et al. (2013) found that the frequency of *Bru*1 was low in Argentine cultivars and US-derived breeding parents, both overall (6.6%) and amongst brown rust-resistant individuals (16.3%). This is perhaps expected, as the germplasm from both Louisiana and Argentina are closely related (Perera et al. 2012). In China, Li et al. (2017) surveyed 101 breeding clones for brown rust response and *Bru*1 genotype. Not only was *Bru*1 highly prevalent amongst the accessions surveyed (47.5%), but it was also present in the majority (62%) of resistant clones. A survey of Brazilian germplasm for *Bru*1 was conducted by Neuber et al. (2017). As for the germplasm from China and Florida, the majority of brown rust-resistant clones (73.5%) tested positive for *Bru*1.

In addition to surveying the parental germplasm at Canal Point, Glynn et al. (2013) improved the dominant PCR marker by converting the assay to real-time PCR. Another improvement of the *Bru*1 assay was made by McCord and Migneault (2016) who converted the CAPS marker into an unlabelled probe melting assay (Fig. 2). Though more sensitive to DNA concentration than the CAPS marker, unlabelled probe melting is a closed-tube process that does not require enzyme digestion or gel electrophoresis post PCR, thereby saving time and reagents. In addition, a droplet digital PCR (ddPCR) assay was developed by McCord (2016) that allows estimation of the dosage of *Bru*1 (Fig. 3). This allows identification of parents with higher *Bru*1 copy numbers, which transmit the locus to a higher portion of their progeny and can also be used to identify individuals in different dosage classes prior to testing for gene dosage effects. At Canal Point, screening for the presence of *Bru*1 in the second clonal stage of the breeding programme (1500–1600 genotypes) is now routinely performed, as is ddPCR-based dosage analysis of *Bru*1-positive genotypes used as parents.

4.2 Sucrose content and biomass yield

As maximum sugar yield per unit area is the most important target in sugarcane agriculture, a number of studies have sought to identify QTL and markers for sucrose content and/or biomass yield. Ming et al. (2001) utilized RFLP markers to develop linkage maps and

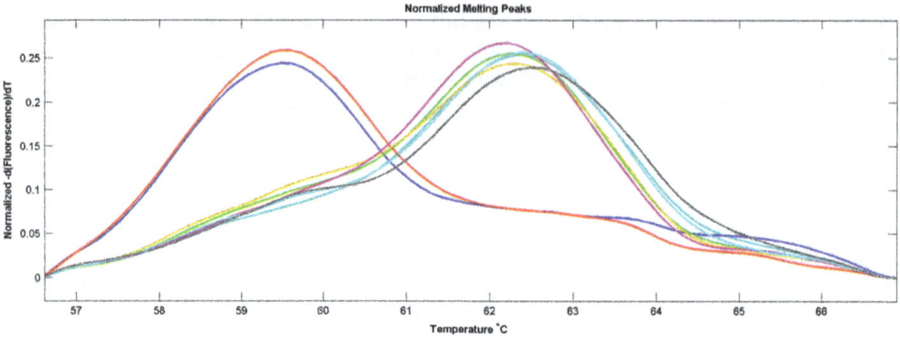

Figure 2 Results of unlabelled probe melting analysis for *Bru*1. The plot displays the negative derivative of fluorescence over temperature and is normalized. The probe (targeting an SNP in the CAPS marker developed by Costet et al. (2012)) melts at a higher temperature (approximately 62°C) in *Bru*1-positive genotypes vs. *Bru*1-negative genotypes (approximately 59.5°C).

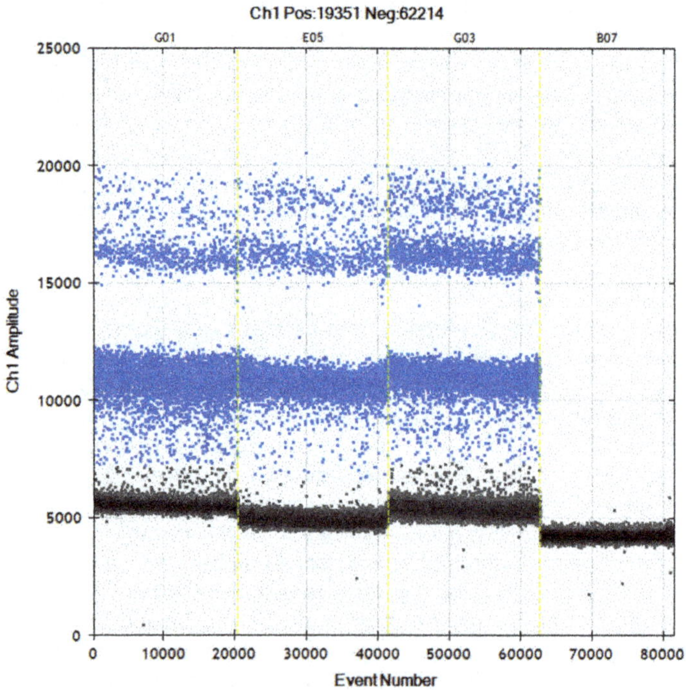

Figure 3 Sample data from a droplet digital PCR (ddPCR) assay for detecting differences in *Bru*1 allele dosage. The grey droplets do not contain any target DNA. The blue droplets at the top of the figure contain both *Bru*1 and the reference marker, while the blue droplets in the middle contain the reference marker only. Samples G01, E05 and G03 are *Bru*1 positive, while sample B07 is a no-template control. Approximately 20 000 individual droplets are present in each sample. Counting the number of positive droplets for each target (*Bru*1 and reference) allows the initial concentration of the targets to be determined. Changes in *Bru*1 concentration relative to the reference indicate changes in allele dosage.

search for QTL associated with sugar content, in this case defined as kilograms of sucrose per tonne of cane. Two populations were analysed; 'Green German' × *S. spontaneum* 'IND 81-146' (n = 264), hereafter the GI population, and *S. spontaneum* 'PIN 84-1' × 'Muntok Java' (n = 239), known as the PM population. Both 'Green German' and 'Muntok Java' were classified as *S. officinarum*, but the accessions used in this study possessed more than 80 chromosomes. A total of 18 QTLs were identified in the GI population. The best-fitting model utilized 14 of these QTLs that together explained 65.5% of trait variation. Although *S. officinarum* and *S. spontaneum* generally have high and low sucrose content, respectively, one QTL from each of the parents was found with an opposite effect, that is, a high-sucrose QTL from 'IND 81-146' and a low-sucrose QTL from 'Green German'. In the PM population, 23 QTLs were identified, and the best-fitting model utilized 22 of them. This model accounted for 68.3% of trait variation. In this population, a single *S. spontaneum* QTL was found with a positive effect on sucrose; no negative-effect QTL were found in 'Muntok Java'. Several of the sugar-related QTLs identified in this study appeared to be in genomic regions near known genes or QTLs in maize that correspond with sugar content in the seed. A follow-on study with the same populations and markers (Ming et al. 2002a) identified QTLs for sucrose content (here defined as the percentage of sucrose in the juice), sucrose yield (influenced by sucrose content, fibre and biomass yield), stalk weight and stalk number (components of biomass yield), fibre content and ash. A total of 102 significant marker-trait associations were detected, most of which (61) were located on linkage groups. The number of QTLs per trait varied considerably. For example, 10 and 24 stalk weight QTLs were found in the GI and PM populations, respectively, accounting for 62.7% and 71.6% of trait variation, respectively. In contrast, only two and one QTL, respectively, were found for stalk number, and these QTLs explained only 13.9% and 6.1% of trait variation. The most significant single QTL was for stalk weight; it was contributed by 'Muntok Java' and accounted for 16.2% of trait variation. Both populations were also evaluated for plant height and flowering behaviour (Ming et al. 2002b), and QTLs were compared with those found for the same traits in sorghum. Five QTLs for plant height originally found in sorghum, as well as a QTL for flowering, were also detected in the PM population. Reasonably large-effect QTLs for plant height were detected in this study, with one from 'Muntok Java' explaining 22.6% of the trait variation. Similarly, Jordan et al. (2004) detected QTLs in sugarcane for stalk number and suckering (vigorous, low-sucrose shoots that emerge later than the main group of stalks) that corresponded to sorghum QTL for tillering and rhizomatousness.

Using a portion (n = 295) of the selfed 'R570' population, and a map based on approximately 1000 AFLP markers, Hoarau et al. (2002) identified 'quantitative trait alleles' or QTA (markers associated with quantitative traits) for plant height, stalk population and diameter and Brix. Of 45 single-marker QTAs detected, only five were significant across both years of evaluation, probably due to strong environment (year was confounded with crop cycle) and genotype × environment effects. The percentage of variation explained by QTA ranged from 3 to 7% (within years). Evaluation of digenic interactions identified four pairs of markers whose interaction was significant across both years for stalk diameter (1 pair), stalk number (2 pairs) and Brix (1 pair), but the percentage of variation explained by these pairs was not reported. In contrast to Ming et al. (2002a), the percentage of variation explained by stalk number QTA, the only trait in common, was considerably higher (44–51% depending on year). Aitken et al. (2005, 2006b) developed a similarly sized mapping population (n = 230) from a cross between *S. officinarum* IJ76-514 and the Australian commercial cultivar 'Q165'. A total of 1054 SSR and AFLP markers were used to generate a map of 'Q165' and identify QTL for Brix and pol (juice sucrose content) contributed by this parent. Sucrose content tends to increase throughout the harvest season and is controlled by environmental and genetic factors. For this reason, data were taken both early and mid-season over two years. QTL identified by single genes (markers) and digenic (epistatic) interactions were explored. For pol, three and six QTLs based on single-dose markers were common in both years at early and mid-season respectively, and for Brix, five and three QTLs were detected, respectively. Analysis of multi-dose markers revealed one multi-season QTL for pol at early and mid-season, and one and three QTL for Brix. Single markers explained from 3% to 9% of variation for either trait. Digenic interactions corresponding to six regions were significant across years for both traits, explaining 11% to 13% of trait variation (which traits were not specified). QTL models including single-, multi-dose and interaction QTLs (including some higher than first-order interactions) explained 51% and 55% of combined (across years) early and mid-season variation for pol, and 58% and 54% of the early and mid-season variation for Brix.

Unlike most non-woody crops, sugarcane is grown as a perennial crop, generally for at least three seasons, but often much longer, especially in tropical regions. The most desirable QTLs are those that are stable across multiple years (and locations), which obviously require more time to generate the phenotypic data. Pastina et al. (2012) evaluated a mapping population over three years (and two locations) and also attempted to improve the analysis by utilizing mixed models (optimized for each trait prior to searching for QTL) to calculate QTL interactions with harvest and/or location. A small population (n = 100) from the Brazilian cross 'SP80-180' × 'SP80-4966' was phenotyped for biomass yield, sugar yield, fibre and pol and genotyped using RFLP and SSR markers. For the QTL detection step, interval mapping (Lander and Botstein 1989) was used along with the classical single marker analysis. Interval mapping can increase the precision of QTL estimates by testing for the presence of QTL at locations in between markers (i.e. intervals). A final multi-QTL model (including QTL and unlinked but trait-associated markers) was then developed for each trait. For all traits, QTLs were detected which were stable across environments, but the majority had significant interactions with harvest and/or location. The percentage variation explained by the various QTLs was not reported.

Likewise, Singh et al. (2013) analysed phenotypic data over a larger number of environments. They utilized SSR markers to search for QTLs for eight sugar and yield-related traits in a population derived from two elite Indian cultivars ('Co86011' × 'CoH70', n = 207). Whereas most QTL mapping studies have looked at clonal plots of each genotype, the authors of this study evaluated the population for Brix at the true seedling stage. Clonal plots were further evaluated in six environments (three locations and two seasons). As for Pastina et al. (2012), composite interval mapping (Zang 1993), a more sophisticated mapping technique, was used to detect QTL. Composite interval mapping is an improvement over interval mapping, in that it can account for the presence of linked QTLs. Five QTLs for Brix were stable across all environments, one of which co-located with a pol QTL, which is not surprising given that the traits are highly correlated. Each Brix QTL explained 6–9% of trait variation. Neither of the two seedling Brix QTL were co-located with clonal Brix QTL, but one was located in the same region as one of the pol QTL; each explained approximately 2.5% of trait variation. Including those QTLs co-located with clonal Brix and seedling Brix, six QTLs for pol were detected, explaining 2.5–8.9% of trait variation. QTLs for the yield- and agronomic-related traits (stalk number, length, diameter, internode length and number of green leaves) tended to have larger effects than the sugar-related QTL, explaining 9–19% of trait variation.

4.3 Disease resistance

After yield, resistance to disease is arguably the next most important target. QTL and/or markers for disease resistance often show large effects, as is the case for *Bru*1. Mudge et al. (1996) developed a linkage map of *S. officinarum* 'LA Purple' × *S. robustum* 'Molokai 5829' (n = 84) using RAPD markers. One marker that was linked at a distance of 7.3 cM to susceptibility to eyespot disease, caused by the fungus *Bipolaris sacchari*, was found. Eyespot susceptibility segregated 1:1 in this population, suggesting similar gene action (simplex dominant) as for *Bru*1. This would have made it a good candidate for marker-assisted selection, but apparently due to the fact that eyespot is a minor disease (Rott et al. 2014), no additional markers have been reported in the literature. In Australia, McIntyre et al. (2005a, 2005b) developed a linkage map from 192 individuals of a biparental cross between 'Q117', a commercial sugarcane cultivar, and 74C42, an elite breeding clone, using approximately 1000 AFLP, RFLP and SSR markers. Several of the RFLP markers were derived from resistance gene analogues, which have been explored in a number of species in efforts to identify disease resistance genes. The population was scored for resistance to brown rust and *Pachymetra* root rot (PRR), caused by the oomycete *Pachymetra chaunorhiza*. Three QTLs (here defined as genomic regions) and one unlinked marker were associated with PRR in the mapping population in two seasons. These QTLs (analysed by single marker analysis) explained 4–16% of trait variation depending on the marker and year. For brown rust, two unlinked markers and approximately seven QTLs were identified, explaining a similar amount of variation (4–18%). With respect to marker-assisted breeding, the authors took the important next step of validating the markers identified in the first analyses. Six markers (actual markers, not QTL) associated with PRR and seven markers associated with brown rust were used to screen 154 sugarcane clones, primarily of Australian origin. Three of the PRR-associated markers and one of the brown rust–associated markers were significant in the validation group (all at $P < 0.05$); two more brown rust markers were significant at $P < 0.1$.

A large effect QTL for yellow spot disease, caused by *Mycovellosiella koepkei*, was found in a segregating biparental population from Mauritius by Aljanabi et al. (2007). The population consisted of 227 individuals from a cross between 'M 137/75' which is resistant, and 'R570', which is susceptible. Segregation analysis of yellow spot resistance suggested the presence of a single dominant gene. Utilizing 557 SSR and AFLP markers present in only the resistant parent and segregating in the progeny, the authors found a major QTL for yellow spot resistance that explained 23.8% of trait variation. One AFLP marker was linked to the QTL at a distance of 14 cM. To date, no further research to develop a diagnostic marker for breeding has been reported.

Possibly the largest QTL mapping effort in sugarcane (apart from *Bru*1), prior to the use of next-generation array and/or sequence-based markers, was that of Raboin et al. (2006). The mapping population consisted of 198 individuals from a cross between 'R570' (used to identify *Bru*1) and 'MQ76-53', an F1 hybrid between a commercial sugarcane cultivar and an *S. spontaneum* clone. A total of 1666 AFLP, SSR and RFLP markers, along with traits such as resistance to brown rust and stalk colour, were evaluated. AFLP markers that were linked (separately) to brown rust (23 cM) and red stalk colour (6.5 cM) were found. Stalk colour (red vs. all others) segregated as a single dominant trait, but diagnostic markers have not been developed, presumably because the trait is not agronomically important. The brown rust resistance locus is distinct from *Bru*1, which also segregated in this cross. The closest linked marker for brown rust resistance explained 35% of trait variation amongst the 90 individuals of the cross that did not contain *Bru*1, indicating the presence of another major effect QTL for brown rust.

4.4 Nutrient use efficiency

In terms of sustainability, nutrient use efficiency is another target for marker-assisted breeding. Whan et al. (2010) analysed a biparental population of 168 individuals derived from *S. officinarum* IJ76-514 × 'Q165' (Aitken et al. 2005, 2007), for nitrogen use efficiency (NUE) under controlled (greenhouse) conditions, using two nitrogen rates (limiting and non-limiting). Using single-marker analysis at a threshold of $P < 0.005$, significant QTLs (based on the presence of more than one linked marker significantly associated with a trait) were found for NUE and several other nitrogen- and growth-related traits. In general, the QTLs identified were unique to the nitrogen treatments. Two QTLs were found for NUE in 'Q165' under non-limiting conditions, explaining 6–7% of trait variation, both of which had a negative effect. Another two QTLs were detected under limiting N; one had a negative effect, explaining 12% of trait variation, and the other had a positive effect and accounted for 9% of trait variation. For the female parent, a total of nine QTL were identified for NUE, explaining 6–8% of trait variation, of which four were detected under non-limiting N (two with positive effects) and five under limiting N (all with negative effects). Several unlinked markers from the female parent were also associated with NUE, one of which had a positive effect on the trait and explained 11% of the variation. QTLs associated with glutamine synthetase (GS) activity, one of the major enzymes involved in nitrogen metabolism, were not linked to those associated with NUE.

4.5 New marker technologies

One of the main challenges with linkage mapping and trait analysis in polyploids is that more markers are required than those in diploid species. The linkage maps developed as part of the QTL analyses reported so far are incomplete, and a number of markers associated with traits were unlinked. In addition, marker density was likely insufficient to allow for more sophisticated QTL mapping approaches such as interval mapping or composite interval mapping. Indeed, these QTL mapping methods were used in only a few cases (Pastina et al. 2012; Singh et al. 2013). Newer array and sequence-based technologies deliver more markers and have been used increasingly for trait mapping in sugarcane. Diversity Arrays Technology (DArT) markers, which are based on hybridization of a labelled reduced representation library to a DNA microarray, were used by Costet et al. (2012b) to increase the marker density of the 'R570' × 'MQ76-53' map. A total of 1161 new markers were added. In addition, the population was evaluated for resistance to *Sugarcane yellow leaf virus* (SCYLV). Single-marker analysis identified one SSR and three AFLP markers linked to SCYLV resistance, of which the most significant explained more than 30% of the trait variation.

Restriction site–associated DNA markers (RAD-seq, Baird et al. 2008) and genotyping by sequencing (GBS, Elshire et al. 2011) are related to DArT markers (and AFLPs) in that the initial steps involve generating a reduced representation of the genome via enzyme digestion, ligation of adapters and amplification of targeted fragments via PCR. The newer technologies extract additional information from the reduced representation library by sequencing the fragments using NGS technologies. Parmessur et al. (2014) reported initial efforts in utilizing RAD-seq technology to map and identify markers for early ripening (sucrose accumulation). A small biparental population (n = 107) was developed from a cross between 'CP 67-412', an early ripening clone from the mainland United States, and 'M 245/76', a late ripening clone from Mauritius. A large number of single-dose SNP markers were identified for each parent (9500 for 'CP 67-412' and 7560 for 'M245/76'), although approximately 40% of the markers had more than 20% missing data. Despite the missing data, a map was generated for each parent, with each map containing more than 3200 markers. Recently, Balsalobre et al. (2017) utilized GBS to identify more than 7000 single-dose markers in a biparental cross. This population consisted of 151 genotypes from a cross between the Brazilian cultivars 'SP80-3280' and 'RB835486'. An additional 629 SSR and target region amplification polymorphism or TRAP markers (Hu and Vick 2003) were utilized in the mapping process. Phenotypic data were collected from two locations over three years (six environments) and included sucrose percentage, Brix, stalk diameter and fibre percentage. Composite interval mapping was used to identify and characterize QTL. Although nearly 7700 single-dose markers were identified, only 993 were actually utilized in mapping. This is a drastic rate of attrition, possibly due (at least in part) to the small population size, limiting the amount of recombination. Despite this reduction in the number of markers, QTLs were detected for all traits that were stable across locations and years. These QTLs explained 2.7–9.2% of trait variation.

5 GWAS/SNP-based mapping and QTL analysis

The other main strategy for identifying trait-linked markers is association mapping. In this case, biparental progeny are not evaluated, but large groups (panels) of individuals are screened. This method can give better resolution than biparental linkage mapping, as many more meiotic events from different backgrounds tend to reduce the distance over which linkage disequilibrium operates. When association mapping is used with large numbers of markers (typically SNPs), it is known as genome wide association mapping, or GWAS.

An early use of GWAS in sugarcane was described by Wei et al. (2010). A modest number (n = 1531) of DArT markers was used in conjunction with an association population. This population (n = 480) consisted of breeding parents, commercial cultivars and randomly selected seedlings. The population was evaluated for one season (plant cane) at three locations. Although association mapping was performed to identify markers for cane yield and sucrose content, a major feature of this study was to measure the effects of population structure, genotype × environment interaction (G × E) and spatial variation on GWAS results, using a mixed-model analysis. Controlling for population structure is necessary in GWAS to eliminate markers that only appear to be associated with a trait, but are instead spurious associations due to genetic relatedness. G × E and spatial variation can influence results in both association and linkage mapping contexts. Population structure (based on pedigree), G × E and spatial variation, all affected GWAS results by inflating the number of significant markers detected and/or missing markers that would have been detected had these effects been accounted for. Overall, G × E had the largest effect, and the authors concluded that utilization of multiple environments and accounting for G × E were necessary in obtaining accurate association mapping results.

Debibakas et al. (2014) were the next to report the results of GWAS in sugarcane. This study focused on resistance to SCYLV, utilizing a panel of 189 genotypes from 26 breeding programmes and 3849 DArT and AFLP markers. The panel was phenotyped in plant cane and second ratoon (i.e. second regrowth of original planting) using two successive trials at the same location in Guadeloupe under natural infection conditions (no reliable mechanical inoculation method has been developed for SCYLV). Data on the percentage of SCYLV-positive leaves and stalks were obtained, and the virus population (titre) in leaves and stalks was estimated based on the intensity of the immunoassay staining. Population structure was accounted for using PCA of marker data as a covariate in a general linear model. In addition, the authors modelled kinship (lower-level family relatedness) utilizing a mixed linear model. Including both kinship and population structure in the model proved to be the best method at reducing inflation of the test statistic. A total of six significant markers were detected, accounting for 9–14% of the trait variation, depending on the marker, structure/kinship model (two different Q matrices were used), trial and sub-trait being measured. Two of these markers were significant in both tissue types and sub-traits, with one being localized to chromosome 2 (of sorghum) and the other to chromosome 5. The marker on chromosome 2 is linked (in sorghum) to genes homologous to *Arabidopsis* genes that are regulated in response to *tobacco mosaic virus* (TMV) infection, and the marker on chromosome 5 is linked (in sorghum) to a gene homologous to an *Arabidopsis* gene associated with aphid feeding.

Gouy et al. (2015) also explored the feasibility of GWAS in sugarcane by screening an association panel of 189 sugarcane genotypes from around the world with a combination of 3327 AFLP, DArT and SSR markers. Thirteen traits were analysed, including agronomic and disease-/pest-related traits, across multiple locations. Population structure was modelled using marker information and either Bayesian clustering analysis using the STRUCTURE programme (Pritchard et al. 2000) or PCA. For association analysis, both general and mixed linear models were used, both with and without population structure information. As for Debibakas et al., accounting for both kinship and population structure proved most effective in reducing spurious associations. Controlling for population structure but not kinship resulted in 23 significant markers, whereas controlling for both identified nine markers, of which only three overlapped with the simpler model. Of the nine significant markers, two were found for Brix, one for flowering intensity, four for brown rust and two for gumming. One of the diagnostic markers for *Bru*1 was included in the screen and was 're-detected', explaining up to 51.5% of trait variation. The variation accounted for by other markers ranged from 8.4% to 13.9%. The authors suggested that the power of the experimental design with respect to quantitative traits, the correlation between population structure and some of the traits measured and the number of markers used limited the number of significant markers detected.

Banerjee et al. (2015) utilized 989 SSR fragments (from 123 primer pairs) to perform GWAS in a small panel (n = 108) of Indian sugarcane clones. Over three environments (three plant-cane crops at a single location), a number of yield-related traits, including sucrose content, stalk length, weight, diameter, population and nodes per stalk were evaluated. As for Gouy et al. (2015), accounting for both population structure and kinship was the most appropriate model for controlling inflation of the test statistic. Stable markers (significant across all years) were found for each trait analysed. Individually, these markers accounted for 5.9–16.4% of the trait variation depending on marker and trait; this is comparable with the results of Gouy et al. The most predictive marker, for stalk population, explained an average of 15.7%. The yearly average amount of variation explained by all markers for a particular trait ranged from 19% for node number to 57% for stalk population. In addition to identifying significant markers for traits of interest, the authors found significant differences in the population structure based on pedigree analysis vs. marker genotypes.

6 Genomic selection

With some exceptions for disease-related markers, markers and/or QTLs identified with important traits in sugarcane, either through linkage mapping or GWAS, have only accounted for a small portion of the variation seen for the trait. Even if a significant portion of trait variation is accounted for by a given set of markers, it is inefficient from a breeding standpoint to attempt selection using markers with small effects. This is because very few individuals will contain enough of the favourable alleles to show a measurable difference in the trait value(s). Rather than selecting the markers with significant but small effects directly, the technique of GS develops a prediction model for each trait based on hundreds to thousands of markers regardless of their individual significance, along with phenotypic data (reviewed in Desta and Ortiz (2014) and Poland and Rutkoski (2016)). This model can then be used for selection by applying it to new genotypes using only their marker information, that is, no phenotypic information is required. If trait heritability and model prediction are good enough, genetic gain can be enhanced by shortening the breeding cycle via the elimination of at least some of the phenotyping involved in a breeding programme. GS has been used extensively and successfully in animal breeding (Hayes et al. 2009; Miar et al. 2015), but in more recent years, has also been applied to crop plants.

The use of GS in sugarcane was initially assessed by Gouy et al. (2013). Two independent selection panels (no overlap), each comprising 167 unique accessions, were each genotyped with 1499 DArT markers. One panel was evaluated in Réunion, and the other in Guadeloupe; each panel was evaluated at multiple locations. Phenotypic data on stalk diameter and population, juice Brix, bagasse content, fibre characteristics (acid detergent fibre, acid detergent lignin and *in vitro* digestibility) and diseases such as smut, brown rust and sugarcane yellow leaf were collected.

The model parameters and performance of GS can depend on the statistical methods used to create the prediction models. A number of methods have been used for GS, which vary in a number of ways, including assumptions of underlying distributions (parametric vs. non-parametric), accommodation for non-additive variance, etc. (Calus 2010). In this example, ridge regression (RR), Bayesian least absolute shrinkage and selector operator (LASSO), reproducing kernel Hilbert spaces (RKHS) and partial least squares regression (PLSR) were tested. In addition to evaluating different methods of GS model development, two methods of analysing the accuracy of GS were evaluated. Both methods utilized the Pearson correlation coefficient between the predicted and observed genetic values for each member of a given panel. For the first method, each panel was divided into five subpanels. Four of the subpanels were used as the training population to predict the values of the fifth subpopulation. This subdivision and prediction was performed randomly 500 times to generate a distribution of correlations for each trait and model. The second was cross-validation, wherein the model developed with the entire Guadeloupe panel as a training population was used to predict the genetic values of the Réunion panel and vice versa. The median Pearson correlation coefficients (within panels) ranged from 0.11 (non-significant) for brown rust in the Guadeloupe panel using PLSR to 0.62 for Brix in the Réunion panel using Bayesian LASSO. Correlations were generally similar across different methods, and correlations were generally higher for the Réunion panel vs. the Guadeloupe panel. Brown rust was evaluated only in those individuals that did not contain *Bru*1; the resulting reduction in panel size may have contributed to the lower accuracy of GS for this trait. For inter-panel cross-validation, correlations ranged from 0.13 (non-significant) for smut, using Réunion as the test panel, to 0.55 for Brix, again using Réunion as the test panel. Correlations were generally higher using Réunion as the test panel, and as for the within-panel assessment, correlations were generally very similar across methods. The results were satisfactory, especially the accuracy (correlation) using training and test panels from different hemispheres. They also suggested that higher marker density could have improved accuracy by detecting strong marker-trait associations, which can be leveraged by Bayesian methods, and by more accurately assessing the genetic diversity present in the panels (which influences estimates of population structure that in turn affect model accuracy).

A much greater marker density was achieved by Costa (2015) who utilized 37 014 SNP markers to genotype a panel of 514 genotypes. These genotypes were derived from 98 half-sib families using parents from the RIDESA sugarcane breeding programme in Brazil and had been through one cycle of selection. Phenotypic data were collected in the plant-cane crop from a single location and included sucrose percentage, tonnes of cane per hectare, tonnes of sugar per hectare and bagasse fibre percentage. As described in Gouy et al. (2013), Costa evaluated several models for GS, including genomic best linear unbiased predictors (GBLUP), Bayesian LASSO, Bayes A, Bayes B and Bayesian ridge regression. The accuracy of GS in this experiment was determined in a manner similar to the within-panel cross-validation method employed by Gouy et al. (2013), in that the panel was randomly subdivided into ten subpanels, nine of which were used as a training set for predicting the values of the tenth subpanel. The routine was repeated ten times (vs. 500 for Gouy et al.). The Pearson correlation between predicted and observed genetic values was termed predictive ability, whereas accuracy was defined (with respect to additive effects) as the ratio of predictive ability to the square root of narrow-sense heritability as defined by Resende (2002) for the trait of interest. As an additional comparison, two more prediction models were also fitted for all traits that included genomic information (with regards to relatedness) and/or classical pedigree information. These are not considered GS models *per se*, but were included to explore in general terms the advantages of using marker information in addition to pedigree information. The use of information from relatives (via pedigree) has been used by mixed-model analysis for many years, particularly in animal breeding.

Correlation coefficients between the predicted and observed genetic values were very similar for the different GS methods used, in agreement with Gouy et al. (2013). Correlations ranged from 0.033 for tonnes of sugar per hectare using the Bayes B method to 0.322 for fibre percentage using Bayesian ridge regression. The correlation coefficients for sucrose percentage were nearly as high as those for fibre, ranging from 0.304 to 0.310. Although the values of the coefficients differ, the high relative ranking of coefficients for sucrose content is also in general agreement with the results (for Brix) of Gouy et al. (2013). The predictive abilities of the different methods were also very similar, and again, the lowest accuracy was for tonnes of sugar per hectare (0.07 using the Bayes B method), and the highest accuracy was for fibre percentage (0.57 using Bayesian ridge regression). When compared with each other, the prediction model using pedigree information had a lower correlation with observed genetic values than the model using pedigree and genomic information for all traits except tonnes of sugar per hectare, indicating that use of markers for estimating relatedness (and genetic covariance) is superior to using pedigree information alone. Regarding the pedigree + genomic model, correlations were higher for sucrose and fibre percentages (0.311 and 0.305, respectively) than for tonnes of cane or tonnes of sugar per hectare (both 0.004, non-significant).

Although many more markers were used by Costa than by Gouy et al., correlation coefficients for comparable traits (sucrose content and fibre percentage in this study, Brix and bagasse percentage in Gouy et al. (2013)) were similar. Although not discussed by Costa, the expected advantage of higher marker density was likely negated by the fact that the phenotypic data were based on a single unreplicated (with respect to the test clones) trial in a single environment. This is of particular importance with respect to the yield traits (tonnes of sugar/cane per hectare) as these traits have lower heritability. On the other hand, Costa suggested that still higher marker densities may be needed for GS in sugarcane, due to the sheer size of the genome. Another suggestion for model improvement was to incorporate models that account for non-additive effects. Gouy et al. (2013) used just such a model (RKHS), but the lower marker density in that study may have masked any benefits of the model.

The accuracies reported by these GS experiments in sugarcane may not be good enough to use GS for the purposes of direct selection of commercial clones. Improved trial design (which leads to more accurate phenotypes) and perhaps higher marker densities will likely lead to higher accuracies. In either case, GS may be useful 'as is' in the selection of parents based on their breeding value, which is the average effect of a parent on its progeny, and is influenced by the additive genetic variance in a population. Recently, researchers in Australia (Wei and Jackson 2016; Jackson and Wei 2017) utilized DArT markers to perform GS in Australian sugarcane germplasm. The set of 384 DArT markers was very small by GS standards, but these markers were pre-selected for their association with cane yield and sugar content in previous experiments (Wei et al. 2010), and in fact did not appreciably affect accuracy than when using a larger number of markers (P. Jackson, pers. comm.). The markers were used to genotype 1135 clones being evaluated in the second stage (i.e. the stage following seedling selection) of the Australian breeding programme. Phenotypic data for cane yield and sucrose content were collected over three seasons in a series of five sub-trials (not all clones were tested in all trials). The methods used for model selection included Bayes A, Bayes B, Bayesian LASSO, GBLUP, RKHS and a simple method of summing all fixed effects of individual markers. Predictions of genetic value were made, but the authors extended their experiments to also predict the breeding value of stage 2 individuals. A total of 94 clones were selected to be used for determination of their breeding value based on progeny testing. These clones were selected by one of the four methods: random selection, phenotypic selection based on genetic effects, genotypic selection based on predicted marker effects and index selection based on phenotypic and marker data. Those clones which flowered were crossed with a number of commercial parents. The families were evaluated for biomass yield and sucrose content at four locations. These data were used to calculate breeding values for the 41 parents that flowered and were used in crosses. Correlations (accuracies) were then calculated between actual breeding values and the values predicted by phenotypic performance of the parents and the values predicted by marker information (GS). Correlations were determined for breeding values for both cane yield and sucrose content singly and for an economic index combining the two breeding values, with extra weight given to cane yield.

Several key results were obtained over the course of these experiments (Table 1). First, the correlation between phenotypic performance of the parents and their actual breeding value (based on progeny testing) was low and non-significant for all traits. Second, the most appropriate GS method for model development was the simple model that summed fixed effects over all markers. The reasons for this are unclear and perhaps not statistically satisfying, but this method showed an improvement of as much as 0.1 units over more elaborate methods (P. Jackson, pers. comm.). Third, the correlations between observed breeding values and genetic values (estimated via genomic prediction) for both cane yield and sucrose content were statistically significant. A significant negative correlation was observed between the breeding value for cane yield and genomic predictions for sucrose content, but this was reduced by eliminating individual markers with negative effects on yield. The non-significant correlations between parental phenotype and breeding value and the significant correlations between genomic predictions and breeding value indicate that selecting parents based on predicted breeding value is superior to selecting them based on phenotypic performance. The authors suggest that breeding cycle time could be greatly shortened by using GS to identify parents at early stages, rather than waiting until clones are nearing release as is traditionally done (Fig. 4).

GS in sugarcane is still in an early stage. In addition to the costs of genotyping, the costs of generating high-quality phenotypic data are significant. Due to the crop's high yields, perennial growth habit and high labour requirements, it is logistically formidable to plant and harvest sugarcane field trials with enough genotypes and sufficient replication and plot sizes. A combination of high marker density (using sequence-based markers) and more robust phenotypic data sets should improve the predictive ability of GS for both clonal and parental selection. In addition, models that incorporate non-additive effects may improve model performance, as Jackson and Wei (2017) showed, in the context of an association mapping experiment, that more significant two-way marker interactions were detected for sucrose content than would be expected by random chance. The complexity of the calculations for non-additive effects is significant, but the development of more elaborate models, either for association mapping or GS, is likely to be a key for allowing markers to be used more extensively in sugarcane breeding.

Table 1 Correlations between estimates of breeding value of 41 parents (where breeding value is determined from progeny performance) for overall economic breeding value (EBV), CCS and cane yield and prior predictors of these breeding values

	EBV	CCS BV	TCH BV
CCS	0.16	0.26	–0.13
TCH	0.24	0.23	0.02
Markers CCS (1)	0.18	0.61***	–0.59***
Markers CCS (2)	0.39**	0.68***	–0.36*
Markers TCH	0.42**	0.20	0.35*

Reproduced with permission from Jackson and Wei (2017).

Predictors were CCS – CCS of the parental clone measured in a field trial; TCH – cane yield of the parental clone measured in a field trial.

Markers CCS (1) – a marker-based function for CCS; Markers CCS (2) – a marker-based function for CCS involving deleting markers with negative impact on yield.

Markers TCH – a marker-based function for cane yield.

*, **, *** refer to statistical significance at $P < 0.05, 0.01, 0.001$, respectively. (Note: TCH = tonnes of cane per hectare; CCS = commercial cane sucrose, per cent).

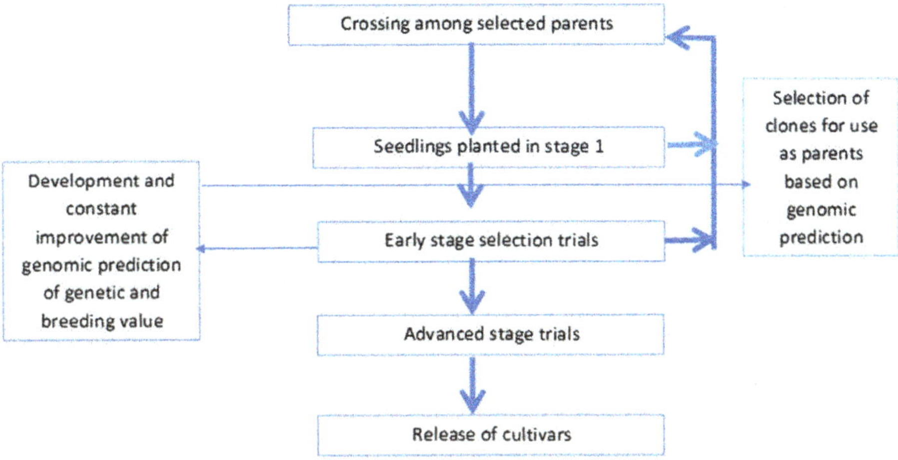

Figure 4 Modified breeding scheme proposed by Wei and Jackson (2016), wherein information from genomic selection/prediction is used to select parents at earlier stages in the breeding cycle, thus enhancing genetic gain. From the original caption, 'Heavy lines refer to flow of clones, and light lines refer to key flows of data'.

7 Future trends and conclusion

Despite the challenges posed by sugarcane, geneticists and breeders have actively sought to use DNA marker technology to enhance breeding efforts. Markers have been used to explore taxonomy, to estimate genetic diversity and to develop unique molecular fingerprints. Numerous studies have been conducted to identify markers linked to important agronomic traits, using a wide array of marker technologies (gel, array and sequence based) and biparental linkage and association-based approaches. Apart from disease resistance (*Bru*1 in particular), the markers that have been found accounted for a relatively small portion of the variation in traits of interest. In part, this should not be surprising, as the two most important traits in sugarcane (biomass yield and sugar content) are quantitatively distributed. More sophisticated QTL mapping and modelling methods, including those that account for non-additive gene action, may yet identify individual markers that can be used directly in breeding. In addition, breeding programmes may choose to validate existing trait-associated markers and deploy them even if effects are relatively small, and if the traits are of sufficiently high value. GS holds promise for sugarcane, especially if training panels are of an appropriate size, phenotyping is robust and perhaps if models are developed to leverage non-additive variance. A genome sequence of sugarcane, when it becomes available, will enhance efforts in QTL mapping, GWAS and GS. Positional information from a genome sequence would aid the development of more accurate and efficient genotyping assays. Modern sugarcane is a young invention, and despite its challenges, we should expect to be able to utilize a combination of the latest marker technologies (including a sequenced genome), sound phenotypic data and new sophisticated models to increase rates of genetic gain in this important crop.

8 Where to look for further information

Future research in marker-assisted breeding of sugarcane is expected in several areas, including: improved models for genomic selection; improved platforms for low cost, high throughput genotyping; biotechnology (including genome editing); improved models for GWAS and QTL mapping that account for the polyploid nature of sugarcane; and sequenced genomes of modern sugarcane and donor species (*S. officinarum* and *S. spontaneum*). Key organizations involved in this research include (but are not limited to): Commonwealth Scientific and Industrial Research Organization (CSIRO); Sugar Research Australia; United States Department of Agriculture, Agricultural Research Service (USDA-ARS); Centre de coopération internationale en recherche agronomique pour le développement (CIRAD); Empresa Brasileira de Pesquisa Agropecuária (EMBPRAPA); and University of Illinois Urbana-Champaign (UIUC).

9 References

Aitken, K. S., P. A. Jackson and C. L. McIntyre. 2005. A combination of AFLP and SSR markers provides extensive map coverage and identification of homo(eo)logous linkage groups in a sugarcane cultivar. *Theor. Appl. Genet.* 110:789–801.

Aitken, K. S., J.-C. Li, P. Jackson, G. Piperidis and C. L. McIntyre. 2006a. AFLP analysis of genetic diversity within *Saccharum officinarum* and comparison with sugarcane cultivars. *Aust. J. Ag. Res.* 57:1167–84.

Aitken, K. S., P. A. Jackson and C. L. McIntyre. 2006b. Quantitative trait loci identified for sugar traits in a sugarcane (*Saccharum* spp.) cultivar × *Saccharum officinarum* population. *Theor. Appl. Genet.* 112:1306–17.

Aitken, K. S., P. A. Jackson and C. L. McIntyre. 2007. Construction of a genetic linkage map for Saccharum officinarum incorporating both simplex and duplex markers to increase genome coverage. *Genome* 50:742–56.

Aljanabi, S. M., Y. Parmessur, H. Kross, S. Dhayan, S. Saumtally, K. Ramdoyal, L. J. C. Autrey and A. Dookun-Saumtally. 2007. Identification of a major quantitative trait locus (QTL) for yellow spot (*Mycovellosiella koepkei*) resistance in sugarcane. *Mol. Breed.* 19:1–14.

Artschwager, E. and E. W. Brandes. 1958. *Sugarcane (Saccharum officinarum* L.*) Agriculture Handbook No. 122*. United States Department of Agriculture, Washington DC.

Asnaghi, C., F. Paulet, C. Kaye, L. Grivet, M. Deu, J. C. Glaszmann and A. D'Hont. 2000. Application of synteny across Poaceae to determine the map location of a sugarcane rust resistance gene. *Theor. Appl. Genet.* 101:962–9.

Asnaghi, C., A. D'Hont, J. C. Glaszmann and P. Rott. 2001. Resistance of sugarcane cultivar R 570 to *Puccinia melanocephala* isolates from different geographic locations. *Plant Dis.* 85:282–6.

Asnaghi, C., D. Roques, S. Ruffel, C. Kaye, J. Y. Hoarau, H. Télismart, J. C. Girard, L. M. Raboin, A. M. Risterucci, L. Grivet and A. D'Hont. 2004. Targeted mapping of a sugarcane rust resistance gene (*Bru*1) using bulked segregant analysis and AFLP markers. *Theor. Appl. Genet.* 108:759–64.

Baird, N. A., P. D. Etter, T. S. Atwood, M. C. Currey, A. L. Shiver, Z. A. Lewis, E. U. Selker, W. A. Cresko and E. A. Johnson. 2008. Rapid SNP discovery and genetic mapping using sequenced RAD markers. *PLoS One* 3:e3376.

Balsalobre, T. W. A., G. S. Pereira, G. R. A. Margarido, R. Gazaffi, F. Z. Barreto, C. O. Anoni, C. B. Cardoso-Silva, E. A. Costa, M. C. Mancini, H. P. Hoffmann, A. P. de Souza, A. A. F. Garcia and M. S. Carneiro. 2017. GBS-based single dosage markers for linkage and QTL mapping allow gene mining for yield-related traits in sugarcane. *BMC Genomics* 18:72–90.

Banerjee, N., A. Siraree, S. Yadav, S. Kumar, J. Singh, S. Kumar, D. K. Pandey and R. K. Singh. 2015. Marker-trait association study for sucrose and yield in contributing traits in sugarcane (*Saccharum* spp. hybrid). *Euphytica* 205:185–201.

Buerstmayr, H., T. Ban and J. A. Anderson. 2009. QTL mapping and marker-assisted selection for Fusarium head blight resistance in wheat: A review. *Plant Breed.* 128:1–26.

Calus, M. P. L. 2010. Genomic breeding value prediction: Methods and procedures. *Animal* 4:157–64.

Collard, B. C. Y. and D. J. Mackill. 2008. Marker-assisted selection: An approach for precision plant breeding in the twenty-first century. *Phil. Trans. R Soc. B* 363:557–72.

Collard, B. C. Y., M. Z. Z. Jahufer, J. B Brouwer and E. C. K. Pang. 2005. An introduction to markers, quantitative trait loci (QTL) mapping and marker-assisted selection for crop improvement: The basic concepts. *Euphytica* 142:169–96.

Costa, P. M. A. 2015. Prediction of breeding values in sugarcane using pedigree and genomic information. Ph.D. thesis, Universidade Federal de Viçosa, Minas Gerais, Brazil. URL: http://www.locus.ufv.br/handle/123456789/7539. Accessed 27 April 2017.

Costet, L., L. Le Cunff, S. Royaert, L. M. Raboin, C. Hervouet, L. Toubi, H. Telismart, O. Garsmeur, Y. Rousselle, J. Pauquet, S. Nibouche, J. C. Glaszmann, J. Y. Hoarau and A. D'Hont. 2012a. Haplotype structure around *Bru*1 reveals a narrow genetic basis for brown rust resistance in modern sugarcane cultivars. *Theor. Appl. Genet.* 125:825–36.

Costet, L., L. M. Raboin, M. Payet, A. D'Hont and S. Nibouche. 2012b. A major quantitative trait allele for resistance to the *Sugarcane yellow leaf virus* (*Luteoviridae*). *Plant Breed.* 131:637–40.

D'Hont, A., Y. H. Lu, P. Feldmann and J. C. Glaszmann. 1993. Cytoplasmic diversity in sugarcane revealed by heterologous probes. *Sugar Cane* 1:12–15.

D'Hont, A., P. S. Rao, P. Feldmann, L. Grivet, N. Islam-Faridi, P. Taylor and J. C. Glaszmann. 1995. Identification and characterization of sugarcane intergeneric hybrids, *Saccharum officinarum* × *Erianthus arundinaceus*, with molecular markers and DNA in situ hybridization. *Theor. Appl. Genet.* 91:320–6.

D'Hont, A., L. Grivet, P. Feldmann, J. C. Glaszmann, S. Rao and N. Berding. 1996. Characterization of the double genome structure of modern sugarcane cultivars (*Saccharum* spp.) by molecular cytogenetics. *Mol. Gen. Genet.* 250:405–13.

D'Hont, A., D. Ison, K. Alix, C. Roux and J. C. Glaszmann. 1998. Determination of basic chromosome numbers in the genus *Saccharum* by physical mapping of ribosomal RNA genes. *Genome* 41:221–5.

Daugrois, J. H., L. Grivet, D. Roques, H. Y. Hoarau, H. Lombard, J. C. Glaszmann and A. D'Hont. 1996. A putative major gene for rust resistance linked with a RFLP marker in sugarcane cultivar 'R570'. *Theor. Appl. Genet.* 92:1059–64.

Debibakas, S., S. Rocher, O. Garsmeur, L. Toubi, D. Roques, A. D'Hont, J. Y. Hoarau and J. H. Daugrois. 2014. Prospecting sugarcane resistance to sugarcane yellow leaf virus by genome-wide association. *Theor. Appl. Genet.* 127:1719–32.

Desta, Z. A. and R. Ortiz. 2014. Genomic selection: Genome-wide prediction in plant improvement. *Trends Plant Sci.* 19:592–601.

Dodds, H. H. 1938. The revolution in sugar-cane varieties in South Africa. *Proc. S. Afr. Sugar Technol. Ass.* 12:6–20.

Elshire, R. J., J. C. Glaubitz, Q. Sun, J. A. Poland, K. Kawamoto, E. S. Buckler and S. E. Mitchell. 2011. A robust, simple genotyping-by-sequencing (GBS) approach for high-diversity species. *PLoS One* 6:e19379.

Garsmeur, O., C. Charron, S. Bocs, V. Jouffe, S. Samain, A. Couloux, G. Droc, C. Zini, J. C. Glaszmann, M. A. Van Sluys and A. D'Hont. 2011. High homologous gene conservation despite extreme autopolyploid redundancy in sugarcane. *New Phytol.* 189:629–42.

Glaszmann, J. C., A. Fautret, J. L. Noyer, P. Feldmann and C. Lanaud. 1989. Biochemical genetic markers in sugarcane. *Theor. Appl. Genet.* 78:537–43.

Glaszmann, J. C., Y. H. Lu and C. Lanaud. 1990. Variation of nuclear ribosomal DNA in sugarcane. *J. Genet. Breed.* 44:191–7.

Glynn, N. C., C. Laborde, R. W. Davidson, M. S. Irey, B. Glaz, A. D'Hont and J. C. Comstock. 2013. Utilization of a major brown rust resistance gene in sugarcane breeding. *Mol. Breed.* 31:323–31.

Gouy, M., Y. Rousselle, D. Bastianelli, P. Lecomte, L. Bonnal, D. Roques, J. C. Efile, S. Rocher, J. Daugrois, L. Toubi, S. Nabaneza, C. Hervouet, H. Telismart, M. Denis, A. Thong-Chane, J. C. Glaszmann, J. Y. Hoarau, S. Nibouche and L. Costet. 2013. Experimental assessment of the accuracy of genomic selection in sugarcane. *Theor. Appl. Genet.* 126:2575–86.

Gouy, M., Y. Rousselle, A. Thong-Chane, A. Anglade, S. Royaert, S. Nibouche and L. Costet. 2015. Genome wide association mapping of agro-morphological traits in sugarcane. *Euphytica* 202:269–84.

Gravois, K. A. and K. P. Bischoff. 2008. New sugarcane varieties to the rescue. *LSU AgCenter.* URL: http://www.lsuagcenter.com/portals/communications/publications/agmag/archive/2008/spring/new-sugarcane-varieties-to-the-rescue. Accessed 7 May 2017.

Grivet, L., A. D'Hont, D. Roques, P. Feldmann, C. Lanaud and J. C. Glaszmann. 1996. RFLP mapping in cultivated sugarcane (*Saccharum* spp.): Genome organization in a highly polyploid and aneuploid interspecific hybrid. *Genetics* 142:987–1000.

Hayes, B. J., P. J. Bowman, A. J. Chamberlain and M. E. Goddard. 2009. Genomic selection in dairy cattle: Progress and challenges. *J. Dairy Sci.* 92:433–43.

Heinz, D. J. 1969. Isozyme prints for variety identification. *Sugarcane Breed. Newslett.* 24:8.

Hoarau, J. Y., B. Offman, A. D'Hont, A. M. Risterucci, D. Roques, J. C. Glaszmann and L. Grivet. 2001. Genetic dissection of a modern sugarcane cultivar (*Saccharum spp.*). I. Genome mapping with AFLP markers. *Theor. Appl. Genet.* 103:84–97.

Hoarau, J. Y., L. Grivet, B. Offmann, L. M. Raboin, J. P. Diorflar, J. Payet, M. Hellmann, A. D'Hont and J. C. Glaszmann. 2002. Genetic dissection of a modern sugarcane cultivar (*Saccharum spp.*). II. Detection of QTLs for yield components. *Theor. Appl. Genet.* 105:1027.

Hu, J. and B. A. Vick. 2003. Target region amplification polymorphism: A novel marker technique for plant genotyping. *Plant Mol. Biol. Rep.* 21:289–94.

Jackson, P. and X. Wei. 2017. Using DNA markers to speed up progress in sugarcane breeding. *Proc. Aust. Soc. Sugar Cane Technol.* 39:296–303.

Jannoo, N., L. Grivet, M. Seguin, F. Paulet, R. Domaingue, P. S. Rao, A. Dookum, A. D'Hont and J. C. Glaszmann. 1999. Molecular investigation of the genetic base of sugarcane cultivars. *Theor. Appl. Genet.* 99:171–84.

Jena, K. K. and D. J. Mackill. 2008. Molecular markers and their use in marker-assisted selection in rice. *Crop Sci.* 48:1266–76.

Jordan, D. R., R. E. Casu, P. Besse, B. C. Carroll, N. Berding, and C. L. McIntyre. 2004. Markers associated with stalk number and suckering in sugarcane collocate with tillering and rhizomatousness QTLs in sorghum. *Genome* 47:988–93.

Joshi, S. V. and E. H. Albertse. 2013. Development of a fingerprinting database and cultivar identification in sugarcane using a genetic analyzer. *Proc. S. Afr. Sugar Technol. Ass.* 86:200–12.

Lander, E. S. and D. Botstein. 1989. Mapping Mendelian factors underlying quantitative traits using RFLP linkage maps. *Genetics* 121:185–99.

Le Cunff, L., O. Garsmeur, L. M. Raboin, J. Pauquet, H. Telismart, A. Selvi, L. Grivet, R. Philippe, D. Begum, M. Deu, L. Costet, R. Wing, J. C. Glaszmann and A. D'Hont. 2008. Diploid/polyploid syntenic shuttle mapping and haplotype-specific chromosome walking toward a rust resistance gene (*Bru*1) in highly polyploid sugarcane (2n ~12x ~115). *Genetics* 180:649–60.

Li, W. F., X. Y. Wang, Y. K. Huang, R. Y. Zhang, H. L. Shan, J. Yin and Z. M. Luo. 2017. Molecular detection of *Bru*1 gene and identification of brown rust resistance in Chinese sugarcane germplasm. *Sugar Tech.* 19:183–90.

Lu, Y. H., A. D'Hont, D. I. T. Walker, P. S. Rao, P. Feldmann and J. C. Glaszmann. 1994. Relationships among ancestral species of sugarcane revealed with RFL using single copy maize nuclear probes. *Euphytica* 78:7–18.

Maccheroni, W., H. Jordão, R. Degaspari and S. Matsuoka. 2007. Development of a dependable microsatellite-based fingerprinting system for sugarcane. *Proc. Intl. Soc. Sugar Cane Technol.* 26:889–900.

McCord, P. H. 2016. Using droplet digital PCR (ddPCR) to detect copy number variation in sugarcane, a high-level polyploid. *Euphytica* 209:439–48.

McCord, P. H. and A. J. Migneault. 2016. Genotyping sugarcane for the brown rust resistance locus *Bru*1 using unlabeled probe melting. *Sugar Tech* 18:401–6.

McIntyre, C. L., R. E. Casu, J. Drenth, D. Knight, V. A. Whan, B. J. Croft, D. R. Jordan and J. M. Manners. 2005a. Resistance gene analogues in sugarcane and sorghum and their association with quantitative trait loci for rust resistance. *Genome* 48:391–400.

McIntyre, C. L., V. A. Whan, B. Croft, R. Magarey and G. R. Smith. 2005b. Identification and validation of molecular markers associated with Pachymetra root rot and brown rust resistance in sugarcane using map- and association-based approaches. *Mol. Breed.* 16:151–61.

Miar, Y., G. Plastow and Z. Wang. 2015. Genomic selection, a new era for pork quality improvement. *Springer Sci. Rev.* 3:27–37.

Michelmore, R. W., I. Paran and R. V. Kesseli. 1991. Identification of markers linked to disease-resistance genes by bulked segregant analysis; a rapid method to detect markers in specific genomic regions by using a segregating population. *Proc. Natl. Acad. Sci. USA* 88:9828–32.

Ming, R., S. C. Liu, P. H. Moore, J. E. Irvine and A. H. Paterson. 2001. QTL analysis in a complex autopolyploid: Genetic control of sugar content in sugarcane. *Genome Res.* 11:2075–84.

Ming, R., Y. W. Wang, X. Draye, P. H. Moore, J. E. Irvine and A. H. Paterson. 2002a. Molecular dissection of complex traits in autopolyploids: Mapping QTLs affecting sugar yield and related traits in sugarcane. *Theor. Appl. Genet.* 105:332–45.

Ming, R., T. A. Del Monte, E. Hernandez, P. H. Moore, J. E. Irvine and A. H. Paterson. 2002b. Comparative analysis of QTLs affecting plant height and flowering among closely-related diploid and polyploid genomes. *Genome* 45:794–803.

Mudge, J., W. R. Andersen, R. L. Kehrer and D. J. Fairbanks. 1996. A RAPD genetic map of *Saccharum officinarum*. *Crop Sci.* 36:1362–6.

Nayak, S. N., J. Song, A. Villa, B. Pathak, T. Ayala-Silva, X. Yang, J. Todd, N. C. Glynn, D. N. Kuhn, B. Glaz, R. A. Gilbert, J. C. Comstock and J. Wang. 2014. Promoting utilization of *Saccharum* spp. Genetic resources through genetic diversity analysis and core collection construction. *PLos One* 9:e110856.

Neuber, C., F. R., Camilo dos Santos, J. B. da Costa, M. Volpin, M. A Xavier, D. Perecin, R. C. V. Burbano, M. G. A. Landell and L. R. Pinto. 2017. Survey of the *Bru*1 gene for brown rust resistance in Brazilian local and basic sugarcane germplasm. *Plant Breed.* 136:182–7.

Pan, Y. B. 2010. Databasing molecular identities of sugarcane (*Saccharum* spp.) clones constructed with microsatellite (SSR) markers. *Am. J Plant Sci.* 1:87–94.

Parco, A. S., M. C. Avellaneda, A. H. Hale, J. W. Hoy, C. A. Kimbeng, M. J. Pontif, K. A. Gravois, N. Baisakh. 2014. Frequency and distribution of the brown rust resistance gene *Bru*1 and implications for the Louisiana sugarcane breeding programme. *Plant Breed.* 133:654–9.

Parmessur, Y., J. Arro, J. Joomun, R. Ming and A. Saumtally. 2014. Application of RADseq to the linkage mapping of sugarcane to identify markers linked to early ripening. In: *Plant and Animal Genome XXII*, San Diego, CA, 12 January 2014. URL: https://pag.confex.com/pag/xxii/webprogram/Paper9365.html. Accessed 10 April 2017.

Pastina, M. M., M. Malosetti, R. Gazaffi, M. Mollinari, G. R. A. Margarido, K. M. Oliveira, L. R. Pinto, A. P. Souza, F. A. van Euwijk and A. A. F. Garcia. 2012. A mixed model QTL analysis for sugarcane multiple-harvest-location trial data. *Theor. Appl. Genet.* 124:835–49.

Perera, M. F., M. E. Arias, D. Costilla, A. C. Luque, M. B. Garcia, C. D. Romero, J. Racedo, S. Ostengo, M. P. Filippone, M. I. Cuenya and A. P. Castagnaro. 2012. Genetic diversity assessment and genotype identification in sugarcane based on DNA markers and morphological traits. *Euphytica* 185:491–510.

Piperidis, G., A. R. Rattey, G. O. Taylor and M. C. Cox. 2004. DNA markers: A tool for identifying sugarcane varieties. *Proc. Aust. Soc. Sugar Cane Technol.* 26.

Poland, J. and J. Rutkoski. 2016. Advances and challenges in genomic selection for disease resistance. *Annu. Rev. Phytopathol.* 54:79–98.

Pritchard, J. K., M. Stephens and P. Donnelly. 2000. Inference of population structure using multilocus genotype data. *Genetics* 155:945–59.

Raboin, L. M., K. M. Oliveira, L. Le Cunff, H. Telismart, D. Roques, M. Butterfield, J. Y. Hoarau and A. D'Hont. 2006. Genetic mapping in sugarcane, a high polyploid, using bi-parental progeny: Identification of a gene controlling stalk colour and a new rust resistance gene. *Theor. Appl. Genet.* 112:1382–91.

Racedo, J., M. F. Perera, R. Bertani, C. Funes, V. González, M. I. Cuenya, A. D'Hont, B. Welin and A. P. Castagnaro. 2013. *Bru1* gene and potential alternative sources of resistance to sugarcane brown rust disease. *Euphytica* 191:429–36.

Resende, M. D. V. 2002. *Biometric Genetics and Statistics in the Breeding of Perennial Crops* (Portuguese). EMBRAPA Informação Tecnológica, Brasilia; EMBRAPA Florestas, Colombo, Paraná, Brazil.

Roach, B. T. 1989. Origin and improvement of the genetic base of sugarcane. *Proc. Aust. Soc. Sugar Cane Technol.* 11:34–47.

Rott, P., J. C. Comstock, R. A. Gilbert and H. S. Sandhu. 2014. Sugarcane eyespot disease. Institute of Food and Agricultural Sciences Extension bulletin SSAGR203, University of Florida. URL: http://edis.ifas.ufl.edu/pdffiles/SC/SC00300.pdf. Accessed 9 March 2017.

Singh, R. K., S. P. Singh, D. K. Tiwari, S. Srivastava, S. B. Singh, M. L. Sharma, R. Singh, T. Mohapatra and N. K. Singh. 2013. Genetic mapping and QTL analysis for sugar yield-related traits in sugarcane. *Euphytica* 191:333–53.

Sreenivasan, T. V., B. S. Ahloowalia and D. J. Heinz. 1987. Cytogenetics, pp. 211–53. In: D. J. Heinz (Ed.), *Sugarcane Improvement through Breeding*. Elsevier, Amsterdam.

Vos, P., R. Hogers, M. Bleeker, M. Reijans, T. van de Lee, M. Hornes, A. Frijters, J. Pot, J. Peleman, M. Kuiper and M. Zabeau. 1995. AFLP: A new technique for DNA fingerprinting. *Nucl. Acids Res.* 23:4407–14.

Wei, X. and P. A. Jackson. 2016. Addressing slow rates of long-term genetic gain in sugarcane. *Proc. Int. Soc. Sugar Cane Technol.* 29:1923–30.

Wei, X., P. A. Jackson, S. Hermann, A. Kilian, K. Heller-Uszynska and E. Deomano. 2010. Simultaneously accounting for population structure, genotype by environment interaction, and spatial variation in marker-trait associations in sugarcane. *Genome* 53:973–81.

Whan, A., N. Robinson, P. Lakshmanan, S. Schmidt and K. Aitken. 2010. A quantitative genetics approach to nitrogen use efficiency in sugarcane. *Func. Plant Biol.* 37:448–54.

Zeng, Z. B. 1993. Theoretical basis for separation of multiple linked gene effects in mapping quantitative trait loci. *Proc. Natl. Acad. Sci. USA* 90:10972–6.

Chapter 6

Genetic improvement of sugarcane by transgenic, intragenic and genome editing technologies

Fredy Altpeter and Ratna Karan, University of Florida, USA

1 Introduction

Sugarcane (*Saccharum* spp. hybrids) is a C4 grass grown in tropical and subtropical regions of the world and contributes to 80% of the world's raw sugar production. Sugarcane is also an important crop for commercial production of biofuel due to its exceptional biomass production and accumulation of large amount of sucrose in the stem internodes. Sugarcane is considered by the US Environmental Protection Agency as a feedstock for production of advanced biofuel due to its superior contribution to reduce the life cycle greenhouse gas production in the fight against global warming and climate change (https://www.epa.gov/renewable-fuel-standard-program/renewable-fuel-annual-standards). Cultivated sugarcane is a highly polyploid (2n = 100–130) interspecific hybrid between *Saccharum officinarum* and *S. spontaneum* with contributions from *S. robustum*, *S. sinense*, *S. barberi* and related grass genera such as *Erianthus*, *Miscanthus* and *Narenga* (Brandes 1958; Purseglove 1972; Daniels and Roach 1987). In modern sugarcane cultivars, the majority of chromosomes are derived from *S. officinarum*, approximately 10% of the chromosomes are from *S. spontaneum* and 10% are recombinant chromosomes between the two species (D'Hont et al. 1996). Sugarcane breeding has developed cultivars with high sugar accumulation, higher yields, disease resistance and improved ratooning ability (Ming et al. 2006). However, due to its narrow genetic base, low

http://dx.doi.org/10.19103/AS.2017.0035.07

fertility and the complexity of the sugarcane genome, it is very challenging to generate advanced sugarcane cultivars that are highly productive under changing environmental and biotic stress conditions (Arruda 2012). Biotechnology including tissue culture, transgenic, intragenic and genome editing approaches will accelerate sugarcane improvement and cultivar development to fulfil the growing demand for renewable fuel and food. The commercial launch of transgenic, insect-resistant sugarcane by Brazil' Centro de Tecnologia Canavieira (CTC) is expected in the next months and will boost the interest in additional transgenic traits for sugarcane improvement. Sugarcane is also a prime target for the production of value-added products via synthetic biology due to its exceptional biomass production and sugar accumulation. Sugarcane biotechnology is enabling this rapidly growing research field which will fuel the emerging bio-economy (reviewed by Baltes and Voytas 2015). Advances in *in vitro* techniques will accelerate the development of desired sugarcane genotypes (Lakshmanan et al. 2005; Taparia et al. 2012a,b) and will continue to provide disease-free clonal propagules for commercial sugarcane production (Snyman et al. 2011). A number of traits were successfully introduced into sugarcane by genetic transformation and tissue culture techniques (reviewed by Lakshmanan et al. 2005; Altpeter and Oraby 2010). Vegetative propagation of sugarcane prevents the segregation of different genotypes that is common in sexual progenies of heterozygous crops. The lack of segregation in vegetative progenies makes sugarcane also an ideal crop for gene stacking by retransformation or co-transformation of several un-linked target genes as typically required for metabolic engineering. Vegetative propagation and crop sterility increase the biosafety of GM sugarcane and minimize the potential of gene flow to related species or nearby crops. Use of biotechnology approaches that may not require or facilitate regulatory approval such as cisgenic and intragenic will accelerate commercialization of biotech products (Schouten et al. 2006; Rommens et al. 2007; Dermawan et al. 2016). However, the technologies still encompass random insertion of genes in the genome of the target crop, causing event-to-event variation in target gene expression (Carmona et al. 2005; Dermawan et al. 2016). Genome editing techniques with designer nucleases such as meganuclease, zinc finger nucleases (ZFNs), transcription activator–like effector nucleases (TALEN) and CRISPR (clustered regularly interspaced short palindromic repeats)-Cas9 (CRISPR-associated) allow the targeting of precise locations within the genome for insertion of target genes, precision nucleotide substitutions for allele replacement or targeted mutagenesis to create desired loss of function alleles (Belhaj et al. 2013; Voytas 2013; Voytas and Gao 2014; Jung and Altpeter 2016a; Kannan et al. 2018). Ongoing efforts in improving crop transformation and genome editing technologies (Altpeter et al. 2016) will empower a genome- to systems-level understanding of plant–environment interactions and accelerate crop improvement and synthetic biology (Shih et al. 2016). Sugarcane processing mills resemble biorefineries and are located in close proximity to sugarcane production fields which minimizes costs associated with feedstock transportation and addition of value streams. Sucrose accumulation in the stems of sugarcane can be redirected for the production of value-added compounds to fuel the emerging bio-based economy.

2 Tissue culture, gene transfer and selectable marker systems

2.1 Tissue culture

Tissue culture is used for micropropagation of elite clones, production of disease-free clones, conservation of useful germplasm as well as production of transgenic, intragenic, genome-edited or somaclonal variants with desirable attributes. *In vitro* cultures of sugarcane are initiated from meristematic explants such as apical meristem, immature leaves, immature inflorescence or axillary buds and are regenerated into plantlets via somatic embryogenesis or organogenesis. The first report of *in vitro* culture of sugarcane dates back to 1964 (Nickell 1964), followed by reports on plant regeneration from callus (Barba and Nickell 1969; Heinz and Mee 1969). Embryogenic callus was induced from immature leaves (Nadar et al. 1978; Ho and Vasil 1983), apical meristems (Ahloowalia and Maretzki 1983) or immature inflorescence (Liu 1993). Growth regulators, mainly auxins and cytokinins, regulate the spatial and temporal expressions of genes to control the genetic programme of the somatic cells to transition between embryo developmental stages (Feher 2015). Different factors have been reported to affect the morphogenic potential of sugarcane explants. Generally auxins like 2,4-dichlorophenoxyacetic acid (2,4-D) or picloram are used to induce embryogenic callus. Shoot regeneration occurs by transferring callus to media without auxin (Ahloowalia and Maretzki 1983) or with a weak auxin (e.g. indole acetic acid) or a low concentration of α-naphthalene acetic acid and a cytokinin-like 6-benzyl amino purine (BAP) or kinetin (Larkin 1982; Irvine et al. 1991). Including BAP in the regeneration media increased the number of shoot formations from sugarcane tissues (Gambley et al. 1994). Media supplementation with thidiazuron resulted in faster shoot regeneration compared to BAP, kinetin or zeatin (Gallo-Meagher et al. 2000; Chengalrayan and Gallo-Meagher 2001). Orientation of the explant on culture media, size and developmental stage of the explant, the concentration and type of growth regulators and illumination of cultures are reported to determine the regeneration potential and regeneration pathway of the explant including direct or indirect organogenesis or embryogenesis (Lakshmanan et al. 2005; Garcia et al. 2007; Snyman et al. 2011). Aftab and Iqbal (1999) described direct initiation of embryos on the cut edges of immature leaves when cultured on media with 2,4-D and casein hydrolysate in the light. Decreasing the 2,4-D concentration from 3 to 1.5 mg L^{-1} reduced the occurrence of mutated plants following indirect embryogenesis (Basnayake et al. 2011). Plants produced through direct embryogenesis or direct organogenesis minimize the chance of somaclonal variation due to the absence of a callus phase. Transposon activation in callus occurs due to the dynamic changes in plant chromatin by altering the DNA methylation pattern (De-Araujo et al. 2005; Singer et al. 2001; Schellenbaum et al. 2008). Genetic mechanisms responsible for controlling the epigenetic barrier to convert somatic cells into totipotent cells are still being explored (De-la-Pena et al. 2015). Direct organogenesis is the preferred approach for micropropagation of sugarcane and typically results in a very low frequency of somaclonal variation (Snyman et al. 2011). However, direct embryogenesis is preferred for minimizing somaclonal variation and avoiding chimerism after genetic transformation (Snyman et al. 2006; Taparia et al. 2012a), as the tissue culture period is very short and the somatic embryos are unicellular in origin.

2.2 Gene transfer and selectable marker systems

Bower and Birch (1992) reported the first stably transformed sugarcane plants, following biolistic transfer of constitutively expressed *npt*II into embryogenic callus and geneticin selection. Transgenic sugarcane plants were also produced by *Agrobacterium* (Arencibia et al. 1998; Enriquez-Obregon et al. 1998) or electroporation-mediated transformation (Arencibia et al. 1995; 1999). Biolistic gene transfer into callus is the most frequently used method for sugarcane transformation (Lakshmanan et al. 2005; Altpeter and Oraby 2010; Birch 2014). Several parameters including direct embryogenesis (Snyman et al.

2006), reducing the DNA amount and using minimal expression cassettes without vector backbone were optimized for generating single copy events with improved efficiency and transgene performance (Taparia et al. 2012a). A direct comparison of *Agrobacterium*-mediated and improved biolistic gene transfer revealed similar transformation efficiency, numbers of single copy events, transgene expression level and stability and better field performance of lines derived from biolistic gene transfer (Jackson et al. 2013; Joyce et al. 2014; Wu et al. 2015).

Alternative selectable marker genes and selection protocols were previously reviewed by Lakshmanan et al. (2005), Altpeter and Oraby (2010) and Birch (2014). The neomycin phosphotransferase (*npt*II, Bower and Birch 1992) is the most frequently used selectable marker gene followed by phosphinothricin acetyltransferase (*bar/pat*, Gallo-Meagher and Irvine 1996; Manickavasagam et al. 2004), phosphomannose isomerase (*manA*, Jain et al. 2007), acetolactate synthase (*als*, Van der Vyver et al. 2013; Dermawan et al. 2016) and hygromycin phosphotransferase (*hpt*, Arencibia et al. 1998) for sugarcane transformation. Including a selective agent during the entire regeneration process but reducing the concentration of the selective agent (e.g. geneticin) during this phase reduces the occurrence of chimerism while improving transformation rates of recalcitrant genotypes (Bower et al. 1996; Basnayake et al. 2011).

3 Transgene expression for the improvement of agronomic traits

Commercialization of transgenic sugarcane requires events with stable expression of transgenes which are uniformly transmitted to vegetative progenies without compromising agronomic performance (Dong et al. 2014). Gene silencing, potentially compromising stable transgene expression in sugarcane, was comprehensively reviewed by Birch (2014). Most of the transgenic sugarcane plants that were evaluated under field conditions were developed by biolistic gene transfer (Snyman et al. 1998; Leibbrandt and Snyman 2003; Gilbert et al. 2005; Gilbert et al. 2009; Hamerli and Birch 2011; Weng et al. 2011; Jung et al. 2013; Noguera et al. 2015; Zhou et al. 2016) and few used *Agrobacterium*-mediated gene transfer (Enríquez-Obregón et al. 1998; Joyce et al. 2014). Interestingly, a greater percentage of events derived by biolistics were similar to the parental cultivar for growth and yield than events derived from *Agrobacterium*-mediated gene transfer (Joyce et al. 2014).

Weed management is important for reaching the high yield potential of sugarcane. A number of studies focused on the development of herbicide-resistant sugarcane. Gallo-Meagher and Irvine (1996) established the use of the *bar* gene as an effective selectable marker for biolistic sugarcane transformation which conferred resistance to bialaphos and the herbicide Ignite. Stability of herbicide resistance was confirmed under greenhouse conditions in vegetative progenies. Enríquez-Obregón et al. (1998) developed *bar* -gene-containing transgenic sugarcane lines by *Agrobacterium*-mediated transformation and confirmed their resistance to the herbicide Basta under greenhouse and small-plot conditions. Snyman et al. (1998) produced transgenic sugarcane by biolistic transformation of the *bar* gene and chose one herbicide-resistant clone for field evaluation. A small-scale field trial using vegetatively propagated clones displayed the herbicide resistance in transgenic sugarcane without any phytotoxic symptoms even after repeated application of Buster (7 L ha^{-1}) on plant cane and two ratoons. A detailed evaluation of the agronomic performance of herbicide-tolerant sugarcane derived from biolistic gene transfer was described by Leibbrandt and Snyman (2003). One transgenic line harbouring nine copies of the *bar* gene was selected for a replicated field trial following the stable expression of the transgene and herbicide tolerance in vegetative progenies. Morphological and agronomic characteristics, such as stalk morphology, fibre content, smut disease resistance and yield, measured in the first ratoon did not significantly differ between transgenic and non-transgenic sugarcane under different weed control conditions. This study revealed that even high transgene copy numbers can lead to good transgene and agronomic performance. Noguera et al. (2015) produced glyphosate-tolerant sugarcane lines by biolistic gene transfer of the *epsps* gene into callus. While all of the transgenic lines maintained stable expression of herbicide resistance under field conditions, many displayed phenotypic aberrations. However a few lines were also identified that were almost indistinguishable from wild type regarding agronomic and industrial characteristics and composition. As callus culture may induce somaclonal variation in transgenic sugarcane, the authors suggest generation of at least 30–40 events with this approach. This will increase the probability of recovering a few transformed lines with similar growth and production characteristics as the parental line. Acetolactate synthase (ALS) is a key enzyme in the biosynthetic pathway of branched-chain amino acids, and also a target for ALS-inhibiting herbicides including imidazolinones and sulfonylureas (Yu and Powles 2014). Van der Vywer et al. (2013) utilized a mutated *als* gene from tobacco under control of the maize ubiquitin promoter and the *Agrobacterium* nos 3'UTR as the selectable marker with chlorsulfuron as the selective agent. Regenerated plants displayed resistance to the herbicide chlorsulfuron under greenhouse conditions.

Diseases caused by viruses, fungi, bacteria, phytoplasmas or nematodes are major contributors to sugarcane yield losses worldwide (Goncalves et al. 2012). Sugarcane producers prefer to control diseases with resistant varieties to minimize costs and environmental impacts of pesticide applications. Therefore, generating disease-resistant cultivars is a very important component of a sugarcane cultivar development programme (Rott et al. 2013). Viruses are one of the world's most important sugarcane diseases with *Sugarcane mosaic virus* (SCMV) and *Sugarcane yellow leaf virus* (SCYLV) being most devastating (Goncalves et al. 2012). Resistance against SCMV, SCYLV or *Sorghum mosaic virus* (SrMV) was achieved by co-suppression, antisense or RNAi suppression of the coat proteins and was confirmed under greenhouse (Zhu et al. 2011; Guo et al. 2015) and/or field conditions (Ingelbrecht et al. 1999; Butterfield et al. 2002; Gilbert et al. 2005, 2009). Ingelbrecht et al. (1999) reported immunity to SrMV strain SCH, following a post-transcriptional gene silencing approach that was based on the co-suppression of the untranslatable coat protein of this strain. However, virus resistance requires a high degree of sequence similarity between the transgene and the virus's endogenous coat protein gene. Therefore, susceptibility was observed to a distantly related strain (Ingelbrecht et al. 1999). Segregation of transgenes and resistance to SrMV were also studied in sexual sugarcane progenies (Butterfield et al. 2002). While herbicide resistance conferred by the *bar* gene was transmitted to progeny, most of the transgenic progenies with the coat protein construct were susceptible to SrMV, which may be a consequence of the complex transgene integration into the original events. Several transgenic accessions with improved agronomic performance and SCMV resistance compared to commercial controls were identified in multi-year field testing of transgenic sugarcane lines with a constitutive expression construct of the untranslatable SCMV strain E coat protein gene

(Gilbert et al. 2005). However, extensive phenotypic variability was observed, leading to yield reduction in most transgenic sugarcane lines. Similarly, transgenic sugarcane with an antisense construct to the untranslatable coat protein of the SCYLV displayed resistance to the virus under replicated multi-year field conditions along with more than 30% reductions in cane and sugar yields, compared to commercial controls (Gilbert et al. 2009). In both cases transgenic lines were derived from biolistic gene transfer into callus (Gilbert et al. 2005, 2009). Shorter tissue culture procedures and optimized biolistic gene transfer parameters are now available for reducing the occurrence of undesired somaclonal mutations (Taparia et al. 2012a,b; Wu et al. 2015; Jung et al. 2016b; Kannan et al. 2018).

Leaf scald in sugarcane is a devastating disease caused by the bacterium *Xanthomonas albilineans*, which systemically infects the xylem vessels and produces low molecular weight toxins known as 'albicidins' (Birch and Patil 1987). The albicidin detoxification gene '*albD*' was cloned from *Pantoea dispersa* (Zhang and Birch 1996), and constitutively expressed in sugarcane (Zhang et al. 1999). Transgenic sugarcane plants expressing '*albD*' gene did not develop leaf scald disease symptoms, and lines with high AlbD enzyme activity also protected the plants against systemic multiplication of the pathogen (Zhang et al. 1999).

Sugarcane is known to host more than 1500 *insects*, but the vast majority have restricted geographic distributions. The new bio-economy may result in expansion of sugarcane production areas and will require an integrated approach to respond to emerging pest problems (Goebel and Sallam 2011). Biotechnology-derived insect resistance will likely play a major role in future pest management strategies. Insect resistance in sugarcane was achieved by overexpression of proteinase inhibitor or *Bacillus thuringiensis* delta-endotoxin (BT) genes and was confirmed under greenhouse (Arencibia et al. 1997; Weng et al. 2006; Kalunke et al. 2009; Arvinth et al. 2010) and/or field conditions (Arencibia et al. 1999; Falco and Silva-Filho 2003; Weng et al. 2011; Gao et al. 2016). In an effort to control dead heart disease in sugarcane, the light-regulated phosphoenolpyruvate carboxylase promoter was used to express BT *cry1Ab* gene in the leaf of stably transformed plants. The transgenic plants expressing BT *cry1Ab* killed the *Scirpophaga nivella* larvae within one hour of its feeding to the plants. Accumulation of endotoxin under the light-regulated promoter in different tissues of field grown transgenic sugarcane varied greatly and was 39% higher in young leaves than in mature leaves, while BT was not detected in the cane juice (Khan et al. 2011). Stem borer–resistant sugarcane lines were reported from field trials following the constitutive expression of a codon optimized *cry1Ac* under constitutive promoters (Weng et al. 2011; Gao et al. 2016). Several events derived from biolistic gene transfer into callus performed as well as untransformed controls in respect of agronomic and industrial traits (Weng et al. 2011; Gao et al. 2016).

Abiotic stress tolerance is of increasing importance due to climate change and resulting variability in weather patterns. Experiments under controlled environmental conditions have evaluated a number of approaches to elevate drought tolerance in sugarcane but still need to be followed up with field testing for validation. Stress-inducible expression of proline in sugarcane improved water-deficit tolerance (Molinari et al. 2007), as well as salinity stress tolerance (Guerzoni et al. 2014) by protecting the photosynthetic apparatus and preventing oxidative damage caused by the imposed stresses. Expression of the transcription factor *AtDREB2A* under stress-inducible Rab17 promoter from maize in transgenic sugarcane led to the up-regulation of genes involved in plant response to drought stress. Under drought stress, transgenic plants displayed a higher relative water content, leaf water potential, higher photosynthetic rates and sucrose accumulation (Reis et al. 2014). Constitutive expression of the arabidopsis vacuolar pyrophosphatase

(*AVP1*) gene, *DREB2* or *heat-shock protein 70* from *Erianthus arundinaceus* also enhanced drought and salinity tolerance, and stacking of *DREB2* with constitutively expressed pea *DNA helicase 45* further enhanced salinity tolerance (Augustine et al. 2015a,b,c,d; Kumar et al. 2014). Constitutive expression of a cell death suppressor gene from Arabidopsis (*AtBI-1*) increased long-term water-stress tolerance in transgenic lines. However, constitutive *AtBI-1* expression increased susceptibility of transgenic sugarcane plants to brown rust and reduced biomass accumulation, suggesting the use of a drought-inducible promoter (Ramiro et al. 2016).

Optimizing nutrient uptake has also great potential for elevating the productivity of sugarcane. Three components of the signalling pathway for potassium uptake from *Arabidopsis thaliana 'AtCBL9, AtCIPK23'* and *'AtAKT1'* were co-overexpressed in sugarcane (*Saccharum* spp.) and resulted in 31–35% elevated potassium content in transgenic plants under potassium deficiency conditions, compared to untransformed controls (Li et al. 2014). Sugarcane transformed with a barley alanine aminotransferase gene under transcriptional control of the rice OsAnt1 promoter displayed improved nitrogen use efficiency and biomass accumulation under limited (0.4 mM) nitrogen conditions in a greenhouse trial (Snyman et al. 2015).

4 Case study: transgenic, intragenic and genome editing approaches to improve bioethanol production from sugarcane

The first generation process for ethanol production utilizes the sucrose which accumulates in the sugarcane stems as source for microbial fermentation. A second generation process is required for converting the abundant amount of cell wall bound sugars present in the lignocellulosic sugarcane biomass residues (bagasse, straw and tops) to ethanol. Some of the commercial ethanol production facilities in Brazil have recently integrated the first- and second-generation technologies for commercial ethanol production (Dias et al. 2014). This is expected to elevate both the sustainability of the process and the biofuel yield per ton of harvested sugarcane biomass (Dias et al. 2014).

4.1 Transgenic approaches

Lignin is a recalcitrance factor for cell wall saccharification (Jørgensen et al. 2007). RNAi suppression of the lignin biosynthetic caffeic acid O-methyltransferase (*COMT*) gene by 67–97% reduced the lignin content by 3.9–13.7%, respectively. The ratio of the lignin monomers syringyl to guaiacyl was reduced from 1.5 in the wild type to values ranging between 1.3 and 0.8 (Jung et al. 2012). Selected *COMT*-RNAi lines evaluated in the field maintained these *COMT* and lignin suppression levels. Moderate reduction in total lignin (6%) improved saccharification efficiency up to 23% without significant difference in biomass yield, total structural carbohydrates or Brix value in comparison to tissue culture–derived or transgenic control plants (Jung et al. 2013). Overexpression of the maize transcription factor ZmMYB42 reduced lignin content by 8–21% and improved saccharification efficiency by up to 43% while displaying similar phenotypes as the transgenic control under greenhouse conditions (Poovaiah et al., 2016).

In planta production of enzymes that hydrolyse lignocellulosic substrates to fermentable sugars can contribute to making cellulosic ethanol a cost-competitive liquid transport fuel (Jorgensen et al. 2007; Harrison et al. 2011; Kim et al. 2017). Harrison et al. (2011) described the accumulation of cellobiohydrolases (CBH I and CBH II) and endoglucanase (EG) in transgenic sugarcane under the control of leaf-specific maize *PepC* promoter and different subcellular targeting signals. EG accumulated to highest levels (up to 0.05% of total soluble protein) with chloroplast targeting (Harrison et al. 2011). Greenhouse-grown mature ratoon events produced equivalent or elevated levels of active CBH I, CBH II and EG compared to primary transgenics (Harrison et al. 2014). Recently, Kim et al. (2017) reported accumulation of the hyperthermostable GH10 xylanase *Xyl10B* in transgenic sugarcane at 1.2% of the total soluble protein using the constitutive maize ubiquitin promoter and subcellular targeting to the chloroplast. Tissues retained most of the xylanase activity after being dried at 35°C and being stored for two months. Xyl10B-producing sugarcane displayed growth and development similar to the non-transformed control under greenhouse conditions. This may be due to the fact that the hyperthermostable enzyme has only residual catalytic activity at temperatures below 70°C. Thin-layer chromatography and high-performance liquid chromatography analyses of hydrolysis products confirmed the superior catalytic activity and thermostability (100°C) of the *in planta*–produced Xyl10B.

4.2 Intragenic approaches

Despite the tremendous benefits of transgenic crops for the grower, consumer and the environment (Brookes and Barfoot 2016a,b; Sainger et al. 2016), the effort, time and cost for regulatory approval have contributed to a dramatic contraction of crop biotechnology innovations (Graff et al. 2009; Fedoroff 2015). The costs of meeting regulatory requirements when bringing one transgenic crop event to market were estimated at $35 100 000 in 2011 (McDougall 2011). Exclusive use of gene components from the target crop or a sexually compatible species should facilitate regulatory approval of biotech crops and may increase the public acceptance of this technology. A cisgenic approach was first discussed by Schouten et al. (2006), which uses the gene and its associated regulatory elements (promoter and 3′UTR) from the host plant itself without rearrangements of individual components. However, an intragenic approach also includes gene components from sexually compatible species as well as recombination of different gene components (Rommens et al. 2007). Development of a selectable marker/selection system for the efficient recovery of cis or intragenic events is a major challenge for this approach. Interestingly, specific mutations in the *als* gene of plants are conferring resistance to sulfonylurea and imidazolinone herbicides (Yu and Powles 2014). Therefore, such mutations can provide the basis of a selectable marker system for the cisgenic or intragenic approach. Dermawan et al. (2016) established an intragenic selection protocol for sugarcane by using a mutated *als* gene as an intragenic selectable marker for sugarcane. The mutated *als* gene and all regulatory sequences such as promoter and 3′ UTR were derived from sexually compatible sorghum and were delivered by biolistics as a minimal expression cassette without vector backbone. Regenerated plants were resistant to the herbicide chlorsulfuron under greenhouse conditions. This selection approach was then combined with an intragenic approach to suppress the lignin biosynthetic 4-coumarate:coenzyme A ligase (*4CL*) gene by RNAi using minimal expression cassettes with sequences of all components from sorghum or sugarcane

(Jung et al. 2016b). RNAi suppression of the *4-CL-1* gene in field-grown intragenic sugarcane lines resulted in up to 16.5% reduction in total lignin content along with altered monolignol ratios compared to non-transformed controls. The saccharification efficiency of mature lignocellulosic biomass was improved by 52% in intragenic sugarcane without compromising biomass accumulation in comparison to the non-transformed commercial cultivar. This intragenic approach is expected to enhance both the profitability and sustainability of bioethanol production from sugarcane.

4.3 Genome editing approaches

Genome editing is a transformative technology for crop improvement (Voytas and Gao 2014) and is expected to meet greater public and regulatory acceptance than transgenic approaches (Wolt et al. 2016). Sequence-specific nucleases (SSN) such as CRISPR/Cas9, TAL effector nucleases, ZFNs and meganucleases can be designed to precisely target one specific DNA sequence within a genome, creating a DNA double-strand break (DSB). Repair of such DSBs in the absence of a repair template typically undergoes the erroneous non-homologous end joining (NHEJ) pathway, resulting in gene knockouts due to frame shift mutations. Alternatively, precise gene replacement can occur by homology directed repair (HDR) if homologous DNA fragments are available. Targeted loss-of-function mutations have resulted in crop improvement including disease resistance in rice and wheat, oil quality in soybean, and storage and processing quality of potato (reviewed by Baltes and Voytas 2015; Weeks et al. 2016). Creation of loss-of-function mutations with SSN in sugarcane is very challenging due to the sequence variation of functionally redundant target genes in the highly polyploid sugarcane genome. The first report on successful genome editing of sugarcane used a TALEN pair targeted to a highly conserved region of the *caffeic acid O-methyltransferase* (*COMT*) gene to reduce lignin content in sugarcane (Jung and Altpeter 2016a). Targeted *COMT* mutations were identified in up to 74% of the lines with stable integration of the TALEN pair. Capillary electrophoresis and high throughput sequencing of the *COMT* polymerase chain reaction (PCR) amplicon indicated that the mutation frequency in different events ranged from 8 to 99% of the *COMT* copies/alleles. Mutations were stably transmitted to vegetative progenies and events with 99% mutation frequency displayed 29–32% reduction of lignin along with significant reduction of syringyl subunit content. Sequencing of long *COMT* amplicons recently revealed that more than 100 copies/alleles were mutated, resulting in drastically improved saccharification efficiency without compromising agronomic performance under field conditions (Kannan et al. 2018). These improved cell wall characteristics compare favourably with the lignin suppression following RNAi of *COMT* and should increase the ethanol yield from sugarcane (Jung et al. 2013; Mohan 2016).

In addition to the targeted 'loss of function' mutagenesis described earlier, precision genome editing is also capable of generating 'gain of function' mutations by precise nucleotide substitutions (Baltes and Voytas 2015). This process is also known as gene targeting (GT) and involves template-mediated and HDR of targeted DNA DSBs. HDR occurs at much lower frequencies than the NHEJ process. NHEJ requires no template and typically is associated with small nucleotide deletions and insertions causing 'loss of function' by shifting the open reading frame of the exon. For efficient recovery of the rare GT events it is desirable to create a phenotype that is selectable at the cellular level in addition to providing benefits for crop production. Van der Vywer et al. (2013) and Dermawan et al. (2016) previously demonstrated with transgenic or intragenic approaches that a mutated *als* gene provides an effective selectable marker for sugarcane as well as a herbicide resistance trait. In contrast to these transgenic and intragenic approaches Oz et al. (2018) recently targeted the endogenous sugarcane ALS gene for precision genome editing using CRISPR-Cas9 to direct DBSs specifically to the ALS gene in addition to a template that introduced one or two herbicide resistance conferring point mutations. Sequencing of the PCR amplicons and herbicide application revealed that the targeted nucleotide substitutions were incorporated precisely and conferred herbicide resistance. This selectable GT technology will now support multiplexed GT and should replace transgenes as selectable markers.

5 Metabolic engineering to elevate sucrose yields and produce alternative sugars or value-added biomaterials

5.1 Sucrose and sucrose isomers

Sucrose accumulation in its stem parenchyma cell vacuoles and the apoplast surrounding these cells (Welbaum and Meinzer 1990; Uys et al. 2007) together with its high biomass yield make sugarcane an attractive target for metabolic engineering to either elevate sucrose yields or sustainably produce value-added biomaterials. Promoters which direct expression preferentially to mature stem are valuable tools to minimize pleiotropic effects in sugarcane (Mudge et al. 2013).

Biotech approaches for *increasing sucrose yield* focus on improved photosynthetic efficiency (reviewed by Ort et al. 2015), transport or metabolic processes enhancing sink strength (reviewed by Wang et al. 2013). Overexpression or suppression of genes encoding sucrose synthesizing or metabolizing enzymes including sucrose phosphate synthase, soluble acid or neutral invertase or pyrophosphate: fructose 6-phosphate 1-phosphotransferase did so far not contribute to elevated sucrose yields in sugarcane (Vickers et al. 2005; Botha et al. 2001, Groenewald and Botha 2008; Rossouw et al. 2010). The rapid emergence of a systems-level understanding of sucrose accumulation and flux control mechanisms is informing future strategies to improve carbon partitioning in the stem of sugarcane (reviewed by Wang et al. 2013; Shih et al. 2016).

Trehalulose (TH) is a sucrose isomer which has higher solubility and acid stability than sucrose and is therefore preferred for the preparation of jam and other highly concentrated sugar solutions. A TH synthase gene from '*Pseudomonas mesoacidophila* MX-45' was expressed in sugarcane under constitutive maize ubiquitin (Ubi-1) promoter and vacuolar targeting signal. TH synthase expressing sugarcane lines accumulated up to 347 mM TH in whole-cane juice, comprising more than 40% of total sugars. The transgenic lines displayed reduced stem diameter and vigour in the first vegetative field generation following regeneration from callus and greenhouse propagation. They grew more vigorously when replanted under field conditions, but the experimental design did not allow to quantify biomass yield reliably (Hamerli and Birch 2011).

Isomaltulose is a sucrose isomer with beneficial properties including lower glycaemic index and low cariogenicity. Sugarcane constitutively expressing bacterial sucrose isomerase with vacuolar targeting signal accumulated isomaltulose at up to 50% of the total sugar without reducing the sucrose content, resulting in doubled total sugar concentrations in harvested juice of greenhouse-grown plants. Characterization of the isomaltulose accumulating lines suggested increased photosynthesis, sucrose transport and sink strength (Wu and Birch 2007). However, in contrast to those greenhouse results, total sugar levels were not increased and isomaltulose concentrations were lower when these lines and elite sugarcane cultivars expressing the same construct were grown under field conditions (Basnayake et al. 2012). The vast majority of the transgenic lines derived from callus cultures showed reduced vigour and/or stem diameter. However, two lines with biomass yields resembling the non-transformed control and isomaltulose accumulation up to 33% of the total sugar content were also identified (Basnayake et al. 2012). Expressing a vacuole-targeted, silencing-resistant version of the sucrose isomerase under a mature stem–specific promoter from sugarcane resulted in the accumulation of isomaltulose of up to 81% of the total sugar content without compromising germination of setts or biomass accumulation (Mudge et al. 2013). This study clearly demonstrated the potential of sugarcane for the efficient production of value-added materials. It also suggests that using a mature stem–specific promoter for metabolic engineering may minimize or prevent any detrimental effects of the transgene expression on agronomic performance.

5.2 Promoters for stem-specific expression in sugarcane

Promoters contain many regulatory cis-elements that allow binding of specific transcription activators or repressors to influence tissue-specific or developmentally regulated gene expression. Most of the studies with transgenic sugarcane involve constitutive or leaf-specific promoters. However, few promoters which direct expression preferentially to the stem have been explored. Promoters directing expression to the parenchyma tissues of maturing stems are of particular interest for biotechnology approaches aimed at enhancing sucrose content or accumulating value-added biomaterials during mature stem development (Moyle and Birch 2013a). Stem-specific promoters from defence and fibre biosynthesis-related genes may facilitate the expression of specific transgenes for enhanced stress responses or biomass composition (Damaj et al. 2010).

Promoters of two highly expressed genes in the vascular tissues of the sugarcane stem and involved in lignification *'DIRIGENT (SHDIR16)'* and *'O-METHYL- TRANSFERASE (SHOMT)'* were subcloned to *uidA* reporter gene for expression analysis (Damaj et al. 2010). Stem-preferred expression in the vascular tissues was directed by these constructs in sugarcane as well as in different monocots such as rice, maize and sorghum. GUS expression under control of *SHDIR16* and *SHOMT* promoters was 34 and 8 times higher in sugarcane stems than under the control of the maize *ubiquitin* promoter, respectively. Histochemical localization of GUS expression suggested that the expression of genes under control of these promoters is directed preferentially to the bundle sheath cells of sclerenchymatous tissues and in cells surrounding the phloem, xylem and protoxylem. GUS expression under these promoters was also induced in stem, root and leaf tissue by treatments with salicylic acid, jasmonic acid and methyl jasmonate, which mediate plant responses to biotic and abiotic stresses (Damaj et al. 2010).

Moyle and Birch (2013a) characterized promoters corresponding to two ESTs (CA081293 and CA260047) that were highly expressed in the maturing stem based on microarray data of Casu et al. (2004). Sequence analysis of corresponding BAC clones identified multiple alleles of *ScLSG* (*Loading Stem Gene*) with diversity in 5′UTR and promoter sequences (Moyle and Birch 2013a). Peak expression of seven *ScLSG* alleles was found in the sucrose loading zone of maturing stem. Within the stem, there was expression in parenchyma, vascular and rind tissues and two alleles had very low expression in non-stem tissues. Moyle and Birch (2013b) further characterized six promoters of *ScLSG* alleles along with their corresponding 5′UTR region by reporter gene activity following subcloning upstream of a silencing-resistant *luciferase* gene. Out of six *ScLSG* promoters analysed in stable transgenic sugarcane lines, four promoters displayed tissue selectivity with high expression of luciferase in the sucrose loading zone and throughout the mature stem in contrast to leaf and root expression.

5.3 Polyhydroxybutyrate

Polyhydroxybutyrate (PHB) is a biodegradable plastic synthesized by bacteria as a storage carbon. To utilize the potential of high biomass sugarcane for sustainable production of biodegradable plastic, Petrasovits et al. (2007), constitutively co-expressed three bacterial PHB biosynthesis genes (*phaA, phaB* and *phaC*) under the maize ubiquitin promoter. PHB accumulation in leaves was highest with subcellular targeting to the plastids [up to 1.88% of leaf dry weight (DW)] and did not reduce biomass. Lower PHB accumulation in stems compared to leaves was due to the difference in the abundance of different cell types, with the high PHB accumulating vascular cells being more abundant in leaves (Petrasovits et al. 2007). PHB accumulation in greenhouse-grown plants increased from 3 to 9 months and was highest in old leaf and lowest in young leaf. The maximum PHB accumulation was 1.77% of leaf DW without affecting plant height, total aerial biomass and sugar content compared to non-transformed controls (Purnell et al. 2007). To further increase the PHB content in sugarcane leaf, Petrasovits et al. (2012) tested four different promoters, and the maize chlorophyll A/B-binding protein promoter (*cab-m5*) facilitated the highest PHB accumulation up to 4.8% of leaf DW. However, the highest PHB-producing lines showed reduced biomass and mild chlorosis (Petrasovits et al. 2012) under greenhouse conditions. Acetyl-CoA is the initial substrate for PHB biosynthesis as well as for the competing fatty acid biosynthesis. Treatment of transgenic PHB-accumulating lines with a chemical inhibitor of acetyl-CoA carboxylase (ACCase) suppressed fatty acid biosynthesis and increased the PHB accumulation by fourfold in immature leaf tissues and by 50–60% if applied to mature plants before harvest (Petrasovits et al. 2013). Chloroplasts of sugarcane mesophyll cells produced a significantly lower level of PHB compared with bundle sheath cells (Petrasovits et al. 2007), thereby limiting the full PHB yield potential of the plant. To increase the PHB accumulation in mesophyll chloroplasts, the ectopic expression of β-ketothiolase was replaced with acetoacetyl-CoA synthase. This pathway engineering raised average PHB accumulation to 6.3% of leaf DW (McQualter et al. 2015). PHB accumulation of up to 1.21% of leaf DW is associated with normal plant development, while that of 1.3% and higher of leaf DW results in a progressive decrease in biomass and increased chlorosis (Petrasovits et al. 2012) or decreased carbohydrate content (McQualter et al. 2015). A systems biology approach and metabolic modelling proposed that the PHB polymer disrupts ATP supply in bundle sheath cells potentially by damaging thylakoid membranes or scattering photosynthetically active radiation (McQualter et al. 2016).

5.4 Triacylglycerol for biodiesel production

The use of renewable fuels such as ethanol from perennial feedstocks such as sugarcane significantly contributes to reduction in greenhouse gases. Plants store lipids as glycerol esters of fatty acids, also known as triacylglycerol (TAG). TAG is one of the most energy-rich forms of reduced carbon available in nature (Thelen and Ohlrogge 2002). The production of lipids in vegetative tissues of high biomass yielding crops has been proposed as a new strategy for increasing energy density and could significantly contribute to sustainable elevation of biodiesel yields (Ohlrogge and Chapman 2011; Huang et al. 2016). Recently, metabolic engineering of sugarcane for accumulation of TAG in vegetative biomass was reported (Zale et al. 2016). Analysis of transgenic sugarcane plants revealed the increase in TAG accumulation in leaf or stem by 1.9% or 0.9% of DW, respectively, by constitutive co-expression of *WRINKLED1*, *diacylglycerol acyltransferase1-2* and *oleosin* and simultaneous co-suppression of *ADP-glucose pyrophosphorylase* and a subunit of the *peroxisomal ABC transporter-1*. Lipid droplets of the mesophyll tissues from transgenic leaves were visible by confocal microscopy. Modelling predicts that biodiesel production from oil-accumulating sugarcane, even at TAG accumulation levels of 2% of DW, is cost-competitive to biodiesel production from soybean (Huang et al. 2016). Continued efforts in elevating lipid accumulation in sugarcane could contribute to large-scale replacement of fossil fuel–derived diesel without unrealistic demands on crop production area (Huang et al. 2016).

6 Future perspectives

Recent advancements in synthetic biology, intragenic and genome editing technologies are expected to accelerate genetic improvement of sugarcane. These technologies will not only advance crop improvement for conventional applications but also facilitate metabolic engineering contributing to new value streams to fuel the emerging bio-economy (Shih et al. 2016). Genome editing and intragenic technologies are expected to meet greater public and regulatory acceptance than transgenic approaches (Wolt et al. 2016). The large-scale application of these powerful technologies is critical for advancing sustainable agriculture and will depend on reducing regulatory costs associated with the commercial release of genetically improved sugarcane. Multiplexed genome editing will enable precision breeding by replacing the inferior alleles with superior alleles without the disruptive meiotic segregation. This will be of tremendous benefits for vegetatively propagated crops with complex polyploid genomes such as sugarcane. Generating elite events without a transgenic footprint will require genome editing protocols which prevent the traditional approach of integration of transgenes (Zhang et al. 2016).

7 Where to look for further information

7.1 A textbook for the novice

Stewart, C. N., Jr., 2016. *Plant Biotechnology and Genetics*, 2nd Edition. John Wiley & Sons Inc, ISBN13: 9781118820124; ISBN10: 1118820126.

7.2 Key societies and professional conferences worth visiting to keep up to date with trends

- Society for In Vitro Biology https://sivb.org/
- American Society of Plant Biologists http://aspb.org

- Crop Science Society of America https://www.crops.org/
- International Association of Plant Biotechnology http://iapb2018.com/
- In Vitro Biology Conference https://sivb.org/meetings/
- Plant Biology Congress http://plantbiology.aspb.org/
- International Plant and Animal Genome Conference http://www.intlpag.org/
- International Congress of Plant Molecular Biology http://www.ipmb2015.org/

7.3 Key journals

- *Molecular Plant* https://www.journals.elsevier.com/molecular-plant/
- *Plant Biotechnology Journal* http://onlinelibrary.wiley.com/journal/10.1111/(ISSN)1467-7652
- *Plant Molecular Biology* http://link.springer.com/journal/11103
- *The Plant Genome* https://dl.sciencesocieties.org/publications/tpg
- *Nature Plants* http://www.nature.com/nplants/
- *Plant Cell Reports* http://link.springer.com/journal/299
- *Plant Cell Tissue and Organ Culture* http://link.springer.com/journal/11240
- *Molecular Breeding* http://link.springer.com/journal/11032
- *Transgenic Research* http://link.springer.com/journal/11248
- *In Vitro Cellular and Developmental Biology* – Plant http://www.springer.com/life+sciences/plant+sciences/journal/11627

8 References

Aftab, F. and Iqbal, J., 1999. Plant regeneration from protoplasts derived from cell suspension of adventive somatic embryos in sugarcane (*Saccharum* spp. Hybrid cv. Col-54 and cv. CP-43/33). *Plant Cell Tissue and Organ Culture*, 56, 155–62.

Ahloowalia, B. S. and Maretzki, A., 1983. Plant regeneration via somatic embryogenesis in sugarcane. *Plant Cell Reports*, 2, 21–5.

Altpeter, F. and Oraby, H., 2010. Sugarcane. In: Kempken, F. and Jung, C. (Eds) *Genetic Modification of Plants*, Springer, Berlin, pp. 453–72.

Altpeter, F., Springer, N. M., Bartley, L. E., Blechl, A., Brutnell, T. P., Citovsky, V., Conrad, L., Gelvin, S. B., Jackson, D., Kausch, A. P., Lemaux, P. G., Medford, J. I., Orozo-Cardenas, M., Tricoli, D., VanEck, J., Voytas, D. F., Walbot, V., Wang, K., Zhang, Z. J. and Stewart, C. N., 2016. Advancing crop transformation in the era of genome editing. *The Plant Cell*, 28(7), 1510–20.

Arencibia, A., Molina, P., Riva, G. and Selman-Housein, G., 1995. Production of transgenic sugarcane (*Saccharum officinarum* L.) plants by intact cell electroporation. *Plant Cell Reports*, 14, 305–9.

Arencibia, A. D., Carmona, E. R., Téllez, P., Chan, M. T., Yu, S. M., Trujillo, L. E. and Oramas, P., 1998. An efficient protocol for sugarcane (*Saccharum* spp. L.) transformation mediated by *Agrobacterium tumefaciens*. *Transgenic Research*, 7(3), 213–22.

Arencibia, A. D., Carmona, E. R., Cornide, M. T., Castiglione, S., O'Relly, J., Chinea, A., Oramas, P. and Sala, F., 1999. Somaclonal variation in insect-resistant transgenic sugarcane (*Saccharum* hybrid) plants produced by cell electroporation. *Transgenic Research*, 8(5), 349–60.

Arencibia, A. D., Vazquez, R. I., Prieto, D., Tellez, P., Carmona, E. R., Coego, A., Hernandez, L., DelaRiva, G. A. and Selman, H. G., 1997. Transgenic sugarcane plants resistant to stem borer attack. *Molecular Breeding*, 3(4), 247–55.

Arruda, P., 2012. Genetically modified sugarcane for bioenergy generation. *Current Opinion in Biotechnoogy*, 23(3), 315–22.

Arvinth, S., Arun, S., Selvakesavan, R. K., Srikanth, J., Mukunthan, N., Ananda Kumar, P., Premachandran, M. N. and Subramonian, N., 2010. Genetic transformation and pyramiding of aprotinin-expressing sugarcane with cry1Ab for shoot borer (*Chilo infuscatellus*) resistance. *Plant Cell Reports*, 29(4), 383–95.

Augustine, S. M., Ashwin Narayan, J., Syamaladevi, D. P., Appunu, C., Chakravarthi, M., Ravichandran, V., Tuteja, N. and Subramonian, N., 2015a. Overexpression of EaDREB2 and pyramiding of EaDREB2 with the pea DNA helicase gene (PDH45) enhance drought and salinity tolerance in sugarcane (*Saccharum* spp. hybrid). *Plant Cell Reports*, 34(2), 247–63.

Augustine, S. M., Ashwin Narayan, J., Syamaladevi, D. P., Appunu, C., Chakravarthi, M., Ravichandran, V., Tuteja, N. and Subramonian, N., 2015b. Introduction of Pea DNA Helicase 45 into sugarcane (*Saccharum* spp. Hybrid) enhances cell membrane thermostability and upregulation of stress-responsive genes leads to abiotic stress tolerance. *Molecular Biotechnology*, 57(5), 475–88.

Augustine, S. M., Cherian, A. V., Syamaladevi, D. P. and Subramonian, N., 2015c. *Erianthus arundinaceus* HSP70 (EaHSP70) acts as a key regulator in the formation of anisotropic interdigitation in sugarcane (*Saccharum* spp. hybrid) in response to drought stress. *Plant and Cell Physiology*, 56(12), 2368–80.

Augustine, S. M., Narayan, J. A., Syamaladevi, D. P., Appunu, C., Chakravarthi, M., Ravichandran, V. and Subramonian, N., 2015d. *Erianthus arundinaceus* HSP70 (EaHSP70) overexpression increases drought and salinity tolerance in sugarcane (*Saccharum* spp. hybrid). *Plant Science*, 232, 23–34.

Baltes, N. J. and Voytas, D. F., 2015. Enabling plant synthetic biology through genome engineering. *Trends in Biotechnology*, 33(2), 120–31.

Barba, R. and Nickell, L. G., 1969. Nutrition and organ differentiation in tissue cultures of sugarcane, a monocotyledon. *Planta*, 89, 299–302.

Basnayake, S. W. V., Moyle, R. and Birch, R. G., 2011. Embryogenic callus proliferation and regeneration conditions for genetic transformation of diverse sugarcane cultivars. *Plant Cell Reports*, 30(3), 439–48.

Basnayake, S. W. V., Morgan, T. C., Wu, L. and Birch, R. G., 2012. Field performance of transgenic sugarcane expressing isomaltulose synthase. *Plant Biotechnology Journal*, 10(2), 217–25.

Belhaj, K., Chaparro-Garcia, A., Kamoun, S. and Nekrasov, V., 2013. Plant genome editing made easy: Targeted mutagenesis in model and crop plants using the CRISPR/Cas system. *Plant Methods*, 9(1), 39.

Birch, R. G. and Patil, S. S., 1987. Correlation between albicidin production and chlorosis induction by *Xanthomonas albilineans*, the sugarcane leaf scald pathogen. *Physiological and Molecular Plant Pathology*, 30, 199–206.

Birch, R. G., Shen, B., Sawyer, B. J. B., Huttner, E., Tucker, W. Q. J. and Betzner, A. S., 2010. Evaluation and application of a luciferase fusion system for rapid in vivo analysis of RNAi targets and constructs in plants. *Plant Biotechnology Journal*, 8(4), 465–75.

Birch, R. G., 2014. Sugarcane Biotechnology: Axenic culture, gene transfer, and transgene expression. In: Moore P. H. and Botha F. C. (Eds) *Sugarcane: Physiology, Biochemistry, and Functional Biology*. John Wiley & Sons Ltd, Hoboken USA, pp. 645–81.

Botha, F., Sawyer, B. and Birch, R., 2001. Sucrose metabolism in the culm of transgenic sugarcane with reduced soluble acid invertase activity. In: Hogarth, D. M. (Ed.) *Proceedings of the International Society of Sugar Cane Technologists XXIV Congress*, Mackay, Australia, Australian Society of Sugar Cane Technologists, pp. 588–91.

Bower, R. and Birch, R. G., 1992. Transgenic sugarcane plants via microprojectile bombardment. *Plant Journal*, 2(3), 409–16.

Bower, R., Elliott, A. R., Potier, B. A. M. and Birch, R. G., 1996. High-efficiency, microprojectile-mediated cotransformation of sugarcane, using visible or selectable markers. *Molecular Breeding*, 2(3), 239–49.

Brandes, E. W., 1958. Origin, classification and characteristics. In: Artschwager, E. and Brandes, E. W. (Eds) *Sugarcane (S. officinarum L.). USDA Agricultural Handbook*, USDA, Washington, DC, pp. 1–35.

Brookes, G. and Barfoot, P., 2016a. Global income and production impacts of using GM crop technology 1996–2014. *GM Crops and Food*, 7(1), 38–77.

Brookes, G. and Barfoot, P., 2016b. Environmental impacts of genetically modified (GM) crop use 1996–2014: Impacts on pesticide use and carbon emissions. *GM Crops and Food*, 7(2), 84–116.

Butterfield, M. K., Irvine, J. E., Valdez Garza, M. and Mirkov, T. E., 2002. Inheritance and segregation of virus and herbicide resistance transgenes in sugarcane. *Theoretical and Applied Genetics*, 104(5), 797–803.

Carmona, E. R., Arencibia, A. D., López, J., Simpson, J., Vargas, D. and Sala, F., 2005. Analysis of genomic variability in transgenic sugarcane plants produced by *Agrobacterium tumefaciens* infection. *Plant Breeding*, 124(1), 33–8.

Chengalrayan, K. and Gallo-Meagher, M., 2001. Effect of various growth regulators on shoot regeneration of sugarcane. *In Vitro Cellular and Developmental Biology – Plant*, 37, 434–9.

Damaj, M. B., Kumpatla, S. P., Emani, C., Beremand, P. D., Reddy, A. S., Rathore, K. S., Buenrostro-Nava, M. T., Curtis, I. S., Thomas, T. L. and Mirkov, E. E., 2010. Sugarcane DIRIGENT and O-METHYLTRANSFERASE promoters confer stem-regulated gene expression in diverse monocots. *Planta*, 231(6), 1439–58.

Daniels, J. and Roach, B. T., 1987. Taxonomy and evolution. In: Heinz, D. J. (Ed.) *Sugarcane Improvement through Breeding*, vol. 11, Elsevier, Amsterdam, the Netherlands, pp. 7–84.

De-Araujo, P. G., Rossi, M., De-Jesus, E. M., Saccaro Jr., N. L., Kajihara, D., Massa, R., De-Felix, M., Drummond, R. D., Falco, M. C. and Chabregas, S. M., 2005. Transcriptionally active transposable elements in recent hybrid sugarcane, *Plant Journal*, 44(5), 707–17.

De-la-Peña, C., Nic-Can, G. I., Galaz-Ávalos, R. M., Avilez-Montalvo, R. and Loyola-Vargas, V. M., 2015. The role of chromatin modifications in somatic embryogenesis in plants. *Frontiers in Plant Science*, 6, 635.

Dermawan, H., Karan, R., Jung, J. H., Zhao, Y., Parajuli, S., Sanahuja, G. and Altpeter, F., 2016. Development of an intragenic gene transfer and selection protocol for sugarcane resulting in resistance to acetolactate synthase-inhibiting herbicide. *Plant Cell Tissue and Organ Culture*, 126(3), 459–68.

D'Hont, A., Grivet, L., Feldmann, P., Rao, S., Berding, N. and Glaszmann, J. C., 1996. Characterisation of the double genome structure of modern sugarcane cultivars (*Saccharum* spp.) by molecular cytogenetics. *Molecular General Genetics* 250(4), 405–13.

Dias, M. O. S., Cavalett, O. and Maciel, R., 2014. Integrated first and second generation ethanol production from sugarcane. *Chemical Engineering Transactions*, 37, 445–50.

Dong, S., Delucca, P., Geijskes, R. J., Ke, J., Mayo, K., Mai, P., Sainz, M., Caffall, K., Moser, T., Yarnall, M., Setliff, K., Jain, R., Rawls, E., Smith-Jones, M. and Dunder, E., 2014. Advances in *Agrobacterium*-mediated sugarcane transformation and stable transgene expression. *Sugar Technology*, 16(4), 366–71.

Enríquez-Obregón, G. A., Vázquez-Padrón, R. I., Prieto-Samsonov, D. L., De la Riva, G. A. and Selman-Housein, G., 1998. Herbicide-resistant sugarcane (*Saccharum officinarum* L.) plants by *Agrobacterium*-mediated transformation. *Planta*, 206(1), 20–7.

Falco, M. C. and Silva-Filho, M. C., 2003. Expression of soybean proteinase inhibitors in transgenic sugarcane plants: Effects on natural defense against *Diatraea saccharalis*. *Plant Physiology and Biochemistry*, 41(8), 761–6.

Fedoroff, N. V., 2015. Food in a future of 10 billion. *Agriculture and Food Security*, 4, 11.

Fehér, A., 2015. Somatic embryogenesis – Stress-induced remodeling of plant cell fate. *Biochem Biophys Acta*, 1849, 385–402.

Gallo-Meagher, M. and Irvine, J. E., 1996. Herbicide resistant transgenic sugarcane plants containing the bar gene. *Crop Science*, 36(5), 1367–74.

Gallo-Meagher, M., English, R. G. and Abouzid, A., 2000. Thidiazuron stimulates shoot regeneration of sugarcane embryogenic callus. *In Vitro Cellular and Developmental Biology – Plant*, 36, 37–40.

Gambley, R. L., Bryant, J. D., Masel, N. P. and Smith, G. R., 1994. Cytokinin-enhanced regeneration of plants from microprojectile bombarded sugarcane meristematic tissue. *Australian Journal of Plant Physiology*, 21(5), 603–12.

Gao, S., Yang, Y., Wang, C., Guo, J., Zhou, D., Wu, Q., Su, Y., Xu, L. and Que, Y., 2016. Transgenic sugarcane with a cry1Ac gene exhibited better phenotypic traits and enhanced resistance against sugarcane borer. *PLoS ONE*, 11(4), 1–16.

Garcia, R., Cidade, D., Castellar, A., Lips, A., Magioli, C., Callado, C. and Mansu, E., 2007. In vitro morphogenesis patterns from shoot apices of sugar cane are determined by light and type of growth regulator. *Plant Cell Tissue and Organ Culture*, 90(2), 181–90.

Gilbert, R. A., Gallo-Meagher, M., Comstock, J. C., Miller, J. D., Jain, M. and Abouzid, A., 2005. Agronomic evaluation of sugarcane lines transformed for resistance to Sugarcane mosaic virus strain E. *Crop Science*, 45(5), 2060–7.

Gilbert, R. A., Glynn, N. C., Comstock, J. C. and Davis, M. J., 2009. Agronomic performance and genetic characterization of sugarcane transformed for resistance to sugarcane yellow leaf virus. *Field Crops Research*, 111(1–2), 39–46.

Goebel, F. R. and Sallam, N., 2011. New pest threats for sugarcane in the new bioeconomy and how to manage them. *Current Opinion in Environmental Sustainability*, 3, 81–9.

Gonçalves, M. C., Pinto, L. R., Creste, S. and Landell, M. G. A. 2012. Virus diseases of Sugarcane a constant challenge to sugarcane breeding in Brazil. *Functional Plant Science and Biotechnology*, 6, 108–16.

Graff, G. D., Zilberman, D. and Bennett, A. B., 2009. The contraction of agbiotech product quality innovation. *Nature Biotechnology*, 27(8), 702–4.

Groenewald, J. H. and Botha, F. C., 2008. Down-regulation of pyrophosphate: Fructose 6-phosphate 1-phosphotransferase (PFP) activity in sugarcane enhances sucrose accumulation in immature internodes. *Transgenic Research*, 17(1), 85–92.

Guerzoni, J. T. S., Belintani, N. G., Moreira, R. M. P., Hoshino, A. A., Domingues, D. S., Filho, J. C. B. and Vieira, L. G. E., 2014. Stress-induced Δ1-pyrroline-5-carboxylate synthetase (P5CS) gene confers tolerance to salt stress in transgenic sugarcane. *Acta Physiologiae Plantarum*, 36(9), 2309–19.

Guo, J., Gao, S., Lin, Q., Wang, H., Que, Y. and Xu, L., 2015. Transgenic sugarcane resistant to sorghum mosaic virus based on coat protein gene silencing by RNA interference. *BioMed Research International*, 2015, 9.

Hamerli, D. and Birch, R. G., 2011. Transgenic expression of trehalulose synthase results in high concentrations of the sucrose isomer trehalulose in mature stems of field-grown sugarcane. *Plant Biotechnology Journal*, 9(1), 32–7.

Harrison, M. D., Geijskes, J., Coleman, H. D., Shand, K., Kinkema, M., Palupe, A., Hassall, R., Sainz, M., Lloyd, R., Miles, S. and Dale, J. L., 2011. Accumulation of recombinant cellobiohydrolase and endoglucanase in the leaves of mature transgenic sugarcane. *Plant Biotechnology Journal*, 9(8), 884–96.

Harrison, M. D., Geijskes, R. J., Lloyd, R., Miles, S., Palupe, A., Sainz, M. B. and Dale, J. L., 2014. Recombinant cellulase accumulation in the leaves of mature, vegetatively propagated transgenic sugarcane. *Molecular Biotechnology*, 56(9):795–802.

Heinz, D. J. and Mee, G. W. P., 1969. Plant differentiation from callus tissue of *Saccharum* species. *Crop Science*, 9, 346–8.

Ho, W. J. and Vasil, I. K., 1983. Somatic embryogenesis in sugarcane (*Saccharum officinarum* L) I. the morphology and physiology of callus formation and the ontogeny of somatic embryos. *Protoplasma*, 118, 169–80.

Huang, H., Long, S. and Singh, S., 2016. Techno-economic analysis of biodiesel and ethanol co-production from lipid –producing sugarcane. *Biofuels Bioproducts and Biorefining*, 10, 299–315.

Ingelbrecht, I., Irvine, J. and Mirkov, T., 1999. Posttranscriptional gene silencing in transgenic sugarcane. Dissection of homology-dependent virus resistance in a monocot that has a complex polyploid genome. *Plant physiology*, 119(4), 1187–98.

Irvine, J. E., Benda, G. T. A., Legendre, B. L. and Machado Jr., G. R., 1991. The frequency of marker changes in sugarcane plants regenerated from callus culture. II. Evidence for vegetative and genetic transmission, epigenetic effects and chimera1 disruption. *Plant Cell Tissue and Organ Culture*, 26, 115–25.

Jackson, M. A., Anderson, D. J. and Birch, R. G., 2013. Comparison of *Agrobacterium* and particle bombardment using whole plasmid or minimal cassette for production of high-expressing, low-copy transgenic plants. *Transgenic Research*, 22(1), 143–51.

Jain, M., Chengalrayan, K., Abouzid, A. and Gallo, M., 2007. Prospecting the utility of a PMI/mannose selection system for the recovery of transgenic sugarcane (*Saccharum* spp. hybrid) plants. *Plant Cell Reports*, 26(5), 581–90.

Jorgensen, H., Kristensen, J. B. and Felby, C., 2007. Enzymatic conversion of lignocellulose into fermentable sugars: Challenges and opportunities. *Biofuels Bioproducts and Biorefining*, 1(2), 119–34.

Joyce, P., Hermann, S., O'Connell, A., Dinh, Q., Shumbe, L. and Lakshmanan, P., 2014. Field performance of transgenic sugarcane produced using *Agrobacterium* and biolistics methods. *Plant Biotechnology Journal*, 12(4), 411–24.

Jung, J. H. and Altpeter, F., 2016a. TALEN mediated targeted mutagenesis of the caffeic acid O-methyltransferase in highly polyploid sugarcane improves cell wall composition for production of bioethanol. *Plant Molecular Biology*, 92, 131–42.

Jung, J. H., Fouad, W. M., Vermerris, W., Gallo, M. and Altpeter, F., 2012. RNAi suppression of lignin biosynthesis in sugarcane reduces recalcitrance for biofuel production from lignocellulosic biomass. *Plant Biotechnology Journal*, 10(9), 1067–76.

Jung, J. H., Vermerris, W., Gallo, M., Fedenko, J. R., Erickson, J. E. and Altpeter, F., 2013. RNA interference suppression of lignin biosynthesis increases fermentable sugar yields for biofuel production from field-grown sugarcane. *Plant Biotechnology Journal*, 11(6), 709–16.

Jung, J. H., Kannan, B., Dermawan, H., Moxley, G. W. and Altpeter, F., 2016b. Precision breeding for RNAi suppression of a major 4-coumarate:coenzyme A ligase gene improves cell wall saccharification from field grown sugarcane. *Plant Molecular Biology*, 92(4–5), 505–17.

Kalunke, R. M., Kolge, A. M., Babu, K. H. and Prasad, D. T., 2009. *Agrobacterium* mediated transformation of sugarcane for borer resistance using Cry 1Aa3 gene and one-step regeneration of transgenic plants. *Sugar Technology*, 11(4), 355–9.

Kannan, B., Jung, J. H. Moxley, G.W., Lee, S.-M. and Altpeter, F., 2018. TALEN mediated targeted mutagenesis of more than 100 COMT copies/alleles in highly polyploid sugarcane improves saccharification efficiency without compromising biomass yield (Plant Biotechnology Journal, accepted for publication).

Khan, M. S., Ali, S. and Iqbal, J., 2011. Developmental and photosynthetic regulation of d-endotoxin reveals that engineered sugarcane conferring resistance to 'dead heart' contains no toxins in cane juice. *Molecular Biology Reports*, 38(4), 2359–69.

Kim, J. Y., Nong, G., Rice, J. D., Gallo, M., Preston, J. F. and Altpeter, F., 2017. *In planta* production and characterization of a hyperthermostable GH10 xylanase in transgenic sugarcane. *Plant Molecular Biology*, 93(4–5). 465–78.

Kumar, T., Uzma, Khan, M. R., Abbas, Z. and Ali, G. M., 2014. Genetic improvement of sugarcane for drought and salinity stress tolerance using arabidopsis vacuolar pyrophosphatase (AVP1) gene. *Molecular Biotechnology*, 56(3), 199–209.

Lakshmanan, P., Geijskes, R. J., Aitken, K. S., Grof, C. L. P., Bonnett, G. D. and Smith, G. R., 2005. Sugarcane biotechnology: The challenges and opportunities. *In Vitro Cellular and Developmental Biology – Plant*, 41(4), 345–63.

Larkin, P. J., 1982. Sugarcane tissue and protoplast culture. *Plant Cell Tissue and Organ Culture*, 1, 149–63.

Leibbrandt, N. B. and Snyman, S. J., 2003. Stability of gene expression and agronomic performance of a transgenic herbicide-resistant sugarcane line in South Africa. *Crop Science*, 43(2), 671–7.

Li, Q., Fan, L., Luo, Q., He, H., Zhang, J., Zeng, Q., Li, Y., Zhou, W., Huang, Z., Deng, H. and Qi, Y., 2014. Co-overexpression of AtCBL9, AtCIPK23 and AtAKT1 enhances K+ uptake of sugarcane under low-K+ stress. *Plant OMICS*, 7(3), 188–94.

Liu, M. C., 1993. Factors affecting induction, somatic embryogenesis and plant regeneration of callus from cultured immature inflorescences of sugarcane. *Journal of Plant Physiology*, 141(6), 714–20.

Manickavasagam, M., Ganapathi, A., Anbazhagan, V. R., Sudhakar, B., Selvaraj, N., Vasudevan, A. and Kasthurirengan, S., 2004. *Agrobacterium*-mediated genetic transformation and development of herbicide-resistant sugarcane (*Saccharum* species hybrids) using axillary buds. *Plant Cell Reports*, 23(3), 134–43.

McDougall, P., 2011. The cost and time involved in the discovery, development and authorisation of a new plant biotechnology derived trait. Crop Life International, pp. 1–24.

McQualter, R. B., Petrasovits, L. A., Gebbie, L. K., Schweitzer, D., Blackman, D. M., Chrysanthopoulos, P., Hodson, M. P., Plan, M. R., Riches, J. D., Snell, K. D., Brumbley, S. M. and Nielsen, L. K., 2015. The use of an acetoacetyl-CoA synthase in place of a β-ketothiolase enhances poly-3-hydroxybutyrate production in sugarcane mesophyll cells. *Plant Biotechnology Journal*, 13(5):700–7.

McQualter, R. B., Bellasio, C., Gebbie, L. K., Petrasovits, L. A., Palfreyman, R. W., Hodson, M. P., Plan, M. R., Blackman, D. M., Brumbley, S. M. and Nielsen, L. K., 2016. Systems biology and metabolic modelling unveils limitations to polyhydroxybutyrate accumulation in sugarcane leaves; lessons for C4 engineering. *Plant Biotechnology Journal*, 14(2), 567–80.

Ming, R., Moore P. H., Wu, K., D' Hont, A., Glaszmann, J. C., Tew, T. L., Mirkow, T. E., Silva, J. D., Jifon, J., Rai, M., Schnell, R. J., Brumbley, S. M., Lakshmanan, P., Comstock, J. C. and Paterson, A. H., 2006. Sugarcane improvement through breeding and biotechnology. *Plant Breeding Review*, 27, 15–118.

Mohan, C., 2016. Genome editing in sugarcane: Challenges ahead. *Frontiers in Plant Science*, *7*, 1542. (doi: 10.3389/fpls.2016.01542).

Molinari, H. B. C., Marur, C. J., Daros, E., De Campos, M. K. F., De Carvalho, J. F. R. P., Filho, J. C. B., Pereira, L. F. P. and Vieira, L. G. E., 2007. Evaluation of the stress-inducible production of proline in transgenic sugarcane (*Saccharum* spp.): Osmotic adjustment, chlorophyll fluorescence and oxidative stress. *Physiologia Plantarum*, 130(2), 218–29.

Moyle, R. L. and Birch, R. G., 2013a. Diversity of sequences and expression patterns among alleles of a sugarcane loading stem gene. *Theoretical and Applied Genetics*, 126(7), 1775–82.

Moyle, R. L. and Birch, R. G., 2013b. Sugarcane loading stem gene promoters drive transgene expression preferentially in the stem. *Plant Molecular Biology*, 82(2), 51–8.

Mudge, S. R., Basnayake, S. W. V, Moyle, R. L., Osabe, K., Graham, M. W., Morgan, T. E. and Birch, R. G., 2013. Mature-stem expression of a silencing-resistant sucrose isomerase gene drives isomaltulose accumulation to high levels in sugarcane. *Plant Biotechnology Journal*, 11(4), 502–9.

Nadar, H., Soepraptopo, S., Heinz, D. J. and Ladd, S. L., 1978. Fine structure of sugarcane (*Saccharum* sp.) callus and the role of auxin in embryogenesis. *Crop Science*, 18, 210–16.

Nickell, L. G., 1964. Tissue and cell cultures of sugarcane – another research tool. *Hawaiian Planters' Record*, 57, 223–9.

Noguera, A., Enrique, R., Perera, M. F., Ostengo, S., Racedo, J., Costilla, D., Zossi, S., Cuenya, M. I., Filippone, M. P., Welin, B. and Castagnaro, A. P., 2015. Genetic characterization and field evaluation to recover parental phenotype in transgenic sugarcane: A step toward commercial release. *Molecular Breeding*, 35, 115.

Ohlrogge, J. and Chapman, K., 2011. The seeds of green energy. *Biochemist*, 33, 34–8.

Ort, D. R., Merchant, S. S., Alric, J., Barkan, A., Blankenship, R. E., Bock, R., Croce, R., Hanson, M. R., Hibberd, J. M., Long, S. P., Moore, T. A., Moroney, J., Niyogi, K. K., Parry, M. A., Peralta-Yahya, P. P., Prince, R. C., Redding, K. E., Spalding, M. H., van Wijk, K. J., Vermaas, W. F., von Caemmerer, S., Weber, A. P., Yeates, T. O., Yuan, J. S. and Zhu, X. G., 2015. Redesigning photosynthesis to sustainably meet global food and bioenergy demand. *Proceddings National Academy of Science USA*, 112, 8529–36.

Oz, T., Karan, R., Meretto A. and Altpeter, F., 2018. CRISPR/Cas9 and template-mediated multi-allelic homologous recombination of the acetolactate synthase gene in sugarcane confers herbicide resistance (submitted for publication).

Petrasovits, L. A., Purnell, M. P., Nielsen, L. K. and Brumbley, S. M., 2007. Production of polyhydroxybutyrate in sugarcane. *Plant Biotechnology Journal*, 5(1), 162–72.

Petrasovits, L. A., Zhao, L., McQualter, R. B., Snell, K. D., Somleva, M. N., Patterson, N. A., Nielsen, L. K. and Brumbley, S. M., 2012. Enhanced polyhydroxybutyrate production in transgenic sugarcane. *Plant Biotechnology Journal*, 10(5), 569–78.

Petrasovits, L. A., McQualter, R. B., Gebbie, L. K., Blackman, D. M., Nielsen, L. K. and Brumbley, S. M., 2013. Chemical inhibition of acetyl coenzyme A carboxylase as a strategy to increase polyhydroxybutyrate yields in transgenic sugarcane. *Plant Biotechnology Journal*, 11(9), 1146–51.

Poovaiah, C. R., Bewg, W. P., Lan, W., Ralph, J. and Coleman, H. D., 2016. Sugarcane transgenics expressing MYB transcription factors show improved glucose release. *Biotechnol Biofuels*, 9, 143.

Purseglove, J. W., 1972. *Tropical Crops: Monocotyledons*. Longman, London.

Ramiro, D. A., Melotto-Passarin, D. M., Barbosa, M. D. A., Santos, F. Dos, Gomez, S. G. P., Massola Júnior, N. S., Lam, E. and Carrer, H., 2016. Expression of Arabidopsis Bax Inhibitor-1 in transgenic sugarcane confers drought tolerance. *Plant Biotechnology Journal*, 14, 1826–37.

Reis, R. R., Andrade Dias Brito da Cunha, B., Martins, P. K., Martins, M. T. B., Alekcevetch, J. C., Chalfun-Junior, A., Andrade, A. C., Ribeiro, A. P., Qin, F., Mizoi, J., Yamaguchi-Shinozaki, K., Nakashima, K., Carvalho, J. de F. C., de Sousa, C. A. F., Nepomuceno, A. L., Kobayashi, A. K. and Molinari, H. B. C., 2014. Induced over-expression of AtDREB2A-CA improves drought tolerance in sugarcane. *Plant Science*, 221–222, 59–68.

Rommens, C. M., Haring, M. A., Swords, K., Davies, H. V. and Belknap, W. R., 2007. The intragenic approach as a new extension to traditional plant breeding. *Trends in Plant Science*, 12(9), 397–403.

Rossouw, D., Kossmann, J., Botha, F. C. and Groenewald, J. H., 2010. Reduced neutral invertase activity in the culm tissues of transgenic sugarcane plants results in a decrease in respiration and sucrose cycling and an increase in the sucrose to hexose ratio. *Functional Plant Biology*, 37, 22–31.

Rott, P. C., Girard, J. C. and Comstock J. C., 2013. Impact of pathogen genetics on breeding for resistance to sugarcane diseases. *Proceedings International Society of Sugarcane Technologists Congress*, 28, PL07.

Sainger, M., Sainger, P. A., Chaudhary, D., Jaiwal, R., Singh, R. P., Dhankher, O. P. and Jaiwal, P. K., 2016. GM crops for developing world in the era of climate change: For increase of farmer's income, poverty alleviation, nutrition and health. In: Jaiwal, P. K., Singh, R. P., Dhanker, O. P. *Genetic Manipulation in Plants for Mitigation of Climate Change*, Springer, New Delhi, India, pp. 223–41.

Schellenbaum, P., Mohler, V., Wenzel, G. and Walter, B., 2008. Variation in DNA methylation patterns of grapevine somaclones (*Vitis vinifera* L.). *BMC Plant Biology*, 8, 78.

Schouten, H. J., Krens, F. A. and Jacobsen, E., 2006. Cisgenic plants are similar to traditionally bred plants: International regulations for genetically modified organisms should be altered to exempt cisgenesis. *EMBO Reports*, 7(8), 750–3.

Shih, P. M., Liang, Y. and Loqué, D., 2016. Biotechnology and synthetic biology approaches for metabolic engineering of bioenergy crops. *Plant Journal*, 87(1), 103–17.

Singer, T., Yordan, C. and Martienssen, R. A., 2001. Robertson'smutator transposonsin *A. thaliana* are regulated by the chromatin-remodeling gene decrease in DNA Methylation (DDM1). *Genes and Development*, 15, 591–602.

Snyman, S., Leibbrandt, N. and Botha, F., 1998. Buster resistant sugarcane. *Proceedings of South African Sugar Technologists Association*, 72, 138–9.

Snyman, S. J., Meyer, G. M., Richards, J. M., Haricharan, N., Ramgareeb, S. and Huckett, B. I., 2006. Refining the application of direct embryogenesis in sugarcane: Effect of the developmental phase of leaf disc explants and the timing of DNA transfer on transformation efficiency. *Plant Cell Reports*, 25(10), 1016–23.

Snyman, S. J., Meyer, G. M., Koch, A. C., Banasiak, M. and Watt, M. P., 2011. Applications of in vitro culture systems for commercial sugarcane production and improvement. *In Vitro Cellular and Developmental Biology – Plant*, 47(2), 234–49.

Snyman, S. J., Hajari, E., Watt, M. P., Lu, Y. and Kridl, J. C., 2015. Improved nitrogen use efficiency in transgenic sugarcane: Phenotypic assessment in a pot trial under low nitrogen conditions. *Plant Cell Reports*, 34(5):667–9.

Taparia, Y., Fouad, W. M., Gallo, M. and Altpeter, F., 2012a. Rapid production of transgenic sugarcane with the introduction of simple loci following biolistic transfer of a minimal expression cassette and direct embryogenesis. *In Vitro Cellular and Developmental Biology – Plant*, 48(1), 15–22.

Taparia, Y., Gallo, M. and Altpeter, F., 2012b. Comparison of direct and indirect embryogenesis protocols, biolistic gene transfer and selection parameters for efficient genetic transformation of sugarcane. *Plant Cell Tissue and Organ Culture*, 111(2), 131–41.

Thelen, J. J. and Ohlrogge, J. B., 2002. Metabolic engineering of fatty acid biosynthesis in plants. *Metabolic Engineering*, 4, 12–21.

Uys, L., Botha, F. C., Hofmeyr, J. H. S. and Rohwer, J. M., 2007. Kinetic model of sucrose accumulation in maturing sugarcane culm tissue. *Phytochemistry*, 68, 2375– 2392.

van der Vyver, C., Conradie, T., Kossmann, J. and Lloyd, J., 2013. In vitro selection of transgenic sugarcane callus utilizing a plant gene encoding a mutant form of acetolactate synthase. *In Vitro Cellular and Developmental Biology – Plant*, 49(2), 198–206.

Vickers, J. E., Grof, C. P. L., Bonnett, G. D., Jackson, P. A. and Morgan, T. E., 2005. Effects of tissue culture,biolistic transformation,and introduction of PPO and SPS gene constructs on performance of sugarcane clones in the field. *Australian Journal of Agricultural Research*, 56, 57–68.

Voytas, D. F., 2013. Plant genome engineering with sequence-specific nucleases. *Annual Review of Plant Biology*, 64, 327–50.

Voytas, D. F. and Gao, C., 2014. Precision genome engineering and agriculture: Opportunities and regulatory challenges. *PLoS Biology*, 12(6), 1–6.

Wang, J., Nayak, S., Koch, K. and Ming, R., 2013. Carbon partitioning in sugarcane (*Saccharum* species). *Frontiers in Plant Science*, 18(4), 201.

Weeks, D. P., Spalding, M. H. and Yang, B., 2016. Use of designer nucleases for targeted gene and genome editing in plants. *Plant Biotechnology Journal*, 14(2), 483–95.

Welbaum, G. E. and Meinzer, F. C., 1990. Compartmentation of solutes and water in developing sugarcane stalk tissue. *Plant Physiology*, 93(3), 1147–53.

Weng, L. X., Deng, H., Xu, J. L., Li, Q., Wang, L. H., Jiang, Z., Hai, B. Z., Li, Q. and Zhang, L. H., 2006. Regeneration of sugarcane elite breeding lines and engineering of stem borer resistance. *Pest Management Science*, 62(2), 178–87.

Weng, L. X., Deng, H. H., Xu, J. L., Li, Q., Zhang, Y. Q., Jiang, Z. De, Li, Q. W., Chen, J. W. and Zhang, L. H., 2011. Transgenic sugarcane plants expressing high levels of modified cry1Ac provide effective control against stem borers in field trials. *Transgenic Research*, 20(4), 759–72.

Wu, H., Awan, F. S., Vilarinho, A., Zeng, Q., Kannan, B., Phipps, T., McCuiston, J., Wang, W., Caffall, K. and Altpeter, F., 2015. Transgene integration complexity and expression stability following biolistic or *Agrobacterium*-mediated transformation of sugarcane. *In Vitro Cellular and Developmental Biology – Plant*, 51(6), 603–11.

Wu, L. and Birch, R. G., 2007. Doubled sugar content in sugarcane plants modified to produce a sucrose isomer. *Plant Biotechnology Journal*, 5(1), 109–17.

Yu, Q. and Powles, S. B., 2014. Resistance to AHAS inhibitor herbicides : Current understanding Resistance to AHAS inhibitor herbicides . *Pest Management Science*, 70, 1340–50.

Zale, J., Jung, J. H., Kim, J. Y., Pathak, B., Karan, R., Liu, H., Chen, X., Wu, H., Candreva, J., Zhai, Z., Shanklin, J. and Altpeter, F., 2016. Metabolic engineering of sugarcane to accumulate energy-dense triacylglycerols in vegetative biomass. *Plant Biotechnology Journal*, 14(2), 661–9.

Zhang, L. and Birch, R. G., 1996. Biocontrol of sugar cane leaf scald disease by an isolate of *Pantoea dispersa* which detoxifies albicidin phytotoxins. *Letters in Applied Microbiology*, 22, 132–136.

Zhang, L., Xu, J. and Birch, R. G., 1999. Engineered detoxification confers resistance against a pathogenic bacterium. *Nature Biotechnology*, 17, 1021–4.

Zhang, Y., Liang, Z., Zong, Y., Wang, Y., Liu, J., Chen, K., Qiu, J.-L. and Gao, C., 2016. Efficient and transgene-free genome editing in wheat through transient expression of CRISPR/Cas9 DNA or RNA. *Nature Communications*, 7, 12617.

Zhou, D., Xu, L., Gao, S., Guo, J., Luo, J., You, Q. and Que, Y., 2016. Cry1Ac Transgenic sugarcane does not affect the diversity of microbial communities and has no significant effect on enzyme activities in rhizosphere soil within one crop season. *Frontiers in Plant Science*, 7, 265.

Zhu, Y. J., McCafferty, H., Osterman, G., Lim, S., Agbayani, R., Lehrer, A., Schenck, S. and Komor, E., 2011. Genetic transformation with untranslatable coat protein gene of sugarcane yellow leaf virus reduces virus titers in sugarcane. *Transgenic Research*, 20(3), 503–12.

Part 2

Progress in understanding and managing diseases, pests and weeds

Chapter 7

Ensuring biosecurity in sugarcane cultivation

Nicole Thompson, Sugar Research Australia, Australia

1 Introduction

Plant biosecurity can be defined as 'a set of measures designed to protect a crop, crops or a subgroup of crops from plant pests at national, regional and individual farm levels' (Anonymous, 2016b). Ensuring sound biosecurity in sugarcane production is extremely important for crop health, for maintaining or improving yields, and for increasing the production. Managing biosecurity includes management of the endemic pests as well as preparedness for exotic pests and incursions. A pest is defined as 'any species, strain or biotype of plant, animal, or pathogenic agent, injurious to plants or plant products' (www.IPPC.int), and this definition includes insect pests, pathogens and parasitic weeds. In this chapter a pest is considered to be endemic if it occurs in the country (an established pest), and exotic if it occurs outside the country, or outside the area of production. Biosecurity has a role to play in preventing and controlling emerging diseases – for a recent review of emerging plant viral diseases see Rodoni (2009).

In this chapter I will refer to the International Plant Protection Convention (IPPC, www.ippc.int) which sets internationally agreed standards for biosecurity and managing biosecurity threats. The role of National Plant Protection Organisations (NPPOs), risk analysis and risk management procedures used to complete a pest risk analysis (PRA) will be discussed. Common surveillance methods and the future of surveillance will be discussed, followed by a case study of a recent survey conducted in Indonesia.

http://dx.doi.org/10.19103/AS.2017.0035.19

Quarantine arrangements for sugarcane usually start prior to its entry into the importing country: import permit conditions are set by the NPPO based on a PRA and industry standards. The chapter will examine effective quarantine arrangements for sugarcane germ plasm exchange with recommendations drawn from the author's experience and research publications on facilities, diagnostics and the challenges related to sugarcane quarantine. The chapter will not discuss the legal requirements for intellectual property rights or genetic resource regulations. A case study showing how Australia prepared for and dealt with the sugarcane smut incursion will be discussed at the end of the chapter.

2 The National Plant Protection Organisation

Each country or region has an NPPO that determines import conditions, based on a PRA. In most cases, the NPPO is a member of the IPPC. The IPPC's mission is 'To secure the cooperation among nations to protect global plant resources from the spread and introduction of pests of plants, in order to preserve food security, biodiversity and to facilitate trade' (Anonymous, 2016c). International Standards for Phytosanitary Measures (ISPMs) are agreed-upon phytosanitary measures, developed by the IPPC, providing guidance on procedures, regulations and treatments that can be used to manage pest and disease risks for international movement of goods. The ISPMs are recognised by the World Trade Organisation and are used by participating governments to determine how to manage the risk of importing sugarcane germ plasm: this can include the requirements for a post-entry (or border) quarantine facility, the conditions for applying phytosanitary measures, the appropriate way to conduct a PRA and appropriate methods for pest surveillance. There are over 177 contracting party signatories of the IPPC, and currently more than 50 ISPMs (see Section 6). This is a good source of extra information about general quarantine requirements, but it should be noted that the information provided by the IPPC are not necessarily those adopted by an NPPO: government guidelines may differ from those of the ISPM.

3 Risk analysis of sugarcane biosecurity

Risk analysis is an important formal step in identifying potential threats and enabling a mechanism to compare and rank these risks. PRA has been defined as 'identifying pests of quarantine concern to an area, determining the pathways and the likelihood of introduction and establishment, and selecting optimum effective pest management options for minimising their introduction and establishment potential' (Singh, 1995).

The IPPC provides a framework for conducting PRA for insect pests, pathogens, plants, parasitic plants and genetically modified organisms. The definition of pests has undergone several iterations in the IPPC to ensure that it encompasses items of quarantine concern. Individual NPPOs can have their definition of a 'pest' with the most notable change being that parasitic weeds are often excluded, or placed as part of a weed risk assessment (Stone and Byrne, 2011). Management of risk will often require commodity-specific biosecurity plans that detail the prevention of pest introduction, risk analysis of pathways for introduction and emergency preparedness (Flood and Day, 2016).

There are formal guidelines for completing a PRA based on the IPPC guidelines for NPPOs. A summary of these procedures from ISPM11-7 are explained below, and full information including updates is available for free from the IPPC website.

ISPM 11 details the PRA procedures and considerations for quarantine pests. The standard provides details for the conduct of PRA to determine if pests are quarantine pests for agricultural crops, as well as for environmental and biological diversity. There are three stages in the process of completing a PRA:

- *Stage 1: Initiating the process.* Identification of the pests and pathways that are of quarantine concern. 'Pathways' are potential methods of introduction of a pest.
- *Stage 2: Risk assessment.* Categorising individual pests to determine that they are quarantine pests. The risk assessment evaluates the probability of the pest entry, establishment and spread, and the potential economic and environmental consequences.
- *Stage 3: Risk management.* Identifying management options to reduce the risks identified in Stage 2 to an acceptable level.

3.1 Stage 1: Initiation

A PRA can be initiated by the identification of a pathway that presents a potential pest hazard, identification of new pests that may require phytosanitary measures or during the review of policies and priorities. Information gathering is very important at this stage to clarify the identity of the pest, its present distribution, host range and so on. The information is likely to come from a number of sources, such as textbooks, pest information websites, journal articles and expert opinions. Any pre-existing PRA should also be considered: it may be out of date, from another NPPO, or for a similar pest but may still have relevant information. At the conclusion of Stage 1 the pest and pathway will have been identified and a decision made whether to continue to Stage 2. It is possible that during the initiation stage the pest will be identified to be not of quarantine concern, so no PRA will be completed.

3.2 Stage 2: PRA

This stage is the most intensive part of the process. There are three broadly related steps in a PRA: pest categorisation, assessment of the probability of introduction and spread and assessment of the economic consequences (including environmental impacts).

Stage 1 will have identified pests that require a PRA. The pest categorisation step involves the determination of the identity of the pest, its presence or absence in the area, the regulatory status, the potential for establishment and spread and their potential economic impact.

The identity of the pest needs to be clearly determined. In sugarcane, there are many resources for the identification of pests, including textbooks (Rott et al., 2000), field guides (Magarey, 2013; Samson et al., 2013), publications and websites. For some sugarcane pests, this remains difficult because either the cause is unknown, or the species is not well defined. This is not a barrier to complete a PRA because as long as the symptoms are consistent and it is transmissible a PRA can be completed. In the case where a vector is involved, the vector can also be considered a pest when it is also associated with the causal organism. In some cases, for example *Perkinsiella* sp., the vector is a pest in its own right as well as being a vector for the Fiji disease virus, thus the vector and vector + virus must be considered separately.

The presence or absence of a pest in the area must be determined, and it is necessary for the pest to be absent from the area under consideration for a PRA to be completed. If it is present in part of the area, then it should be under official regulatory control (e.g. local quarantine boundaries) and the PRA should only be conducted for the area where the pest is absent.

The potential for establishment and spread is determined by comparing the ecological/climatic conditions required for the pest to thrive with those in the area under assessment. This includes the prevalence of primary host species, alternative host species and vectors. Expert opinion and data should be sought for obtaining information about pest preferences, climate conditions and so on. Since this is acknowledged to be a prediction as much empirical evidence should be gathered as possible. During this process case histories and case studies of comparable pests, including reproductive rate, method of survival and differing cultural practices, can be considered. Yield loss estimates can be used to gauge the importance of the pest, and should take into account the presence of other hosts and the natural environment.

At the end of the second stage there should be a clear indication that the pest has the potential to be a quarantine pest with significant economic consequences, and the PRA should continue.

Pest introduction is comprised of both the entry and its establishment. The probability of introduction will depend on the pathway, and all relevant pathways should be investigated. The greater the number of pathways, the greater the probability of entry of a pest. Examples of pathways for sugarcane pests include variety exchange through established import permit schemes, illegal importation, movement of other host plants and natural spread. These all have different risks and management protocols, so should all be assessed.

An assessment for the potential economic consequences should be done in terms of quantitative monetary data. Information can be gained from the areas where the pest is already prevalent, or has been introduced. In modern PRA, both direct and indirect pest effects are characterised. Direct effects include crop losses (yield and quality); cost of control measures; frequency, type and amount of damage; rate of spread and environmental effects. Indirect pest effects include changes to producer costs (includes control costs), changes to exporter markets (less common for sugarcane, but is possible), effects of the control measures to environment, eradication or containment costs and social effects. In order for this analysis to be completed an agricultural economist is likely required. A cost estimate in terms of monetary data will provide a direct comparison between pests, and the sources of information, assumptions and methods of analysis should be specified.

A conclusion on the probability of their introduction, spread and economic (including environmental) consequences should be expressed in terms that are most suitable for interpretation and comparison, either in a quantitative or qualitative ($) form.

A decision matrix can be used to assess the data. The weighting of the matrix is done by the assessors, using expert opinion where necessary to fill knowledge gaps. This usually allows a 'ranking' of pests from most to least important.

An example of the decision-making process is adapted in Annex 1 from the Plant Health Australia guidelines (Anonymous, 2013).

3.3 Stage 3: pest risk management

The conclusions from Stage 2 can be used to decide whether risk management is required, and the strength of management required to bring the risk to an appropriate level. Pest risk management identifies ways to react to the risk and evaluates the efficiency of these actions, to identify the most appropriate options. Uncertainty that is identified through Stages 1 and 2 should be acknowledged and the management process designed in proportion to the risk. This may include an unknown pathogen, unknown mode of spread or unknown yield loss.

The appropriate level of risk is determined by the NPPO and government, and this differs widely around the world. Information from the first two stages of this process is used in making this decision. If the risk is determined to be unacceptable, then the first step is to determine and select appropriate risk management options. These can include: phytosanitary measures to eliminate the pest before introduction; inspection of the consignment and/or testing for pest freedom; pre-entry or post-entry quarantine system; conditions or treatments, restrictions on end use; or prohibition of part of a consignment. The pest risk management options can be enforced on the exporting or importing country or both. This may require the exporting country to provide evidence of pest freedom through surveillance, testing or treatment. The importing country may also be required to increase surveillance, be prepared for eradication programmes or conduct tests as part of the risk management strategies.

Phytosanitary certificates and compliance strategies are by far the most common strategy for risk management of sugarcane pathogens when moving sugarcane germ plasm around the world, and most are based on established industry standards (Frison and Putter, 1993). Export certification includes phytosanitary certificates that provide official assurance that a consignment is 'considered to be free from the quarantine pests specified by the importing contracting party and to confirm with the current phytosanitary requirements of the importing contracting party' (ISPM 12). These are issued by the NPPO in the exporting country to comply with import permit conditions to manage risk in the importing country.

If there are no management options that satisfactorily reduce the risk to an acceptable level, the commodity can be prohibited. Each country has its own methods for appealing this prohibition, and there has been successful reversal of decision when new information or new treatment methods become available. No phytosanitary treatment should be considered permanent, and new methods and information continue to be assessed on a regular basis.

3.4 Final PRA

The final documentation produced by a PRA as specified by the IPPC includes:

- Purpose of the PRA
- Pest, pest list, pathways, PRA area, endangered area
- Sources of information
- Categorised pest list
- Conclusions of risk assessment
    - Probability
    - Consequences
- Risk management
    - Options identified
    - Options selected

This documentation may be required to uphold any decisions and to show the decision-making process.

An example of recent PRA for sugarcane is within the Australian Sugar Industry Biosecurity Plan (Anonymous, 2016a). The document is put together using joint government and industry resources and is revised every five years.

4 Surveying

The IPPC defines two main kinds of surveillance: general surveillance and specific surveys. They have different objectives and use different methods.

General surveillance is defined in ISPM 6 as 'a process whereby information on particular pests which are of concern for an area is gathered from many sources, wherever it is available and provided for use by the NPPO'. The main purpose of general surveillance is to determine pest status in an area, provide early detection of an incursion and provide information to the NPPO to determine pest distribution lists. This information feeds back into PRA and can influence quarantine restrictions and also determine if a specific survey is required. Usually these general surveillance surveys are 'desktop' surveys and compile information from many sources, including expert opinion.

Specific surveys are defined as 'procedure by which NPPOs obtain information on pests of concern on specific sites in an area over a defined period of time'. These surveys are usually focused on a pest or commodity and can include detection, delimiting or monitoring surveys. These usually involve fieldwork and coordination with other bodies. Information from these specific surveys feed back into general surveys and PRA.

Which survey approach is adopted will depend on the pest, commodity, the NPPO and the resources available. For this reason, many field-based surveys are done by research organisations in conjunction with local governments. Prioritisation is generally given to quarantine pests that are easily identified and that affect a commodity that has a high agricultural value and/or requires surveys for export. However, in the case of an incursion there is usually an urgency for detection and delimiting surveys are part of an effective incursion management plan.

4.1 Survey design

Depending on whether the purpose of the survey is to look for a pest of unknown status, to gather data about a known pest population or to determine if the pest is confined to one area they are classified as detection, monitoring or delimiting surveys. They can use a combination of technical methods including sampling (collection and testing), trapping (e.g. pan traps) and visual observation. General information about survey design, methods and considerations is available from the IPPC 'Plant Pest Surveillance' guidelines.

The pest or pathogen, the crop and the environmental conditions also influence survey design. Information from recent incursions has been used to determine the best method for conducting surveys for some crops (e.g. citrus canker in Australia caused by *Xanthomonas axonopodis* (Potts et al., 2013)), while for some other incursions, delimiting survey using probabilistic maps has been used effectively for analysis of a plant invasion (Hauser et al., 2016) and specific spatial arrangements for surveys (Berec et al., 2015). It would be useful to conduct future research to analyse recent pest and disease incursions in sugarcane to determine the best methods for surveillance. It is likely that the use of technology will further enhance survey capability to deliver accurate and informative results.

4.2 Future of surveillance

Remote sensing (e.g. Abdel-Rahman et al., 2010; Everingham et al., 2007; Johansen et al., 2014; Grisham et al., 2010) and the use of drones (UAV) is an active area of research in sugarcane pest and disease analysis. These technologies may allow easier surveillance to take place, for general or specific purposes because of the difficulty in gaining access within a sugarcane crop during parts of the year. This is an important area for sugarcane technological development.

Sample collection technologies are also evolving, with the use of molecular methods of detection, FTA cards for rapid sample collection (Grund et al., 2010; Wilson et al., 2016) and in-field detection techniques being evaluated (e.g. *Sugarcane white leaf phytoplasma* detection kit from Innova Biotechnology; www.innovabiotechnology.com). This could give a more rapid result for delimiting surveys, or for incursion management.

4.3 Case study: survey for sugarcane streak mosaic virus in Indonesia

Sugarcane streak mosaic virus (SCSMV) is a pathogen of high biosecurity risk to Australia. Sugar Research Australia (SRA) and the Indonesian Sugar Research Institute are working on a collaborative project to determine the distribution of SCSMV in Indonesia. Three delimiting surveys of three different regions over three years are being done, and the survey plan is described below.

The survey team consisted of at least two plant pathologists and two entomologists plus local assistance (Fig. 1). The survey was scheduled to be conducted when the cane reaches an appropriate height for SCSMV visual screening (4–5 months), and in the dry season to allow easy access. Sites of commercial sugarcane plantation were pre-chosen for their location and specific fields chosen based on the known susceptible varieties, or where this was not available, based on accessibility. Each site had the variety, crop age, crop class and GPS location recorded on a standardised recording sheet. The survey team was delegated to several tasks during the survey: one member of the team conducted three

Figure 1 Indonesian Sugar Research Institute and Sugar Research Australia teams conducting sugarcane streak mosaic virus survey in Sumatra, Indonesia in August 2016.

random transects (usually with the help of a local assistant) of 10–20 m each. For each transect the total number of stools was recorded, and the presence of pests and diseases recorded. (It should be noted that the normal number of transects done is usually 10, however, because this survey was attempting to visit many sites, it was reduced to three for brevity.) Two members of the team (the pathologists) inspected the area, looking only for SCSMV – when an infected stool was found, samples were taken for future molecular diagnostic testing. The pathologists also recorded and sampled other pests and diseases that were found, and made note of pests and diseases recorded previously in the area, but were not present during the survey. The remaining survey members took insect specimens from within the field for identification and testing. Each field was visited for 30 min to 1 h with up to eight sites visited per production area. The specimens were collected and are to be analysed by molecular methods to determine the distribution of SCSMV and identify any possible vectors. The survey has collected information on other endemic pests, and has given some information on disease freedom from exotic pests as well.

The results of the delimiting surveys will give information regarding the distribution of SCSMV in Indonesia. The data will be used with other project findings to develop monitoring and management programmes in Indonesia and for incursion management plan and PRA for Australia.

5 Effective quarantine arrangements

5.1 Facilities

There are two main types of quarantine facilities for sugarcane import and export that will be discussed in this section: These have been defined as the 'through' facilities and the 'endpoint' facilities.

5.1.1 Visacane

The biggest and most used 'through' facility for sugarcane germ plasm is Visacane located in Montpellier, France, away from other sugarcane production areas (Guinet et al., 2015). Visacane is CIRAD's (Centre de coopération internationale en recherche agronomique pour le développement) sugarcane quarantine unit: it is a comprehensive service that facilitates exchange of sugarcane varieties, acting as an import and export quarantine service in compliance with plant breeders' intellectual property rights. Visacane is integrated in a plant pathology research unit and facilitates the safe movement of sugarcane germ plasm to 30 partners worldwide (Guinet et al., 2015). Over the last ten years, Visacane has imported 1033 varieties from 16 producing countries and exported more than 7200 batches of cuttings of disease-free varieties to sugarcane centres worldwide. From 2006 to 2015 major sugarcane diseases were detected in 22% of the materials (Guinet et al., 2016).

5.1.2 'Endpoint' facilities

Most of the other sugarcane quarantine facilities are 'endpoint' facilities. These are defined in the present chapter as facilities that source varieties for use in that country's industry: the varieties are brought into the country for exclusive use in the local breeding programme, commercial production, germ plasm collections and so on. Endpoint facilities can be run by government, private companies or a combination of both. This depends on the requirements and arrangements by the NPPO for each country and is regulated by the relevant government departments.

An example of a completely government-run unit is in the United States by USDA-APHIS (United States Department of Agriculture – Animal and Plant Health Inspection Service). The quarantine is facilitated completely by APHIS, with importing, plant observation and pathogen testing being coordinated by the Plant Germplasm Quarantine Program in Beltsville, Maryland.

An example of a joint facility is in Australia. The import conditions are determined by the Australian Government, but SRA has been approved to conduct quarantine and pathogen tests (Thompson et al., 2012).

An example of a private procedure is one where there are no government-enforced importing conditions. These are usually found where sugarcane is not a major crop, usually if the germ plasm is being imported for laboratory or glasshouse testing. This is a rare, special case and it is up to the private importer to negotiate pathogen freedom for the materials.

5.2 Sugarcane quarantine

Sugarcane quarantine for germ plasm transfer is managed very similarly around the world. It is usually based on the technical guidelines produced by the FAO/IBPGR in 1993 (Frison and Putter, 1993). Despite these guidelines being 25 years old, they are still relevant for designing basic methods to prepare and ship sugarcane setts, tissue culture plantlets and true seeds. What is absent from these guidelines are specific country requirements, importer requirements and specific pathogen testing procedures that have been implemented.

Recently, the documentation of sugarcane quarantine practices has been published for several importers: notably Australia (Thompson et al., 2012) and Visacane (Guinet et al., 2015). In addition, information from facilities around the world has been used to give the general quarantine procedures and pathogen testing methods that are detailed below as important factors to consider.

5.2.1 Closed facility requirements

The first year (or cycle) of quarantine is done in an enclosed facility. The general considerations for these are:

- Isolated from commercial sugarcane crops.
  - Recommend at least 100 km, or a natural barrier.
  - Examples: inside urban zone, island, across a mountain range.
- Isolated from other sugarcane and other plants.
  - Plants under quarantine should be housed separately from plants not under quarantine.
  - Ability to keep consignments separate.
- Insect-proof.
  - Ability to treat for insects and identify common pests.
- Washing facilities.
  - Can range from wash-in/wash-out to footbath only, depending on local requirements.

- Ability to maintain sugarcane for multiple growth cycles.
  - More than one chamber.
- Treatment facilities.
  - Hot water facility, tissue culture, etc.
- Ability to conduct pathogen testing/disease screening.
  - Diagnostic laboratory.
- Ability to have regular inspections of the plants by trained staff.
  - At least weekly to detect any unusual symptoms.
  - Unusual symptoms may indicate a disease of unknown aetiology.

5.2.2 Open facility requirements

The second year (or cycle) is often completed in an enclosed facility as above, however, in some circumstances an open quarantine plot is used. In these cases, infection by endemic pests must be considered as a possibility, so considerations are:

- Isolation from commercial sugarcane crops.
- Isolation from other sugarcane plants and other plants.
- Use of guard rows.
- Regular inspection for pests and diseases.
- Weed, pest and disease control.

5.2.3 Disease and pathogen testing laboratory requirements

- Each pathogen is diagnosed separately using bioassays, serological or nucleic acid-based methods. A comparison between Visacane and SRA is shown in Table 2.
- Methods include ELISA (enzyme-linked immunosorbent assay), TBIA (tissue blot immune assay), PCR (polymerase chain reaction), RT-PCR (reverse transcription-PCR), qPCR (quantitative PCR), RT-qPCR, metagenomics, sequencing, etc.
- All tests must be validated by the diagnostic laboratory using appropriate positive controls.
- All tests must be done at the appropriate time for disease expression and/or detection. This should be validated for each pathogen.
- Appropriate tissues must be used for diagnostics.
- All tests must have appropriate positive and negative controls.
  - Includes controls for the quality of the extracted nucleic acid (such as endogenous or exogenous controls).

5.3 Basic method for germ plasm exchange

Individual practices for germ plasm exchange may vary based on the phytosanitary measures prescribed on the import permit, or requested by the importer, and these should be followed for the first instance. More general methods described below have been adapted from Frison and Putter (1993): in the most part, the practices are very similar.

Materials should be obtained from a nursery or growth area that is free from major pests and diseases. The germ plasm should be sourced from a plant crop or first ratoon plant that has undergone a cold soak in running water for 40–48 h followed by long hot water treatment (50°C for 3 h; CS-LHWT). All implements used in the preparation of the materials should be sterilised in 70% w/v ethanol or other suitable disinfectant.

5.3.1 Setts

For setts, they should be cleared of all dirt, debris, leaves and leaf sheaths and cut into appropriate lengths (recommend 3-bud setts) and given a short hot water treatment at 50°C for 30 min–1 h before dipping in an appropriate fungicide. The ends of the setts can be dipped in low-melting point paraffin wax to slow the degradation of the setts, and they should be labelled directly with the variety name. The setts are shipped with all the appropriate documentation as determined by the import permit and NPPO. Upon receipt and determining that there are no insects or other contaminants, the setts can be cut into one-bud setts and given a short hot water treatment. Pre-germination can be done in germination cabinets, or they can be planted directly into the quarantine facility.

5.3.2 Tissue cultures

Tissue cultures should be derived from shoot tip or buds – they should not be derived from callus culture to minimise off-types. No antibiotics or charcoal should be added to the final shipment media, and the vessels should be clear to allow inspection of the consignment. Some shipments are done successfully in ziplock bags with minimal amounts of media, and others in plastic culture vessels, and others on tissue culture plates. The plantlets are shipped with all the appropriate documentation as determined by the import permit and NPPO.

Once received and after determining that there are no contaminants, the tissue culture plants can be de-flasked and planted into appropriate potting mix.

5.3.3 Quarantine of clonal materials

Any excess materials, including dunnage, containers and so on, should be disposed of as quarantine waste as per the NPPO facility guidelines.

Varieties are maintained in the facility for at least one growth cycle. During this time testing for disease freedom (Table 1) and regular observations are conducted. If a pathogen under quarantine restriction is identified, then the directions of the NPPO must be followed: in some cases, elimination can be done by several different treatments. If a pathogen not under NPPO restriction is found then elimination and re-testing is suggested. If the pathogen cannot be eliminated, it is recommended that the variety be discarded, to ensure that there is no importing of new variants of the pathogen. Local discretion can be made whether to discard or keep the variety if the pathogen is considered a minor risk.

Common elimination treatments include CS-LHWT for bacterial and fungal pathogens, and apical meristem tip culture for virus and phytoplasma (Chatenet et al., 2001). These elimination treatments are not 100% effective, so it is usual practice to place any treated material back into the first year of quarantine cycle, and tested again for disease freedom.

It is important to keep plants in the two growth periods or cycles of quarantine separate. This is to stop any potential disease spread from untested plants to tested ones. A common

Table 1 A comparison of diagnostic methods used at two sugarcane quarantine facilities for a range of high-risk pathogens

		Diagnostic method and reference	
Disease	Pathogen	Sugar Research Australia*	Visacane*
Leaf scald	*Xanthomonas albilineans*	Bio-PCR (Wang et al., 1999)	Isolation (Davis et al., 1994b)
Ratoon stunting disease	*Leifsonia xyli* subsp. *xyli*	ELISA (Croft et al., 1994) PCR (Pan et al., 1998)	TBIA (Davis et al., 1994a)
Fiji leaf gall	Fiji disease virus	RT-PCR (Smith and Van de Velde, 1994)	RT-PCR (Smith and Van de Velde, 1994)
Mosaic	Sugarcane mosaic virus Sorghum mosaic virus	ELISA (Agdia PathoScreen Potyvirus ELISA kit) RT-PCR (Smith and Van de Velde, 1994; Langeveld et al., 1991; Alegria et al., 2003; Geering et al., 2004)	RT-PCR (Marie-Jeanne et al., 2000)
Streak mosaic	Sugarcane streak mosaic virus	RT-PCR (Damayanti and Putra, 2011)	RT-PCR (Li et al., 2008)
Streak	Sugarcane streak virus, Sugarcane streak Egypt virus, Sugarcane streak Reunion virus, Maize streak virus	PCR (Rybicki and Hughes, 1990; James et al., 2004)	Unpublished PCR (CIRAD)
Ramu stunt	Ramu stunt virus (unassigned)	RT-PCR (Braithwaite et al., 2007)	RT-PCR (Braithwaite et al., 2007)
Yellow leaf	Sugarcane yellow leaf virus	RT-PCR (Girard et al., 2010)	Leaf TBIA and RT-PCR (Girard et al., 2010; Girard et al., 2012) Stalk TBIA (Rassaby et al., 2004)
Grassy shoot White leaf	Grassy shoot phytoplasma White leaf phytoplasma	Nested PCR and amplicon sequencing (Deng and Hiruki, 1991; Tran-Nguyen et al., 2000; Schneider et al., 1995)	Nested PCR (Deng and Hiruki, 1991; Smart et al., 1996; Lorenz et al., 1995; Padovan et al., 1995)
Red leaf mottle	Peanut clump virus	RT-PCR (Bragard and Delfosse, pers comm)	RT-PCR (Bragard and Delfosse, pers comm)

* Thompson et al. (2012) and author information.

* Guinet et al. (2015).

practice is to keep consignments separate and to test all plants at the same time. If one variety in a consignment is positive, all plants in the consignment should be re-tested to ensure no spread of the disease. This is especially important for diseases that can be spread easily in a glasshouse, or if there are potential vectors present at any time during the quarantine.

Once all varieties in a consignment have tested negative, one stalk is selected for CS-LHWT and replanting. It is important to choose the stalk that was tested for disease freedom to eliminate the possibility that two stalks have come from two different plants. The variety is replanted into Cycle 2, which is commonly the 'propagation' phase or an open quarantine phase. Observation of the ratoon is done for as long as possible: this depends on the facility and NPPO direction. The replant is also observed for any unusual symptoms, and tests repeated if necessary or directed. The second cycle can take on different forms depending on the operator, for example at the SRA quarantine the plants are grown for 4–5 months and then propagated by tissue culture for mass release into the SRA variety selection programme. At Visacane, these varieties are grown for 9–12 months and then ready for exporting to the requestor and so must meet import requirements of the next country (Guinet et al., 2015).

Each NPPO will have different arrangements and different practices to determine when a plant is deemed to have 'passed' quarantine and can be released into the country. In Australia, for example, the 'release' from quarantine restriction occurs immediately after pathogen testing and treatment has been completed. The propagation phase is therefore not under quarantine control.

5.3.4 True seed

Sugarcane true seed exchange is of growing importance, and many countries are starting to use true seed to allow local selection of exotic germ plasm. Using true seed can maximise the genetic potential from parents, as it allows the selection of varieties to be done under local conditions, and any resulting varieties can be widely distributed domestically without restrictions (Hoy et al., 2015). There are very few studies done on the transmission of diseases through seed, with no evidence of pathogen transmission through the seed (Maroon-Lango et al., 2013). Quarantine conditions for seed also generally follow the guidelines of Frison and Putter (1993). In brief, the seed is de-fuzzed, treated with a broad-spectrum fungicide and inspected for any contaminants. The seed is then shipped with documentation as per the import permit and NPPO. Upon arrival, the seed is inspected, then grown in containment for a minimum of three months. It is recommended that disease screening occur at this stage to focus on viruses, as these are the most likely to be transmitted through seed. Close observation and pathogen testing should occur throughout the quarantine period, and abnormal findings investigated. Depending on the NPPO direction, the seedlings can be transferred to an open quarantine and inspected regularly throughout the growth cycle. Usually the seedlings are directly released from quarantine restriction after completion of the testing and inspection.

5.4 Pathogen diagnostics

The main pathogens that NPPOs focus on for quarantine diagnostics are those that cause exotic diseases. Endemic pathogens are rarely present on import permits. In some cases, specified pathogens and diseases may be from other crops, or have no diagnostic test readily available. In most cases, observation for disease symptoms is required by the NPPO if there is no diagnostic test available, and other diagnostic laboratories are required for some tests. Some NPPOs have a standard requirement for sugarcane germ plasm, and others have completed PRA for each sending country. The import permits and conditions can vary widely between countries.

Most sugarcane quarantine disease-testing facilities test for exotic pathogens, endemic pathogens, and do routine observations to ensure a clean source of materials. Table 1 shows a comparison of two facilities: Visacane and SRA, taken from publications and updated with new information for SRA.

The specific methods used for diagnostics vary; however, the pathogens tested for are the same, and each test has been validated for use in the laboratory. Other regions test for different pathogens – for example, USDA-APHIS tests for Sugarcane striate mosaic virus (Thompson and Randles, 2001); however, these are generally more minor pathogens that have only been found in a limited range and so can be easily excluded by the exporter. Also notably absent are fungal pathogens, oomycetes, some bacterial pathogens and leaf diseases. The treatments that setts or tissue culture plants undergo prior to exporting will eliminate most of these diseases, and they will be found by observation.

6 Dealing with disease/pest incursions: a case study

Sugarcane smut incursion in Australia

Sugarcane smut (*Sporisorium scitamineum*) was recognised as a major pest risk to Australia during the worldwide outbreaks throughout the twentieth century. A PRA was done by BSES on behalf of the Australian Government in 1997 (Croft and Magarey, 1997a), and contingency plans (Croft and Magarey, 1997b) and a sugar industry biosecurity plan developed for incursion management. Sugarcane smut was identified for the first time in Australia in July 1998 in the Ord River Irrigation Area (ORIA), located in northern Western Australia (Riley et al., 1999). The ORIA was a relatively small sugarcane production area (~1% of Australia's production) established in 1995 and located more than 1500 km away from the major sugarcane-producing areas on the east coast of Australia (Fig. 2). The most likely source of the smut was wind-blown spores from Indonesia, and it spread rapidly in the plantation (Croft and Braithwaite, 2006). The initial incursion could not be eradicated, and containment of smut in the ORIA was attempted by ploughing out heavily infested fields, removing susceptible varieties, imposing strict quarantine restrictions on movement of plants and machinery between ORIA and the eastern sugarcane production areas (Croft and Braithwaite, 2006). The eastern sugarcane production area also began to prepare for an incursion by surveillance, education, monitoring and accelerating research into sugarcane smut. Research areas included determining the likely impact of sugarcane smut by screening Australian varieties for resistance both in the ORIA and in offshore screening in Indonesia (Croft et al., 2000); increasing imports of germ plasm with high resistance to sugarcane smut; determining the genetic variability of sugarcane smut (Croft and Braithwaite, 2006); and having a diagnostic test available and ready to deploy (James et al., 2007).

Sugarcane smut was first found on the east coast of Australia in June 2006 in the Childers district (Table 2). This triggered an Emergency Plant Pest Response under the then current PLANTPLAN, which was coordinated by the Queensland Department of Primary Industries and Fisheries (QDPI&F), BSES and CANEGROWERS. The response included communications, surveillance, quarantine declarations and enforcement (Croft et al., 2008).

The identification of a characteristic whip was diagnostic of the disease; however, confirmation by another method was required, so PCR-based diagnostics was completed and morphological characterisation of the spores done. The PCR was done within two days of the initial finding, and the morphological characterisation done shortly after (Croft et al., 2008).

Figure 2 Map of the sugar industry in Australia, reproduced with permission from http://asmc.com.au/industryoverview/. The inset shows the Ord river irrigation area (ORIA) in Western Australia and its location relative to the major sugar producing regions on the Australian east coast.

Table 2 Timeline of the sugarcane smut incursion in Australia

1997	Pest risk analysis and contingency plan developed Offshore resistance screening started (1998–2007)
July 1998	Sugarcane smut identified in ORIA, Western Australia Ploughing-out and use of more resistant varieties in ORIA Monitoring of smut in ORIA Intensified state-to-state quarantine Surveys of Eastern Australia Education of sugar industry and public Molecular characterisation of smut in ORIA Intensification of smut resistance screening programmes in ORIA Updating of contingency and biosecurity response plans
8 June 2006	Sugarcane smut identified near Childers, Queensland
9 June 2006	Sugarcane smut confirmed by pathologist on-farm at Childers samples sent for analysis to laboratory OCPPO and Qld Chief PPO notified of possible incursion
10 June 2006	PCR diagnostics completed and confirmed sugarcane smut by BSES Incursion formally confirmed and notified to OCPPO and QCPPO Meeting of response managers (QDPI&F, BSES, CANEGROWERS)
12 June 2006	Morphological confirmation of smut spores by QDPI&F
14 June 2006	Consultative Committee on Emergency Plant Pests (CCEPP) meeting to ask for industry response plan (IRP) from the response managers
26 June 2006	CCEPP met to discuss IRP
June 2006	Smutbuster programme initiated to accelerate breeding of smut-resistant sugarcane varieties Smut resistance screening commences in Bundaberg region
7 July 2006	CCEPP met to request revision of IRP and economic analysis for benefits of eradication to be decided
18 July 2006	CCEPP met to discuss progress
26 July 2006	NMG (National Management Group, representing all Australian states) met to consider revised IRP and request further revisions
16 Aug 2006	NMG met to consider final version of IRP The NMG were unable to come to a unanimous decision, so cost-sharing arrangements were not activated and no funds were available for grower re-imbursement or destruction costs The response managers resolved to continue the emergency response, and agreed-upon compensation to growers
7 Nov 2006	Sugarcane smut detected north of Mackay Disease determined to be widespread and not able to be eradicated
Nov 2006	Spore trapping begun for (1) epidemiology and (2) early detection of smut
15 Dec 2006	Sugarcane smut detected in Herbert region
22 Nov 2007	Sugarcane smut detected in Maryborough region
Dec 2007	Sugarcane smut detected in Proserpine region
Sept 2008	Sugarcane smut detected in Northern regions
Oct 2008	Sugarcane smut detected in Burdekin region

Nov 2009	Sugarcane smut detected in NSW New Sugar Industry Biosecurity Plan developed (V2)
Dec 2009	Sugarcane smut found at Woodford
2010–present	Sugarcane smut widespread in most districts Widespread distribution of resistant varieties through accelerated resistant screening programme Reviews of emergency response completed 2016: New Sugar Industry Biosecurity Plan developed (V3)

Delimiting surveillance was done using extensive trace-back to identify plant sources and machinery movements. The surveillance in the initial region was very intensive, and predicted 99% probability of detecting disease if 10% of the row in a farm contained an infected plant. The surveillance was targeted to inspect 110 rows per property in susceptible varieties. Outside of the initial infection area, surveillance was less intense, but still predicted a 95% confidence interval for detecting disease present on 10% of farms (Croft et al., 2008).

In November 2006, smut infection was confirmed north of Mackay and it was concluded that the disease was established and widespread, therefore eradication was unlikely to be successful. Surveillance continued with the use of local productivity services, BSES and mill staff on the ground and deploying spore traps for early detection (Magarey et al., 2008, 2009). The disease spread to other regions over the next few years, and a campaign to limit crop losses was started by limiting the planting of susceptible varieties, ploughing-out susceptible varieties and replacement with more resistant varieties. Predictions were made for estimated crop losses, and these proved to be quite accurate on a wide scale (Antony et al., 2010).

Some production areas were able to replace their susceptible varieties before smut became a major problem in their area, therefore yield losses in those areas were lessened (Magarey et al., 2010, 2011). In some areas, however, the reliance on a limited number of varieties has caused other problems, with reports on increases of other diseases in those areas.

Sugarcane smut resistance screening has become part of the ongoing plant improvement programme, and there was an enhanced effort in the breeding programme, for example the 'Smutbuster' programme (Bhuiyan et al., 2013), which is now starting to produce more varieties. In the 10 years since the sugarcane smut incursion into Eastern Australia, and now, it has become another disease that has been managed. The management strategy has not deviated from what was predicted: clean planting materials, resistant varieties and fungicide treatment at planting has become the standard practice for controlling sugarcane smut in Australia.

Sugarcane smut was the first test of the Australian Government emergency response under the EPPD, and many reviews of the response, the contingency plans and the operations have been completed (Antony et al., 2010; Willcox et al., 2008). Changes to the plans have been made in response to identified problems; however, some seem to have lapsed from their initial urgency. Many of the operational problems faced during the incursion are expected to rise again, and recommendations for a 'trial run' or 'mock incursion' are suggested to take place in the near future. A new Sugar Industry Biosecurity Plan has been developed (Anonymous, 2016a) and it includes updated PRA for pests and diseases. The plans are in place to respond to the next incursion.

7 Future trends and conclusion

There are many challenges involved in quarantine, from facility management to government policy to controlling for all diseases.

Emerging diseases and first reports

These can be a problem when the reports are not timely or well described. PRA are reassessed when first reports are published and restrictions are put in place to protect the importing country. First reports need careful assessment and follow-up information including delimiting surveys in order for an accurate PRA to inform import conditions.

Some published first reports include diseases detected in quarantine and if the diseases are intercepted and eliminated, then this should be clearly stated. In addition, feedback should be provided to the exporting country, and NPPOs often have guidelines and recommendations for interception, first reports and incursion management that need to be considered.

Developing and updating protocols for detection

There are hundreds of different methods for diagnosing sugarcane pathogens: the importance of validation on 'real-world' positive control samples by the diagnostic laboratory is paramount. If the laboratory is changing from one test to another, the tests should be run side by side to show that the new test is 'better' than the old test, for example, has higher specificity, more sensitivity and so on. If required, the test must also be approved by the NPPO.

Availability of the positive controls for test development and validation is important, and researchers should collaborate on this area.

New methods of detection for unknown pathogens

Metagenomics and other next-generation sequencing tools have the potential to change the face of diagnostics. One challenge related to current diagnostics is the highly specific nature of the majority of tests developed, leading to pathogen variants not being detected. New and uncharacterised pathogens also will go undetected. A method being investigated is a metagenomics approach targeting all viruses infecting sugarcane (virome) and using sophisticated sequencing and bioinformatics to identify unknown targets (Roossinck et al., 2015; Candresse et al., 2014; Guinet et al., 2015). With the complete sequencing of the sugarcane genome still ongoing, this could provide new methods for quarantine diagnostics in the future.

8 Where to look for further information

The International Plant Protection Convention website (https://ippc.int/en/) has a range of relevant documents including those listed in Table 3.

Table 3 List of some relevant ISPMs

Doc #	Title	Publication date
ISPM 01	Phytosanitary principles for the protection of plants and the application of phytosanitary measures in international trade	14 January 2016
ISPM 02	Framework for pest risk analysis	14 January 2016
ISPM 04	Requirements for the establishment of pest-free areas	14 January 2016
ISPM 05	Glossary of phytosanitary terms (as adopted by Commission on Phytosanitary Measures (CPM)-11)	20 May 2016
ISPM 06	Guidelines for surveillance	14 January 2016
ISPM 07	Phytosanitary certification system	14 January 2016
ISPM 08	Determination of pest status in an area	14 January 2016
ISPM 09	Guidelines for pest eradication programmes	14 January 2016
ISPM 10	Requirements for the establishment of pest-free places of production and pest-free production sites	14 January 2016
ISPM 11	Pest risk analysis for quarantine pests	14 January 2016
ISPM 12 (as revised by CPM-9, 2014)	Phytosanitary certificates	14 January 2016
ISPM 21	Pest risk analysis for regulated non-quarantine pests	14 January 2016
ISPM 22	Requirements for the establishment of areas of low pest prevalence	14 January 2016
ISPM 28	Phytosanitary treatments for regulated pests	14 January 2016
ISPM 29	Recognition of pest-free areas and areas of low pest prevalence	14 January 2016
ISPM 32	Categorisation of commodities according to their pest risk	14 January 2016
ISPM 34	Design and operation of post-entry quarantine stations for plants	14 January 2016

9 Appendix

Annex 1

Pest risk assessment charts adapted from Plant Health Australia (Anonymous, 2013).

We will use an example pathogen to complete the PRA charts. In this case, a systemic fungal disease with fragile spores (short viability) to enter a geographically nearby area that is highly susceptible to the pathogen. However, there are cultural practices in place inside the at-risk area that may limit the spread.

1 Entry, Establishment and Spread Potential assessed using definitions below

Define the *entry potential*:

Entry potential	Definition
Negligible	The probability of entry is extremely low given the combination of all known factors including the geographic distribution of the pest, quarantine practices applied, probability of pest survival in transit and pathways for pest entry and distribution to a suitable host.
Low	The probability of entry is low, but clearly possible given the expected combination of factors described above.
Medium	Pest entry is likely, given the combination of factors described above.
High	Pest entry is very likely and potentially frequent given the combination of factors described above.
Unknown	The pest entry potential is unknown or very little of value is known.

For the example pathogen: Entry potential is 'Medium'. The fragile spores mean that natural spread is probably low, however, the geographical area is nearby.

Define the *establishment potential:*

Establishment potential	Definition
Negligible	The pest has limited potential to survive and become established given the combination of all known factors.
Low	The pest has the potential to survive and become established in approximately one-third or less of the range of hosts. The pest could have a low probability of contact with susceptible hosts.
Medium	The pest has the potential to survive and become established in between approximately one-third and two-thirds of the range of hosts.
High	The pest has potential to survive and become established throughout most or all of the range of hosts. Distribution is not limited by environmental conditions that prevail. Based upon its current world distribution, and known conditions of survival, it is likely to survive wherever major hosts are grown.
Unknown	The establishment potential of the pest is unknown or very little of value is known.

For the example pathogen: Establishment potential is 'High'. We know that the pathogen is widespread in similar regions.

Define the *spread potential*:

Spread potential	Definition
Negligible	The pest has very limited potential for spread given the combination of dispersal mechanisms, availability of hosts, vector presence, industry practices and geographic and climatic barriers
Low	The pest has the potential for natural or assisted spread to susceptible hosts yet is hindered by a number of the above factors

Medium	The pest has an increased likelihood of spread due to the above factors
High	The natural spread of the pest to most production areas is largely unhindered and assisted spread is also difficult to manage
Unknown	The spread potential is unknown or very little of value is known

For the example pathogen: The spread potential is 'Low'. This should be determined by a panel of experts and the areas of concern debated.

2 Combine descriptive likelihoods using matrix below to obtain one 'rating' for entry, establishment and spread. Combine in order of entry, establishment and spread.

Likelihood	High	Medium	Low	Negligible
High	High	Medium	Low	Negligible
Medium	Medium	Low	Low	Negligible
Low	Low	Low	Extremely low	Negligible
Negligible	Negligible	Negligible	Negligible	Negligible

For the example pathogen: Combine the 'Medium' entry with 'High' establishment to get 'Medium' for entry and establishment, then combine with 'Medium' spread potential. The rating for entry, establishment and spread is therefore 'Low'. At this stage, it is useful to check the assumptions and information used to gain the individual ratings, especially is expert opinion does not think that the ratings are accurate.

3 Assess the economic impact. Use as much data as possible to make a conclusion.

Economic impact	Definition
Negligible	There are very minor, often undetectable, impacts on production with insignificant changes to host longevity, crop quality, production costs or storage ability. There are no restrictions to market access.
Very low	There are minor, yet measurable, impacts on production including host longevity, crop quality, production costs or storage ability. There are no restrictions to market access.
Low	There are measurable impacts to production including host mortality, reduction in yield, production costs, crop quality, storage losses and/or minimal impacts on market access.
Medium	There are significant impacts on production with either host mortality, reduction in yield, production costs, crop quality, storage losses and/or moderate impacts on market access.
High	There are severe impacts on production including host mortality and significant impacts on either crop quality or storage losses, and/or severe impacts on market access.
Extreme	There is extreme impact on standing crop at all stages of maturity, with high host mortality or unmanageable impacts to crop production and quality, and/or extreme, long-term impacts on market access.
Unknown	The economic potential of the pest is unknown or very little of value is known.

For the example pathogen: The economic impact is deemed to be 'High' due to potential complete crop losses, but there are management tools already available.

4 Combine the 'entry, establishment and spread' matrix result with the economic impact to determine an overall risk rating. Specialists should be used to provide information and to 'logic check' the assessment.

Economic consequence rating	Combined likelihood rating				
	High	Medium	Low	Extremely low	Negligible
Extreme	Extreme risk	Extreme risk	High risk	Medium risk	Low risk
High	High risk	High risk	Medium risk	Low risk	Very low risk
Medium	Medium risk	Medium risk	Low risk	Very low risk	Negligible risk
Low	Low	Low	Very low risk	Negligible risk	Negligible risk
Very low	Very low risk	Very low risk	Negligible risk	Negligible risk	Negligible risk
Negligible	Negligible risk	Negligible risk	Negligible risk	Negligible risk	Negligible risk

For the example pathogen: The likelihood rating of 'Low' is combined with the economic consequence of 'High' to give a 'Medium' risk overall. This is very useful for ranking and comparing pests, and in most cases a PRA using this matrix will not be done in isolation.

10 References

Abdel-Rahman, E. M., Ahmed, F. B., Van Den Berg, M. and Way, M. J. 2010. Potential of spectroscopic data sets for sugarcane thrips (Fulmekiola serrata Kobus) damage detection. *International Journal of Remote Sensing*, 31, 4199–216.

Alegria, O. M., Royer, M., Bousalem, M., Chatenet, M., Peterschmitt, M., Girard, J. C. and Rott, P. 2003. Genetic diversity in the coat protein coding region of eighty-six sugarcane mosaic virus isolates from eight countries, particularly from Cameroon and Congo. *Archives of Virology*, 148, 357–72.

Anonymous. 2013. *Pest Risk Assessment for Use in Industry Biosecurity Plans.* Updated July 2013. Canberra, Australia: Plant Health Australia.

Anonymous. 2016a. *Biosecurity Plan for the Sugarcane Industry: A Shared Responsibility Between Government and Industry.* Plant Health Australia, Canberra, Australia.

Anonymous. 2016b. *Government and Plant Industry Cost Sharing Deed in Respect of Emergency Plant Pest Responses (EPPRD).* Plant Health Australia, Canberra, Australia.

Anonymous. 2016c. *International Plant Protection Convention* [Online]. FAO. Available: https://www.ippc.int/[Accessed 22 August 2016 2016].

Antony, G., Magarey, R. and Milford, B. 2010. Sugarcane smut after three years: A policy retrospective. *Proceedings of the Australian Society of Sugar Cane Technologists*, 32,38–49.

Berec, L., Kean, J. M., Epanchin-Niell, R., Liebhold, A. M. and Haight, R. G. 2015. Designing efficient surveys: Spatial arrangement of sample points for detection of invasive species. *Biological Invasions*, 17, 445–59.

Bhuiyan, S., Croft, B. and Cox, M. 2013. Breeding for sugarcane smut resistance in Australia and industry response: 2006–2011. *Proceedings of the Australian Society of Sugar Cane Technologists*, 35, 1–9.

Braithwaite, K., Croft, B. and Magarey, R. 2007. Progress in identifying the cause of Ramu stunt disease of sugarcane. *Proceedings of the Australian Society of Sugar Cane Technologists*, 29, 235–41.

Candresse, T., Filloux, D., Muhire, B., Julian, C., Galzi, S., Fort, G., Bernardo, P., Daugrois, J.-H., Fernandez, E., Martin, D. P., Varsani, A. and Roumagnac, P. 2014. Appearances can be deceptive: Revealing a hidden viral infection with deep sequencing in a plant quarantine context. *Plos One*, 9 (7), doi:10.1371/journal.pone.0102945.

Chatenet, M., Delage, C., Ripolles, M., Irey, M., Lockhart, B. and Rott, P. 2001. Detection of Sugarcane yellow leaf virus in quarantine and production of virus-free sugarcane by apical meristem culture. *Plant Disease*, 85, 1177–80.

Croft, B., Irawan and Berding, N. 2000. Screening Australian sugarcane clones for smut reaction in Indonesia: initial results. *Proceedings of the Australian Society of Sugar Cane Technologists*, 22, 170–7.

Croft, B. and Braithwaite, K. 2006. Management of an incursion of sugarcane smut in Australia. *Australasian Plant Pathology*, 35, 113–22.

Croft, B., Greet, A., Leaman, T., Teakle, D. and Egan, B. 1994. RSD diagnosis and varietal resistance screening in sugarcane using the EB-EIA technique. *Proceedings of the Australian Society of Sugar Cane Technologists*, 16, 143–51.

Croft, B., Magarey, R., Allsopp, P., Cox, M., Willcox, T., Milford, B. and Wallis, E. 2008. Sugarcane smut in Queensland: arrival and emergency response. *Australasian Plant Pathology*, 37, 26–34.

Croft, B. J. and Magarey, R. C. 1997a. *Pest Risk Analysis of Sugarcane for the Northern Australia Quarantine Strategy*. BSES Limited, Brisbane, Australia.

Croft, B. J. and Magarey, R. C. 1997b. *Sugarcane Smut - A Contingency Plan for the Australian Sugar Industry*. BSES Limited, Brisbane, Australia.

Damayanti, T. A. and Putra, L. K. 2011. First occurence of sugarcane streak mosaic virus infecting sugarcane in Indonesia. *Journal of General Plant Pathology*, 77, 72–4.

Davis, M., Dean, J., Miller, J. and Shine, J. 1994a. A method to screen for resistance to ratoon stunting disease of sugarcane. *Sugar Cane*, 6, 9–16.

Davis, M., Rott, P., Baudin, P. and Dean, J. 1994b. Evaluation of selective media and immunoassays for detection of Xanthomonas albilineans, causal agent of sugarcane leaf scald disease. *Plant Disease*, 78, 78–82.

Deng, S. and Hiruki, C. 1991. Amplification of 16S rRNA gnes from culturable and nonculturable mollicutes. *Journal of Microbiological Methods*, 14, 53–62.

Everingham, Y., Lowe, K. H., Donald, D., Coomans, D. and Markley, J. 2007. Advanced satellite imagery to classify sugarcane crop characteristics. *Agronomy for Sustainable Development*, 27, 111–17.

Flood, J. and Day, R. 2016. Managing risks from pests in global commodity networks – policy perspectives. *Food Security*, 8, 89–101.

Frison, E. A. and Putter, C. A. J. (Eds). 1993. *FAO/IBPGR Technical Guidelines for the Safe Movement of Sugarcane Germplasm*. Rome: Food and Agriculture Organization of the United Nations, Rome/International Board for Plant Genetic Resources.

Geering, A., Thomas, J. and Persley, D. 2004. *A National Diagnostic Protocol for the Identification of Maize Dwarf Mosaic Virus*, QDEEDI, prepared for Plant Health Australia as part of National Diagnostic Protocols Initiative.

Girard, J., Fernandez, E., Daugrois, J., Roques, D., Roumagnac, P. and Rott, P. 2010. Genetic diversity of Sugarcane yellow leaf virus in a sugarcane selection plot in Guadeloupe (FWI). In: *Proceedings of the 27th Congress of the International Society of Sugar Cane Technologists (ISSCT)*, Veracruz, Mexico.

Girard, J. C., Fernandez, E., Daugrois, J. H., Roques, D., Roumagnac, P. and Rott, P. 2012. Genetic diversity of Sugarcane yellow leaf virus in a sugarcane selection plot in Guadeloupe (FWI). *International Sugar Journal*, 114, 104–8.

Grisham, M. P., Johnson, R. M. and Zimba, P. V. 2010. Detecting Sugarcane yellow leaf virus infection in asymptomatic leaves with hyperspectral remote sensing and associated leaf pigment changes. *Journal of Virological Methods*, 167, 140–5.

Grund, E., Darissa, O. and Adam, G. 2010. Application of FTA® Cards to Sample Microbial Plant Pathogens for PCR and RT-PCR. *Journal of Phytopathology*, 158, 750–7.

Guinet, I., Girard, J. C., Roumagnac, P., Daugrois, J. H., Fernandez, E. and Rott, P. 2015. Visacane: An innovative quarantine tool for the exchange of pest and disease free sugarcane germplasm. *International Sugar Journal*, 117, 890–6.

Guinet, I., Bousquet, J. F., Darroussat, M. J., Fernandez, E. and Daugrois, J. H. 2016. Visacane, an overview of the distribution of disease-free sugarcane varieties for the last 10 years. In: *Proceedings of the 29th Congress of the International Society of Sugar Cane Technologists (ISSCT)*, 2016, Chiang Mai, Thailand.

Hauser, C. E., Giljohann, K. M., Rigby, M., Herbert, K., Curran, I., Pascoe, C., Williams, N. S., Cousens, R. D. and Moore, J. L. 2016. Practicable methods for delimiting a plant invasion. *Diversity and Distributions*, 22, 136–47.

Hoy, J. W., Hale, A. and Kimbeng, C. 2015. Sugarcane true seed importation: A new genetic resource? *Sugar Journal*. October 2015, 24–25.

James, A., Braithwaite, K., Magarey, R. and Croft, B. 2007. Laboratory support systems for the sugarcane smut effort. *Proceedings of the Australian Society of Sugar Cane Technologists*, 29, 176–82.

James, A., Spall, V., Magarey, R. and Croft, B. 2004. Biotechnology for disease screening and quarantine in Australia. *Proceedings of the Australian Society of Sugar Cane Technologists*, 26.

Johansen, K., Robson, A., Samson, P., Sallam, N., Chandler, K., Eaton, A., Derby, L. and Jennings, J. 2014. Mapping canegrub damage from high spatial resolution satellite imagery. *Proceedings of the Australian Society of Sugar Cane Technologists*, 36, 62–70.

Langeveld, S. A., Dore, J. M., Memelink, J., Derks, A. F. L. M., Van Der Vlugt, C. I. M., Asjes, C. J. and Bol, J. F. 1991. Identification of potyviruses using the polymerase chain reaction with degenerate primers. *Journal of General Virology*, 72, 1531–42.

Li, R., Mock, R., Huang, Q., Abad, J., Hartung, J. and Kinard, G. 2008. A reliable and inexpensive method of nucleic acid extraction for the PCR-based detection of diverse plant pathogens. *Journal of Virological Methods*, 154, 48–55.

Lorenz, K., Schneider, B., Ahrens, U. and Seemüller, E. 1995. Detection of the apple proliferation and pear decline phytoplasmas by PCR amplification of ribosomal and nonribosomal DNA. *Phytopathology*, 85, 771–6.

Magarey, R., Braithwaite, K., Bade, G., Croft, B. and Lonie, K. 2008. Spore trap detection of sugarcane smut: 12 months experience and outcomes. In: *Proceedings of the Australian Society of Sugar Cane Technologists*, 30, pp. 72–86.

Magarey, R., Bade, G., Braithwaite, K., Croft, B. and Lonie, K. 2009. Smut spore trapping studies conducted in Australian east coast production areas in late 2007–2008. *Proceedings of the Australian Society of Sugar Cane Technologists*, 31, 158–65.

Magarey, R., Bull, J., Sheahan, T. and Denney, D. 2010. Yield losses caused by sugarcane smut in several crops in Queensland. *Proceedings of the Australian Society of Sugar Cane Technologists*, 32, 347–54.

Magarey, R., Sheahan, T., Sefton, M., Kerkwyk, R. and Bull, J. 2011. Yield losses from sugarcane smut: Indications from Herbert mill data and from individual crop assessments. *Proceedings of the Australian Society of Sugar Cane Technologists*, 33, 1–9.

Magarey, R. C. 2013. *Diseases of Australian Sugarcane: Field Guide*. Brisbane: Sugar Research Australia.

Marie-Jeanne, V., Ioos, R., Peyre, J., Alliot, B. and Signoret, P. 2000. Differentiation of poaceae potyviruses by reverse transcription-polymerase chain reaction and restriction analysis. *Journal of Phytopathology-Phytopathologische Zeitschrift*, 148, 141–51.

Maroon-Lango, C. J., Hoy, J. W., Comstock, J. C., Grisham, M. P., Mock, R., Hale, A., Afghan, S., Croft, B. J., Cueva, F. D. L., Hoffmann, H., Kennedy, A., Orozco, H., Saumtally, S., Victoria, J. and Viswanathan, R. 2013. Current knowledge and practices related to seed transmission of sugarcane pathogens and movement of seed. *Journal - American Society of Sugar Cane Technologists*, 33, 20–9.

Padovan, A., Gibb, K., Bertaccini, A., Vibio, M., Bonfiglioli, R., Magarey, P. and Sears, B. 1995. Molecular detection of the Australian grapevine yellows phytoplasma and comparison with grapevine yellows phytoplasmas from Italy. *Australian Journal of Grape and Wine Research*, 1, 25–31.

Pan, Y. B., Grisham, M. P., Burner, D. M., Damann, K. E. and Wei, Q. 1998. A polymerase chain reaction protocol for the detection of Clavibacter xyli subsp xyli, the causal bacterium of sugarcane ratoon stunting disease. *Plant Disease*, 82, 285–90.

Potts, J. M., Cox, M. J., Barkley, P., Christian, R., Telford, G. and Burgman, M. A. 2013. Model-based search strategies for plant diseases: A case study using citrus canker (Xanthomonas citri). *Diversity and Distributions*, 19, 590–602.

Rassaby, L., Girard, J. C., Lemaire, O., Costet, L., Irey, M., Kodja, H., Lockhart, B. and Rott, P. 2004. Spread of Sugarcane yellow leaf virus in sugarcane plants and fields on the island of Réunion. *Plant Pathology*, 53, 117–25.

Riley, I., Jubb, T., Egan, B., Croft, B., Singh, V. and Kumar, V. 1999. First outbreak of sugarcane smut in Australia. In: *Proceedings of the XXIII ISSCT Congress, New Delhi, India, 22–26 February,* 1999. Volume 2., 1999. Sugar Technologists' Association of India, New Delhi, India, pp. 333–7.

Rodoni, B. 2009. The role of plant biosecurity in preventing and controlling emerging plant virus disease epidemics. *Virus Research*, 141, 150–7.

Roossinck, M. J., Martin, D. P. and Roumagnac, P. 2015. Plant virus metagenomics: Advances in virus discovery. *Phytopathology*, 105, 716–27.

Rott, P., Bailey, R. A., Comstock, J. C., Croft, B. J. and Saumtally, A. S. 2000. *A Guide to Sugarcane Diseases*. Montpellier: CIRAD and ISSCT.

Rybicki, E. P. and Hughes, F. L. 1990. Detection and typing of Maize streak virus and other distantly related geminiviruses of grasses by polymerase chain reaction amplification of a conserved viral sequence. *Journal of General Virology*, 71, 2519–26.

Samson, P., Sallam, N. and Chandler, K. 2013. *Pests of Australian Sugarcane: Field Guide*. Brisbane: Sugar Research Australia.

Schneider, B., Cousin, M., Klinkong, S. and Seemüller, E. 1995. Taxonomic relatedness and phylogenetic positions of phytoplasmas associated with diseases of faba bean, sunnhemp, sesame, soybean, and eggplant/Taxonomische Verwandtschaft und phylogenetische Stellung von Phytoplasmen, die Krankheiten bei Pferdebohne, Crotalaria, Sesam, Sojabohne und Aubergine hervorrufen. *Zeitschrift für Pflanzenkrankheiten und Pflanzenschutz/Journal of Plant Diseases and Protection*, 102(3), 225–32.

Singh, S. 1995. Pest Risk Analysis and its implications for pest and disease exclusion in international germplasm exchange. In: Croft, B. J., Piggin, C. M., Wallis, E. S. and Hogarth, D. M. (Eds) *Sugarcane Germplasm Conservation and Exchange*. Canberra, ACT, Australia: Australian Centre for International Agricultural Research.

Smart, C. D., Schneider, B., Blomquist, C. L., Guerra, L. J., Harrison, N. A., Ahrens, U., Lorenz, K. H., Seemueller, E. and Kirkpatrick, B. C. 1996. Phytoplasma-specific PCR primers based on sequences of the 16S-23S rRNA spacer region. *Applied and Environmental Microbiology*, 62, 2988–93.

Smith, G. R. and Van De Velde, R. 1994. Detection of sugarcane mosaic virus and Fiji disease virus in diseased sugarcane using the polymerase chain reaction. *Plant Disease*, 78, 557–61.

Stone, L. M. and Byrne, M. 2011. Comparing the outputs of five weed risk assessment models implemented in Australia: Are there consistencies across models? *Plant Protection Quarterly*, 26, 29–35.

Thompson, N. and Randles, J. W. 2001. The genome organisation and taxonomy of Sugarcane striate mosaic associated virus. *Archives of Virology*, 146, 1441–51.

Thompson, N., Wilson, E. and Croft, B. 2012. Sugarcane quarantine disease screening in Australia. *International Sugar Journal*, 114, 577–83.
Tran-Nguyen, L., Blanche, K. R., Egan, B. and Gibb, K. S. 2000. Diversity of phytoplasmas in northern Australian sugarcane and other grasses. *Plant Pathology*, 49, 666–79.
Wang, Z. K., Comstock, J. C., Hatziloukas, E. and Schaad, N. W. 1999. Comparison of PCR, BIO-PCR, DIA, ELISA and isolation on semiselective medium for detection of Xanthomonas albilineans, the causal agent of leaf scald of sugarcane. *Plant Pathology*, 48, 245–52.
Willcox, T. G., Croft, B. J., Sallam, M. N., Allsopp, P. G. and Milford, B. J. 2008. A review of contingency planning and the emergency response to the incursion of sugarcane smut in Queensland. *Proceedings of the Australian Society of Sugar Cane Technologists*, 30, 43–51.
Wilson, E. J., Thompson, N., Magarey, R. C., Echadian, E. M., Kristini, A. and Hidayat, S. H. 2016. Australian collaboration for development of detection methods for Sugarcane streak mosaic virus. *Proceedings of the Australian Society of Sugar Cane Technologists*, poster presentation.

Chapter 8

Viral metagenomics and sugarcane pathogens

Philippe Roumagnac, CIRAD, France; Dimitre Mollov, USDA-ARS, USA; and Jean-Heinrich Daugrois and Denis Filloux, CIRAD, France

1 Introduction

1.1 Plant viral metagenomics: a brief history

Metagenomics, also referred to as environmental or community genomics, is the genomic analysis of environmental samples that contain mixtures of mostly unknown microscopic life forms, many of which are practically unculturable. Although the term 'metagenomics' first appeared in a 1998 article of Jo Handelsman and coworkers (Handelsman et al., 1998), the origins of the discipline can be traced back to the work of Norman Pace and colleagues in 1980s and early 1990s (Schmidt et al., 1991).

During the last decade metagenomics concepts and techniques, first used by fungal and microbial ecologists, have been applied to studies of viral ecology and pathology. However, while for an ecologist one sample is a very small fraction of the ecosystem, for a pathologist one sample is an organism hosting billions of microbes and viruses (Sender et al., 2016). In 2002, a seminal study using a shotgun cloning-based method to sequence aquatic viral communities demonstrated the potential of viral metagenomics to study viromes, that is, the collection of viruses associated with the targeted organism (Breitbart et al., 2002). Metagenomics-based approaches were further promoted by rapid advances in nucleic acid sequencing technologies (next-generation sequencing, NGS) to study viromes at scales ranging from individual

http://dx.doi.org/10.19103/AS.2017.0035.20

organisms to entire communities (Delwart, 2007; Edwards and Rohwer, 2005). The advent of metagenomics-based approaches has led to the discovery and characterization of new viruses infecting animals (including humans) and plants (Lipkin, 2010; Delwart, 2007; Roossinck, 2012).

The feasibility of plant viral metagenomics became apparent in 2009, when two pioneering studies demonstrated that virus-derived small interfering RNA (siRNA) could be used to detect both known and previously uncharacterized plant viruses within infected plants (Donaire et al., 2009; Kreuze et al., 2009). Plant viral metagenomics quickly became popular, and many recent studies have discovered at least 86 unknown plant viruses (Roossinck et al., 2015). In 2010, a spatial plant virus metagenomics approach – also termed 'ecogenomics' – developed by M. Roossinck and coworkers, yielded geographically tagged cDNA sequences from known and unknown virus species (Roossinck et al., 2010). Besides describing many new viruses from uncultivated plants, ecogenomics studies have revealed that few wild plants infected by viruses displayed obvious evidence of virus infections at the time of sampling (Bernardo et al., 2013b; Muthukumar et al., 2009).

The use of plant viral metagenomics also has other practical applications, including studying of viral diversity and evolution and conducting viral surveillance, diagnostic, and large-scale molecular epidemiological assessments, (Bernardo et al., 2013a). While plant viral metagenomics undoubtedly plays a key role in discovering plant viruses and identifying both known and unknown plant viruses (Roossinck et al., 2015), the potential utility of this approach in plant virus diagnostic is still being debated (Massart et al., 2014). Ecogenomics is also likely to improve our knowledge about spatial and temporal distributions of different plant viruses at the agro-ecosystem level, and to help identify ecological parameters that influence variations in virus distribution patterns.

In this chapter, we first describe the plant virus metagenomics methods that were developed in the last decade. We then focus on the use and application of plant virus metagenomics to identify, detect and study both known and newly emerging sugarcane pathogens. Finally we discuss challenges and opportunities of using the most up-to-date technology in plant pathology.

1.2 Metagenomics approaches

Several plant viral metagenomics approaches have been developed in the last decade and at least 66 studies about metagenomics of plant viruses and viroids have been published since 2009 (Wu et al., 2015). Metagenomics approaches are typically based on nucleic acid extraction, followed by a sequence-independent amplification step and the use of an NGS platform and bioinformatics pipelines (Fig. 1). These metagenomics-based approaches are differentiated according to the classes of nucleic acids that are targeted: (i) total RNA or DNA, (ii) double-stranded RNAs (dsRNA), (iii) virion-associated nucleic acids (VANA) purified from virus-like particles and (iv) virus-derived siRNA.

Total RNA/DNA extraction, combined with sequence-independent amplification or whole transcriptome amplification, followed by high-throughput NGS have effectively helped discover viruses and identify the probable causal agent of viral diseases (Adams et al., 2013b; Al Rwahnih et al., 2009). Recently, this approach based on total RNA extraction was, for example, successfully applied to assemble and characterize a virus provisionally named Ramu stunt virus (RmSV), associated with the Ramu stunt disease of sugarcane (Mollov et al., 2016). Ecogenomics studies based on dsRNA extraction have provided both a first estimation of the prevalence of RNA plant viruses at the agro-ecosystem level and the discovery of many new viruses (Roossinck et al., 2010). However,

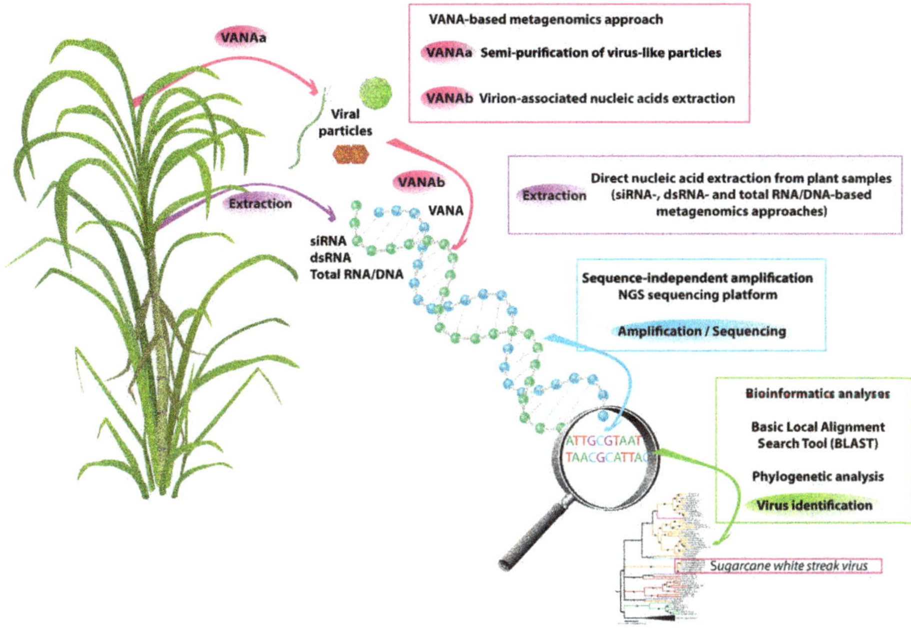

Figure 1 Flow diagram of the steps in metagenomics-based approaches, including the first step of nucleic acids extraction (pink and purple arrows) followed by a sequence-independent amplification step and the use of an NGS platform (blue arrow) and finally the use of bioinformatics pipelines (green arrow).

this approach is time consuming and misses negative sense viruses and thus has been seldom used to detect plant viruses. VANA-based approaches are based on methods that partially purify viral particles from complex samples, with a subsequent extraction of viral nucleic acids. This additional step (in comparison with the other metagenomics-based approaches, Fig. 1) leads to the removal of at least some host nucleic acids. VANA-based approaches have estimated the prevalence of plant viruses at the agro-ecosystem level as ranging from 25.8% to 35.9% in Camargue, France, and in the Western Cape regions of South Africa (Bernardo, 2014). In addition, these studies increased our knowledge about plant virus biodiversity, leading to the discovery of several new viruses using VANA-based approaches and their further characterization (Bernardo et al., 2013b; Roumagnac et al., 2015; Scheets et al., 2011; Thapa et al., 2012). Several viral metagenomics-based studies have targeted short interfering RNAs (siRNAs) generated by the host plant RNA silencing machinery (Roossinck et al., 2015, 22). These small 21 and 24 ntRNAs result from the cleavage of dsRNA found within a plant cell, such as transient dsRNA formed during the replication cycle of RNA viruses. The extraction, sequencing and assembly of these siRNAs have been used to detect both known and unknown plant viruses (Candresse et al., 2014; Kreuze et al., 2009; Roossinck et al., 2015). This approach has also led to the detection of integrated and transcriptionally active endogenous viral elements (EVEs) (Filloux et al., 2015b; Rajeswaran et al., 2014).

We will now describe three recent studies that have used viral metagenomics-based approaches to (i) solve long-standing etiologic enigmas, (ii) identify known and potentially new sugarcane viruses in a sugarcane quarantine context and (iii) assess the diversity of a known sugarcane virus (*Sugarcane yellow leaf virus*).

2 Detecting the multi-segmented Ramu stunt virus

Ramu stunt, known since the 1980s, is a devastating disease of sugarcane that currently only occurs in Papua New Guinea. It is an international quarantine disease that impacts the economic trade of germplasm and hinders cultivar development. The causal organism is a virus (Braithwaite et al., 2007), whose genomic characterization was not completed until recently (Mollov et al., 2016). Symptoms associated with Ramu stunt include light green to yellow streaks varying from small specks to streaks covering the entire length of the leaf (Fig. 2). The genomic sequence of RmSV was obtained by whole transcriptome RNA Seq approach using total RNA extracted from symptomatic sugarcane cv. Ragnar. The total RNA was used to prepare a cDNA library, which was sequenced as 100 bp pair-end reads on an Illumina HiSeq 2500 platform. The sequencing run generated more than 37 million reads. Contigs (a contig is a set of overlapping reads that together represent a consensus region of DNA) were adjusted to include only consensus assemblies between 500 and 10 000 nt. These analyses produced over 18 000 contigs, of which six were identified as related to plant viruses. The six contigs were assigned as RNA 1 through RNA 6 and their total nucleotide count represented the 16 kb RmSV genome. Phylogenetic analysis using the replicases of RmSV and members of *Tenuivirus*, *Phlebovirus*, and *Tospovirus* suggested that RmSV is closely related to the genus *Tenuivirus* (Mollov et al., 2016). In this study, the whole transcriptome RNA Seq approach has proved to be efficient for the assembly and the genome characterization of RmSV, the likely cause of Ramu stunt disease of sugarcane (Mollov et al., 2016).

Figure 2 Symptomatic leaves associated with Ramu stunt virus

3 Discovery of a new sugarcane mastrevirus using viral metagenomics

Besides identifying viruses causing known or emerging diseases (Adams et al., 2013a, 2014; Al Rwahnih et al., 2009; Mollov et al., 2016), metagenomics-based approaches have been used to detect multiple viruses infecting individual quarantined sugarcane plants (Candresse et al., 2014). The siRNA- and VANA-based approaches revealed a mixed infection of *Sugarcane streak Egypt* mastrevirus with another uncharacterized mastrevirus, *Sugarcane white streak virus* (SWSV) (Candresse et al., 2014): The two accessions in the CIRAD sugarcane quarantine programme were known to be infected with the former (by other techniques), but the novel SWSV had escaped detection for the last two decades. SWSV has now been detected in several sugarcane-producing countries, including Sudan, Barbados, Guadeloupe Island and Réunion Island. This study highlights the tremendous potential to use NGS in quarantine and certification programmes to substantially reduce the number of non-intercepted viruses that may accompany the international exchange of plant material. One robust approach might utilize a quarantine screening with a first-stage sequence-independent NGS, followed by sequence-dependent secondary assays. The metagenomics-based screening allows the detection and identification of both known and unknown viruses within a single plant. Virus-specific assays (e.g. PCR/RT-PCR) can then be used to confirm the presence of newly discovered viruses and implement more widespread testing for them.

4 Study of the genetic diversity of *Sugarcane yellow leaf virus*

In 2012, the VANA-based metagenomics approach was used to analyse individually the virome of 300 sugarcane varieties collected from the Biological Resource Centre for Tropical Plants (CRB-PT) located on Guadeloupe Island. Besides characterizing viromes, this study also assessed the diversity of *Sugarcane yellow leaf virus* (SCYLV), which is a phloem-restricted RNA virus (family *Luteoviridae*, genus *Polerovirus*), causing yellow leaf disease. SCYLV is transmitted by aphids and propagated by infected cuttings. Yellow leaf disease emerged two decades ago in Guadeloupe, where it has been observed in several sugarcane clones since 1994 (Daugrois et al., 1999). Since its introduction in Guadeloupe, the virus prevalence has continuously increased in the CRB-PT sugarcane collection and 119 of 189 (63%) CRB-PT varieties tested using RT-PCR in 2010 were infected (Debibakas et al., 2014). Based on phylogenetic analyses of both partial and complete SCYLV genome sequences, SCYLV isolates have been classified into nine different genotypes: BRA from Brazil (Abu et al., 2006), CHN1 (Wang and Zhou, 2010), CHN2 (Wang et al., 2012) and CHN3 (Lin et al., 2014) from China, CUB from Cuba (Abu et al., 2006), HAW from Hawaii (ElSayed et al., 2011), IND from India (Chinnaraja et al., 2013), PER from Peru (Abu et al., 2006), and REU from Réunion Island (Abu et al., 2006). These isolates cluster in three main groups: Group1 – BRA-PER-CHN2-CHN3-HAW; Group2 – CUB-CHN1-IND; and Group3 – REU.

Sugarcane leaf samples were randomly collected from 300 CRB-PT sugarcane varieties and were processed using the 454 pyrosequencing VANA-based metagenomics approach (Filloux et al., 2015a). Mapping the 454 VANA-reads against SCYLV reference genomes

Table 1 Identified genotype in the target regions of ORF1 and ORF3 for each variety and total VANA 454 reads obtained for the variety sample assigned to SCYLV

	Target sequences assigned to genotypes BRA IND REU		
Varieties	ORF1	ORF3	Total reads assigned to SCYLV
B41227	IND	IND	811
B64277	IND	IND	1485
B6623	BRA	BRA	138
B70533	IND+REU	REU	921
B711001	REU	REU	1188
B7410	BRA	BRA	1254
B76359	BRA	BRA	931
B85356	REU	REU	779
Biscuit	IND	IND	342
BBZ85102	IND	IND	305
BJ6902	IND	IND	188
BR6202	IND	IND	624
CB4069	BRA+REU	REU	237
CB4789	BRA	BRA	1017
CO775	REU	REU	699
CO740	BRA+REU	BRA+REU	627
CP52-43	BRA	BRA	2375
CP62-258	REU+IND	BRA+REU	538
CP69-1062	BRA	BRA	576
CR74-250	BRA	BRA	660
CR83323	IND	IND	310
F175	BRA	BRA	2684
Formosa	Bra+IND	Bra+IND	197
FR91699	IND	IND	202
FR91253	REU	REU+IND	137
IAC64257	REU	REU	2038
LF61-5	REU	REU	2151
M292/70	REU	REU	871
M907/61	REU	REU	395
M1227/62	REU	REU	313
Mex68200	BRA	BRA	146

Varieties	Target sequences assigned to genotypes BRA IND REU		Total reads assigned to SCYLV
	ORF1	ORF3	
My515	IND	IND	104
N17	REU	REU	555
NCo334	IND	IND	1243
NG57238	REU	REU	1495
PHIL66-7	BRA	BRA	424
POJ100	BRA	BRA	348
PS41	REU	REU	1013
Q128	REU+IND	REU	1356
R575	REU+IND	REU+IND	1090
R576	REU	REU	1406
ROC1	BRA	BRA	364
ROC3	BRA	BRA	470
ROC7	IND	REU	660
SP783137	IND	IND	128
TC2	IND	IND	3045
TUC74-1	REU	REU	796
TUC75-2	BRA	BRA	2530

revealed that SCYLV-related reads were present in 134 samples, for an infection rate (45%) similar to the previous study on the CRB-PT collection (Debibakas et al., 2014). The diversity of partial ORF1 (positions 1602–1742; genome REU-YL1b, accession number AM072755) and partial ORF3 (positions 3936–4176; genome REU-YL1b, accession number AM072755) was analysed for 48 positive samples using reads mapping against representative complete genomes of each of the SCYLV groups: SCYLV-BRA-YL1 (accession number: AM072750), SCYLV-IND1 (Y236971) and SCYLV-REU-YL1b (AM072755). The viromes of the 48 included 14 with group1 isolates, 12 with group2 isolates, 13 with group3 isolates, and 9 with evidence of mixed SCYLV genotypes (Table 1).

Further analysis of the partial ORF1 and ORF3 consensus sequences among the nine mixed infection plants revealed that eight contained several potential co-infecting ORF alleles (Table 1), and one (ROC7, Table 1) probably contained a recombinant SCYLV isolate. We further amplified, cloned and Sanger-sequenced a fragment encompassing ORF2 to ORF5 from the putative recombinant SCYLV isolate from ROC7 sugarcane varieties (Table 1) and confirmed that this viral isolate harbours a chimeric genomic organization of the ORF2/ORF5 region. RDP4 software (Martin et al., 2015) subsequently confirmed evidence of recombination pattern for the ROC7 ($P = 1.3\ 10^{-67}$) isolate. This recombinant presumably arose from a SCYLV CUB isolate (major parent) and a SCYLV REU isolate (minor parent). While it is well documented that SCYLV is prone to recombination (ElSayed et al., 2012; Lin et al., 2014), this is the first report of recombinant SCYLV isolates in Guadeloupe.

Large scaffolds covering at least 80% of publicly available SCYLV genomes were further assembled from 23 out of the 48 infected varieties, including 22 varieties infected by a single non-recombinant genotype and the ROC7 variety putatively infected by the CUB-REU recombinant described above (Table 1). Phylogenetic analysis demonstrated that SCYLV isolates from Guadeloupe unambiguously branch within the three SCYLV groups (Fig. 3). As expected, the recombinant ROC7 isolate branches deep along the group2 lineage (Fig. 3).

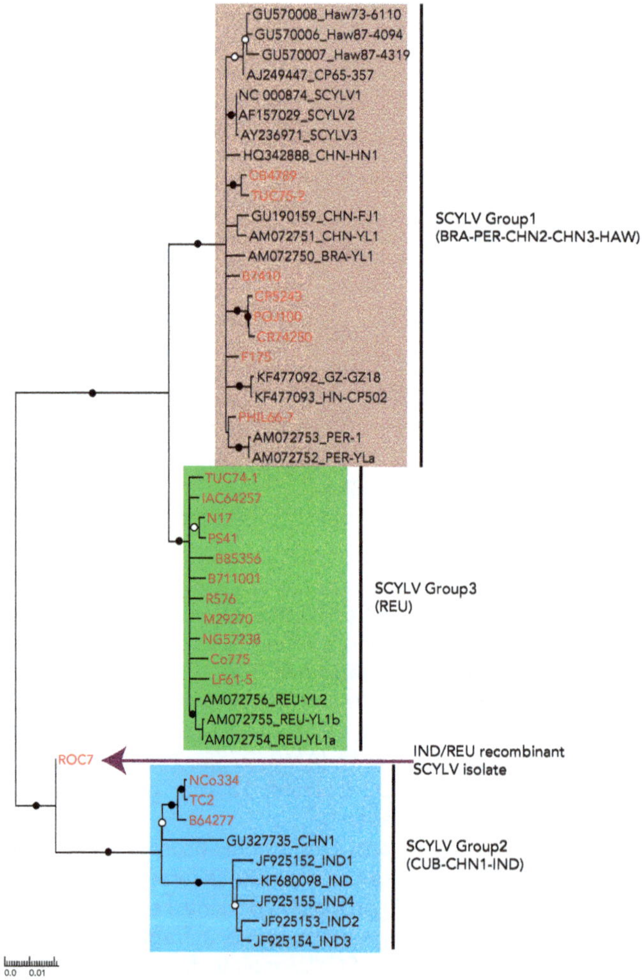

Figure 3 Maximum likelihood tree of SCYLV reconstructed with the current publicly available whole genome of SCYLV isolates and a GTR+G+I nucleotide substitution model (selected as best fit by the Molecular Evolutionary Genetics Analysis (MEGA) software). Five hundred bootstraps replicates were used to test the support of branches. Branches associated with a filled dot have bootstrap support above 90% whereas those with an unfilled dot have bootstrap support above 70%. All branches with less than 50% bootstrap support have been collapsed. SCYLV isolates from Guadeloupe are indicated in red.

This study demonstrated the potential for using metagenomics-based approaches to assess the diversity of sugarcane viruses. These metagenomics-based studies can create a reliable inventory of viruses that are infecting a given host; however, they still only give hints about the diversity of the viruses. Based on these results, in-depth analyses of the viral diversity must be performed using classical molecular methods.

5 Computational identification of plant viruses from metagenomics-based datasets

Using metagenomics for virus discovery and identification requires stepwise computational approaches that include sequence cleaning, *de novo* sequence assembly and/or sequence mapping to known reference sequences. One of the main challenges is to accurately *de novo* assemble metagenomics reads that usually consist of a mixture of DNA from different organisms (viruses, bacteria, fungi or eukaryotic organisms) (Ghurye et al., 2016; Rose et al., 2016). One consequence of the presence of reads amplified from many organisms is the difficulty in identifying true biological differences from sequencing errors (Ghurye et al., 2016). Several approaches have been developed for assembling metagenomics dataset – for example, the greedy method, the Overlap-Layout-Consensus approach and the de-Bruijn graph approach, the last of which has been widely used (Ghurye et al., 2016). This assembly method is based on relationship between substrings of fixed length k (k-mers) derived from the reads (Ghurye et al., 2016). Since *de novo* sequence assembly is most frequently used in viral metagenomics, alignment-based approaches, such as the Basic Local Alignment Search Tool (BLAST), are an essential tool (Altschul et al., 1990). This computational algorithm compares nucleotide or protein sequence similarities to a library or international database of known sequences. Based on the similarity scores, BLAST identifies sequences in the reference database that align with the metagenomics-based query read. This indicates the most likely taxonomic position of the organism from which the read or contig was amplified.

Because large volumes of reads are generated by metagenomics-based approaches, it has been challenging to transform these raw DNA/RNA sequence datasets into biological conclusions. Millions or even billions of sequence reads can be generated from a single NGS protocol and few scientific groups have been able to exploit fully these datasets and tackle their biological issues because of the limitations in computational tools and most scientists' insufficient proficiency in using them. The exponentially increasing metagenomics-based sequence data creates numerous challenges that will need attention. One is the problem of archiving NGS and other big data, which can reach gigabyte and even terabyte levels for a single project. Another is the analysis of metagenomic sequences that will require computational programmes that are increasingly specialized for the clustering of metagenomic sequence reads if the data are to be used effectively. These two challenges will require a high degree of data sharing, community organization and collaboration.

6 Challenges of metagenomics research

6.1 The challenge of 'unknown' reads and contigs

While metagenomics-based approaches are effective for discovering previously unknown viruses and investigating etiologic enigmas, sometimes only a small percentage of the reads can be correlated to known virus species using alignment-based approaches such as BLAST. The majority of the reads and contigs (up to 70% generated during some environmental viral metagenomic studies) have no detectable identity to sequences within public databases (Brum et al., 2016; Rosario and Breitbart, 2011). These uncharacterized sequences have been termed 'viral dark matter' (Brum et al., 2016; Rosario and Breitbart, 2011). In addition, many reads and contigs that are assigned to virus genera or virus families share low amino acid similarities (<50%) to known viral proteins, suggesting that they represent uncharacterized viral species (Rosario and Breitbart, 2011). While it is now hypothesized that viruses represent the largest reservoir of untapped genetic information in the biosphere, our capabilities to identify viruses remain limited. This challenge with the 'viral dark matter' stresses the need to improve several aspects of metagenomics-based studies. Sharing metagenomics datasets by connecting local databases or creating a centralized database for submitting and/or retrieving viral metagenomic data would improve the rate of identification of viruses and probably reduce the volume of 'viral dark matter'. Better characterization of whole viral genomes and their availability in a centralized database would also help identify novel viruses. It is interesting to note that 25 000 bacterial and archaeal host genomes are available in NCBI RefSeq (January 2015), whereas only approximately 1 500 viral genomes are deposited in this international database (Roux et al., 2015). Finally, we probably have to look beyond alignment-based approaches for identifying and classifying metagenomics-based reads. Alternatively, coding sequences (CDS) can be predicted and annotated with domain family homologs, and databases such as the Conserved Domain Database (CDD) (Marchler-Bauer et al., 2015) and Pfam (Finn et al., 2016) can be scanned using position-specific scoring matrices (PSSMs) or hidden Markov models (HMMs) (Lobb et al., 2015). Finally, recent studies using novel approaches, such as comparison of metagenomes using shared k-mers, protein clusters (PCs) and viral populations, have shown that virus-related sequences can be retrieved from the dark matter stock produced by alignment-based approaches (Brum and Sullivan, 2015).

6.2 The ecological role of cryptic viruses

Recent studies have demonstrated that several virus species have mutualistic relationships with their hosts (Roossinck, 2011). Whereas some viruses are essential for the survival of their hosts, others give their hosts a competitive edge in the natural world. It is noteworthy that several studies have shown that both harmless plant viruses and acute plant viruses can reduce the impact of abiotic stress on their host plant (Davis et al., 2015; Marquez et al., 2007; Westwood et al., 2013; Xu et al., 2008).

A seminal study revealed that a panic grass (*Dichanthelium lanuginosum*) commonly found in the geothermal soils of Yellowstone National Park can grow in the high soil temperatures of this region because it is colonized by a fungal endophyte (*Curvularia protuberata*), which is infected by a dsRNA virus (*Curvularia thermal tolerance virus*, (Marquez et al., 2007)). The discovery of this 'ménage-à-trois' (one plant, one fungus, one virus), which is vital for the plant survival in soils with temperatures over 50°C (Marquez et al., 2007), had shed light on the indirect mutualistic relationships of plant viruses with their hosts. Other studies have further shown that acute viral infections can confer tolerance to plant abiotic stresses, such as drought or cold tolerance (Xu et al., 2008). These results highlight the need to further investigate the role of plant viruses in mutualistic relationships with their host, and perhaps even revise our view of plant viruses as being just pathogens. While metagenomics-based studies will provide valuable tools to help decipher plant microbiota, biologists will have to include a holistic view of remarkably complex and specific interactions that can lead to the emergence or the inhibition of plant virus pathogens.

6.3 The challenge of determining causation

Metagenomics-based studies have revealed an unprecedented extraordinary diversity of microorganisms living in a given ecological niche. The description of these complex living microbial communities has challenged the concept of the Koch's postulates, which historically has represented the ultimate standard for establishing the causal nature of an infectious agent. Besides questioning our concepts of viruses as always being pathogens (see section on mutualism) or our capacity to decipher the actual diversity of viruses living in a specific niche (see section on the dark matter), metagenomics-based studies imply complex interactions among multiple microbes, rather than a single pathogen, may frequently cause disease symptoms (Vayssier-Taussat et al., 2014). In certain cases, the methodological shift from disease aetiology to metagenomics is likely to create a paradigm shift in the demonstration of disease causation and to challenge plant sanitation rules (see the following section).

While several propositions suggesting new criteria for microbiota research have emerged during the last decade (Mai et al., 2016; Mokili et al., 2012), it remains difficult to use them to set new universal postulates to determine the causal pathways for microbiota contributions to human, animal and plant health and disease.

6.4 The challenge of sanitation and legal aspects

The international movement of sugarcane germplasm is essential to genetic diversity preservation and sugarcane quality improvement (Maroon-Lango et al., 2013). Frequent and global movement of plant material includes significant risks of transmitting exotic pathogens (Wilkinson et al., 2011), especially viruses that are not detectable by visual examination, for example, most sugarcane cultivars do not exhibit disease symptoms when infected by SCYLV under European or American quarantine settings. This has increased the establishment of strict quarantine and import regulations in many countries. Utilizing NGS technologies to screen germplasm may be advantageous to both regulatory agencies and growers because the use of NGS approaches can speed up quarantine and certification programmes, which will result in reduced quarantine costs and in a faster release of certified plant material (Hadidi et al., 2016, Martin et al., 2016). However, the policy aspects of implementing NGS for quarantine may be challenging. There are many important questions to be addressed: Who is going to process the NGS samples? Data quality control may need to be regulated and strict standards need to be applied for data validation. Once a pathogen is identified, the level of sequence completion sufficient to be considered a positive identification (i.e. the question of whether partial sequences are enough or the whole genome needs to be assembled) needs to be determined. The benefits discussed earlier of using NGS for virus strain differentiation could concurrently pose challenges. If certain strains are present in the region, but more 'exotic' strains are detected on imported material, then how would this finding affect the regulatory considerations for that particular pathogen/strain?

Many unknown pathogens are currently being discovered by NGS (Candresse et al., 2014; Mollov et al., 2016; Roossinck et al., 2015). The biology of these newly discovered viruses and their association to diseases will have to be studied in order to make decisions about safe movement of their host-plant propagation. Martin et al. (2016) have for instance recently proposed that a minimum of graft transmissibility assays should be carried out for newly discovered viruses, viroids, and phytoplasmas expected to be graft transmissible. How to regulate these recently discovered viruses may therefore be a lengthy process that is challenging to both regulatory officials and stakeholders. If organizations apply therapy to germplasm infected with viruses for which little biological data exists before it is released, new protocols aimed to 'clean up' material typically via tissue culture and/or chemical and/or heat therapy and/or cryotherapy may need to be developed, validated and accepted by regulatory agencies. There is also a potential for significant backlog in therapy throughput. These hurdles may add up to many years of delayed germplasm release. Economical, biological and ecological aspects of the newly discovered/novel virus issues should be evaluated and specific pest risk assessments obtained.

The advantages of applying NGS technologies in plant virology are numerous and the technical progress in this field may be too fast for policy makers to adapt. Multiple agencies in several countries (Australia, Canada, France, New Zealand, and the United States) are adopting or have adopted the use of NGS, in addition to the legal diagnostic tests, in quarantine programmes. These agencies have formed internal working groups to discuss the application of NGS to Plant Protection and Quarantine policy and operations (Hadidi et al., 2016). Better coordination between governments to share data and standards should be the starting point in adopting the technology when used to regulate import and export of plant material. As of today there are no international policies regarding NGS use and trading partners need a clear guidance of the regulatory requirements and fiscal responsibilities.

6.5 The challenge of sharing datasets

Open-data sharing within the scientific community is essential for the future advancement of virology. Sharing datasets between labs allow scientists to better characterize new viruses, examine the validity of emerging publications, and make timely and accurate diagnoses. In plant virology, data sharing has played an important role in the classification of new viruses. Viruses are primarily classified using genomic data. For this reason, phylogenetic grouping of viruses is largely dependent on access to other viral genomes through public genomic databases. Large-scale use of these databases has led to significant advancement in standardization of viral taxonomy and phylogenetics (King et al., 2012; Simmonds et al., 2017). Public access databases have also helped create a more uniform standard of data quality and presentation. For that purpose, *The Genomic Standards Consortium* was established by the scientific community to develop an official standardized approach to genomic data collection and exchange (Field et al., 2008, 2014). While researchers are now able to better deposit and disseminate their genomic data, they can also interpret and validate large datasets using analytical tools provided by the international public databases, such as the NCBI Viral Genomes Resource (Brister et al., 2015).

Individually, researchers rely only on complex, incomplete and often small-scale metagenomics or pathogen population analyses to examine pathogen diversity, pathogen-host interactions, pathogen epidemiology, etc. In addition, these elaborate, multifaceted datasets are usually difficult and expensive to reproduce in a timely manner. It is therefore of utmost importance to make these 'local' datasets available to researchers to improve and develop new diagnostic tools and protocols that help detect and manage plant diseases. Rapid data sharing is especially important during outbreaks and epidemics to assemble large datasets that can be used to reconstruct the routes of the pathogen invasion or disentangle the genetic structure of the spreading pathogen (Kallas and O'Connor, 2016).

Despite the many benefits, data sharing can be a topic many scientists avoid. Ambiguity in data ownership laws and policies leaves many researchers reluctant to share datasets before publication. This can delay the development of disease management strategies that are vital to agriculture. Fortunately, as mentioned above, the field of genomics has pioneered a new era of data sharing through large-scale data repositories such as GenBank (NCBI) which make data accessible to the public while also protecting the rights of researchers (Brister et al., 2015).

Other complications may arise when public sector researchers collaborate with corporations because guidelines sharing is often lacking. Hence, when a corporation funds a scientific research, there are usually no clear and uniform policies that dictate who owns the data. Such collaborations must be worked out independently and cases often arise where the corporation does not wish to release certain data to the public. Creating policies that align the interests of scientists, corporations, and the community is a challenge that is still to be resolved.

7 Future application of metagenomics to plant diagnostics, myth or reality?

The advent of metagenomic technology has undoubtedly revolutionized the field of plant pathology and made it possible to identify new viruses by sequencing total nucleic acids, including small RNAs, from plant tissue without prior knowledge of the viral sequence (Prabha et al., 2013; Roossinck et al., 2015). However, there are still many limitations to use these approaches to perform reliable, affordable and portable diagnostics assays (Massart et al., 2014). For instance, inter-laboratory studies are needed to assess the portability of NGS-based protocols and to define standardized detection sensitivity thresholds. Consequently, the gold standard tests for diagnosis of plant viruses are still largely based on PCR-based methodologies (Massart et al., 2014).

The first step in performing a metagenomic analysis is the construction of a metagenomic library. This step alone can be very costly, tedious, and time consuming. Creating a high-quality metagenomic library requires precise methods of nucleic acid extraction, purification, and processing (Lam et al., 2015). Although these methods have been widely commercialized, the diversity of plant hosts and viral pathogens often requires specialized laboratory techniques and processing (Lam et al., 2015). One possible solution to this challenge is a platform where metagenomic libraries can be shared and accessed publicly. However, many administrative and policy obstacles would need to be resolved before this goal is attainable. The second step in performing a metagenomic analysis is to choose the *ad hoc* NGS approach to be used (Illumina HiSeq or MiSeq, Roche 454FLX titanium XL+, Pacific Biosciences, etc.). The challenge of diagnostics-oriented metagenomics-based studies is to balance the required sequencing depth and coverage (Sims et al., 2014), the cost of the NGS approach and the goal of the study. For example, different approaches may be warranted to obtain a high sequencing depth and coverage of few plant viruses infecting few plant samples (Kreuze et al., 2009; Mollov et al., 2016) versus to reliably detect all plant viruses present in hundreds or thousands of plant samples at low sequencing depth and coverage (Bernardo et al., 2016; Roossinck et al., 2010). While future experiments are likely to benefit from lower sequencing costs, routinely using NGS-based approaches in a diagnostics perspective is currently prohibitively expensive for many service programmes.

Another challenge lies in the bioinformatics analysis. The sheer volume of data can be difficult to analyse at all levels, from contig assembly, to annotation matching to a known sequence and *de novo* sequence alignments. This means that contigs that belong to a well-annotated taxon will be easier to read than sequences belonging to taxon under-represented in public databases (Massart et al., 2014). Fortunately, inexpensive and rapid sequencing tools like Sanger sequencing have fuelled the growth of bioinformatics tools and public databases such as GenBank (NCBI) that allow researchers to better analyse sequences.

Finally, as mentioned in the section 'The challenge of determining causation', metagenomics can reveal microbial diversity but does not necessarily show which pathogens cause disease symptoms, either individually or synergistically. However, as researchers and pathologists learn more about virus–host interactions, making accurate diagnoses will become easier. Despite these on-going challenges, metagenomics is almost certain to become a promising approach to better diagnose and characterize plant viruses in the near future. Undoubtedly, as this technology improves and becomes more affordable, its applications in the field of plant pathology will prove very useful.

8 Acknowledgements

The authors thank Dr. Gary Kinard for critical review of this chapter. PR is funded by an EU grant (N° PIOF-GA-2013-622571).

9 References

Abu, A. Y., Rassaby, L., Royer, M., Borg, Z., Braithwaite, K. S., Mirkov, T. E., Irey, M. S., Perrier, X., Smith, G. R. and Rott, P. 2006. Yellow leaf of sugarcane is caused by at least three different genotypes of *Sugarcane yellow leaf virus*, one of which predominates on the Island of Reunion. *Arch.Virol.*, 151, 1355–71.

Adams, I. P., Glover, R., Souza-Richards, R., Bennett, S., Hany, U. and Boonham, N. 2013a. Complete genome sequence of arracacha virus B: A novel cheravirus. *Arch. Virol.*, 158, 909–13.

Adams, I. P., Miano, D. W., Kinyua, Z. M., Wangai, A., Kimani, E., Phiri, N., Reeder, R., Harju, V., Glover, R., Hany, U., Souza-Richards, R., Nath, P. D., Nixon, T., Fox, A., Barnes, A., Smith, J., Skelton, A., Thwaites, R., Mumford, R. and Boonham, N. 2013b. Use of next-generation sequencing for the identification and characterization of Maize chlorotic mottle virus and Sugarcane mosaic virus causing maize lethal necrosis in Kenya. *Plant Pathol.*, 62, 741–9.

Adams, I. P., Rai, S., Deka, M., Harju, V., Hodges, T., Hayward, G., Skelton, A., Fox, A. and Boonham, N. 2014. Genome sequence of vanilla distortion mosaic virus infecting *Coriandrum sativum*. *Arch. Virol.*, 159(12), 3463–5.

Al Rwahnih, M., Daubert, S., Golino, D. and Rowhani, A. 2009. Deep sequencing analysis of RNAs from a grapevine showing Syrah decline symptoms reveals a multiple virus infection that includes a novel virus. *Virology*, 387, 395–401.

Altschul, S. F., Gish, W., Miller, W., Myers, E. W. and Lipman, D. J. 1990. Basic local alignment search tool. *J. Mol. Biol.*, 215, 403–10.

Bernardo, P. 2014. *Ecologie, diversité et découverte de phytovirus à l'échelle de deux agro-écosystèmes dans un cadre spatio-temporel à l'aide de la géo-métagénomique*. PhD thesis, University of Montpellier II.

Bernardo, P., Albina, E., Eloit, M. and Roumagnac, P. 2013a. [Pathology and viral metagenomics, a recent history]. *Med Sci (Paris)*, 29, 501–8.

Bernardo, P., Golden, M., Akram, M., Naimuddin,Nadarajan, N., Fernandez, E., Granier, M., Rebelo, A. G., Peterschmitt, M., Martin, D. P. and Roumagnac, P. 2013b. Identification and characterisation of a highly divergent geminivirus: Evolutionary and taxonomic implications. *Virus Res.*, 177, 35–45.

Bernardo, P., Muhire, B., Francois, S., Deshoux, M., Hartnady, P., Farkas, K., Kraberger, S., Filloux, D., Fernandez, E., Galzi, S., Ferdinand, R., Granier, M., Marais, A., Monge Blasco, P., Candresse, T., Escriu, F., Varsani, A., Harkins, G. W., Martin, D. P. and Roumagnac, P. 2016. Molecular characterization and prevalence of two capulaviruses: Alfalfa leaf curl virus from France and Euphorbia caput-medusae latent virus from South Africa. *Virology*, 493, 142–53.

Braithwaite, K. S., Croft, B. J. and Magarey, R. C. 2007. Progress in identifying the cause of Ramu stunt disease of sugarcane. *Proc. Aust. Soc. Sugar Cane Technol.*, 29, 235–41.

Breitbart, M., Salamon, P., Andresen, B., Mahaffy, J. M., Segall, A. M., Mead, D., Azam, F. and Rohwer, F. 2002. Genomic analysis of uncultured marine viral communities. *Proc. Natl. Acad. Sci. U. S. A.*, 99, 14250–5.

Brister, J. R., Ako-Adjei, D., Bao, Y. and Blinkova, O. 2015. NCBI viral genomes resource. *Nucleic Acids Res.*, 43, D571–7.

Brum, J. R., Ignacio-Espinoza, J. C., Kim, E. H., Trubl, G., Jones, R. M., Roux, S., Verberkmoes, N. C., Rich, V. I. and Sullivan, M. B. 2016. Illuminating structural proteins in viral 'dark matter' with metaproteomics. *Proc. Natl. Acad. Sci. U. S. A.*, 113, 2436–41.

Brum, J. R. and Sullivan, M. B. 2015. Rising to the challenge: Accelerated pace of discovery transforms marine virology. *Nat. Rev. Microbiol.*, 13, 147–59.

Candresse, T., Filloux, D., Muhire, B., Julian, C., Galzi, S., Fort, G., Bernardo, P., Daugrois, J. H., Fernandez, E., Martin, D. P., Varsani, A. and Roumagnac, P. 2014. Appearances can be deceptive: Revealing a hidden viral infection with deep sequencing in a plant quarantine context. *PLoS One*, 9, e102945.

Chinnaraja, C., Viswanathan, R., Karuppaiah, R., Bagyalakshmi, K., Malathi, P. and Parameswari, B. 2013. Complete genome characterization of Sugarcane yellow leaf virus from India: Evidence for RNA recombination. *Eur. J. Plant Pathol.*, 135, 335–49.

Daugrois, J.-H., Jean-Baptiste, I., Lockhart, B. E. L., Irey, M., Chatenet, M. and Rott, P. 1999. First report of Sugarcane yellow leaf virus in the French West Indies. *Plant Dis.*, 83, 588.

Davis, T. S., Bosque-Perez, N. A., Foote, N. E., Magney, T. and Eigenbrode, S. D. 2015. Environmentally dependent host-pathogen and vector-pathogen interactions in the Barley yellow dwarf virus pathosystem. *J. Appl. Ecol.*, 52, 1392–401.

Debibakas, S., Rocher, S., Garsmeur, O., Toubi, L., Roques, D., D'hont, A., Hoarau, J. Y. and Daugrois, J. H. 2014. Prospecting sugarcane resistance to Sugarcane yellow leaf virus by genome-wide association. *Theor. Appl. Genet.*, 127, 1719–32.

Delwart, E. L. 2007. Viral metagenomics. *Rev.Med.Virol.*, 17, 115–31.

Donaire, L., Wang, Y., Gonzalez-Ibeas, D., Mayer, K. F., Aranda, M. A. and Llave, C. 2009. Deep-sequencing of plant viral small RNAs reveals effective and widespread targeting of viral genomes. *Virology*, 392, 203–14.

Edwards, R. A. and Rohwer, F. 2005. Viral metagenomics. *Nat. Rev. Microbiol.*, 3, 504–10.

Elsayed, A. I., Boulila, M., Komor, E. and Zhu, Y. J. 2012. Putative recombination signature and significance of insertion/deletion events in the RNA-dependent RNA polymerase coding region of sugarcane yellow leaf virus. *Biochimie*, 94, 1764–72.

Elsayed, A. I., Weig, A. R. and Komor, E. 2011. Molecular characterization of Hawaiian Sugarcane yellow leaf virus genotypes and their phylogenetic relationship to strains from other sugarcane-growing countries. *Eur. J. Plant Pathol.*, 129, 399–412.

Field, D., Garrity, G., Gray, T., Morrison, N., Selengut, J., Sterk, P. et al. 2008. The minimum information about a genome sequence (MIGS) specification. *Nat. Biotechnol.*, 26, 541–7.

Field, D., Sterk, P., Kottmann, R., De Smet, J. W., Amaral-Zettler, L., Cochrane, G., Cole, J. R., Davies, N., Dawyndt, P., Garrity, G. M., Gilbert, J. A., Glockner, F. O., Hirschman, L., Klenk, H. P., Knight, R., Kyrpides, N., Meyer, F., Karsch-Mizrachi, I., Morrison, N., Robbins, R., San Gil, I., Sansone, S., Schriml, L., Tatusova, T., Ussery, D., Yilmaz, P., White, O., Wooley, J. and Caporaso, G. 2014. Genomic standards consortium projects. *Stand Genomic. Sci.*, 9, 599–601.

Filloux, D., Dallot, S., Delaunay, A., Galzi, S., Jacquot, E. and Roumagnac, P. 2015a. Metagenomics approaches based on virion-associated nucleic acids (VANA): An innovative tool for assessing without a priori viral diversity of plants. *Methods Mol. Biol.*, 1302, 249–57.

Filloux, D., Murrell, S., Koohapitagtam, M., Golden, M., Julian, C., Galzi, S., Uzest, M., Rodier-Goud, M., D'hont, A., Vernerey, M. S., Wilkin, P., Peterschmitt, M., Winter, S., Murrell, B., Martin, D. P. and Roumagnac, P. 2015b. The genomes of many yam species contain transcriptionally active endogenous geminiviral sequences that may be functionally expressed. *Virus Evol.*, 1, 1–17.

Finn, R. D., Coggill, P., Eberhardt, R. Y., Eddy, S. R., Mistry, J., Mitchell, A. L., Potter, S. C., Punta, M., Qureshi, M., Sangrador-Vegas, A., Salazar, G. A., Tate, J. and Bateman, A. 2016. The Pfam protein families database: Towards a more sustainable future. *Nucleic Acids Res.*, 44, D279–85.

Ghurye, J. S., Cepeda-Espinoza, V. and Pop, M. 2016. Metagenomic assembly: Overview, challenges and applications. *Yale J. Biol. Med.*, 89, 353–62.

Hadidi, A., Flores, R., Candresse, T. and Barba, M. 2016. Next-generation sequencing and genome editing in plant virology. *Front. Microbiol.*, 7, 1325.

Handelsman, J., Rondon, M. R., Brady, S. F., Clardy, J. and Goodman, R. M. 1998. Molecular biological access to the chemistry of unknown soil microbes: A new frontier for natural products. *Chem. Biol.*, 5, R245–9.

Kallas, E. G. and O'Connor, D. H. 2016. Real-time sharing of zika virus data in an interconnected world. *JAMA Pediatr.*, 170, 633–4.

King, A. M. Q., Adams, M. J., Carstens, E. B. and Lefkowitz, E. J. 2012. Virus Taxonomy. Classification and Nomenclature of Viruses. Ninth Report of the International Committee on Taxonomy of Viruses. In: King, A. M. Q., Adams, M. J., Carstens, E. B. and Lefkowitz, E. J. (eds.). Amsterdam, Boston, Heidelberg, London, New York, Oxford, Paris, San Diego, San Francisco, Singapore, Sydney, Tokyo: Elsevier.

Kreuze, J. F., Perez, A., Untiveros, M., Quispe, D., Fuentes, S., Barker, I. and Simon, R. 2009. Complete viral genome sequence and discovery of novel viruses by deep sequencing of small RNAs: A generic method for diagnosis, discovery and sequencing of viruses. *Virology*, 388, 1–7.

Lam, K. N., Cheng, J., Engel, K., Neufeld, J. D. and Charles, T. C. 2015. Current and future resources for functional metagenomics. *Front. Microbiol.*, 6, 1196.

Lin, Y. H., Gao, S. J., Damaj, M. B., FU, H. Y., Chen, R. K. and Mirkov, T. E. 2014. Genome characterization of *Sugarcane yellow leaf virus* from China reveals a novel recombinant genotype. *Arch. Virol.*, 159, 1421.

Lipkin, W. I. 2010. Microbe hunting. *Microbiol. Mol. Biol. Rev.*, 74, 363–77.

Lobb, B., Kurtz, D. A., Moreno-Hagelsieb, G. and Doxey, A. C. 2015. Remote homology and the functions of metagenomic dark matter. *Front. Genet.*, 6, 234.

Mai, V., Prosperi, M. and Yaghjyan, L. 2016. Moving microbiota research toward establishing causal associations that represent viable targets for effective public health interventions. *Ann. Epidemiol.*, 26, 306–10.

Marchler-Bauer, A., Derbyshire, M. K., Gonzales, N. R., Lu, S., Chitsaz, F., Geer, L. Y., Geer, R. C., HE, J., Gwadz, M., Hurwitz, D. I., Lanczycki, C. J., LU, F., Marchler, G. H., Song, J. S., Thanki, N., Wang, Z., Yamashita, R. A., Zhang, D., Zheng, C. and Bryant, S. H. 2015. CDD: NCBI's conserved domain database. *Nucleic Acids Res.*, 43, D222–6.

Maroon-Lango, C. J., Hoy, J. W., Comstock, J. C., Grisham, M. P., Mock, R., Hale, A. L., Afghan, S., Croft, B. J., Dela Cueva, F., Hoffmann, H., Kennedy, A., Orozco, H., Saumtally, S., Victoria, J. and Viswanathan, R. 2013. Current knowledge and practices related to seed transmission of sugarcane pathogens and movement of seed. *J. Am. Soc. Sugar Cane Technol.*, 33, 20–9.

Marquez, L. M., Redman, R. S., Rodriguez, R. J. and Roossinck, M. J. 2007. A virus in a fungus in a plant: Three-way symbiosis required for thermal tolerance. *Science*, 315, 513–15.

Martin, D. P., Murell, B., Golden, M., Khoosal, A. and Muhire, B. 2015. RDP4: Detection and analysis of recombination patterns in virus genomes. *Virus Evol.*, 1, vev003.
Martin, R. R., Constable, F. and Tzanetakis, I. E. 2016. Quarantine regulations and the impact of modern detection methods. *Annu. Rev. Phytopathol.*, 54, 189–205.
Massart, S., Olmos, A., Jijakli, H. and Candresse, T. 2014. Current impact and future directions of high throughput sequencing in plant virus diagnostics. *Virus Res.*, 188, 90–6.
Mokili, J. L., Rohwer, F. and Dutilh, B. E. 2012. Metagenomics and future perspectives in virus discovery. *Curr. Opin. Virol.*, 2, 63–7.
Mollov, D., Maroon-Lango, C. and Kuniata, L. 2016. Detection by next generation sequencing of a multi-segmented viral genome from sugarcane associated with Ramu stunt disease. *Virus Genes.*, 52, 152–5.
Muthukumar, V., Melcher, U., Pierce, M., Wiley, G. B., Roe, B. A., Palmer, M. W., Thapa, V., Ali, A. and Ding, T. 2009. Non-cultivated plants of the Tallgrass Prairie Preserve of northeastern Oklahoma frequently contain virus-like sequences in particulate fractions. *Virus Res.*, 141, 169–73.
Prabha, K., Baranwal, V. K. and Jain, R. K. 2013. Applications of next generation high throughput sequencing technologies in characterization, discovery and molecular interaction of plant viruses. *Indian J. Virol.*, 24, 157–65.
Rajeswaran, R., Seguin, J., Chabannes, M., Duroy, P. O., Laboureau, N., Farinelli, L., Iskra-Caruana, M. L. and Pooggin, M. M. 2014. Evasion of short interfering RNA-directed antiviral silencing in musa acuminata persistently infected with six distinct banana streak pararetroviruses. *J. Virol.*, 88, 11516–28.
Roossinck, M. J. 2011. The good viruses: Viral mutualistic symbioses. *Nat. Rev. Microbiol.*, 9, 99–108.
Roossinck, M. J. 2012. Plant virus metagenomics: Biodiversity and ecology. *Annu. Rev. Genet.*, 46, 359–69.
Roossinck, M. J., Martin, D. P. and Roumagnac, P. 2015. Plant virus metagenomics: Advances in virus discovery. *Phytopathology*, 105, 716–27.
Roossinck, M. J., Saha, P., Wiley, G., Quan, J., White, J., Lai, H., Chavarria, F., Shen, G. and Roe, B. 2010. Ecogenomics: Using massively parallel pyrosequencing to understand virus ecology. *Mol. Ecol.*, 19, 81–8.
Rosario, K. and Breitbart, M. 2011. Exploring the viral world through metagenomics. *Curr. Opin. Virol.*, 1, 1–9.
Rose, R., Constantinides, B., Tapinos, A., Robertson, D. L. and Prosperi, M. 2016. Challenges in the analysis of viral metagenomes. *Virus Evol.*, 2, vew022.
Roumagnac, P., Granier, M., Bernardo, P., Deshoux, M., Ferdinand, R., Galzi, S., Fernandez, E., Julian, C., Abt, I., Filloux, D., Mesleard, F., Varsani, A., Blanc, S., Martin, D. P. and Peterschmitt, M. 2015. Alfalfa leaf curl virus: An aphid-transmitted geminivirus. *J. Virol.*, 89, 9683–8.
Roux, S., Hallam, S. J., Woyke, T. and Sullivan, M. B. 2015. Viral dark matter and virus-host interactions resolved from publicly available microbial genomes. *Elife*, 4, e08490.
Scheets, K., Blinkova, O., Melcher, U., Palmer, M. W., Wiley, G. B., Ding, T. and Roe, B. A. 2011. Detection of members of the *Tombusviridae* in the Tallgrass Prairie Preserve, Osage County, Oklahoma, USA. *Virus Res.*, 160, 256–63.
Schmidt, T. M., Delong, E. F. and Pace, N. R. 1991. Analysis of a marine picoplankton community by 16S ribosomal-RNA gene cloning and sequencing. *J. Bacteriol.*, 173, 4371–8.
Sender, R., Fuchs, S. and Milo, R. 2016. Revised estimates for the number of human and bacteria cells in the body. *PLoS Biol.*, 14, e1002533.
Simmonds, P., Adams, M. J., Benko, M., Breitbart, M., Brister, J. R., Carstens, E. B., Davison, A. J., Delwart, E., Gorbalenya, A. E., Harrach, B., Hull, R., King, A. M. Q., Koonin, E. V., Krupovic, M., Kuhn, J. H., Lefkowitz, E. J., Nibert, M. L., Orton, R., Roossinck, M. J., Sabanadzovic, S., Sullivan, M. B., Suttle, C. A., Tesh, R. B., Van Der Vlugt, R. A., Varsani, A. and Zerbini, F. M. 2017. Consensus statement: Virus taxonomy in the age of metagenomics. *Nat. Rev. Micro.*, 15, 161-68.
Sims, D., Sudbery, I., Ilott, N. E., Heger, A. and Ponting, C. P. 2014. Sequencing depth and coverage: Key considerations in genomic analyses. *Nat. Rev. Genet.*, 15, 121–32.

Thapa, V., Melcher, U., Wiley, G. B., Doust, A., Palmer, M. W., Roewe, K., Roe, B. A., Shen, G., Roossinck, M. J., Wang, Y. M. and Kamath, N. 2012. Detection of members of the *Secoviridae* in the Tallgrass Prairie Preserve, Osage County, Oklahoma, USA. *Virus Res.*, 167, 34–42.

Vayssier-Taussat, M., Albina, E., Citti, C., Cosson, J. F., Jacques, M. A., Lebrun, M. H., Le Loir, Y., Ogliastro, M., Petit, M. A., Roumagnac, P. and Candresse, T. 2014. Shifting the paradigm from pathogens to pathobiome: New concepts in the light of meta-omics. *Front. Cell Infect. Microbiol.*, 4, 29.

Wang, M. Q., Xu, D. L., Li, R. and Zhou, G. H. 2012. Genotype identification and genetic diversity of *Sugarcane yellow leaf virus* in China. *Plant Pathol.*, 61, 986–93.

Wang, M. Q. and Zhou, G. H. 2010. A near-complete genome sequence of a distinct isolate of *Sugarcane yellow leaf virus* from China, representing a sixth new genotype. *Virus Genes*, 41, 268–72.

Westwood, J. H., Mccann, L., Naish, M., Dixon, H., Murphy, A. M., Stancombe, M. A., Bennett, M. H., Powell, G., Webb, A. A. and Carr, J. P. 2013. A viral RNA silencing suppressor interferes with abscisic acid-mediated signalling and induces drought tolerance in Arabidopsis thaliana. *Mol. Plant Pathol.*, 14, 158–70.

Wilkinson, K., Grant, W. P., Green, L. E., Hunter, S., Jeger, M. J., Lowe, P., Medley, G. F., Mills, P., Phillipson, J., Poppy, G. M. and Waage, J. 2011. Infectious diseases of animals and plants: An interdisciplinary approach. *Philos. Trans. R Soc. Lond. B Biol. Sci.*, 366, 1933–42.

Wu, Q., Ding, S. W., Zhang, Y. and Zhu, S. 2015. Identification of viruses and viroids by next-generation sequencing and homology-dependent and homology-independent algorithms. *Annu. Rev. Phytopathol.*, 53, 425–44.

Xu, P., Chen, F., Mannas, J. P., Feldman, T., Sumner, L. W. and Roossinck, M. J. 2008. Virus infection improves drought tolerance. *New Phytol.*, 180, 911–21.

Chapter 9

Progress in understanding fungal diseases affecting sugarcane: red rot

R. Viswanathan, A. Ramesh Sundar, R. Selvakumar and P. Malathi, ICAR-Sugarcane Breeding Institute, India

1 Introduction

Red rot is a disease of sugarcane stalks caused by the fungal pathogen *Colletotrichum falcatum*. It is one of the oldest recorded sugarcane diseases. After its first description from Java (Went, 1893), the disease was reported in Australia, the West Indies, Hawaii and the mainland United States (Viswanathan, 2010). In India, Barber (1901) reported severe occurrence of the disease. Butler (1906) subsequently conducted detailed studies of the disease, especially on the pathogen and epidemiology, and named the disease 'red rot'. Worldwide, the disease has been recorded in about 77 countries. However, it occurs most severely in South Asian countries, especially India, Nepal, Pakistan, Myanmar and Bangladesh, as well as in Thailand and Vietnam. The disease also causes limited damage to cane cultivation in the United States (mainly in Louisiana), Brazil and Nigeria (Singh and Singh, 1989; Viswanathan, 2010).

Epidemics of red rot have been very common since 1901, causing havoc on the cultivation of *Saccharum officinarum* and *S. barberi* clones before the era of nobilization to increase disease resistance. After the introduction of hybrid varieties, disease epidemics continued and caused significant damage to sugarcane cultivation in different countries, becoming responsible for the removal of many elite varieties in the field (Viswanathan, 2010). However, the disease was successfully managed by releasing disease-resistant varieties after each epidemic, and such efforts have reduced the threat of the disease in

http://dx.doi.org/10.19103/AS.2017.0035.21

different countries. The use of new sugarcane varieties also led to the emergence of new pathogenic variants of *C. falcatum*, and as a result, the disease has remained a problem in many countries (Singh and Singh, 1989).

Developing red rot-resistant varieties has been a major management strategy to deal with the disease in most countries. Detailed studies have been conducted to standardize disease screening techniques, which have helped to identify resistant sources and inheritance of red rot resistance (Abbott, 1938; Agnihotri, 1983; Chona, 1980; Duttamajumder, 2002; Viswanathan, 2010). Subsequently, the mechanism of red rot resistance has been studied in detail with a focus on the biochemical basis of resistance (Godshall and Lonergan, 1987; Viswanathan and Samiyappan, 1999). Diversity in *C. falcatum* was also studied to establish pathogenic, cultural, morphological and immunogenic variations (Padmanaban et al., 1996; Viswanathan et al., 2003a).

Molecular tools have been employed to understand genetic variation of the pathogen (Malathi et al., 2010; Mohanraj et al., 2002; Suman et al., 2005; Viswanathan, 2010) and more recently to understand host resistance, the *C. falcatum* genome and pathogenicity-related genes (Malathi and Viswanathan, 2012; Muthumeena et al., 2013; Sathyabhama et al., 2015). Recently, Viswanathan et al. (2016a) reported the complete genome and transcriptome of the pathogen using next-generation sequencing. This has opened new avenues to understand the pathogenicity of *C. falcatum*, development of new variants and possible genetic manipulation of the pathogen through a functional approach.

Other than disease management through varietal resistance, new research in areas such as biological control to manage the disease in sugarcane has also been carried out (Ramesh Sundar et al., 2009; Viswanathan and Samiyappan, 2002a). Efficacy of fungicides against the pathogen has been improved with effective delivery mechanism (Malathi et al., 2016; Viswanathan et al., 2016b). This chapter reviews the economic impact of red rot, diagnosis, pathogen variability, application of molecular tools to understand pathogen biology, varietal resistance, screening methods and new management approaches including fungicide delivery.

2 Symptoms and economic impact

During germination, *C. falcatum* infection causes death in germinating buds or shoots (Fig. 1). Disease symptoms can be clearly seen during the maturity phase of the crop. Affected stalks exhibit extensive foliage discolouration and drying, accompanied by death of stalks alone or in stools (Fig. 2). The diseased canes exhibit discoloured lesions of varying intensities on rind tissue (Fig. 3). Internally, the disease is characterized by extensive rotting of stalk tissues with reddish discolouration. Subsequently, the affected tissue turns muddy or dull red. Further, characteristic white spots perpendicular to the long axis of the stalk occur throughout the affected internode tissues (Fig. 4). These white spots in the affected tissues clearly differentiate red rot from other stalk diseases such as wilt, stalk rot and pineapple diseases. Typically, *C. falcatum* infects stalk tissues alone and does not infect lamina tissue. However, after systemic infection, the pathogen in the stalk progresses through the whorl region and causes mid-rib lesions of variable sizes (Fig. 5). Mid-rib lesions can also be caused by the inoculum carried through humid air or rain drops. The lamina tissue remains free from symptoms.

Figure 1 Death of germinated shoots due to primary sources of *C. falcatum* inoculum.

Figure 2 Death of entire sugarcane plant due to red rot.

Since the pathogen infects cane stalks, even a limited infection can impact both cane yield and juice quality (Viswanathan and Samiyappan, 1999). During epidemics, extensive losses to cane production have been recorded in India (Fig. 6) and other Asian countries. In Thailand, red rot occurs in combination with wilt and is referred to as red rot wilt. The disease can cause losses in cane yield of 34.6–73.7% in the plant cane crop and up to 100% in ratoon crops. Severe epidemics of red rot wilt were reported in the 1990s with estimated losses of US$20 million. About 28.5% losses in cane yield, rising to 82.7% when disease severity increased to 75%, was reported in Pakistan. In India, yield losses of

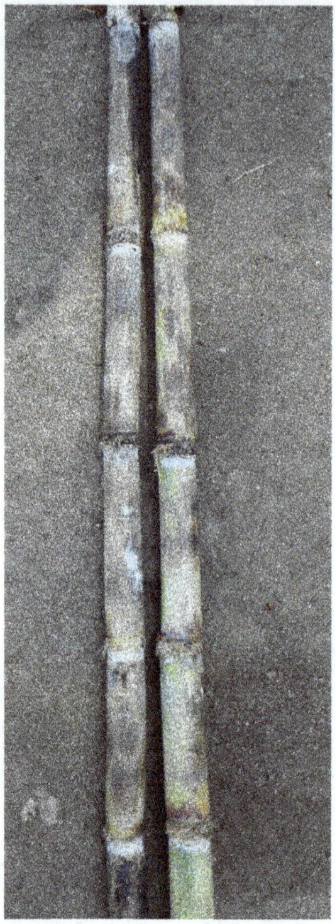

Figure 3 Extensive patchy rind discolouration in red rot-affected canes.

up to 100% occurred in different regions during severe epidemics (Viswanathan, 2010). Such epidemics caused substantial losses to the sugar industry and farmers, estimated between US$80 and US$100 million annually during the epidemics in India (Singh and Singh, 1989).

The pathogen produces large quantities of invertase enzymes which break down sucrose molecules into glucose and fructose. Apart from affecting the reducing sugars, disease-affected cane tissue shows higher levels of total soluble salts and gum, as well as a simultaneous decrease in pH, sucrose and cane juice purity (Singh and Singh, 1989). In Pakistan, a 30–87% loss in extraction of juice and a 30–74% loss in recoverable sugar were reported in red rot-affected cane stalks; similarly, respective losses of 4.8–40.8% and 7.8–43.3% were reported from India. The disease-affected canes showed an increase of 19.2–30.9% in reducing sugars depending on cane variety and the severity of red rot infections (Viswanathan, 2010). Studies of Viswanathan and Samiyappan (1999) revealed that the pathogen infection reduced brix and purity substantially with a 25–75% reduction in sucrose content. Currently, the disease occurs in varying intensities in sugarcane grown in the states of India and continues to cause economic

Figure 4 Rotting of internal tissues of affected canes with reddish discolouration and characteristic white spots.

Figure 5 Leaves in crown exhibit *C. falcatum* infection as mid-rib lesions.

Figure 6 Red rot epidemic in sugarcane: the severely affected field shows complete destruction of canes over a large area in the field.

damage (Viswanathan, 2017; Viswanathan and Rao, 2011). Detailed reviews on the impact of the disease have been published (Singh and Singh, 1989; Viswanathan and Samiyappan, 1999).

3 Disease transmission

Went (1893) identified the pathogen causing red rot as *C. falcatum*. Both anamorphic (Glomerellaceae, Sordariomycetes, Ascomycota) and teleomorphic (*Glomerella tucumanensis* (Speg.) Arx & Muller) stages of the fungus have been reported. The pathogen produces abundant fruiting bodies or 'acervuli' on the affected tissues of the stalk rind, node, bud or leaf mid-rib. In addition, the fungus forms chlamydospores on the surface of the stalk/leaf or within the cane tissue. The fruiting bodies of teleomorph 'perithecia' are found profusely on dead or dying leaf lamina, mid-rib or the leaf sheath (Duttamajumder, 2002). The pathogen is primarily spread through infected setts (stem cuttings) and occasionally, inoculum surviving in the soil in the infected crop debris may also serve as a primary source of infection for young crops. Rain can act as a secondary means of transmission of the disease by washing the inoculum from acervuli on the stalk surface. The inoculum carried by rainwater reaches adjacent plants and initiates infection. Similarly, rain splash also passes inoculum from infected plants to neighbouring plants. The pathogen can also, to a limited extent, become airborne. Acervuli produced on mid-rib lesions and rind/node can play a crucial role in the secondary transmission of the pathogen (Agnihotri, 1983). The pathogen adopts a hemibiotrophic lifestyle like other *Colletotrichum* spp. and the rice blast pathogen *Magnaporthe grisea*.

The fungus enters the cane stalk through various channels. Different parts of the node such as bud scales, root primordia, growth ring and leaf scar serve as portals of entry for the pathogen in growing or mature stalks and initiate infection (Agnihotri, 1983). It was also shown that dormant infections can occur in leaf scar tissue and on the bud scales, which could remain in that condition for long periods of time. Appressoria are observed on the epidermis covering this area.

Abbott (1938) established that *C. falcatum* infection in the growing stalks occurs primarily through the tunnels of the moth borer *Diatraea saccharalis* Fabricius (Lepidoptera: Crambidae) in Louisiana and southern Florida and in some varieties, through the root primordia. Subsequently, Ogunwolu (1983) reported the borer *D. saccharalis* damage accentuated yield effects from red rot, both in the plant and first ratoon crops in Louisiana. Similarly, in Nigeria, the damage caused by this borer facilitates the infection of cane by *C. falcatum* by creating easy access for the fungus into the stem (Olufolaji, 1987). *Eldana saccharina* Walker (Lepidoptera: Pyralidae) and *Sesamia calamistis* Hampson (Lepidoptera: Noctuidae) borings were found to be important points of entry for *C. falcatum* in South Africa (Trenor and Bailey, 1989). However, studies of Chona (1950) have clearly shown that borers play little part in originating the disease or its subsequent spread under Indian conditions. Later studies also revealed that the stalk borer *Chilo auricilius* infestation does not play a conspicuous role in accelerating red rot infection, but showed a possible role in secondary transmission of the pathogen (Viswanathan, 2010).

Apart from borer holes, growth cracks also provide avenues for the entry of *C. falcatum* into cane stalks. Chona (1950) reported that red rot infection in varieties such as Co 233, Co 396, Co 548 and BO 4 took place rapidly through growth cracks. Edgerton (1959) also reported similar findings with Co 290. However, Agnihotri (1983) indicated that pathogen infection takes place only when the conidia land on the split before the sealing process starts.

4 Diagnosis and disease management

4.1 Detection of the red rot pathogen

Infections carried by the seed canes serve as primary sources for disease spread in sugarcane, hence precise diagnosis for the pathogen infection in seed canes is important to prevent spread in the field. Traditional diagnostic techniques are not always reliable since initial infection of nodal and mid-rib tissues is not immediately visible. Enzyme-linked immunosorbent assay (ELISA) techniques have been developed to provide better detection (Viswanathan et al., 2000; Bukhari et al., 2012). These assays showed that tissues such as buds, roots and leaf scars are frequently colonized by the pathogen (Viswanathan et al., 2000). Molecular assays have been developed with actin and calmodulin primers for *C. falcatum*. Their specificity in detecting *C. falcatum* was validated in comparison to other *Colletotrichum* spp. Probe-based diagnostics was standardized for large-scale application using these highly specific primers. These assays were highly useful in diagnosis of *C. falcatum* in symptomless tissue/mixed state of infection and to monitor the pathogen survival in soil (Malathi et al., 2014).

Recently, a loop-mediated isothermal amplification (LAMP) assay was developed for the detection of *C. falcatum* with good sensitivity (Chandra et al., 2015). This assay was five times more sensitive than PCR assays in detecting *C. falcatum* in infected cane tissue. Additionally, it is rapid and visual judgement of colour change is done within an hour. Since the process does not require post-amplification processing, the assay is found to be simple, economical and applicable in laboratories with limited equipment (Chandra et al., 2015). These sensitive assays will help in developing field diagnostic assay kits for clean seed production. Early diagnosis of *C. falcatum* infection in seed canes ensures supply of healthy planting material in the field. Use of such kits for disease monitoring in the field will also help in taking disease management decisions in plant crops before starting ratoon crops.

4.2 Breeding for resistance

A key method for combatting red rot is developing varieties with improved resistance. In 1912, the Sugarcane Breeding Institute (SBI) at Coimbatore initiated the 'nobilization' process by crossing noble (*Saccharum officinarum*) and wild (*S. spontaneum* in particular) cane to derive economic traits from the former and disease resistance from the latter. *S. spontaneum* has been found to have a large number of clones with red rot resistance in contrast to much lower resistance in *S. officinarum*. Resistance in *S. barberi*, *S. sinense* and *S. robustum* to red rot were also low, although resistance was better in *S. barberi* than in the other two species (Viswanathan, 2010).

Research into the inheritance of resistance genes has suggested that the dominant genes from *S. officinarum* might be involved in masking the effect of resistance genes of *S. spontaneum* (Azab and Chilton, 1952). There have been contradictory reports that either there is no association between the degree of resistance of the progenies and parents (Chona, 1980) or that crosses involving resistant parents tend to have more resistant progenies than those of crosses involving susceptible parents (Yin et al., 1996). More recent studies have found evidence of both race-specific (vertical resistance) as well as race-non-specific (horizontal resistance) (Babu et al., 2010; Viswanathan, 2010).

This breeding programme led to the development of many outstanding varieties, which have dominated sugarcane cultivation in Asia, Africa and many other countries (Viswanathan, 2010). During the early part of the last century, the disease was threatening the sugar industry in Louisiana, USA, hence systematic resistance screening, parent selection and cultivation of resistant varieties were followed to contain the disease (Abbott, 1938). Disease severity has significantly been reduced by systematic replacement of susceptible varieties with resistant varieties. These breeding programmes have relied on conventional hybridization, either by the inter-specific or inter-varietal crossing of the chosen parents. Repeated back crossing to the commercial parent without an intervening generation of selfing has proved cost effective, though time-consuming in transferring the desired traits. Other techniques such as mutation breeding and somaclonal variation through tissue culture have not yet proved successful (Viswanathan, 2010).

Reliable disease screening methods are required for successful breeding and selection programmes. Chona (1954) developed a plug inoculation method to screen sugarcane genotypes for red rot resistance. Further work on the critical parameters governing red rot resistance in sugarcane led to the development of a 0–9 rating scale to assess red rot resistance which has been adopted by different research groups across the globe (Fig. 7) (Srinivasan and Bhat, 1961; Viswanathan, 2010). More recently, a new 'nodal swabbing method' has been developed which causes less injury (and therefore a more natural way of infection) than the plug method (Fig. 8). The field screening for red rot in sugarcane takes at least one year and this process has limitations of land, labour and other resources. To screen a large number of clones for red rot in the early stages of a breeding cycle, a rapid screening procedure in a temperature- and humidity-controlled chamber was developed (Fig. 9). This method has been found to be rapid and suitable for large-scale screening of sugarcane clones for red rot resistance (Viswanathan et al., 1998).

Figure 7 Distinct red rot symptoms caused by *C. falcatum* as reddish lesions covering many internodes with white spots in a susceptible cultivar after inoculation by plug method (right). The resistant cultivar does not exhibit any symptoms except dark lesions in the inoculated internode (left). Arrows indicate the point of inoculation.

4.3 Chemical control

In addition to resistant varieties, disease management has depended on chemical control. Vegetative propagation favours transmission of *C. falcatum* through planting materials in sugarcane. In addition, the pathogen propagules surviving in the soil may serve as another primary source of pathogen inoculum. Attempts have therefore been made to contain disease development from sett and soil-borne inoculum of fungal pathogens during the initial growth period of sugarcane by sett treatment with fungicides. Earlier studies indicated that fungicides such as carbendazim, benomyl, vitavax, areton and thiophanate methyl were effective *in vitro* against *C. falcatum*. However, they did not give satisfactory control of the disease under field conditions (Agnihotri, 1983). Later studies revealed that fungicidal efficacy can be improved by prolonged soaking of setts in the fungicide solution (Malathi et al., 2004). Overnight soaking of setts in a fungicide solution was found to be highly effective than mere soaking of setts for one hour for the

Figure 8 *C. falcatum* infection after nodal swabbing method caused death of inoculated canes. Cotton swabbed nodes indicate the pathogen-inoculated nodes.

Figure 9 Screening sugarcane clones for red rot resistance under controlled condition testing. (a) Pathogen-inoculated canes kept inside the chamber for disease development. (b) Disease development is seen as lesions of varying sizes on the rind. Resistant types remain symptom free after ten days of incubation.

management of soil-borne infection by *C. falcatum*. Thiophanate methyl was found to be more effective than carbendazim to manage red rot under field conditions (Malathi et al., 2004). The large amount of seed materials required for commercial planting makes overnight soaking impractical. For this reason, a method to rapidly diffuse fungicides inside the setts through a mechanized treatment was tested and found to be comparable to overnight soaking. It would be possible to upscale this method for commercial planting. While thiophanate methyl was suitable for this treatment, azole fungicides were phytotoxic at elevated concentrations. Combination of thiophanate methyl and azoxystrobin each at 500 ppm were effective against *C. falcatum*, both in overnight soaking and mechanized sett treatment (Malathi et al., 2017). This method of fungicide treatment was confirmed to manage red rot under field conditions in disease-endemic areas (Viswanathan et al., 2016b) and this opens the possibility of managing the disease effectively using fungicides, which was not possible earlier under field conditions.

4.4 Biological control

There has also been research into biological control of red rot. The efficacy of plant growth promoting rhizobacteria (PGPR) belonging to *Pseudomonas* spp. against *C. falcatum* has been established and their application in the field standardized (Viswanathan and Samiyappan, 2002a). The efficacy of *Trichoderma* against *C. falcatum* under field conditions has also been reported (Viswanathan, 2010). Another development in biocontrol application was use of biosurfactants which exhibit antifungal activities and induce immunity in plants. Recently, Hassan et al. (2015) reported the efficacy of surfactin-producing antagonistic *Bacillus* sp. against *C. falcatum* and their potential to induce systemic resistance in sugarcane. The *Bacillus* sp. had a longer shelf life in filter cake formulation of sugarcane and field application reduced red rot severity in the treated canes and resulted in higher yield. Similarly, Goswami et al. (2015) investigated on rhamnolipid bio-surfactant produced by *Pseudomonas aeruginosa* DS9 for its antifungal activities against *C. falcatum*. RL-DS9, the rhamnolipid, exhibited 87% inhibition of conidial germination under *in vitro* conditions and reduced the disease severity at 100 $\mu g\ mL^{-1}$ in the field. These recent studies indicate utility of bio-surfactant molecules from antagonistic microbes in managing red rot and promising results are expected from large-scale use of the molecules.

4.5 Induction of systemic resistance

Apart from bioagents, inducer molecules such as benzo-(1,2,3)-thiadiazole-7-carbonic acid *S*-methyl ester (BTH) and salicylic acid have been found to induce systemic resistance against red rot in susceptible sugarcane cultivars under field conditions. Pre-treatment with these inducers significantly reduced the disease development in the stalks and reduced the disease severity upon pathogen challenge (Ramesh Sundar et al., 2009). Recently, detailed studies were made to isolate and characterize inducer/elicitor molecules from *C. falcatum*. *In vitro* secretome analysis of *C. falcatum* by two-dimensional gel electrophoresis coupled with MALDI TOF/TOF/MS identified molecular signatures of *C. falcatum* such as ceratoplatanin protein (EPL1) and HYP2 and they are hypothesized to play the roles of elicitor and elicitor and effector (dual role), respectively (Ashwin et al., 2017). This has opened up the possibilities of identifying elicitor molecules from *C. falcatum* which can induce resistance in sugarcane. It was found that if disease development was arrested at an early stage of crop growth, the risk of subsequent disease development was significantly reduced. Delivery of the effective fungicides and/or liquid formulations of *Pseudomonas* through micro-irrigation offer a way to manage disease at a later stage in crop development (Viswanathan, 2012). Recently, nano-formulations of chitosan were shown to be efficient against the rice fungal pathogen *Pyricularia grisea* by eliciting induced resistance (Sathiyabama and Manikandan, 2016). Studies conducted at ICAR-SBI revealed antifungal activities of chitosan against *C. falcatum*. Additionally, chitosan-coated systemic acquired resistance (SAR) inducer nanoparticles were synthesized and their nano-formulations are being assessed for their stability and effectiveness.

5 Understanding host–pathogen interactions

5.1 Pathogenic variation of *C. falcatum*

It is important to note that *C. falcatum* displays substantial variation in pathogenicity. New variants have reduced the life of many elite varieties in the field. Under field conditions, mutation and parasexual recombination play a major role in the origin of new pathogenic strains. The phenomenon 'heterokaryosis' contributes to increasing the pathogen virulence (Abbott, 1938; Chona, 1980; Padmanaban et al., 1996, Srinivasan, 1962; Viswanathan et al., 2003a).

Padmanaban et al. (1996) identified a set of 13 host differentials from *Saccharum* spp. and hybrid cultivars to characterize pathogenic variation in *C. falcatum*. Later *C. falcatum* isolates were designated into 11 pathotypes (CF01 to CF11). These pathotypes have been used to screen sugarcane cultivars for resistance to red rot (Viswanathan, 2010). After removal of susceptible varieties from cultivation, the pathogen slowly adapts to the new cultivars and gains virulence over a period of time. Recently, predominant *C. falcatum* pathotype CF06 originated from popular cultivar CoC 671 in tropical India, but CF06 lost its virulence after elimination of its original host variety. Meanwhile a new pathotype (CF12) with high virulence has appeared in this region (Viswanathan, 2017). Such development of new pathotypes occur from time to time depending on the host varieties deployed in the field (Viswanathan, 2010). Molecular grouping of *C. falcatum* isolates has been attempted using RAPD profiles (Mohanraj et al., 2002; Suman et al., 2005), rDNA (Malathi et al., 2010) and vegetative compatibility grouping (Viswanathan et al., 2003a), though no clear correlation has been found. Malathi et al. (2010) studied variability in a set of *C. falcatum* pathotypes at cultural, genetic, pathogenic and molecular levels and found existence of two groups without any correlation between these traits. Further characterization of about 80 pathotypes collected throughout India for cultural, pathogenic variation and molecular variability for conserved gene sequences revealed the existence of less genetic variability in comparison to phenotypic variability (Malathi et al., 2014). Apparent low genetic variability between *C. falcatum* isolates could be due to single nucleotide mutations that were not revealed by the methods used so far to investigate genetic differences in *C. falcatum*.

5.2 Identification of red rot resistance mechanism in sugarcane

Genetics of sugarcane has gained low attention as compared to other crops mainly due to its highly hetereozygous, polyploidy and frequently aneuploid nature, complex genome, poor fertility and long breeding cycle (Gupta et al., 2010). Although inheritance to red rot resistance in sugarcane is not clearly understood, work has been recently carried out to decipher the mechanisms of host defence using genomic and proteomic tools.

In previous decades, different biochemical features imparting resistance in a particular genotype have been reported such as gum formation (Srinivasan and Bhat, 1961) as well as phenols and polyphenol oxidase (Godshall and Lonergan, 1987; Viswanathan, 2012). Prominent induction of peroxidase (PO) and other phenylpropanoid enzymes was found in resistant cultivars after pathogen infection (Viswanathan, 2012). It was shown that 3-deoxyanthocyanidin compounds act as sugarcane phytoalexins and are probably involved in resistance to *C. falcatum* (Ganesh Kumar et al., 2015). Apart from phytoalexins, PR proteins have also been associated with red rot resistance (Viswanathan and Samiyappan, 1999; Viswanathan et al., 2005). These studies suggested a possible role of these enzymes and phytoalexins in red rot resistance and induced resistance mediated by fluorescent pseudomonads (Viswanathan and Samiyappan, 2002a,b; Viswanathan et al., 2003b).

Host–pathogen interaction studies involving differential display and subtractive libraries revealed new information on the differentially expressed transcripts expressed during incompatible interactions (Gupta et al., 2010; Muthumeena et al., 2013; Prathima et al., 2013; Rahul et al., 2016; Sathyabhama et al., 2015). As of now we postulate that pathogen-associated molecular patterns (PAMPs) generated during the attempted invasion of *C. falcatum* are recognized by pattern recognition receptor and R-genes in the host, and this leads to various downstream signalling processes including activation of calcium influx channels, activation of H_2O_2 production, induction of transcription factors such as Myb and WRKY, induction of R-gene expression, production of phytoalexins and expression of defence-related genes. The calcium-dependent protein kinase activated by the influx of Ca^{2+} ions, in the presence of 14-3-3 protein, leads to the generation of reactive oxygen species (ROS), and the ROS production is further aided by the H_2O_2 generated by the SOD/NADPH complex. Rapid generation of ROS further strengthens the magnitude of downstream signalling processes. The collective action of all the above components will result in a defence signal of magnitudes well above the threshold level of hypersensitive response and ultimately cell death (Viswanathan, 2012). Our present understanding points towards a predominantly hypersensitive reaction/cell death-oriented response exerted by resistant sugarcane genotypes upon attempted pathogen invasion and this will be helpful in studies involving characterization of individual components of sugarcane and *C. falcatum* interactions. Further studies are also required to identify key switch taking place during biotrophy to necrotropic interaction during the infection. Developing a stalk reference proteome in sugarcane (Ramesh Sundar et al., 2010) led to establishing a defence proteome for red rot resistance (Viswanathan, 2012). Hameed et al. (2015) identified few resistant gene analogues in red rot-resistant varieties.

Singh et al. (2016) studied genome-wide association mapping of resistance to different races of red rot, and identified putative candidate genes possibly involved in host–pathogen interaction based on synteny between the sorghum and sugarcane genomes. They identified four marker–trait associations (MTAs) for resistance to red rot. They found several genes encoding important plant defence-related proteins to be localized to the vicinity of these MTAs and further studies could identify potential markers for their application in marker-assisted sugarcane breeding. Three of these four markers identified in this study were mapped to positional candidate genes related to plant defence locations of the sorghum genome and these markers may be of high value in marker-assisted selection for red rot resistance in sugarcane. This study also identified proteins with LRR and ring finger as functional domains, cytochrome P450, MAP kinase-4 and legume lectins as possible candidate genes coding for red rot resistance. The identification of candidate genes for red rot resistance may contribute to the understanding of biological mechanisms determining the partial and polygenically inherited resistance to this disease.

5.3 Mechanism of induced resistance

Different metabolites or hydrolytic enzymes of antagonistic *Pseudomonas* and *Trichoderma* strains have been found to contribute to suppression of *C. falcatum*, and induced resistance and growth promotion in sugarcane (Viswanathan, 2010). A high-molecular-weight elicitor was isolated from the mycelial walls of *C. falcatum* and the partially purified elicitor-induced accumulation of phenolics and defence-related enzymes in sugarcane leaves and suspension-cultured cells (Ramesh Sundar et al., 2002). SAR against *C. falcatum* linked to the inducer molecule BTH has also been found (Ramesh Sundar et al., 2009). Recently, the molecular basis of SAR-responsive transcripts in sugarcane were also documented through genomic tools. The study revealed a major role played by cinnamic acid 4-hydroxylase, 4-coumarate:coenzyme A ligase, chalcone synthase and chalcone reductase in the regulation of phytoalexin synthesis and R-genes such as NBS-LRR genes and basal layer antifungal peptide genes in SAR (Nagarathinam et al., 2014). Besides, preliminary investigations on SAR-induced sugarcane indicated an abundant expression of PR proteins. Meanwhile, the role of signalling pathways during the activation of systemic resistance by inducers and molecular signatures of *C. falcatum* (PAMPs) is being studied (Ramesh Sundar, unpublished). The pattern of gene expression will elucidate signalling pathways that respond differentially with respect to host genotype and SAR/pathogen-triggered immunity as observed in other plants.

5.4 Characterization of pathogenicity genes

While studying host resistance mechanisms to *C. falcatum* in sugarcane, considerable efforts were made to understand pathogenicity in the fungal pathogen through genomic tools. Various attempts have been made to identify pathogenicity genes in *C. falcatum* (Karunakaran et al., 2008; Malathi and Viswanathan, 2012). Malathi and Viswanathan (2012) identified metabolites, melanin and hydrolytic enzymes as pathogenicity factors in *C. falcatum*. Also, expression of a number of pathogenicity gene homologs reported in other *Colletotrichum* spp. was confirmed in *C. falcatum* (Scindiya et al., 2017). In addition to understanding pathogenicity mechanism through genomic approaches, there were attempts to identify key candidate target genes of *C. falcatum* involved in pathogenicity and to suppress their expression. Antifungal activities of hydrolytic enzymes from sugarcane and antagonistic *Trichoderma* have been demonstrated against *C. falcatum* (Malathi et al., 2012; Viswanathan et al., 2003b,c). Further studies of interactive proteomics involving sugarcane, *C. falcatum* and antagonistic *Trichoderma* identified the proteins specifically expressed for pathogenicity in *C. falcatum* and defence in sugarcane. These studies identified a set of candidate genes/proteins of *C. falcatum* suppressed by *T. harzianum* in sugarcane stalks during the interaction (Elamathi et al., 2017). Recently, there has been interesting work on a recombinant 'sugarcane wound-inducible protein' SUGARWIN2 against *C. falcatum* by Franco et al. (2014). The study showed that the protein altered *C. falcatum* morphology by increasing vacuolization, points of fractures and a leak of intracellular material, leading to germling apoptosis. The specificity of SUGARWIN2 towards *C. falcatum* and its action has been proposed to be part of a strategy against the fungal pathogen that colonizes the plant after insect borer *Diatraea saccharalis* attack.

These approaches will identify key target candidate genes expressed during *C. falcatum* pathogenesis and potential antifungal genes to suppress their activities. Further, the target genes in *C. falcatum* can also be suppressed through RNAi strategies.

5.5 *C. falcatum* genome and transcriptome

Recently, efforts were made to sequence the genome and transcriptome of *C. falcatum*. The genome consisted of 12 270 genes while its transcriptome revealed 13 320 genes, the notable closely related species were *C. graminicola* and *C. sublineola* which cause anthracnose in maize and sorghum, respectively (Viswanathan et al., 2016a). The *C. falcatum* secretory proteins were found to have large number of CAZy families, esterases, peptidases, proteases, cytochrome P450, transporters, proteinases and transcription factors (Prasanth et al., 2016). These secretory proteins may actively suppress host defence and allow invasive hyphae spread throughout to initiate infection. Comparative analysis of the genes involved in biotrophy–necrotrophy transition and membrane transporters revealed that the number of transporters encoded by *C. falcatum* is significantly higher compared to that encoded by several other *Colletotrichum* spp. infecting other crops (O'Connell et al., 2012). The genomic and transcriptomic analyses have brought new vistas in understanding the functions of effector proteins in fungal pathogenicity during the interaction. Identifying the key genes which interact with host interface using *in silico* techniques will further help to understand the pathogenicity mechanism completely.

6 Summary and future trends

Although heritability of red rot resistance is complex, there has been great success in developing many red rot-resistant varieties through conventional bi-parental crossing and varietal selection. Precise screening methods for testing sugarcane progenies under controlled conditions make it possible to determine red rot resistance at an early stage in the breeding and selection programme. This has accelerated identification of the promising varieties. The identification of the molecular markers for red rot resistance in sugarcane also promises to have application in early progeny selection. Currently, red rot-resistant progenies possess resistance to one or few pathotypes of *C. falcatum* and further studies need to be done to broaden resistance from other sources such as *S. spontaneum* and *Erianthus* spp. with potential resistant genes.

C. falcatum exhibits variation in cultural, morphological, pathogenic and molecular traits. Detailed analysis of the genome and transcriptome data of *C. falcatum* through comparative genomics has helped to identify pathogenicity or virulence genes, including unique genes in *C. falcatum* that are responsible for its particular effects on sugarcane. In future, candidate genes coding for, for example, wound-inducible proteins could be tested for expression in sugarcane to minimize damage caused by *C. falcatum* using transgenic techniques.

Apart from varietal resistance, alternate strategies like induced systemic resistance and SAR show promise. Further, going for nano-formulations is also a promising approach to deliver inducers to manage the disease in a sustainable way. Improved fungicide delivery to setts has been shown to contain disease development in young crops. Once the young crop is protected, secondary spread of the pathogen is reduced significantly.

7 Where to look for further information

Further information on sugarcane red rot can be had by referring to earlier publications such as Duttamajumdar (2002), Singh and Singh (1989), Viswanathan (2010), Viswanathan et al. (2003) and Viswanathan and Rao (2011).

8 Acknowledgements

The research work on red rot was supported by different funding agencies viz. the Department of Biotechnology; the Department of Science & Technology; the Indian Council of Agricultural Research (Outreach project), New Delhi; and ICAR-SBI, Coimbatore.

9 References

Abbott, E. V. (1938). *Red Rot of Sugarcane*. Tech. Bull. USDA, 641, p. 96.

Agnihotri, V. P. (1983). *Diseases of Sugarcane*. Oxford & IBH Publishing Co., New Delhi, p. 363.

Ashwin, N. M. R., Leonard Barnabas, E., Ramesh Sundar, A., Muthumeena, M., Malathi, P. and Viswanathan, R. (2017). Disease suppression effects of resistance inducing agents against red rot of sugarcane. *Eur. J. Plant Pathol.* DOI: 10.1007/s10658-017-1181-1.

Azab, Y. E. and Chilton, S. J. P. (1952). Studies on the inheritance of red rot disease of sugarcane. *Phytopathology*, 42, 282.

Babu, C., Natarajan, U. S., Shanthi, R. M., Govindaraj, P., Sunder, A. R. and Viswanathan, R. (2010), Inheritance of red rot resistance in sugarcane (*Saccharum* sp. hybrids). *Sugar Tech*, 12, 167–71.

Barber, C. A. (1901), Sugarcane diseases in Godawari and Ganjam Districts. *Madras Dept. Land Records and Agri. Bull.*, 512(43), 181–94.

Bukhari, K. A. I. M., Nithya, K., Valluvaparidasan V., Paranidharan V. and Velazhahan R. (2012). Detection of Colletotrichum falcatum causing red rot of sugarcane by enzyme-linked immunosorbent assay. *Arch. Phytopathol. Plant Protect.*, 45, 823–30.

Butler, E. J. (1906). Fungal diseases of sugarcane in Bengal. *Mem. Dept. Agri. India Bot. Ser.* 1(3), 2–24.

Chandra, A., Keizerweerd, A. T., Que, Y. X. and Grisham, M. P. (2015). Loop-mediated isothermal amplification (LAMP) based detection of *Colletotrichum falcatum* causing red rot in sugarcane. *Mol. Biol. Rep.*, 42, 1309–16,

Chona, B. L. (1950). Studies on the diseases of sugarcane in India. III Sources and mode of red rot infection. *Indian J. Agric. Sci.*, 20, 363–85.

Chona, B. L. (1954). Relative resistance of sugarcane varieties to red rot. *Indian J. Agric. Sci.*, 24, 301–15.

Chona, B. L. (1980), Red rot of sugarcane and sugar industry – a review. *Indian Phytopathol.*, 33, 191–207.

Duttamajumdar, S. K. (2002). A century of red rot disease of sugarcane in India. In *Sugarcane Crop Management* (Eds: S. B. Singh, G. P. Rao and S. Eswaramoorthy). SCI TECH Publishing LLC, Houston, Texas, USA, pp. 52–108.

Edgerton, C. W. (1959). *Sugarcane and its Diseases*. Louisiana State University Press, La, USA, pp. 69–100.

Elamathi, E., Malathi, P., Viswanathan, R. and Ramesh Sundar, A. (2017). Molecular analysis of the suppression of *Colletotrichum falcatum* by *Trichoderma harzianum* in sugarcane. *J. Plant Pathol.* 99, 211–18. DOI: http://dx.doi.org/10.4454/jpp.v99i1.3800.

Franco, F. P., Santiago, A. C., Henrique-Silva, F., de Castro, P. A., Goldman, G. H., Moura, D. S. and Silva-Filho, M. C. (2014). The sugarcane defense protein SUGARWIN2 causes cell death in *Colletotrichum falcatum* but not in non-pathogenic fungi. *PLoS ONE*, 9(3), e91159. DOI: 10.1371/journal.pone.0091159.

Ganesh Kumar, V., Viswanathan, R., Malathi, P., Nandakumar, M. and Ramesh Sundar, A. (2015), Differential induction of 3-deoxyanthocyanidin phytoalexins in relation to *Colletotrichum falcatum* resistance in sugarcane. *Sugar Tech*, 17, 314–21.

Godshall, M. A. and Lonergan, T. A. (1987), The effect of sugarcane extracts on the growth of the pathogenic fungus *Colletotrichum falcatum. Physiol. Mol. Plant Pathol.*, 30, 19–42.

Goswami, D., Borah, S. N., Lahkar, J., Handique, P. J. and Deka, S. (2015). Antifungal properties of rhamnolipid produced by *Pseudomonas aeruginosa* DS9 against *Colletotrichum falcatum. J. Basic Microbiol.*, 55, 1–10.

Gupta, V., Raghuvanshi, S., Gupta, A., Saini, N., Gaur, A., Khan, M. S., and Tyagi, A. K. (2010). The water deficit stress and red rot related genes in sugarcane. *Funct. Integr. Genom.*, 10, 207–14.

Hameed, U., Pan, Y. B. and Iqbal, J. (2015), Genetic analysis of resistance gene analogues from a sugarcane cultivar resistant to red rot disease. *J. Phytopathol.*, 163, 755–63.

Hassan, M. N., Namood-e-Sahar, Shah, S. Z., Afghan, S. and Hafeez, F. Y. (2015). Suppression of red rot disease by *Bacillus* sp. based biopesticide formulated in non-sterilized sugarcane filter cake. *BioControl*, 60, 691–702.

Karunakaran, M., Nair, N. V., Rho, H.-S., Choi, J., Kim, S. and Lee, Y.-H. (2008). *Agrobacterium tumefaciens*-mediated transformation in *Colletotrichum falcatum* and *C. acutatum. J. Microbiol. Biotechnol.* 18, 234–41.

Malathi, P, Padmanaban, P, Viswanathan, R, Mohanraj, D, and Ramesh Sundar, A. (2004), Efficacy of Thiophanate methyl against red rot of sugarcane. *Acta Phytopathol. Entomol. Hungarica*, 39(1–3), 39–47.

Malathi, P. and Viswanathan, R. (2012). Identification of pathogenicity determinants in *Colletotrichum falcatum* using wild and mutant cultures. *Sugar Tech*, 14, 383–90.

Malathi, P., Viswanathan, R. and Jothi R. (2006). Specific adaptation of *Colletotrichum falcatum* pathotypes to sugarcane cultivars. *Sugar Tech*, 8(1), 54–8.

Malathi, P., Viswanathan, R. and Padmanaban, P. (2012). Identification of antifungal proteins from fungal and bacterial antagonists against *Colletotrichum falcatum* causing sugarcane red rot. *J. Biol. Control*, 26, 49–54.

Malathi, P., Viswanathan, R. and Ramesh Sundar A. (2014). Phylogenetics and diagnostics of *Colletotrichum falcatum* isolates causing red rot of sugarcane in India. Proceedings of National Symposium on Innovative and Ecofriendly Research Approach for Plant Disease Management, Indian Society of Mycology and Plant Pathology, PDKV, Akola, 8–10 January 2014, p. 208.

Malathi, P., Viswanathan, R. and Ramesh Sundar, A. (2017). Mechanized means of sett treatment: an effective way of delivering fungicides for the management of red rot in sugarcane. *Sugar Tech*, 19, 176–82, doi: 10.1007/s12355-016-0444-z.

Malathi, P., Viswanathan, R., Ramesh Sundar, A., Prakasam, N., Padmanaban, P., Jothi, R., Renuka Devi, S. R. and Poongothai, M. (2010). Variability among *Colletotrichum falcatum* pathotypes used for screening red rot resistance in sugarcane. *Sugar Cane Intern.*, 28(2), 47–52.

Mohanraj, D., Kumaresan, S. and Sreenivasan, T. V. (2002). Molecular characterization of isolates of the sugarcane red rot pathogen *Colletotrichum falcatum* Went. *Indian Phytopathol.*, 55, 147–51.

Muthumeena, M., Ashwin, R., Ramesh Sundar, A., Malathi, P. and Viswanathan, R. (2013). Expression profiling of transcription factors (TFs) in sugarcane X *Colletotrichum falcatum* interaction. *J. Plant Biochem. Biotechnol.*, 22, 286–94

Nagarathinam, S., Ashwin, R., Ramesh Sundar, A., Malathi, P. and Viswanathan, R. (2014). Molecular profiling of systemic acquired resistance (SAR)-responsive transcripts in sugarcane challenged with *Colletotrichum falcatum. Appl. Biochem. Biotechnol.*, 174, 2839–50.

O'Connell, R. J., Thon, M. R., Hacquard, S., Amyotte, S. G., Kleemann, J., Torres, M. F. and Vaillancourt, L. J. (2012). Lifestyle transitions in plant pathogenic *Colletotrichum* fungi deciphered by genome and transcriptome analyses. *Nat Gene.*, 44, 1060–5.

Ogunwolu, E. C. (1983). Some relationship between crop damage inflicted by the sugarcane borer, *Diatraea saccharalis* (F) and selected plant diseases. Ph. D thesis, Louisiana State University, Baton Rouge, Louisiana, USA, 99p.

Olufolaji, D. B. (1987). The sugar cane borer and red rot disease in Nigeria. *Sugar Cane*, (Autumn suppl.), 21–3.

Padmanaban, P., Mohanraj, D., Viswanathan, R., Rao, M. M., Prakasam, N., Jothi, R. and Alexander, K. C. (1996). Differential interaction of sugarcane clones to pathotypes of *Colletotrichum falcatum* Went. *Sugar Cane*, 4, 16–20.

Prasanth, C. N., Viswanathan, R., Malathi, P., Sundar, A. R. and Tiwari, T. (2016). Identification of candidate secretory effector proteins (CSEPS) genes from *Colletotrichum falcatum* and their role in host-pathogen interaction by comparative modelling. *Proc. Intern. Soc. Sugar Cane Technol.*, 29, 1747–50.

Prathima, P. T., Raveendran, M., Kumar, K. K., Rahul, P. R., Ganesh Kumar, V., Viswanathan, R., Ramesh Sundar, A., Malathi, P., Sudhakar, D. and Balasubramanian, P. (2013). Differential regulation of defense-related gene expression in response to red rot pathogen *Colletotrichum falcatum* infection in sugarcane. *Appl. Biochem. Biotechnol.*, 171, 488–503.

Rahul, P. R., Ganesh Kumar, V., Viswanathan, R., Ramesh Sundar, A., Malathi, P., Naveen Prasanth, C. and Prathima, P. T. (2016). Defense transcriptome analysis of sugarcane and *Colletotrichum falcatum* interaction using host suspension cells and pathogen elicitor. *Sugar Tech*, 18, 16–28.

Ramesh Sundar, A., Nagarathinam, S., Ganesh Kumar, V., Rahul, P. R., Raveendran, M., Malathi, P., Agrawal, G. K., Rakwal, R. and Viswanathan, R. (2010). Sugarcane proteomics: Establishment of a protein extraction method for 2-DE in stalk tissues and initiation of sugarcane proteome reference map. *Electrophoresis*, 31, 1959–74.

Ramesh Sundar, A., Velazhahan, R. and Vidhyasekaran, P. (2002). A glycoprotein elicitor isolated from *Colletotrichum falcatum* induces defense mechanisms in sugarcane leaves and suspension-cultured cells. *J. Plant Dis. Protect.*, 109, 601–11.

Ramesh Sundar, A., Viswanathan, R. and Nagarathinam, S. (2009). Induction of systemic acquired resistance (SAR) using synthetic signal molecules against *Colletotrichum falcatum* in *Saccharum officinarum*. *Sugar Tech*, 11, 274–81.

Sathyabhama, M., Viswanathan, R., Nandakumar, M., Malathi, P. and Ramesh Sundar, A. (2015). Understanding sugarcane defence responses during the initial phase of *Colletotrichum falcatum* pathogenesis by suppression subtractive hybridization (SSH). *Physiol. Mol. Plant Pathol.*, 91, 131–40.

Sathiyabama, M. and Manikandan, A. (2016). Chitosan nanoparticle induced defense responses in finger millet plants against blast disease caused by *Pyricularia grisea* (Cke.) Sacc. *Carbohydrate Polymers*, 154, 241–6.

Scindiya, M., Malathi, P., Kaverinathan, K., Viswanathan, R. and Ramesh Sundar, A. (2017). Molecular characterization of pathogenicity gene homologs in *Colletotrichum falcatum* causing red rot in sugarcane. *Sugar Tech*, DOI: 10.1007/s12355-017-0512-z.

Singh, K. and Singh, R. P. (1989). Red rot. In *Diseases of Sugarcane: Major Diseases* (Eds.: C. Ricaud, E. T. Egan, A. G. Gillaspie, Jr. and C. G. Hughes). Elsevier, Amsterdam, pp. 169–88.

Singh, R. K., Banerjee, N., Khan, M. S., Yadav, S., Kumar, S., Duttamajumder, S. K., Lal, R., Patel, J. D., Guo, H., Zhang, D. and Paterson, A. H. (2016). Identification of putative candidate genes for red rot resistance in sugarcane (*Saccharum* species hybrid) using LD-based association mapping. *Mol. Genet. Genom.*, DOI: 10.1007/s00438-016-1190-3.

Srinivasan, K. V. (1962). Some observations on variation in the red rot pathogen, *Glomerella tucumanensis* (Speg.) Arx and Muller. *Proc. Intern. Soc. Sugar Cane Technol.*, 11, 795–802.

Srinivasan, K. V. and Bhat, N. R. (1961). Red rot of sugarcane – criteria for grading resistance. *J. Indian Bot. Soc.* 40, 566–77.

Suman, A., Lal, S., Shasany, A. K., Gaur, A. and Singh, P. (2005). Molecular assessment of diversity among pathotypes of *Colletotrichum falcatum* prevalent in sub-tropical Indian sugarcane. *World J. Microbiol. Biotechnol.*, 21, 1135–40.

Trenor, K. L. and Bailey, R. A. (1989). A preliminary report on the incidence of red rot in the South African sugarcane industry. Proc. Ann. Cong. South African Sugar Technol.' Assoc. 111–115.

Viswanathan, R. (2010). *Plant Disease: Red Rot of Sugarcane*. Anmol Publications Pvt Ltd, New Delhi, 301p.
Viswanathan, R. (2012). Molecular basis of red rot resistance in sugarcane. In *Sugarcane Pathology. Functional Plant Science* and *Biotechnology 6 (Special Issue 2)* (Eds: R. Viswanathan and A. R. Sundar). Global Science Books, Ikenobe, Japan, pp. 40–50.
Viswanathan, R. (2017). Pathogen virulence in sugarcane red rot pathogen versus varieties in cultivation: Classical case of loss in virulence in the pathotype CF06 (Cf671). *Sugar Tech*, 19, 293–99. Doi: 10.1007/s1235501604586.
Viswanathan, R., Malathi, P. and Naik, R. (2016b). A sett treatment device for fungal disease management and healthy nursery programme in sugarcane. *Proc. Ann. Conv. Sugar Technol. India*, 77, 236–53.
Viswanathan, R., Malathi, P. and Padmanaban, P. (2003a). Variation in sugarcane red rot pathogen *Colletotrichum falcatum* Went. In *Frontiers of Fungal Diversity in India* (Eds: G. P. Rao, C. Manoharachari, D. J. Bhat, R. C. Rajak and T. N. Lakhanpal). International Book Distributing Co, Lucknow, pp. 639–67.
Viswanathan, R., Malathi, P., Ramesh Sundar, A., Aarthi, S., Premkumari, S. M. and Padmanaban, P. (2005). Differential induction of chitinases and thaumatin-like proteins in sugarcane in response to infection by *Colletotrichum falcatum* causing red rot disease. *J. Plant Dis. Protect.*, 112, 537–42.
Viswanathan, R., Nandakumar, R. and Samiyappan, R. (2003b). Involvement of pathogenesis-related proteins in *Pseudomonas* spp. induced systemic resistance against *Colletotrichum falcatum* in sugarcane. *J. Plant Diseases Protect.*, 110, 524–37.
Viswanathan, R., Naveen Prasanth, C., Malathi, P. and Ramesh Sundar, A. (2016a). Draft genome sequence of *Colletotrichum falcatum* – a prelude on screening of red rot pathogen in sugarcane. *J Genomics*, 4, 1–3.
Viswanathan, R., Padmanaban, P. and Mohanraj, D. (1998). Comparison of three testing methods for the evaluation of red rot (*Colletotrichum falcatum* Went) in sugarcane. *Indian J. Agric. Sci.*, 68, 226–30.
Viswanathan, R., Padmanaban, P., Mohanraj, D. and Jothi R. (2000). Indirect ELISA technique for the detection of the red rot pathogen in sugarcane (*Saccharum* spp. hybrid) and resistance screening. *Indian J. Agric. Sci.*, 70, 308–11.
Viswanathan, R., Ramesh Sundar, A. and Premkumari, S. M. (2003c). Mycolytic enzymes of extracellular enzymes of antagonistic microbes to *Colletotrichum falcatum*, red rot pathogen of sugarcane. *World J. Microbiol. Biotechnol.*, 19, 953–9.
Viswanathan, R. and Rao, G. P. (2011), Disease scenario and management of major sugarcane diseases in India. *Sugar Tech*, 13, 336–53.
Viswanathan, R. and Samiyappan, R. (1999). Red rot disease in sugarcane: a major constraint for the Indian sugar industry. *Sugar Cane*, 5, 9–15.
Viswanathan R. and Samiyappan, R. (2002a), Induced systemic resistance by fluorescent pseudomonads against red rot disease of sugarcane caused by *Colletotrichum falcatum*. *Crop Protect.*, 21, 1–10.
Viswanathan, R. and Samiyappan, R. (2002b). Role of oxidative enzymes in the plant growth promoting rhizobacteria (PGPR) mediated induced systemic resistance against *Colletotrichum falcatum* in sugarcane. *J. Plant Dis. Protect.*, 109, 88–100.
Went, F. A. F. C. (1893), Het rood snot. *Arch. Java-Suikerindus*, 1, 265–82.
Yin, Z., Hoy, J. W. and Milligan, S. B. (1996). Evaluation and heritability of resistance to sugarcane red rot. *Phytopathology*, 86, 662–7.

Chapter 10

Progress in understanding fungal diseases affecting sugarcane: smut

Claudia B. Monteiro Vitorello, Patricia Dayane Carvalho Schaker, Juliana Benevenuto, Natália de Sousa Teixeira e Silva and Sintia Silva de Almeida, University of São Paulo, Brazil

1 Introduction

Smut fungi are host-specific pathogens affecting many angiosperm clades, although they most commonly occur on Poaceae family members (Begerow et al., 2014). True smut fungi belong to the Ustilaginomycetes class in the Basidiomycota phylum and comprise more than 1,650 species (Toh and Perlin, 2016). In general, smut pathogens are found along with the geographic distribution of their hosts (Begerow et al., 2014). Common to all smut species is the need to undergo sexual cross to form dikaryotic hyphae to infect the host plant. If components of such pathway are disrupted, virulence is severely reduced (Bölker, 2001; Bakkeren et al., 2008; Yan et al., 2016b,c). On the contrary, haploid cells can grow saprophytically and are not able to cause disease. Disease symptoms vary amongst hosts, but they all share fungal sporogenesis as a primary characteristic of the main symptom.

http://dx.doi.org/10.19103/AS.2017.0035.33

The intimate biotrophic interaction of the fungus with its host makes smut fungi a group of very specialized pathogens. Host specialization occurred after the host jump probably occurred during the evolution and led to smut fungi speciation (Begerow et al., 2014).

Sugarcane smut was first described in 1877 in South Africa, reported on a *Saccharum sinense* clone known as 'China cane' (reviewed by several authors including Luthra et al. (1940), Waller (1969), Lee-Lovick (1978) and Bailey (1979)). Reports in India of smut causing severe problems date back to the 1930s, but smut has probably always been in India on susceptible genotypes of *Saccharum barberi* and in the wild grass *Saccharum spontaneum* (Ferreira and Comstock, 1989; Croft and Braithwaite, 2006). The disease reached Argentina in 1940 and soon was identified in Brazil, Paraguay and Bolivia (Bergamin Filho et al., 1987; Rago et al., 2009). Around the same time, Kenya reported the first smut epidemic peak (Waller, 1969). The disease was then observed in Hawaii and in the Caribbean after 1970 (Byther et al., 1971; Lee-Lovick, 1978). Finally, sugarcane smut was reported in Western Australia in 1998 (Croft and Braithwaite, 2006; Magarey et al., 2010). The sugarcane origin centre in Papua New Guinea was considered free of smut until the recent report (August 2016) on commercial crops (Tom et al., 2017). Currently, smut is present in nearly all sugarcane-growing countries, except Fiji (Croft and Braithwaite, 2006; Sundar et al., 2012; Tom et al., 2017) (Fig. 1). The severity of the disease varies widely, from negligible proportions to serious losses (10–80%), depending on the environmental conditions, prevalence of pathogen races and the tolerance level of the sugarcane varieties (Whittle, 1982; Hoy et al., 1986; Sundar et al., 2012). Losses in productivity are related to tillering, reduced cane diameter, reduced sucrose content and increased fibre content (Martinez et al., 2000; Wada et al., 2016).

After this historical perspective of the disease, this chapter will present an overview of data gathered over the past few years using 'omics' techniques (genomics, transcriptomics, proteomics and metabolomics) in combination with functional studies performed to identify candidate genes involved in the interaction. This pathosystem has been investigated considering the responses of sugarcane-resistant and susceptible genotypes to the pathogen as well as the pathogen biology, genetic diversity and strategies used to

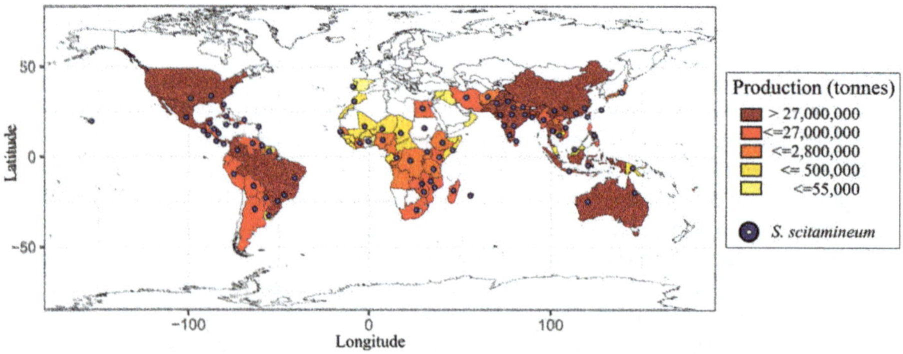

Figure 1 Distribution of smut disease in sugarcane-growing countries. Sugarcane production map was adapted from FAOSTAT 2014 data (http://www.fao.org/faostat) and *Sporisorium scitamineum* (disease agent) distribution according to Plantwise (http://www.plantwise.org/). The recent finding of smutted sugarcane in Papua New Guinea (Tom et al., 2017) is included in the graphic.

infect and colonize sugarcane tissues. Each subsection will discuss different aspects of the knowledge generated for this pathosystem.

2 Characteristics of the disease

The biotrophic fungus *Sporisorium scitamineum* (Syd.) (Piepenbring et al., 2002) (Syn: *Ustilago scitaminea* H. and P. Sydow) is the causal agent of sugarcane smut. The characteristic disease symptom is the emergence of black whip structures from shoot apical meristem or meristems of lateral buds of infected stalks. Whips are largely brown to black in colour assuming various formats (Fig. 2). The whips are composed of a core of pithy sugarcane tissues surrounded by fungal sori responsible for teliospore formation (Piepenbring, 1996; Marques et al., 2016). The infected sugarcane tissues lose their regular structure and vessel organization (Lloyd and Pillay, 1980; Marques et al., 2016; Carvalho et al., 2016). Other general symptoms related to smut are tillering and poor cane formation, and occasionally leaf and stem galls (Croft and Braithwaite, 2006).

Figure 2 Disease symptoms of sugarcane smut. (a) Single whip-like structure from sugarcane shoot apical meristem; (b) three whips emerging from sugarcane lateral tillers (red arrows); (c) tumour-like gall developed over a leaf midrib; (d) basal enlargement of a whip (red arrow); (e) longitudinal section of the whip shown in (d); (f) inoculated (left) and control (right) plants showing stalk diameter differences after whip emission (120 DAI); (g) tillering of infected plants; (h) single culm healthy plants; (i) gall formation in the base of sugarcane stalk.

Sporisorium scitamineum infects cane meristem tissues, grows systemically and induces the formation of reproductive structures in the apical meristem region of the plant (whip) (Lloyd and Pillay, 1980; Sundar et al., 2012; Marques et al., 2016; Carvalho et al., 2016). Teliospores are formed and matured within these structures. The fungal dispersion mainly occurs through teliospores carried by wind or water (Sundar et al., 2012). The life cycle of the pathogen includes teliospore germination on the surface of sugarcane buds (Taniguti et al., 2015). Following germination, diploid teliospores undergo meiosis to produce sporidial cells (haploid). Two sexually compatible sporidial cells may fuse to originate an infective dikaryotic hypha, which plays a vital role in infecting sugarcane (Fig. 3). Sexual compatibility depends on the type of sexual reaction (*mating type*). The mating type is controlled by two loci, *a* and *b*, each harbouring conserved genes (Albert and Schenck, 1996; Bakkeren et al., 2008). The *a* locus encompasses genes encoding a pheromone and a membrane receptor. Compatible mates carry alleles encoding complementary pairs of the receptor and pheromone proteins. The *b* locus harbours genes encoding two subunits of a heterodimeric transcription factor in which the subunits must be encoded by different alleles to produce a functional protein (Gillissen et al., 1992; Kronstad and Staben, 1997; Bölker, 2001; Singh et al., 2004). Subsequently, the dikaryotic hyphae differentiate appressorium to penetrate plant tissues (Peters et al., 2017). During the infection process, environmental and host-specific signals are perceived and transmitted by intracellular signalling pathways. The heterodimeric transcriptional factor regulates sexual mating and/or filamentous growth potentially via modulating glucose metabolism and Hog1-mediating oxidative response (Yan et al., 2016b). This set of events usually

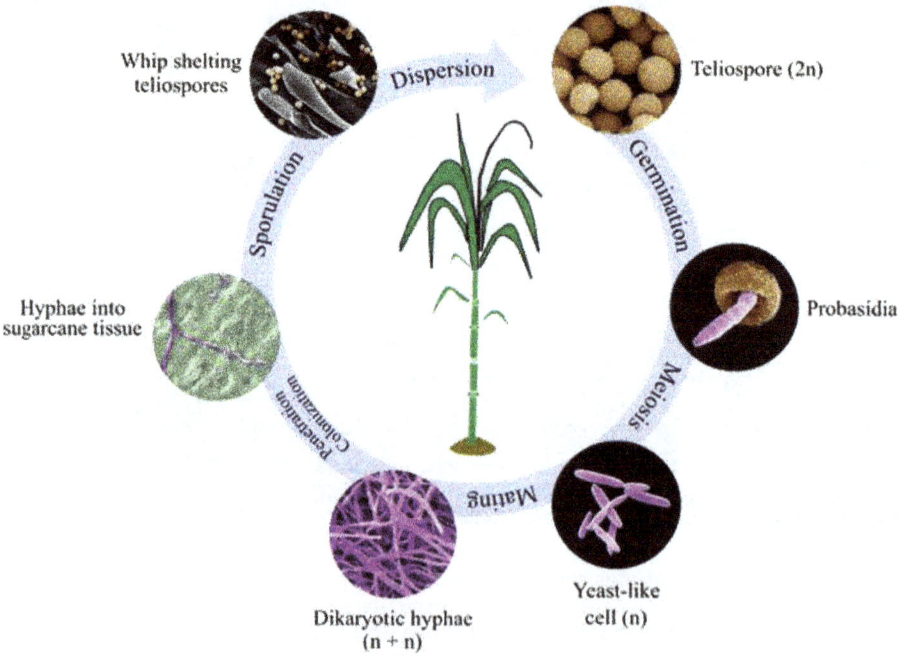

Figure 3 *Sporisorium scitamineum* in various developmental stages and within host tissues.

occurs between 6 and 48 h after teliospore deposition in the sugarcane surface (Sundar et al., 2012; Peters et al., 2017). It has been described that some sugarcane-resistant varieties can be colonized by the fungus and only eventually induce whip development (Carvalho et al., 2016). Rates and patterns of sugarcane tissue colonization differ between susceptible and tolerant/resistant genotypes (Lloyd and Pillay, 1980; Marques et al., 2016; Carvalho et al., 2016; Peters et al., 2017).

3 Disease control and pathogen detection

Effects of climate changes and green harvesting practices will likely impact the incidence of sugarcane diseases (Zhao and Li, 2015). Smut is one of the diseases which will mostly be affected due to increasing global temperature and the maintenance of spores through ratoon crops in non-burnt fields (Waller, 1969). It is described that high temperatures (25–30°C) and dry conditions favour disease spread and progress (Zekarias et al., 2011), along with ratooning practices, which induce the development of symptoms in latently infected plants (Akalach and Touil, 1996; Croft and Braithwaite, 2006).

Because smut occurs in all cane-producing regions with diverse edaphoclimatic conditions, various strategies may be adopted to prevent infection, and reduce or control the severity of the disease. Amongst them are the selection of resistant varieties (Singh et al., 2005a; Croft et al., 2008; Nalawade et al., 2012; Lemma et al., 2015), use of fungicides (Wada, 2003; Bhuiyan et al., 2015b), heat treatment of stalks (Gupta, 1978), crop rotation, soil flooding (Abdou et al., 1990) and roguing (Lee-Lovick, 1978; Rott and Comstock, 2015). Currently, the use of pre-sprouted seedlings with a phytosanitary certificate and seedlings from micropropagation methods are the alternatives to prevent smut and other diseases affecting sugarcane cultivation.

The selection for resistant varieties is still the most effective way to control smut disease. Sugarcane genotypes are evaluated for resistance by submitting buds to artificial inoculation, usually by immersing them in a spore suspension. Strategies to select resistant varieties often include evaluation of the presence of protective scales over meristem regions that will impair and/or delay germination and fungal penetration (Gloria et al., 1995; Peters et al., 2017). These protective scales present trichomes and/or phenolic compounds. The percentage of whipped plants is taken as a measure of disease susceptibility in a given clonal population (Ladd et al., 1974). The whip counting starts six weeks after inoculation and continues up to 10 months (Lemma et al., 2015). The rating scale ranges from 0 (immune) to 9 (very susceptible) (Latiza et al., 1980).

Early diagnosis of the infected plants and early screening of the resistant varieties are important management tools. Currently, the detection of smut before whip emission is possible using techniques of molecular biology and microscopy (Alexander and Ramakrishnan, 1980; Lloyd and Pillay, 1980; Sinha et al., 1982; Lloyd and Naidoo, 1983; Albert and Schenck, 1996; Singh et al., 2004; Dalvi et al., 2012; Kavitha et al., 2014). The first molecular detection method used the sequence of the mating-type *bE* gene as a template for polymerase chain reaction (PCR) amplification (Albert and Schenck, 1996). The improved quantitative method (TaqMan probe) used the same target gene and was able to detect the fungus at the initial stage (12 h) of infection with a potential to investigate seed cane and seedlings (Su et al., 2013a, 2016b). Recently, the rapid and cost-effective LAMP method (loop-mediated isothermal amplification) was described using the sequence of the effector *pep1* gene as a target and detected smut as soon as one day after inoculation (Su et al., 2016d). Fungal colonization throughout the time intervals has also been investigated as a tool to classify resistance in sugarcane genotypes (Su et al., 2016b).

4 Pathogenic and molecular variability of *Sporisorium scitamineum*

The knowledge of the amount and distribution of the genetic variation within and amongst pathogen populations is useful to optimize breeding strategies and screen cultivar genotypes for resistance (McDonald and Linde, 2002). The sensibility of this pathosystem to environmental conditions makes it very difficult to differentiate the degree of aggressiveness amongst fungal isolates and to evaluate the tolerance level of sugarcane varieties against the disease. Under stress conditions, even resistant-classified varieties may present symptoms after smut infection.

Few breakdowns have been reported for cane smut, and the presence of pathogen races is controversial (Ferreira and Comstock, 1989). Two pathogenic races of sugarcane smut (named A and B) were reported in Hawaii in 1976, when the emergence of race B severely threatened the Hawaiian sugar industry (Comstock et al., 1983). In 2001, another new smut race isolated in the Maui Island was reported, affecting one of the major Hawaiian commercial varieties previously considered completely resistant (Schenck, 2003). In Taiwan, three pathogenic races of sugarcane smut were reported (races 1–3) (Lee et al., 1999). Five races were reported in Pakistan (Muhammad and Kausar, 1962), two races in Brazil (Silva and Sanguino, 1978) and two races (X and Y) in Nigeria (Abo and Okusanya, 1996). Luzaran et al. (2012) and Nzioki et al. (2010) suggested the existence of smut races in the Philippines and Kenya, respectively. Bhuiyan et al. (2015a) also reported a pathogenic variation between two spore populations in Australia. Given the variations as observed in the reaction of the same sugarcane cultivar in different countries and the unequal methods applied to define races, Grisham (2001) coordinated a broad race typing for sugarcane smut. A set of 11 cultivars were distributed to ten countries, at 14 locations, for testing the disease reaction to local populations of *Sporisorium scitamineum*. Variability amongst fungal populations was found, but strong evidence for distinct races was observed only in Taiwan (Grisham, 2001).

In addition to the attempts to identify pathological variations in *Sporisorium scitamineum* populations, worldwide efforts have been invested towards the molecular genetic analysis to assess the genetic variation. The main findings on sugarcane smut pathogen variability are summarized in Table 1. These data can provide insights into the evolutionary potential of pathogen populations for circumventing plant resistance genes or counteract fungicides. It can also be useful for inferring phylogenetic relationships, and for the characterization of pathogen races.

High variability was found in Southeast Asian populations and an Asian origin for *S. scitamineum* was proposed (Braithwaite et al., 2004; Raboin et al., 2007). However, the choice of the molecular marker affected the amount of the variation detected, probably due to the genomic region targeted for the analysis (Shen et al., 2016; Benevenuto et al., 2016). Cluster analyses of AFLP data distinguished the three Taiwanese races

Table 1 Molecular variability studies in *Sporisorium scitamineum*. Molecular marker abbreviations are defined below the table

Locality	Number of isolates	Molecular approach	Main findings	Reference
Mainland China	35	ISSR and RAPD	Two main groups and most isolates genetically similar to race3.	Shen et al. (2012)
Mainland China	23	RAPD and SRAP	Molecular variability associated with geographical origin of the isolates.	Que et al. (2012)
Mainland China	100	ISSR and SP-SRAP	Three main groups associated with geographical distribution.	Xu et al. (2014)
Mainland China	23	ITS	Low variation detected amongst isolates, no support for race ascription.	Zhang et al. (2015)
Mainland China	90	SCoT	SCoT is a high polymorphic marker, three main groups, clustering by geographical origin.	Shen et al. (2016)
Hawaii	37	AFLP	Low variability. Clusters did not separate the two Hawaiian races.	Schenck et al. (2004)
Philippines	96	SSR	Three major cluster groups, no association with geographical origin. Variance within populations higher than amongst populations.	Luzaran et al. (2012)
Egypt	12	Differential display of chitinase genes	Distinct band pattern amongst nearly all isolates	Fattah et al. (2010)
Brazil and Argentina	22	telRFLP, AFLP, ITS	Two main groups, no geographical association. High variation at telomeric regions.	Benevenuto et al. (2016)
South Africa, Réunion Island, Hawaii and Guadeloupe	7	RAPD and ITS	Low variability. The Réunion Island isolate was the most distantly related.	Singh et al. (2005b)
Worldwide (13 countries)	38	AFLP	Low level of variation. Southeast Asian isolates are the most divergent group. AFLPs distinguished the three Taiwanese races.	Braithwaite et al. (2004)
Worldwide (15 countries)	142	SSR	Homozygosity at nearly all loci and low variability outside Asia.	Raboin et al. (2007)

ISSR: Inter-Simple Sequence Repeat.
RAPD: Random Amplified Polymorphic DNA.
SRAP: Sequence-Related Amplified Polymorphism.
SP-SRAP: Single Primer-Sequence-Related Amplified Polymorphism.
ITS: Internal Transcribed Spacer.
SCoT: Start Codon-Targeted.
AFLP: Amplified Fragment Length Polymorphism.
SSR: Simple Sequence Repeats.
telRFLP: telomere-associated Restriction Fragment Length Polymorphism.

(Braithwaite et al., 2004; Raboin et al., 2007). On the other hand, AFLP markers did not separate the two Hawaiian races (Schenck et al., 2004). An association between molecular variability and geographical origin was found for Chinese populations (Que et al., 2012; Xu et al., 2014; Shen et al., 2016), whilst South American and Philippine strains did not cluster isolates according to their geographical distribution (Luzaran et al., 2012; Shen et al., 2016; Benevenuto et al., 2016). The absence of a clear geographic structure at local and global levels suggested a human-mediated dispersal of *Sporisorium scitamineum* through the exchange of infected plant material (Braithwaite et al., 2004; Raboin et al., 2007; Shen et al., 2016; Benevenuto et al., 2016). The lack of variability and the close genetic relationship amongst African and American strains imply that all those strains belong to a single lineage (Braithwaite et al., 2004; Raboin et al., 2007). As the same combination of alleles was also found in some Asian populations, the dispersal of a unique lineage from Asia to America and Africa was proposed (Raboin et al., 2007). Also, by using co-dominant microsatellite markers applied to DNA isolated from the dikaryotic hyphae of *Sporisorium scitamineum* strains, Raboin et al. (2007) found homozygous loci for all strains (except for one locus in one isolate from Philippines). This result suggested that selfing is the predominant reproductive mode in *Sporisorium scitamineum*.

Although many independent studies have addressed the molecular variability of *Sporisorium scitamineum* isolates, the genetic basis underlying the variability and its role in pathogenic variations are still poorly understood. Through advancements and cost-effectiveness of high-throughput sequencing technologies, the genomes of four strains of *Sporisorium scitamineum* are now available (Table 2). Dutheil et al. (2016) compared the genome of the South African strain against the Brazilian and Chinese genomes. The South African strain was more closely related to the Brazilian strain (0.3% nucleotide difference in 10 kb windows on an average) than to the Chinese strain (2.6% nucleotide difference in 10 kb windows on an average). Future studies including a higher number of isolates and addressing the nature and location of these single nucleotide polymorphisms as well as the structural variations promise to bring new knowledge into light and improve the discussion about the presence of pathogenic races.

Table 2 *Sporisorium scitamineum* strains for which the whole genome had been sequenced

Strains	Locality	Sugarcane variety	Assembly size (Mb)	Scaffolds + contigs	Protein-coding genes	References
2014001	China	ROC22	19.77	58 + 321	6,636	Queet al. (2014b)
SSC39B	Brazil	IACSP98-2053	19.98	23 + 3	6,677	Taniguti et al. (2015)
BSES15/17	Australia	Q117	19.89	449 + 5,331	*na	Taniguti et al. (2015)
Sscl8	South Africa	*na	19.63	47	6,693	Dutheil et al. (2016)

*na = not available.

5 Insights into *Sporisorium scitamineum* biology based on genomic and transcriptomic data

The first report on the *Sporisorium scitamineum* genome was published by Que et al. (2014b) in which the authors analysed a Chinese strain using the whole-genome shotgun approach. Taniguti et al. (2015) achieved the complete genome sequence (from telomere to telomere of each individual chromosome) of a Brazilian strain of *Sporisorium scitamineum* by using a combination of Illumina and Pacbio sequence reads. Both genomes along with an Australian isolate were compared and few polymorphisms were identified (Taniguti et al., 2015). The haploid genome comprises 19.98 Mb distributed in 23 chromosomes. A total number of 6,677 protein-coding genes were predicted and only 1.24% of the genome corresponded to repetitive elements (Fig. 4). Genomic comparisons amongst four other smut species infecting distinct crop hosts (*Sporisorium scitamineum, Sporisorium reilianum, U. maydis and U. hordei*) showed that 85% of the orthologous gene groups were shared amongst them and that each species had their own set of specific genes (Taniguti et al., 2015). For *Sporisorium scitamineum,* 438 unique genes were identified and may have a role in the adaptation to the sugarcane host (Taniguti et al., 2015).

In the *Sporisorium scitamineum* genome, the two mating-type loci are arranged in a bipolar system: *a* and *b* loci are linked at a distance of 59 Kbp apart on chromosome 2 (Fig. 4) (Taniguti et al., 2015). Differences in the genomic arrangement of *a* and *b* loci imply differences in the mating behaviour as tetra- or bipolar system. Hence, during *Sporisorium scitamineum* meiosis, the mating-type loci segregate as one locus, that is, one spore could give rise to only two different mating-type cells. Bipolar systems favour a higher level of selfing compared with the tetrapolar system in which *a* and *b* loci are unlinked and segregate independently (Fraser and Heitman, 2003). This mating-type loci arrangement in the smut genomes has a direct effect on the fungal biology and pathogen evolution since it is required to initiate sexual development and pathogenicity (Bakkeren et al., 2008).

To predict the potentially secreted proteins encoded by the *Sporisorium scitamineum* genome, Taniguti et al. (2015) considered the presence of signal peptide and the

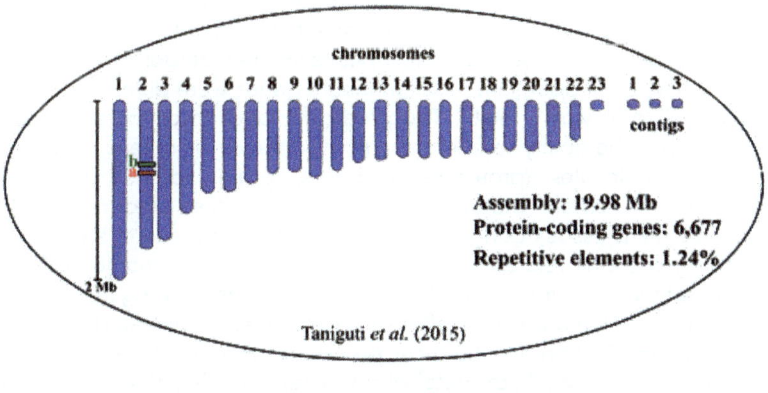

Figure 4 Schematic representation of *Sporisorium scitamineum* genomic features. Red and green bars on chromosome 2 represent the mating-type loci *a* and *b*, respectively.

absence of transmembrane domains and glycosylphosphatidylinositol (GPI anchors) in the encoded proteome. Secreted proteins are of interest to understand the molecular communication between the pathogen and its host. Based on the predictions, the fungal secretome comprised 305 proteins, including 54 carbohydrate-active enzymes (CAZymes) and 70 effector candidates. The number of CAZymes within this set of proteins involved in cell wall breakdown is considered less as compared to necrotrophic pathogens, but similar to other biotrophic fungi (Zhao et al., 2013). The small number of these cell wall degrading enzymes (CWDEs) is supposed to avoid major damage to the plant tissues, since biotrophic fungi depend on living host tissues to feed on and complete their life cycles. It is noteworthy that the *Sporisorium scitamineum* genome encodes a diverse set of CWDEs, including those related to the breakdown of lignin, pectin, cellulose, hemicellulose and cuticle (Taniguti et al., 2015).

Effector candidates were defined after filtering the secretome by size (<30 kDa) and cysteine content (>1.5%) (Taniguti et al., 2015). Using a less stringent approach to identify candidate effector proteins, Dutheil et al. (2016) found ten significant clusters of candidate effectors in the South African *Sporisorium scitamineum* genome. The clustered effector genes were significantly closer to uncharacterized interspersed repeats than other genes in the genome, suggesting that effector clusters are compartmentalized in the genome (Taniguti et al., 2015; Dutheil et al., 2016). The non-random distribution of effectors at the proximity of repeat-rich compartments in many filamentous plant pathogens (known as 'two-speed genome' model) likely serves as a cradle for adaptive evolution (Dong et al., 2015). Other 148 secreted proteins of unknown functions were predicted in which 26% are uniquely encoded by the *Sporisorium scitamineum* genome (Taniguti et al., 2015). All of them are candidates to be further characterized.

Some studies were initiated in that direction based on the genomic similarity between *Sporisorium scitamineum* and other smut fungi. Some of the well-characterized effector proteins of *U. maydis* are also probably relevant in *Sporisorium scitamineum* colonization and proliferation in sugarcane (Barnabas et al., 2017b). The orthologue genes encoding six putative effectors (Pep1, Cmu1, Tin2, Srt1, Stp1 and Pit1) significantly increased the expression three days after inoculation (DAI) and were expressed in distinct patterns throughout the infection cycle. The results of this study also indicate that some of the candidate effectors could be conserved amongst other closely related smut fungi. Pep1 is one of the most studied and probably a common effector in different pathogenic fungi, even across the monocot–dicot border of host plants (Hemetsberger et al., 2015).

Approximately 13% of all *Sporisorium scitamineum* genes were differentially expressed *in planta* compared to *in vitro* growth. The predicted function of these genes showed that the pathogen has the ability to overcome plant defence by means of detoxification of plant defence molecules (proteases inhibitors, pisatin demethylase, benzoate 4-monooxygenase, chorismate mutase, superoxide dismutase and catalase) and to grow and colonize plant tissues using an arsenal of nutrient uptake-related genes, such as those encoding proteins related to the transport of sugar, nicotinic acid and small peptides. Comparative transcriptional profiles of the pathogen growing *in vitro* and during interaction in plants 5 DAI and 200 DAI (after the whip emission) revealed that transporter activity and signal transduction are the most prevalent functions in samples 5 DAI, and hydrolase activity and carbohydrate metabolism in samples 200 DAI. Additionally, 131 fungal genes were expressed only *in planta* compared to *in vitro* growth, some of which were distributed in ten putative pathogenic islands (Taniguti et al., 2015). The recently published *Sporisorium scitamineum in vitro* secretome analysis (Barnabas et al., 2017a) identified ten proteins exclusively secreted in culture mimicking cane plant conditions. Of these, five identified proteins were related to nutrient acquisition and use.

The genome of *Sporisorium scitamineum* was also searched using the pathogen–host interaction database (PHI-base) and 192 genes were annotated as candidate virulence factors (Que et al., 2014b). They play diverse roles in the redox of short-chain compounds, cell wall breakdown energy metabolism and transport. Gene expression profiles of these 192 genes during the course of infection (0, 12, 24, 48, 96 and 120 hpi) proved that 52% of them were expressed in the process of infection and 31 genes were expressed in all the stages analysed. These 31 genes mainly encoded proteins related to energy metabolism and redox of short-chain compound-related enzymes. Noteworthy is the continued increase of cyclophilin B at all stages reaching its maximum at 120 hpi (a 388-fold change). Cyclophilin B possesses cis-trans peptidyl-prolyl isomerase activity and is probably an important determinant of virulence. A metabolome analysis also identified a toxin produced by the fungus throughout plant colonization whose function remains to be established (Schaker et al., 2017).

A successful biotrophic pathogenic fungus is able to mislead the host from its cell wall chitin recognition. The *Sporisorium scitamineum* genome encodes for a chitin deacetylase enzyme that is predicted to alter chitin to chitosan in order to avoid triggering host defences (Taniguti et al., 2015). In addition, plant chitinases are not capable of hydrolysing chitosan. The accumulation of *Cmu1* protein is probably another strategy to surpass host defence (Djamei et al., 2011; Barnabas et al., 2016). This protein seems to channelize chorismate to the phenylpropanoid pathway (PP) in order to reduce its availability for the salicylic acid (SA) biosynthesis, which is involved in defence mechanisms.

Molecular tools to study *Sporisorium scitamineum* biology are available, which facilitate studies to narrow down candidate genes related to pathogenicity. Efficient protocols for genetic transformation and gene knock out were described along with fluorescent-marker-tagged mutants to follow colonization, pathogen proliferation and fungal biology within plant cells and tissues (Yan et al., 2016a,c; Carvalho et al., 2016; Lu et al., 2017).

6 Sugarcane responses to pathogen invasiveness and development

Sugarcane resistance to smut is a multifactor-inheritable feature in which various components may contribute in different stages of the plant–pathogen interaction (Lloyd and Naidoo, 1983; Hector et al., 1995; McNeil et al., 2011; Sundar et al., 2015). Smut resistance seems to involve pre-formed (physical and chemical barriers) and post-formed (molecular and physiological) mechanisms, and each genotype has a particular combination of these features. Genotypes with multiple mechanisms are more likely to survive in changing environments and/or in the cases of new pathogen race emergence (Bhuiyan et al., 2013). Here, we have compiled the available information of candidate genes and biological processes associated with sugarcane responses.

Amongst the pre-formed resistance mechanisms are: 1) the number of bud scales with lignified tissues to impair fungal entrance; 2) the abundance of trichomes, which can have a water-repelling effect (Waller, 1970; Gloria et al., 1995) and 3) the bud biochemical composition (James, 1973). For instance, high glycoside-flavonoid compounds (Lloyd and Pillay, 1980; Lloyd and Naidoo, 1983; Fontaniella et al., 2002; Millanes et al., 2005; Legaz et al., 2011) can promote cell aggregation and inhibition of cell polarity prior to teliospore germination (Millanes et al., 2008; Sanchez-Elordi et al., 2016). However, pre-formed defences may be easily lost due to bud damage and/or sprouting, and are highly dependent on environmental conditions that have an effect on bud morphology (Waller, 1970).

Regarding post-formed defences, most studies focus primarily on particular mechanisms activated upon pathogen perception. More recent studies on the global response by means of transcriptomic analysis (Thokoane and Rutherford, 2001; Borrás-Hidalgo et al., 2005; LaO et al., 2008; Wu et al., 2013; Que et al., 2014a; Huang et al., 2015; Schaker et al., 2016); proteome (Que et al., 2011; Barnabas et al., 2016; Su et al., 2016c) and metabolome (Schaker et al., 2017) provide a more comprehensive view of the interaction.

Smut-resistant genotypes can perceive the pathogen earlier than the susceptible ones, triggering an extensive transcriptional reprogramming (Que et al., 2014a; Peters et al., 2017). The first mechanism to activate defence response relies on PAMP (*Pathogen-Associated Molecular Patterns*) recognition by host receptors. The resistant sugarcane variety N52/219 activates the expression of an S-receptor-like-kinase involved in chitin perception and of a Pto-like kinase (Thokoane and Rutherford, 2001), which can lead to increased levels of chitinase (Esh, 2014; Que et al., 2014a). Other pathogenesis-related (PR) proteins, such as β-1,3-glucanase, are also known to be involved in the fungal cell wall breakdown, and the antifungal thaumatin was found to have a positive correlation with smut resistance (Heinze et al., 2001; Su et al., 2013b, 2016a; Que et al., 2014a; Peng et al., 2017). A very elegant experiment showed that a resistant variety may produce an arginase to induce agglutination of teliospores to impair their germination. However, the teliospores also secrete a fungal arginase that accelerates their own germination. The fungal arginase had a higher affinity for teliospore binding than that of the plant arginase, reverting their effect in agglutination (Sánchez-Elordi et al., 2015). The authors concluded that the binding of fungal arginase to smut teliospores activates a signal transduction cascade that enhances germination. Moreover, the pathogen recognition in resistant varieties leads to the activation of MAP kinase signalling pathway potentially to activate defence mechanisms (Wu et al., 2013; Que et al., 2014a).

Other receptors encoding cytoplasmic NBS-LRR-like proteins (*Nucleotide-Binding Site - Leucine-Rich Repeat*) are responsive to smut infection in both resistant (Borrás-Hidalgo et al., 2005; LaO et al., 2008; Que et al., 2011) and susceptible sugarcane genotypes (Schaker et al., 2016). In a compatible interaction (susceptible genotype), a prematurely increased expression of an RGA (*Resistance Gene Analogue*), homologue to a BAM1 receptor-like kinase (RGA482), was observed (Rossi et al., 2003). BAM1 is related to maintain meristem identity, suggesting that signalling to the late whip emergence is activated soon after inoculation in susceptible genotypes (Schaker et al., 2016).

The production of reactive oxygen species (ROS) is one of the earliest defence responses to smut (Su et al., 2014, 2016b; Peters et al., 2017). Resistant plants increased H_2O_2 accumulation and reduced the activity of antioxidant enzymes, which coincided with fungal developmental phases within plant tissues: promycelium outgrowth, appressorium formation and bud colonization (Peters et al., 2017). It is proposed that this mechanism ensures triggering of defence responses to resist fungal infection and it is not observed in susceptible plants (Peters et al., 2017). However, in susceptible plants, ROS scavenging was described after the whip emission (Barnabas et al., 2016; Schaker et al., 2016).

Oxidative burst is known to activate the PP. PP intermediates are precursors of antimicrobial metabolites, such as flavonoids, and are also related to lignin synthesis. Differential gene expression and protein accumulation related to modifications of the plant cell wall were positively correlated with smut resistance (Que et al., 2014a; Peters et al., 2017). Enzyme functions such as those of phenylalanine ammonia-lyase (PAL), cinnamyl alcohol dehydrogenase and peroxidase (de Armas et al., 2007; Santiago et al., 2009, 2010, 2012; Esh, 2014; Que et al., 2014a; Su et al., 2016c) and increased PAL derivatives such as p-coumaric, syringic, ferulic and caffeic acids (de Armas et al., 2007; Santiago et al., 2009) were all increased in resistant plants. High levels of conjugated polyamines to phenolic compounds are also detected in infected plants (Legaz et al., 1998; Piñon et al., 1999). It has been proposed that phenolics are produced as a linear response of resistance acquisition against smut infection (Lloyd and Naidoo, 1983).

After the whip emission, plants also respond to the fungal presence by modifying intermediates of the PP (Barnabas et al., 2016; Schaker et al., 2016). Surprisingly, the activation of PP in this case seems to rely on tyrosine accumulation (Schaker et al., 2017). Transcripts encoding to a bifunctional PAL (phenylalanine/tyrosine ammonia-lyase – PTAL), shikimate O-hydroxycinnamoyltransferase, cinnamoyl-CoA reductase and peroxidase were all accumulated in plants after whip emergence (Schaker et al., 2016, 2017). Besides, proteomics revealed complementary results in which plants after whip emergence showed increased amounts of PTAL, caffeic acid 3-*O*-methyltransferase, caffeoyl-CoA *O*-methyltransferase, isoflavone reductase and cinnamyl alcohol dehydrogenase (Barnabas et al., 2016).

Other cell wall components, such as xylose, cellobiose and rhamnose were less abundant in meristems of plants developing whips, suggesting meristem loosening to allow whip growth (Schaker et al., 2016, 2017). A similar molecular pattern is described for flowering meristems that lose their capacity to synthesize secondary cell wall (Moore and Berding, 2014). Additionally, the whip emission is accompanied by increased transcriptional levels of a neutral alkaline invertase (cytosolic) and a soluble acid invertase (vacuolar), and increased activity in paths related to glycolysis, tricarboxylic acid cycle and pentose phosphate pathway (PPP) (Schaker et al., 2016, 2017). These data contribute to explain the reduction of sucrose content in smutted plants. It seems that sugarcane metabolism shifted the allocation of carbon to pathways unrelated to sucrose storage, such as starch accumulation (Schaker et al., 2017). Finally, the whip emergence is characterized by the overexpression of MADS-box family of genes, which is also described to control pathways associated with flowering (Schaker et al., 2016).

The hormonal status in both compatible and incompatible interactions is affected. Increased levels of SA, ethylene (ET), jasmonate and gibberellin (GA) were correlated with resistance (Borrás-Hidalgo et al., 2005; LaO et al., 2008; Que et al., 2014a; Huang et al., 2015; Su et al., 2016c). On the other hand, auxin (AU) was the most relevant hormone in compatible interactions, where it may serve as a negative regulator of the immune system and related to the changes of the meristem functions (Hector et al., 1995; Schaker et al., 2016). Transcriptional responses associated with other hormones were detected in smutted plants, such as ET, CYT (cytokinin), BR (brassinosteroid) and GA, which probably contribute to the activation of PPP, and disease symptoms of tillering and poor cane formation (Schaker et al., 2016).

Early identification of resistant genotypes based on molecular tools for sugarcane is an arduous task. Modern varieties of sugarcane are interspecific hybrids with complex and highly polyploid genomes. Even so, the identification of multiple molecular markers associated with smut resistance was successful towards some sugarcane genotypes (Esh and Nars, 2014; Khan et al., 2017).

Recently, Su et al. (2016b) combined physiological and biochemical methods with the pathogen proliferation index (TaqMan qRT-PCR) for the early selection of resistant genotypes after artificial inoculation in controlled indoor experiments. Based upon the changes in the activities of enzymes related to ROS metabolism (POD, a peroxidase of the pathogen-associated protein 9 subfamily; SOD; APX; CAT; MDA) and the PR protein β-1,3-glucanase in sugarcane buds, six of the nine evaluated genotypes displayed a strong correlation with incidences of field smut disease (YZ03-258,YZ01-1413, GT02-467, ROC22, YZ03-103 and FN39).

7 Conclusion

Various studies undertaken over the past years helped improve our knowledge on the disease and uncovered potential targets to be used in the selection process of new smut-resistant varieties. So far, we know that sugarcane undergoes transcriptional changes very early after inoculation. The so-called early phase of infection in more resistant genotypes is characterized by a strong oxidative burst as well as changes related to the antioxidant systems to delay or impair fungal proliferation. ROS probably signs in to increase chitinases, glucanases, thaumatin activities and other antifungal metabolites, tissue lignification and hormonal changes via MAPK signalling (Fig. 5). On the other hand, whip development in susceptible genotypes is related to changes in carbon partitioning, tissue lignification via PTAL overexpression and breakdown of starch and hormonal changes, mainly AU. Also, whip development and flowering probably share some of the intrinsic molecular mechanisms involving the MADS-box family of transcriptional factors. Research groups of various countries using different sugarcane genotypes best adapted to their own conditions have reached similar conclusions regarding the general mechanisms of smut disease and resistance. This probably means that regardless of the genotypes, we record similar responses and must be close to select good targets to be incorporated in breeding programmes. It is also clear, however, that the pathogen displays a potential to adapt, as well as genomic variability and plasticity to be selected according to changes imposed by the selection of new resistant varieties. We need to account for climate changes and attribute changes in sugarcane agricultural practices for the future and include these variables for a comprehensive outlook of the pathosystem. It seems that devoting efforts to test some of the hypotheses raised in combination by various authors are in order to prevent and/or control breakthroughs of the disease. The most significant challenge now is to understand the genome of sugarcane, which is a very complex interspecific hybrid and is highly polyploid at various levels of aneuploidy. The identification of alleles, paralogues, homologues, homeologues and recombinant chromosomes for specific combinations of characteristics in distinct genotypes yet remains to be exploited. Various researchers have dedicated their efforts in that direction and we hope for a complete set of the sugarcane genomes (Souza et al., 2011; Dal-Bianco et al., 2012; de Setta et al., 2014; Okura et al., 2016) to be deciphered in the near future.

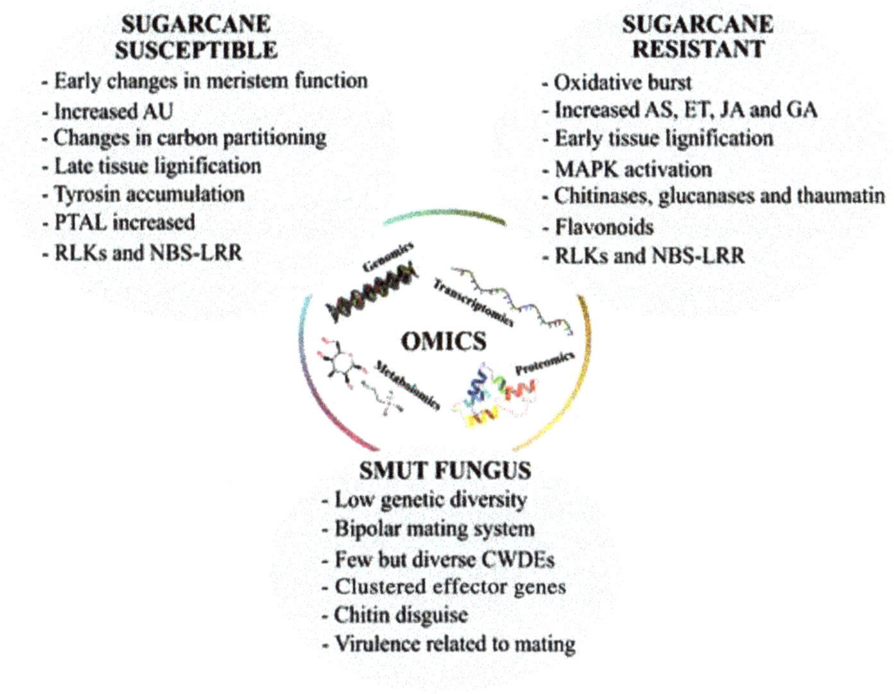

Figure 5 Main findings concerning the molecular cross-talk between *Sporisorium scitamineum* and sugarcane genotypes (susceptible and resistant). Most of the results were obtained using omics techniques and require functional validation.

8 Future trends

Much effort has been made in sugarcane-cultivating countries to improve knowledge related to smut disease. This chapter explored the work developed in different countries especially Australia, Brazil, China, India and the United States and some countries of Africa within universities and research institutes. Topics were related to the molecular cross-talking between the host and the pathogen, the understanding of pathogen genetic variability and distribution, and pathogen biology and description of races, all of which may ultimately contribute to improving breeding strategies.

Sugarcane is the fifth most cultivated plant in the world with an enormous potential to produce biomolecules including ethanol, one of the most common forms of biofuel today. Modern sugarcane varieties are hybrids of *Saccharum officinarum* and *Saccharum spontaneum* that combine the capacity to accumulate high sugar content, rhizome robustness and the ability to resist diseases. This raises a potential issue considering sugarcane as an energy crop. Breeding programmes for the development of energy cane may improve the genetic contribution of this ancestral species in order to increase fibre content and biomass of sugarcane varieties (Matsuoka et al., 2014; Carvalho-Netto et al., 2014). Because productivity needs to be scaled up to accommodate the future demands for food and energy without increasing cultivated areas, for which to accomplish, successful management of diseases is a major concern amongst the growers. The scientific community addresses this concern, which is clearly evident with the ever increasing research publications dealing with smut. A more integrative approach (system biology) and a strong collaboration amongst research groups at the global level with different expertise are possible ways to improve the use of all the information gathered from various sources. The identification of a key set of effectors and their use as a tool for the early detection of resistant plants is probably another good strategy for assisting breeding programmes. It would be even better to compare sugarcane responses to all fungal pathogens threatening sugarcane cultivation to find a common ground to investigate. Sugarcane smut and other fungal diseases will remain as challenging issues in the agriculture scenario for years to come.

9 Acknowledgements

The authors wish to recognize the support of FAPESP (grant numbers: [CBMV] 2015/07112-4, [PDCS] 2013/25599-2, [JB] 2014/21802-0 and [NSTS] 2016/16376-8) and CNPq (grant numbers [SSA]166088/2015-3 and [CBMV] 443827/2014-1). The authors are very grateful to investigators whose work led to the construction of this review and apologize to those whose work has not been cited due to space limitations.

10 References

Abdou, Y. A., Moursy, M. A., Abdel-Fattah, M. N. D. and Mansour, I. M. 1990. Effects of temperature and certain cultural practices on longevity of teliospores of *Ustilago scitaminea*. *Bulletin of the Faculty of Agriculture, University of Cairo* 41: 511–20.

Abo, M. and Okusanya, B. A. 1996. Incidence and variability reaction of sugarcane smut (*Ustilago scitaminea* Syd.) isolates in greenhouse and laboratory tests in Nigeria. *Discovery and Innovation* 8: 227–31.

Akalach, M. and Touil, B. 1996. Occurrence and spread of sugarcane smut caused by *Ustilago scitaminea* in Morocco. *Plant Disease* 80: 1363.

Albert, H. H. and Schenck, S. 1996. Amplification from a homolog of the bE mating-type gene as a sensitive assay for the presence of *Ustilago scitaminea* DNA. *Plant Disease* 80: 1189–92.

Alexander, K. and Ramakrishnan, K. 1980. Infection of the bud, establishment in the host and production of whips in sugarcane (*Ustilago scitaminea*). *Proceedings of the International Society of Sugar Cane Technologists* 17: 1452–55.

Bailey, R. A. 1979. Possibilities for the control of sugarcane smut (*Ustilago scitaminea*) with fungicides. *Proceedings of The South African Sugar Technologists' Association*, 137–42.

Bakkeren, G., Kämper, J. and Schirawski, J. 2008. Sex in smut fungi: structure, function and evolution of mating-type complexes. *Fungal Genetics and Biology* 45: 15–21.

Barnabas, L., Ashwin, N. M. R., Kaverinathan, K., Trentin, A. R., Pivato, M., Sundar, A. R., Malathi, P., Viswanathan, R., Carletti, P., Arrigoni, G., et al. 2017a. In vitro secretomic analysis identifies putative pathogenicity-related proteins of *Sporisorium scitamineum* - the sugarcane smut fungus. *Fungal Biology* 121: 199–211.

Barnabas, L., Ashwin, N. M. R., Kaverinathan, K., Trentin, A. R., Pivato, M., Sundar, A. R., Malathi, P., Viswanathan, R., Rosana, O. B., Neethukrishna, K., et al. 2016. Proteomic analysis of a compatible interaction between sugarcane and *Sporisorium scitamineum*. *Proteomics* 16: 1111–22.

Barnabas, L., Ashwin, N. M. R., Ramesh Sundar, A., Malathi, P., Viswanathan, R.. 2017b. Putative orthologs of *Ustilago maydis* effectors screened from the genome of sugarcane smut fungus - *Sporisorium scitamineum. Australasian Plant Pathology* 46: 147–56.

Begerow, D., Schäfer, A. M., Kellner, R., Yurkov, A., Kemler, M., Oberwinkler, F. and Bauer, R. 2014. Ustilaginomycotina. In: David J. McLaughlin and Joseph W. Spatafora (Eds) *Systematics and Evolution*. Berlin, Heidelberg: Springer Berlin Heidelberg, pp. 295–329.

Benevenuto, J., Longatto, D. P., Reis, G. V., Mielnichuk, N., Palhares, A. C., Carvalho, G., Saito, S., Quecine, M. C., Sanguino, A., Vieira, M. L. C., et al. 2016. Molecular variability and genetic relationship among Brazilian strains of the sugarcane smut fungus (M. Bidochka, Ed.). *FEMS Microbiology Letters* 363: fnw277.

Bergamin Filho, A., Amorim, L., Cardoso, C. O. N., Sanguino, A., Irvine, J. E. and Silva, W. M. 1987. Carvão da cana-de-açúcar e sua epidemiologia. *Boletim Técnico Copersucar* Special ed., Piracicaba: Copersucar, pp. 1–23.

Bhuiyan, S. A., Croft, B. J., Deomano, E. C., James, R. S. and Stringer, J. K. 2013. Mechanism of resistance in Australian sugarcane parent clones to smut and the effect of hot water treatment. *Crop and Pasture Science* 64: 892.

Bhuiyan, S. A., Croft, B. J., Stringer, J. K. and Deomano, E. C. 2015a. Pathogenic variation in spore populations of *Sporisorium scitamineum*, causal agent of sugarcane smut in Australia. *Plant Disease* 99: 93–9.

Bhuiyan, S. A., Croft, B. J. and Tucker, G. R. 2015b. New method of controlling sugarcane smut using Flutriafol fungicide. *Plant Disease* 99: 1367–73.

Bölker, M. 2001. *Ustilago maydis* - A valuable model system for the study of fungal dimorphism and virulence. *Microbiology* 147: 1395–401.

Borrás-Hidalgo, O., Thomma, B. P. H. J., Carmona, E., Borroto, C. J., Pujol, M., Arencibia, A. and Lopez, J. 2005. Identification of sugarcane genes induced in disease-resistant somaclones upon inoculation with *Ustilago scitaminea* or *Bipolaris sacchari*. *Plant Physiology and Biochemistry* 43: 1115–21.

Braithwaite, K. S., Bakkeren, G., Croft, B. J. and Brumbley, S. M. 2004. Genetic variation in a worldwide collection of the sugarcane smut fungus *Ustilago scitaminea*. *Proceedings of the Australian Society of Sugar Cane Technologists* 26: 48–58.

Byther, R. S., Steiner, G. W. and Wismer, C. A. 1971. New Sugarcane diseases reported in Hawaii. *Sugarcane Pathologists' Newsletter* 7: 18–21.

Carvalho, G., Quecine, M. C., Longatto, D. P., Peters, L. P., Almeida, J. R., Shyton, T. G., Silva, M. M. L., Crestana, G. S., Creste, S. and Monteiro-Vitorello, C. B. 2016. *Sporisorium scitamineum* colonisation of sugarcane genotypes susceptible and resistant to smut revealed by GFP-tagged strains. *Annals of Applied Biology* 169: 329–41.

Carvalho-Netto, O. V., Bressiani, J. A., Soriano, H. L., Fiori, C. S., Santos, J. M., Barbosa, G. V., Xavier, M. A., Landell, M. G. and Pereira, G. A. 2014. The potential of the energy cane as the main biomass crop for the cellulosic industry. *Chemical and Biological Technologies in Agriculture* 1: 20.

Comstock, J., Ferreira, S. and Tew, T. 1983. Hawaii's approach to control of sugarcane smut. *Plant Disease* 67: 452–7.

Croft, B. J., Berding, N., Cox, M. C. and Bhuiyan, S. 2008. Breeding smut-resistant sugarcane varieties in Australia: progress and future directions. *Proceedings of the Australian Society of Sugar Cane Technologists* 30: 125–34.

Croft, B. J. and Braithwaite, K. S. 2006. Management of an incursion of sugarcane smut in Australia. *Australasian Plant Pathology* 35: 113–22.

Dal-Bianco, M., Carneiro, M. S., Hotta, C. T., Chapola, R. G., Hoffmann, H. P., Garcia, A. A. F. and Souza G. M. 2012. Sugarcane improvement: How far can we go? *Current Opinion in Biotechnology* 23:265–70.

Dalvi, S. G., Vasekar, V. C., Yadav, A., Tawar, P. N., Dixit, G. B., Threetha Prasad, D. and Deshmukh, R. B. 2012. Screening of promising sugarcane somaclones for agronomic traits, and smut resistance using PCR amplification of Inter Transcribed Region (ITS) of *Sporisorium scitaminae*. *Sugar Technolgy* 14: 68–75.

de Armas, R., Santiago, R., Legaz, M.-E. and Vicente, C. 2007. Levels of phenolic compounds and enzyme activity can be used to screen for resistance of sugarcane to smut (*Ustilago scitaminea*). *Australasian Plant Pathology* 36: 32.

de Setta, N., Monteiro-Vitorello, C. B., Metcalfe, C. J., Cruz, G. M. Q., Del Bem, L. E., Vicentini, R., Nogueira, F. T. S., Campos, R. A., Nunes, S. L., Turrini, P. C. G. and et al. 2014. Building the sugarcane genome for biotechnology and identifying evolutionary trends. *BMC Genomics* 15: 540.

Djamei, A., Schipper, K., Rabe, F., Ghosh, A., Vincon, V., Kahnt, J., Osorio, S., Tohge, T., Fernie, A. R., Feussner, I., et al. 2011. Metabolic priming by a secreted fungal effector. *Nature* 478: 395–8.

Dong, S., Raffaele, S. and Kamoun, S. 2015. The two-speed genomes of filamentous pathogens: Waltz with plants. *Current Opinion in Genetics & Development* 35: 57–65.

Dutheil, J. Y., Mannhaupt, G., Schweizer, G., Sieber C. M.K., Münsterkötter, M., Güldener, U., Schirawski, J. and Kahmann, R. 2016. A tale of genome compartmentalization: the evolution of virulence clusters in smut fungi. *Genome Biology and Evolution* 8: 681–704.

Esh A. 2014. The activity of pathogenesis related proteins in smut resistant and susceptible sugarcane (GT54-9) mutants induced by gamma radiation. *Advances in Plants & Agriculture Research* 1: 1–12.

Esh, A., Nars, M. 2014. Molecular markers for smut resistance in mutated population of sugarcane clone GT-54-9 through Amplified Fragment Length Polymorphism (AFLP). *EC Agriculture* 11: 7–14.

Fattah, A. I., Alamri, S, Abou-Shanab, R. A. I. and Hafez, E. E. 2010. Fingerprinting of *Ustilago scitaminea* (Sydow) in Egypt using differential display technique: chitinase gene the main marker. *Research Journal of Agriculture and Biological Sciences* 6: 8–13.

Ferreira, S. A. and Comstock, J. C. 1989. Smut. In: Ricaud, C., Egan, B. T., Gillaspie A. G. and Hughes C. G. (Eds) *Diseases of Sugarcane*. Elsevier, Amsterdam, The Netherlands, pp. 211–29.

Fontaniella, B., Márquez, A., Rodríguez, C. W., Piñón, D., Solas, M. T., Vicente, C. and Legaz, M. E. 2002. A role for sugarcane glycoproteins in the resistance of sugarcane to *Ustilago scitaminea*. *Plant Physiology and Biochemistry* 40: 881–9.

Fraser, J. A. and Heitman, J. 2003. Fungal mating-type loci. *Current biology : CB* 13: R792–5.

Gillissen, B., Bergemann, J., Sandmann, C., Schroeer, B., Bölker, M. and Kahmann, R. 1992. A two-component regulatory system for self/non-self recognition in *Ustilago maydis*. *Cell* 68: 647–57.

Gloria, B. A. da, Albernas, M. C. C. and Amorim, L. 1995. Structural characteristics of buds of sugarcane cultivars with different levels for resistance to smut. *Journal of Plant Diseases and Protection* 102: 502–8.

Grisham, M. 2001. An international project on genetic variability within sugarcane smut. *Proceedings of the International Society of Sugar Cane Technologists* 24: 459–61.

Gupta, S. C. 1978. Control of diseases by hot water treatment of sugarcane seed material. *Indian Sugar Crops Journal* 5: 28–9.

Hector, E., Prada, F. de, Rodriguez, R. (National SCRI (Cuba)). 1995. Experimental evidence for the presence of different smut resistance mechanisms in sugarcane. *Proceedings of the International Society of Sugar Cane Technologists* 2: 565–73.

Heinze, B., Thokoane, L., Williams, N., Barnes, J. and Rutherford, R. 2001. The smut-sugarcane interaction as a model system for the integration of marker discovery and gene isolation. *Proceedings of The South African Sugar Technologists' Association* 75: 104–7.

Hemetsberger, C., Mueller, A. N., Matei, A., Herrberger, C., Hensel, G., Kumlehn, J., Mishra, B., Sharma, R., Thines, M., Hückelhoven, R., et al. 2015. The fungal core effector Pep1 is conserved across smuts of dicots and monocots. *New Phytologist* 206: 1116–26.

Hoy, J., Hollier, C., Fontenot, D. and Grelen, L. 1986. Incidence of sugarcane smut in Louisiana and its effect on yield. *Plant Disease* 70: 59–60.

Huang, N., Zhang, Y. Y., Xiao, X. H., Huang, L., Wu, Q. B., Que, Y. X., Xu, L. P. 2015. Identification of smut-responsive genes in sugarcane using cDNA-SRAP. *Genetics and Molecular Research* 14: 6808–18.

James, G. L. 1973. Smut spore germination in sugarcane internode surfaces. *Proceedings of The South African Sugar Technologists' Association* 47: 179–80.

Kavitha, M., Sundar, A. R., Padmanaban, P., Viswanathan, R. and Malathi, P. 2014. Comparative study on early detection of sugarcane smut (*Sporisorium scitamineum*) by polymerase chain reaction (PCR) and microscopy. *African Journal of Biotechnology* 13: 4635–38.

Khan, M., Pan, Y. and Iqbal, J. 2017. Development of an RAPD-based SCAR marker for smut disease resistance in commercial sugarcane cultivars of Pakistan. *Crop Protection* 94: 166–72.

Kronstad, J. W. and Staben, C. 1997. Mating type in filamentous fungi. *Annual Review of Genetics* 31: 245–76.

Ladd, S., Heinz, D. and Meyer, H. 1974. Control of sugarcane (*Saccharum* sp.) smut disease (*Ustilago scitaminea*) through breeding and selection of resistant clones. *Proceedings of the International Society of Sugar Cane Technologists* 15: 36–44.

LaO, M., Arencibia, A., Carmona, E., Acevedo, R., Rodríguez, E., León, O., Santana, I. 2008. Differential expression analysis by cDNA-AFLP of Saccharum spp. after inoculation with the host pathogen *Sporisorium scitamineum. Plant Cell Reports* 27: 1103–11.

Latiza, A. S., Ampusta, D. C. and Rivera, J. R. 1980. Reaction of sugarcane clones to strain B of *Sporisorium scitamineum. Proceedings of the International Society of Sugar Cane Technologists* 2: 1456–62.

Lee, C., Yuan, C. and Liang, Y. 1999. Occurrence of a new pathogenic race of culmicolous smut of sugarcane in Taiwan. *Proceedings of the International Society of Sugar Cane Technologists* 23: 406–7.

Lee-Lovick, G. 1978. Smut of sugarcane - *Ustilago scitaminea. Review of Plant Pathology* 57: 181–8.

Legaz, M. E., Armas, R. D., Pinon, D. and Vicente, C. 1998. Relationships between phenolics-conjugated polyamines and sensitivity of sugarcane to smut (*Ustilago scitaminea*). *Journal of Experimental Botany* 49: 1723–28.

Legaz, M. E., Santiago, R., de Armas, R., Alarcón, B., Díaz, E. M., Sánchez-Elordi, E., Sacristán, M., Fontaniella, B., Millanes, A. M., Legaz, M. E., Santiago, R., Armas, R. De, Millanes, A. M., Blanch, M. and Vicente, C. 2011. Molecular defence responses of sugarcane (*Saccharum officinarum* L.) to smut (*Sporisorium scitamineum* (Syd.). In: A. Mendez-Vilas (Ed.) *Science Against Microbial Pathogens: Communicating Current Research and Technological Advances*. Piepenbr & Oberw, Badajoz, Spain, pp. 1244–50.

Lemma, A., Hagos, H., Zekarias, Y. and Tekle, A. 2015. Study on the reaction of sugarcane genotypes (CIRAD-2011) to sugarcane smut (*Sporisorium scitamineum*) in the Ethiopian sugarcane plantations. *Advances in Crop Science and Technology* 3: 1–3.

Lloyd, H. L. and Naidoo, G. 1983. Chemical-assay potentially suitable for determination of smut resistance of sugarcane cultivars. *Plant Disease* 67: 1103–05.

Lloyd, H. L. and Pillay M. 1980. The development of an improved method for evaluating sugarcane for resistance to smut. *Proceedings of The South African Sugar Technologists' Association* 54: 168–72.

Lu, S., Shen, X. and Chen, B. 2017. Development of an efficient vector system for gene knock-out and near in-cis gene complementation in the sugarcane smut fungus. *Scientific Reports* 7: 3113.

Luthra, J., Suttar, A. and Sandhu, S. 1940. Experiments on the control of smut of sugarcane. *Proceedings in Indian Academy of Sciences* 12: 118–28.

Luzaran, R. T., Cueva, F. M. dela, Cumagun, C. J. R., Velasco, L. R. I. and Dalisay, T. U. 2012. Variability of sugarcane smut pathogen, *Ustilago scitaminea* Sydow in the Philippines. *Philippine Journal of Crop Science* 37: 38–51.

Magarey, R., Denney, D., Sheahan, T. and Bull, J. 2010. The Australian sugarcane smut epidemic: Epidemiological considerations and predictions for the final strages. *Proceedings of the Australian Society of Sugar Cane Technologists* 32: 375–87.

Marques, J. P. R., Appezzato-da-Glória, B., Piepenbring, M., Massola, N. S., Monteiro-Vitorello, C. B. and Vieira, M. L. C. 2016. Sugarcane smut: Shedding light on the development of the whip-shaped sorus. *Annals of Botany* 119: 815–27.

Martinez, M., Medina, I., Naranjo, S., Rodriguez, C., de Armas, R., Pinon, D., Vicente, C. and Legaz, M. 2000. Changes of some chemical parameters, involved in sucrose recovery from sugarcane juices, related to the susceptibility or resistance of sugarcane plants to smut (*Ustilago scitaminea*). *International Sugar Journal* 50: 341–6.

Matsuoka, S., Kennedy, A. J., Santos, E. G. D. dos, Tomazela, A. L., Rubio, L. C. S. and Rubio, L. C. S.. 2014. Energy cane: Its concept, development, characteristics, and prospects. *Advances in Botany* 2014: 1–13.

McDonald, B. A. and Linde, C. 2002. Pathogen population genetics, evolutionary potential, and durable resistance. *Annual Review of Phytopathology* 40: 349–79.

McNeil, M. D., Hermann, S., Jackson, P. A. and Aitken, K. S. 2011. Conversion of AFLP markers to high-throughput markers in a complex polyploid, sugarcane. *Molecular Breeding* 27: 395–407.

Millanes, A.-M., Fontaniella, B., Legaz, M.-E. and Vicente, C. 2005. Glycoproteins from sugarcane plants regulate cell polarity of *Ustilago scitaminea* teliospores. *Journal of Plant Physiology* 162: 253–65.

Millanes, A.-M., Vicente, C. and Legaz, M.-E. 2008. Sugarcane glycoproteins bind to surface, specific ligands and modify cytoskeleton arrangement of *Ustilago scitaminea* teliospores. *Journal of Plant Interactions* 3: 95–110.

Moore, P. H. and Berding, N. 2014. Flowering. In: Moore, P. H. and Botha, F. C. (Eds) *Sugarcane – Physiology, Biochemistry and Functional Biology*. John Wiley & Sons, Inc., Ames, Iowa, USA, pp. 379–410.

Muhammad, S. and Kausar, A. 1962. Preliminary studies on the genetics of sugarcane smut, *Ustilago scitaminea* Sydow. *Biologia, Lahore* 8: 65–74.

Nalawade, S. V., Indi, D. V., Pandhare, R. A. and Pawar, S. M. 2012. Evaluation of sugarcane genotypes for resistance to whip smut caused by *Ustilago scitaminea*. *Ecology, Environment and Conservation* 18: 551–3.

Nzioki, H., Jamoza, J., Olweny, C. and Rono, J. 2010. Characterization of physiologic races of sugarcane smut (Ustilago scitaminea) in Kenya. *African Journal of Microbiology Research* 4: 1694–97.

Okura, V. K., de Souza, R. S. C., de Siqueira Tada, S. F. and Arruda, P. 2016. BAC-Pool Sequencing and Assembly of 19 Mb of the Complex Sugarcane Genome. *Frontiers in Plant Science* 7: 342.

Peng, Q., Su, Y., Ling, H., Ahmad, W., Gao, S., Guo, J., Que, Y. and Xu, L. 2017. A sugarcane pathogenesis-related protein, ScPR10, plays a positive role in defense responses under *Sporisorium scitamineum*, SrMV, SA, and MeJA stresses. *Plant Cell Reports* 36, 1427–40.

Peters, L. P., Carvalho, G., Vilhena, M. B., Creste, S., Azevedo, R. A. and Monteiro-Vitorello, C. B. 2017. Functional analysis of oxidative burst in sugarcane smut-resistant and -susceptible genotypes. *Planta* 245: 749–64.

Piepenbring, M. 1996. Smut fungi (*Ustilaginales* and *Tilletiales*) in Costa Rica. *Nova Hedwigia* 113: 1–155.

Piepenbring, M., Stoll, M. and Oberwinkler, F. 2002. The generic position of *Ustilago maydis, Ustilago scitaminea*, and *Ustilago esculenta* (*Ustilaginales*). *Mycological Progress* 1: 71–80.

Piñon, D., de Armas, R., Vicente, C. and Legaz, M. E. 1999. Role of polyamines in the infection of sugarcane buds by Ustilago scitaminea spores. *Plant Physiology and Biochemistry* 37: 57–64.

Que, Y.-X., Lin, J.-W., Song, X.-X., Xu, L.-P. and Chen, R.-K.. 2011. Differential gene expression in sugarcane in response to challenge by fungal pathogen Ustilago scitaminea revealed by cDNA-AFLP. *Journal of Biomedicine and Biotechnology* 2011: 1–10.

Que Y., Su, Y., Guo, J., Wu, Q. and Xu, L. 2014a. A global view of transcriptome dynamics during *Sporisorium scitamineum* challenge in sugarcane by RNA-seq. *PLoS ONE* 9: e106476.

Que, Y., Xu, L., Lin, J., Chen, R. and Grisham, M. P. 2012. Molecular variation of *Sporisorium scitamineum* in Mainland China revealed by RAPD and SRAP Markers. *Plant Disease* 96: 1519–25.

Que, Y., Xu, L., Wu, Q., Liu, Y., Ling, H., Liu, Y., Zhang, Y., Guo, J., Su, Y., Chen, J., et al. 2014b. Genome sequencing of *Sporisorium scitamineum* provides insights into the pathogenic mechanisms of sugarcane smut. *BMC genomics* 15: 1–19.

Raboin, L.-M., Selvi, A., Oliveira, K. M., Paulet, F., Calatayud, C., Zapater, M.-F., Brottier, P., Luzaran, R., Garsmeur, O., Carlier, J., et al. 2007. Evidence for the dispersal of a unique lineage from Asia to America and Africa in the sugarcane fungal pathogen *Ustilago scitaminea*. *Fungal Genetics and Biology* 44: 64–76.

Rago, A. M., Casagrande, M. V. and Massola Júnior, N. S. 2009. Variabilidade patogênica de *Ustilago scitaminea* no estado de São Paulo. *Summa Phytopathologica* 35: 93–97.

Rossi, M., Araujo, P. G., Paulet, F., Garsmeur, O., Dias, V. M., Chen, H., Van Sluys, M.-A. and D'Hont, A. 2003. Genomic distribution and characterization of EST-derived resistance gene analogs (RGAs) in sugarcane. *Molecular Genetics and Genomics* 269: 406–19.

Rott, P. and Comstock, J. C. 2015. *Sugarcane Smut Disease*. Institute of Food and Agricultural Sciences, University of Florida, Gainesville, Florida, USA, pp. 1–2.

Sanchez-Elordi, E., Morales De Los Rios, L., Diaz, E. M., Avila, A., Legaz, M. E. and Vicente, C. 2016. Defensive glycoproteins from sugarcane plants induce chemotaxis, cytoagglutination and death of smut teliospores. *Journal of Plant Pathology* 98: 493–501.

Sánchez-Elordi, E., Ríos, L. M. D. L., Vicente, C. and Legaz, M.-E. 2015. Sugar cane arginase competes with the same fungal enzyme as a false quorum signal against smut teliospores. *Phytochemistry Letters* 14: 115–22.

Santiago, R., Alarcón, B., de Armas, R., Vicente, C. and Legaz, M. E. 2012. Changes in cinnamyl alcohol dehydrogenase activities from sugarcane cultivars inoculated with *Sporisorium scitamineum* sporidia. *Physiologia Plantarum* 145: 245–59.

Santiago, R., de Armas, R., Blanch, M., Vicente, C. and Legaz, M.-E. 2010. *In vitro* effects of caffeic acid upon growth of the fungi *Sporisorium scitamineum*. *Journal of Plant Interactions* 5: 233–40.

Santiago, R., de Armas, R., Fontaniella, B., Vicente, C. and Legaz, M.-E. 2009. Changes in soluble and cell wall-bound hydroxycinnamic and hydroxybenzoic acids in sugarcane cultivars inoculated with *Sporisorium scitamineum* sporidia. *European Journal of Plant Pathology* 124: 439–50.

Schaker, P. D. C., Palhares, A. C., Taniguti, L. M., Peters, L. P., Creste, S, Aitken, K. S., Van Sluys, M-A, Kitajima, J. P., Vieira, M. L. C. and Monteiro-Vitorello, C. B. 2016. RNAseq transcriptional profiling following whip development in sugarcane smut disease. *PLoS ONE* 11: e0162237.

Schaker, P. D. C., Peters, L. P., Cataldi, T. R., Labate, C. A., Caldana, C. and Monteiro-Vitorello, C. B. 2017. Metabolome dynamics of smutted sugarcane reveals mechanisms involved in disease progression and whip emission. *Frontiers in Plant Science* 8: 882.

Schenck, S. 2003. New race of sugarcane smut on Maui. *Hawaii Agriculture Research Center* 69: 1–4.

Schenck, S., Crepeau, M. W., Wu, K. K., Moore, P. H., Yu, Q. and Ming, R. 2004. Genetic diversity and relationships in native hawaiian *Saccharum officinarum* Sugarcane. *Journal of Heredity* 95: 327–31.

Shen, W., Xi, P., Li, M., Liu, R., Sun, L., Jiang, Z. and Zhang, L. 2012. Genetic diversity of Ustilago scitaminea Syd. in Southern China revealed by combined ISSR and RAPD analysis. *African Journal of biotechnology* 11: 11693–703.

Shen, W., Xu, G., Luo, M. and Jiang, Z. 2016. Genetic diversity of *Sporisorium scitamineum* in mainland China assessed by SCoT analysis. *Tropical Plant Pathology* 41: 288–96.

Silva, W. D. da and Sanguino, A. 1978. Evaluating reaction of American cane varieties to *Ustilago scitaminea* in Brazil. *Sugarcane Pathologists' Newsletter* 21: 10–11.

Singh, N., Somai, B. M. and Pillay, D. 2004. Smut disease assessment by PCR and microscopy in inoculated tissue cultured sugarcane cultivars. *Plant Science* 167: 987–94.

Singh, N., Somai, B. M. and Pillay, D. 2005a. In vitro screening of sugarcane to evaluate smut susceptibility. *Plant Cell, Tissue and Organ Culture* 80: 259–66.

Singh, N., Somai, B. M. and Pillay, D. 2005b. Molecular profiling demonstrates limited diversity amongst geographically separate strains of *Ustilago scitaminea*. *FEMS Microbiology Letters* 247: 7–15.

Sinha, O. K., Singh, K. and Misra, S. R. 1982. Stain technique for detection of smut hyphae in nodal buds of sugarcane. *Plant Disease* 66: 932–3.

Souza, G. M., Berges, H., Bocs, S., Casu, R., D'Hont, A., Ferreira, J. E., Henry, R., Ming, R., Potier, B., Van Sluys, M.-A., et al. 2011. The Sugarcane Genome Challenge: Strategies for Sequencing a Highly Complex Genome. *Tropical Plant Biology* 4: 145–56.

Su, Y., Guo, J., Ling, H., Chen, S., Wang, S., Xu, L., Allan, A. C. and Que Y. 2014. Isolation of a novel peroxisomal catalase gene from sugarcane, which is responsive to biotic and abiotic stresses. *PLoS ONE* 9: e84426.

Su, Y., Wang, S., Guo, J., Xue. B., Xu, L. and Que, Y. 2013a. A TaqMan real-time PCR assay for detection and quantification of *Sporisorium scitamineum* in Sugarcane. *The Scientific World Journal* 2013: 1–9.

Su, Y., Xu, L., Xue, B., Wu, Q., Guo, J., Wu, L. and Que, Y. 2013b. Molecular cloning and characterization of two pathogenesis-related β-1,3-glucanase genes ScGluA1 and ScGluD1 from sugarcane infected by *Sporisorium scitamineum. Plant Cell Reports* 32: 1503–19.

Su, Y., Wang, Z., Liu, F., Li, Z., Peng, Q., Guo, J., Xu, L. and Que, Y. 2016a. Isolation and characterization of ScGluD2, a new sugarcane beta-1,3-Glucanase D family gene induced by *Sporisorium scitamineum*, ABA, H2O2, NaCl, and CdCl2 Stresses. *Frontiers in Plant Science* 7: 1348.

Su, Y., Wang, Z., Xu, L., Peng, Q., Liu, F., Li, Z. and Que, Y. 2016b. Early selection for smut resistance in sugarcane using pathogen proliferation and changes in physiological and biochemical indices. *Frontiers in Plant Science* 7: 1–10.

Su, Y., Xu, L., Wang, Z., Peng, Q., Yang, Y., Chen, Y. and Que, Y. 2016c. Comparative proteomics reveals that central metabolism changes are associated with resistance against *Sporisorium scitamineum* in sugarcane. *BMC Genomics* 17: 800.

Su, Y., Yang, Y., Peng, Q., Zhou, D., Chen, Y., Wang, Z., Xu, L., Que, Y. 2016d. Development and application of a rapid and visual loop-mediated isothermal amplification for the detection of *Sporisorium scitamineum* in sugarcane. *Scientific Reports* 6: 23994.

Sundar, A. R., Ashwin, N. M. R., Barnabas, E. L., Malathi, P. and Viswanathan, R. 2015. Disease resistance in sugarcane - an overview. *Scientia Agraria Paranaensis* 14: 200–12.

Sundar, A., Barnabas, E., Malathi, P., Viswanathan, R., Sundar, A. R., Barnabas, E. L., Malathi, P. and Viswanathan, R. 2012. A mini-review on smut disease of sugarcane caused by *Sporisorium scitamineum*. In: Mworia, J. K. (Ed.) *Botany*. InTech, Rijeka, Croatia, pp. 107–28.

Taniguti, L. M., Schaker, P. D. C., Benevenuto, J., Peters, L. P., Carvalho, G., Palhares, A., Quecine, M. C., Nunes, F. R. S., Kmit, M. C. P., Wai, A., et al. 2015. Complete genome sequence of *Sporisorium scitamineum* and biotrophic interaction transcriptome with sugarcane. *PLoS ONE* 10: e0129318.

Thokoane, L. N. and Rutherford, R. S.. 2001. cDNA-AFLP differential display of sugarcane (*Saccharum* spp. hybrids) genes induced by challenge with the fungal pathogen *Ustilago scitaminea* (sugarcane smut). *Proceedings of The South African Sugar Technologists' Association* 75: 104–7.

Toh, S. S. and Perlin, M. H.. 2016. Resurgence of Less-Studied Smut Fungi as Models of Phytopathogenesis in the Omics Age. *Phytopathology* 106: 1244–54.

Tom, L., Braithwaite, K. and Kuniata, L. S. 2017. Incursion of sugarcane smut in commercial cana crops at Ramu, Papua New Guinea. *Proceedings of the Australian Society of Sugar Cane Technologists* 39: 377–84.

Wada, A. C. 2003. Control of sugarcane smut disease in Nigeria with fungicides. *Crop Protection* 22: 45–9.

Wada, A. C., Anaso, A. B. and Bassey, M. S. 2016. Sugar cane whip smut (*Sporisorium scitamineum* Syd) caused field sucrose and juice quality losses of two sugar cane varieties in Nigeria. *International Journal of Plant & Soil Science* 10: 1–11.

Waller, J. M. 1969. Sugarcane smut (Ustilago scitaminea) in Kenya. *Transactions of the British Mycological Society* 52: 139–51.

Waller, J. M. 1970. Sugarcane smut (*Ustilago scitaminea*) in Kenya. *Transactions of the British Mycological Society* 54: 405–14.

Whittle, A. M. 1982. Yield loss in sugar-cane due to culmicolous smut infection. *Tropical agriculture* 59: 239–42.

Wu, Q., Xu, L., Guo, J., Su, Y. and Que, Y. 2013. Transcriptome profile analysis of sugarcane responses to *Sporisorium scitaminea* infection using solexa sequencing technology. *BioMed Research International* 2013: 1–9.

Xu, L., Lu, Y., You, Q., Liu, X., Grisham, M. P., Pan, Y. and Que, Y. 2014. Biogeographical variation and population genetic structure of *Sporisorium scitamineum* in Mainland China: insights from ISSR and SP-SRAP markers. *The Scientific World Journal* 2014: 1–13.

Yan, M., Cai, E., Zhou, J., Chang, C., Xi, P., Shen, W., Li, L., Jiang, Z., Deng, Y. Z., Zhang, L.-H. 2016a. A dual-color imaging system for sugarcane smut fungus *Sporisorium scitamineum*. *Plant Disease* 100: 2357–62.

Yan, M., Dai, W., Cai, E., Deng, Y. Z., Chang, C., Jiang, Z. and Zhang, L.-H. 2016b Transcriptome analysis of Sporisorium scitamineum reveals critical environmental signals for fungal sexual mating and filamentous growth. *BMC Genomics* 17: 354.

Yan, M., Zhu, G., Lin, S., Xian, X., Chang, C., Xi, P., Shen, W., Huang, W., Cai, E., Jiang, Z., et al. 2016c. The mating-type locus b of the sugarcane smut Sporisorium scitamineum is essential for mating, filamentous growth and pathogenicity. *Fungal Genetics and Biology* 86: 1–8.

Zekarias, Y., Mashilla Dejene, G. and Tegegn, F. Y. 2011. Importance and status of sugarcane smut (*Ustilago scitaminea*) in the Ethiopian sugar estates. *Ethiopian Journal of Agricultural Sciences* 21: 35–46.

Zhang, Y. Y., Huang, N., Xiao, X. H., Huang, L., Liu, F., Su, W. H. and Que, Y. X. 2015. Molecular variation of *Sporisorium scitamineum* in Mainland China revealed by internal transcribed spacers. *Genetics and Molecular Research* 14: 7894–7909.

Zhao, D. and Li, Y-R. 2015. Climate change and sugarcane production: Potential impact and mitigation strategies. *International Journal of Agronomy* 2015: 1–10.

Zhao, Z., Liu, H., Wang, C. and Xu, J.-R. 2013. Comparative analysis of fungal genomes reveals different plant cell wall degrading capacity in fungi. *BMC Genomics* 14: 274.

Chapter 11

Progress in understanding fungal diseases affecting sugarcane: rusts

R. Stuart Rutherford, South African Sugarcane Research Institute (SASRI) and University of KwaZulu-Natal, South Africa

1 Introduction

Rust fungi (Uredinales or Pucciniales) are a globally distributed order of obligate parasitic fungi occurring on vascular plants, comprising *c.* 7800 species in c. 166 genera and 14 families (Kirk et al., 2008; Aime et al., 2014). The majority of species occur in the genera *Puccinia* and *Uromyces*, for which there are over 5000 and 1500 taxon names, respectively. The disease caused by these fungi is also referred to as 'rust'. The typical symptom is a 'rust-coloured' (often orange or yellow) pustule on a plant leaf. Economically, rust fungi cause enormous crop losses, most notably in wheat which is affected by three globally distributed rust species (Hovmøller et al., 2010).

There are currently three known rusts of sugarcane – brown rust caused by *Puccinia melanocephala* H & P. Syd., orange rust caused by *Puccinia kuehnii* (W. Krüger) E. J. Butler and tawny rust caused by *Macruropyxis fulva* L. A. Martin, S. A. McFarl. and L. A. Castl. Of these, only brown rust was of any economic importance prior to the year 2000, having spread around the world up to the late 1970s. The new century has seen the worldwide spread of orange rust, probably initiated by massive spore production by a newly virulent pathotype on a widely grown cultivar in Australia.

Whilst some crop pathogenic rusts are studied extensively, the rust species of native or naturalized vegetation generally have not been. *Macruropyxis fulva*, a previously undescribed rust of native *Miscanthidium capense* in southern Africa, has been afforded

http://dx.doi.org/10.19103/AS.2017.0035.34

the opportunity to extend its host range to closely related *Saccharum*, but to date has not spread out of the Southern African region.

Global trade and tourism expansion is increasing the likelihood of world-wide plant pathogen epiphytotics (Mills et al., 2011). Human disturbance of ecosystems, altering spatial and temporal distribution of remnant wild vegetation patches embedded in an agricultural landscape, can drive the emergence of new crop pathogens and their subsequent specialization on the crop host (Stukenbrock and McDonald, 2008; Giraud et al., 2010; Papaix et al., 2015). Of particular relevance to sugarcane is that widely distributed and genetically homogeneous crop monocultures are conducive for rapid propagation across large areas following local pathogen emergence, accidental introduction or arrival by other means (Ali et al., 2014).

Due to climate change, a higher probability of extreme storm events could increase spore dispersal and the dispersal of plant debris including spores of rusts (Pachauri and Reisinger, 2007; Singh et al., 2011). Little is known about the potential effect of changing atmospheric composition or physical climate on the biology of sugarcane rusts, or on the host–pathogen interaction. Generally, increased CO_2 and temperature does affect plant–pathogen interactions, and pathogens can be directly influenced by changes in rainfall, relative humidity and plant canopy architecture (Chakraborty et al., 2011; Pangga et al., 2011; Noctor and Mhamdi, 2017). For example, sugarcane orange rust in Florida was more severe in the 2012 and 2013 growing seasons than other years due to warmer winters (Zhao et al., 2015a; Zhao and Li, 2015).

For most crops affected by rusts, future risks include: continued global movement of plants and their rusts; new rust races evolving faster; reduced effectiveness and durability of rust resistances; the evolution of new host–rust relationships; reduced effectiveness of fungicides; and increased crop loss due to rusts (Chakraborty et al., 2011).

In the light of these risks, this chapter summarizes current knowledge of sugarcane rusts in terms of relevant background biology, pathogen dispersal, disease identification, environmental conditions favouring disease, crop loss, host-plant resistance and control measures.

2 Biology of rusts

2.1 Life cycle

Rusts are exclusively obligate biotrophic and pathogenic fungi of vascular plants. Many rust species exhibit complicated life cycles with variable host plant specificity (Anikster, 1989). A common feature in many rusts is a heteroecious life cycle, requiring two exclusive and unrelated host plant taxa for their completion, as illustrated in the well-documented example of the stem rust *Puccinia graminis* Pers.:Pers. f. sp. *tritici* Eriks. E. Henn, on species of *Berberis* (barberry) and both *Triticum* (wheat) and *Hordeum* (barley).

Rust fungi have been divided into three categories depending upon the number of spore states in their life cycle: macrocyclic, demicyclic and microcyclic (Cummins and Hiratsuka, 2003). Macrocyclic rust fungi produce five spore states [teliospores, basidiospores, spermatia (pycniospores), aeciospores and urediniospores], which occur on either a single host (autoecious) or two hosts (heteroecious). Heteroecious rust fungi disperse between the alternate hosts by means of basidiospores (from telial host to aecial host) and aeciospores (from aecial host to telial host). In heteroecious rusts, sexual recombination takes place on the alternate host, whilst asexual (clonal) reproduction takes place typically on a grass host (Leonard and Szabo, 2005) (Fig. 1). Absence of the uredinial spore, Stage II, is classified as demicyclic, and absence of both aecial and uredinial spores, Stages I and II, is classified as microcyclic (Helfer, 2014).

Teliospores of *Puccinia melanocephala* and *Puccinia kuenhii* are known, although in the latter case only Butler (1914) has recorded his observation. Basidiospores arising after teliospore germination do not initiate infection on sugarcane. However, no alternate hosts are known for *Puccinia melanocephala* and *Puccinia kuenhii* (Purdy et al., 1983). Teliospores have not been observed for *M. fulva*. However, the discovery of this rust on the grass *Miscanthidium capense* in the high Drakensberg Mountains of South Africa suggests that teliospores may be produced in response to the cold dry winters of that region.

Hemileia vastatrix Berk. & Broome, the causal agent of coffee leaf rust, appears similar to sugarcane rusts in that urediniospores function as the dominant propagules, whilst the role of basidiospores produced during germination of teliospores *in situ* on coffee leaves remains uncertain, except that they do not infect coffee (Agrios, 2005). An unproven hypothesis is that *H. vastatrix* is heteroecious, capable of completing its life cycle on

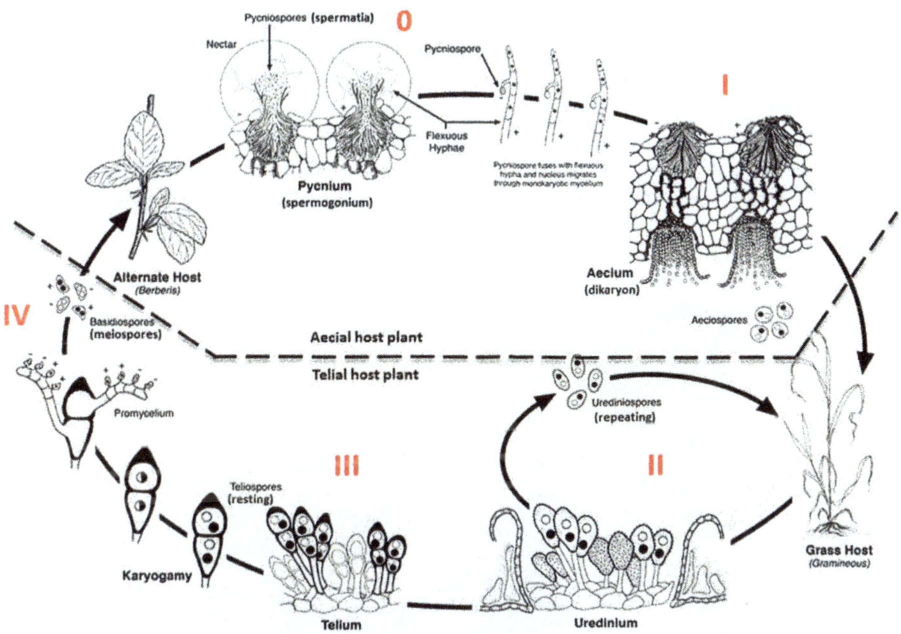

Figure 1 Life cycle of the heteroecious macrocyclic rust *Puccinia graminis* f. sp. *tritici*. Stages 0 and I are on the aecial host, and stages II and III are on the telial host; stage IV develops on germination of teliospores. Meiosis producing haploid (n) spores follows immediately after the fusion of nuclei (karyogamy, 2n) in germinating teliospores. Plasmogamy leading to dikaryotic (n + n) cells, follows after the transfer of pycniospores (spermatia) to compatible receptive hyphae in the pycnium (spermogonium). Adapted from (https://www.ars.usda.gov/ARSUser Files/50620500/Cerealrusts/PgLifecycle.jpg); courtesy of USDA ARS (Accessed 20 July 2017).

an as yet unidentified aecial host plant (Ritschel, 2005). Alternatively, *H. vastatrix* may represent an early diverging autoecious rust, in which the teliospores are vestigial and non-functional. Urediniospores would then have multiple functions: dispersal, infection, survival and even sexual recombination (Evans, 1993; Carvalho et al., 2011).

Sugarcane rusts may therefore be autoecious (with urediniospores, and teliospores only), but with the role of teliospores not being demonstrated. Similarly, the neotropical rust fungus *Prospodium tuberculatum* (Speg.) Arthur, which was released in Australia in 2001 for the classical biological control of the invasive alien weed *Lantana camara*, cycles solely through urediniospores on *L. camara* throughout most of its native range. However, at high altitude extremes, this rust has retained the ability to produce teliospores (Ellison et al., 2006).

2.2 Dispersal

Most relevant to sugarcane rusts, urediniospores are the repeating spores on the telial host. They are produced in vast numbers and represent the spore stage with the highest dispersal potential, generally being considered as the agents of rust epiphytotics (Nascimento et al., 2012). Spore sizes are c. 15–30 μm diameter and many species have urediniospores with thickened cell walls and fine spines, promoting long-distance dispersal in air currents (Brown, 1997; Wang et al., 2010) and by attaching themselves to animals or other moving objects (hair, clothing or luggage) (Fig. 2).

Sugarcane brown rust was almost certainly transported by cyclonic winds from Cameroon to the Dominican Republic in 1978 (Purdy et al., 1985). Mims and Mims (2003) suggested that major fires, which can produce smoke plumes three kilometres or more high, could be much more effective in launching spores into the troposphere than thermals generated by the sun heating the ground. They speculated that the transport of brown rust to the Dominican Republic may have been initiated by burning of sugarcane at harvest in Cameroon. Africa

Figure 2 Profuse urediniospores of sugarcane tawny rust adhering to clothing.

is also the likely source of introduction of orange rust into the Western hemisphere, though further investigation to confirm this hypothesis is required (Saumtally et al., 2011).

Accidental spore transport via human travel may also contribute to the intercontinental dispersal of rust pathogens. The rapid, global spread of the wheat yellow rust (aka stripe rust) fungus *Puccinia striiformis* Westend. f. sp. *tritici* Erikss. is an example, being first introduced into Australia in 1979 from Europe, and again when the same strains of the pathogen that first appeared in North America in 2000 were detected in Western Australia just two years later (Wellings, 2007; Hovmøller et al., 2008).

The dispersal potential of the final, teliospore, stage of rusts is limited, in that many species produce immobile teliospores. The larger size of teliospores reduces their wind dispersal potential. Teliospores are normally thick-walled, long-lived, often dark-coloured and relatively resistant to desiccation and UV radiation (Brown, 1997). As actively growing rusts cannot survive unfavourable conditions on dead plant tissue, resting teliospores are normally produced at the end of the growing season.

2.3 Genetic variability

A significant challenge for resistance breeding is the 'boom and bust' cycle resulting from the emergence of new pathogen races that overcome single major resistance genes. A predominant evolutionary mechanism involves changes in which single virulence phenotypes are acquired sequentially by new races within a rust lineage. This is most likely due to mutations occurring within single Avirulence (*Avr*) genes which acquire virulence in terms of a particular major resistance (*R*) gene in the plant. Among sugarcane rusts, *Puccinia melanocephala* is known to possess the ability to repeatedly adapt and overcome host plant major gene resistance (Hoy et al., 2014).

Sexual reproduction on an alternate host is also known to generate high virulence diversity (Rouse et al., 2009). In response to wheat stem rust outbreaks in the early 1900s, a barberry eradication programme was conducted throughout the North American wheat belt for nearly fifty years. As a result, the number of epiphytotics and diversity of the *Puccinia graminis* f. sp. *tritici* population was hugely reduced (Roelfs, 1982). Since then, only one epiphytotic in 1974 has been reported (Leonard and Szabo, 2005).

In the absence of sexual recombination, as may be the case in sugarcane rusts, somatic hybridization, which involves fusion of dikaryotic (N + N) hyphae from different isolates and either nuclear exchange or parasexual recombination and exchange of chromosomes, is another mechanism that may introduce variability (Park and Wellings, 2012). Also, in the apparently clonal rust species *H. vastatrix*, Carvalho et al. (2011) have described hidden sexual reproduction disguised within the urediniospore. This cryptosexuality may be a common event in rust fungi and could explain why new physiological races arise often and rapidly in *H. vastatrix*.

3 Host plant resistance

3.1 Non-host resistance

Any individual plant species can only be infected by a limited number of potential pathogens. All plants are therefore non-hosts to the majority of pathogens, which highlights the effectiveness of non-host resistance (NHR) (Heath, 1991).

Understanding the basis of NHR in plants that are outside a given pathogen's host range has attracted considerable attention, partly because this form of immunity is durable, broad-spectrum and displayed by all members of a plant species to all isolates of a potential pathogenic species (Nurnberger and Lipka, 2005). However, the likelihood that a potential pathogen can infect two plant species in a habitat increases considerably with decreasing phylogenetic distance between the plant species. Part of this basic NHR may be due to the absence of appropriate chemical and physical signals that are required for recognition of a potential host.

Plant surfaces are important ecologically as they represent the first point of contact with other organisms. It has been suggested that plant surfaces contain signals influencing the germination of biotrophic fungal spores and that these are the first stage in determining host specificity.

Wax from the Poaceae is generally dominated by n-alkanols. Tulloch et al. (1980) found a degree of chemical uniformity within the tribe Triticeae; either alcohols, principally C26 hexacosanol (*Hordeum*) and C28 octacosanol (*Triticum*), or β-diketones (other genera) were the major constituents of the wax, whilst hydrocarbons and esters were always minor constituents.

Octacosanal has been suggested to be a morphogenetically active component for *Puccinia graminis* f. sp. *tritici*, that is, involved in host plant recognition and infection structure differentiation in epicuticular wax of wheat (Reisige et al., 2006). Octacosanol (alcohol)-to-octacosanal (aldehyde) ratios have been shown to vary between sugarcane cultivars (Rutherford and van Staden, 1996).

Chemical differences can also lead to physical differences, for example, β-diketones form hollow wax nanotubes, while primary alcohols form flat plates (Koch et al., 2006). β-diketones increase light reflectance and result in a glaucous phenotype that may modulate leaf temperature (Shelton, 2012), and may decrease leaf wettability, both factors being known to affect rust infection (Statler and Nordgaard, 1980; Seymour, 1985). Additional factors that affect leaf wettability include the number, size and pattern of trichomes; stomatal density; the shape of epidermal cells and leaf water status (Wang et al., 2015).

Certain rust pathogens that initially germinate on the surface of leaves have thigmotropic sensing mechanisms that measure physical features of stomatal guard cell anatomy, e.g. the height of lips of the guard cells. When the fungal hypha encounters a lip of the correct height, it undergoes a developmental programme resulting in the formation of appressoria and infection pegs that enter the stomata and begin colonization of the leaf interior. Allen et al. (1991) found that *Puccinia melanocephala* developed appressoria on artificial ridges optimally within a height range of 0.4–0.8 μm. Appressorium formation can therefore be affected by various thigmotropic and/or chemotropic stimuli depending upon rust species. The absence of appropriate stimuli contributes towards NHR.

From the appressorium, an infection peg is inserted between the stomatal guard cells, and a substomatal vesicle is produced within the leaf apoplast (Fig. 3). Infection hyphae emerge from this vesicle and ramify through the apoplastic space, inserting (or attempting to insert) haustoria into adjacent leaf cells. Sotomayor et al. (1983) found that *Puccinia melanocephala* did not develop beyond the infection hypha stage in the non-host wheat, or the haustorial mother cell stage in the non-host oats. In both cases, contact between a fungal cell and a plant cell resulted in autofluorescence. In barley, Dracatos et al. (2014) showed that non-penetrating contact between *Puccinia graminis* f. sp. *avenae* and non-host barley cells also resulted in localized deposition of autofluorescent material at contact sites, without plant cell death.

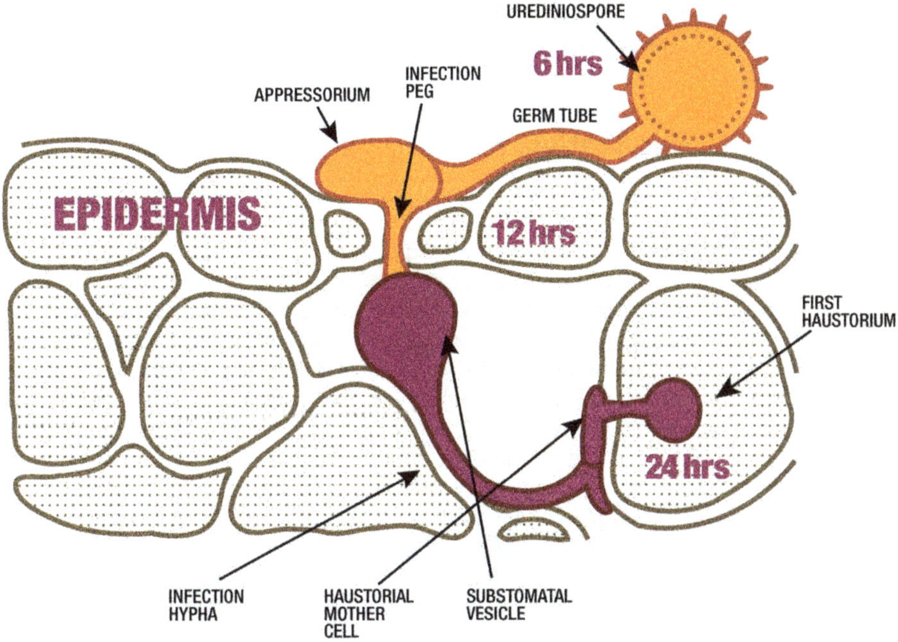

Figure 3 Typical rust infection stages.

This response suggests the possibility that effector production (fungal *avr* gene), and recognition thereof (plant *R*-gene), extends beyond the plant–haustoria interface. For example, the barley *Rpg1* resistance gene encodes a kinase constitutively expressed in all tissues with elevated levels in the epidermis. It appears to respond to components present in avirulent urediniospores that may be released early in infection (Nirmala et al., 2011). From the 1940s to 1989, *Rpg1* conferred broad resistance to all known races of *Puccinia graminis* f sp. *tritici*.

Once pre-formed barriers are overcome, the potential attacker must then also overcome any induced defence. The front line of pathogen-induced defence is triggered by evolutionarily conserved pathogen-derived molecules, called pathogen-associated molecular patterns (PAMPs), which are recognized by non-specific host pattern-recognition receptors (HPRRs) at the plasma membrane. HPRRs, termed 'receptor-like kinases' (RLKs), consist of an extracellular ligand-binding domain, a single trans-plasma membrane domain, and a cytosolic protein kinase domain. This generic recognition system detects evolutionarily conserved PAMP molecules such as β-glucan and chitin from fungi (Nurnberger et al., 2004; Kaku et al., 2006).

In addition to PAMPs, microbial enzyme degradation products of the plant cell wall also elicit responses. Typical 'cell wall signals' are relatively low-molecular weight microbial degradation products such as pectin-derived oligogalacturonides. This class of elicitors has been termed 'damage associated molecular patterns' (DAMPs) because of their origin as metabolites from cell damage (Zipfel, 2009).

Putative HPRRs, including cell-wall-bound receptor kinase, RLKs, and chitin binding receptor kinase, have been detected in the interaction between sugarcane and *Puccinia melanocephala* (Carmona et al., 2004).

3.2 Race-specific (qualitative) host resistance

The evolution of induced plant immunity has been proposed to follow a zigzag interplay between PAMP-triggered immunity (PTI), effector-triggered susceptibility (ETS) and effector-triggered immunity (ETI) (Jones and Dangl, 2006).

PTI can be overcome by pathogen effector molecules that trigger ETS by altering the plant's physiological state to benefit pathogen colonization, or by suppressing host plant defences that are activated by PAMP detection (Speth et al., 2007). In ETS, a detectable PAMP/DAMP-induced resistance can remain to a greater or lesser extent depending upon the efficiency of ETS (Nurnberger and Lipka, 2005). This type of resistance has been referred to as 'basal resistance', which is quantitative in nature (Boller and Felix, 2009).

The transition from PAMP/DAMP-triggered immunity (PTI) to ETS can lead to a second group of responses known as gene-for-gene resistance or ETI. ETI is initiated through recognition of pathogen effectors (avirulence – *avr* gene products) by resistance (R) genes of the host plant (Jones and Dangl, 2006). *R*-genes encode receptors such as those containing an N terminal coiled-coil structure, a central nucleotide-binding (NB) site, and a leucine-rich repeat (LRR) region. Such NB-LRR major *R*-genes recognize fungal effectors and lead to a range of resistance responses usually associated with a hypersensitive response (HR), a cell death process that occurs at the site of attempted haustorial ingress (Fig. 3). The HR is effective against obligate biotrophic pathogens such as rusts which must obtain nutrients from living host tissue (Mur et al., 2008).

Qualitative (major, race-specific) resistance is used extensively in the breeding of graminaceous grain crops in the tribe Triticeae (sub-family Pooideae). On the other hand, only a few major resistance genes, such as the *Rp* genes for resistance to maize common rust (Ramakrishna et al., 2002), have been reported in the tribe Andropogoneae (sub-family Panicoideae) to which sugarcane belongs to.

So far, only two putative major *R*-genes have been found in sugarcane, both conferring resistance to brown rust (Daugrois et al., 1996; Raboin et al., 2006). However, the first of these, brown rust 1 (*Bru1*), is apparently effective against all races of brown rust, suggesting that it is not an NB-LRR major *R*-gene that recognizes a race-specific pathogen effector. Fourteen genes have been identified in the *Bru1* region (Garsmeur et al., 2011).

3.3 Race-nonspecific (quantitative) host resistance

Quantitative resistance (QR) is usually mediated by multiple genes (i.e. polygenic), or quantitative resistance loci (QRLs), each one of which encodes a partial increase in resistance (Poland et al., 2009; St Clair, 2010; Roux et al., 2014; Niks et al., 2015). QR is also considered to be horizontal, that is, effective against all pathogen races, or race-nonspecific. Consequently, QR is said to be more durable than qualitative resistance (Parlevliet, 2002), such that the pathogen requires multiple mutation and/or recombination events to overcome the resistance. Adaptation to resistance is expected to result in a gradual erosion of the resistance efficacy, rather than a sudden breakdown (Van Den Berg et al., 2014).

While QR is often said to be conferred by QRLs of relatively small effect, some QRLs can have such large effects that they blur the boundary between qualitative and QR e.g. *Lr34* in wheat (Krattinger et al., 2009). In another example, the slow-rusting gene *Yr36*, cloned from an ancestral wheat accession, confers high-temperature adult plant partial resistance to wheat stripe rust, *Puccinia striiformis* f. sp. *tritici* (Fu et al., 2009). *Yr36*-dependent resistance manifests itself as a rapid formation of autofluorescent cells around a *Puccinia striiformis* infection site prior to cell penetration.

A component of QR may be conferred by weak ETI triggered by altered or 'defeated' *R*-genes (Li et al., 1999; Fukuoka et al., 2014). McIntyre et al. (2005) amplified sugarcane NB-LRR fragments from cDNA and used these fragments to produce eight RFLP markers that were significantly associated with QR in the progeny from parents lacking any apparent major gene resistance. Each marker was associated with less than 10% of the variation in rust resistance. This is consistent with multigenic determination of rust resistance in this population, and also with the notion of altered or defeated *R*-genes being involved in QR.

HPRRs, the receptors that are part of innate defence, may also play a role in QR. In *Arabidopsis thaliana*, the receptor kinase flagellin sensing 2 (FLS2) confers recognition of bacterial flagellin (flg22) and activates defence. Vetter et al. (2012) showed extensive variation in flg22 perception among plant genotypes which correlated with differences in FLS2 protein abundance. In addition, components of pathways downstream of this PAMP perception likely contributed to the observed quantitative variation.

Corwin et al. (2016) suggested that thousands of causal genes can underlie variation in QR. In maize, the majority of potential small-effect causal genes affected resistance mechanisms downstream of perception and signalling (Kump et al., 2011; Belcher et al., 2012). These included genes for defensins, other pathogenesis-related proteins and enzymes involved in secondary metabolism.

Secondary metabolite phytoanticipins (pre-formed), which are stored as inactive molecules that can be activated and released following biotic attack, contribute towards QR. Leme et al. (2014) suggested that higher levels of the pre-formed flavonoid luteolin-8-C-glucoside may be related to resistance to orange rust in sugarcane. However, studies with sugarcane varieties inoculated with *Puccinia kuehnii* would be necessary to test this hypothesis.

Allelic variation in genes controlling plant secondary chemistry can influence QR (Rowe and Kliebenstein, 2008). Likewise, allelic differences in ATP-binding cassette (ABC) transporters have also been implicated in QR (Krattinger et al., 2009). Carmona et al. (2004) and Oloriz et al. (2012) detected differential ABC transporter gene expression in brown rust-resistant somaclonal variants and chemically induced sugarcane mutants, compared to susceptible 'parent' genotypes.

3.4 Breeding for resistance

The largest proportion of genetic resistance used by sugarcane breeders appears to be quantitative, similar to disease resistance in maize (Balint-Kurti and Johal, 2009). Sugarcane is naturally an outcrossing species, which makes the genetic architecture controlling quantitative traits more complex than in self-pollinating plants like rice and wheat (Wallace et al., 2014). Maize and sugarcane breeders may have sufficient diversity available to allow them bring together multiple small-effect QRLs to achieve effective QR.

Pyramiding QRLs leads to increased QR. Fukuoka et al. (2015) tested the blast resistance provided by four independent QRLs in rice by introgressing different combinations into the same genetic background. They showed that multiple different QRL combinations provided increased levels of resistance.

Another strategy involves the combination of major *R*-genes with QR factors in order to preserve the durability of resistant cultivars (Quenouille et al., 2014). Those authors found that QRLs controlling QR can have a pleiotropic effect on the durability of major *R*-genes.

4 Control of rusts

4.1 Chemical

With the potential presence of three rust species attacking sugarcane, the task of plant breeding for resistance becomes more difficult. Fungicides represent an alternative control measure that can be integrated with host plant resistance. In the past, fungicides had been considered to be neither practical nor economical for rust control in sugarcane (Ryan and Egan, 1989; Comstock and Raid, 1994). However, this situation is changing as recent years have seen the registration of more efficacious fungicides for rust control.

The fungicides of choice for the control of rusts are triazoles (Sterol demethylation inhibitors – SBI(I) – Fungicide Resistance Action Committee (FRAC) class 3), morpholines (Sterol reductase and isomerase inhibitors – SBI(II) – FRAC 5), carboxamides (Succinate dehydrogenase inhibitors – SDHI – FRAC 7) and strobilurins (Quinone-outside inhibitors – QoIs – FRAC 11).

Properties of the different fungicidal modes of action vary within and between classes and can imply when the fungicide can be applied. According to Hewitt (1998), preventive action occurs when the fungicide is applied on healthy tissues (pre-penetration), curative action when it is applied after the penetration of the fungus, but before the appearance of symptoms, and the eradicant action occurs when the fungicide is applied after symptom manifestation. In addition, an antisporant prevents the fungus from producing spores or decreases spore production without stopping the vegetative growth of the pathogen.

Triazoles, carboxamides and strobilurins are all known to display varying degrees of acropetal systemic movement in plants (Godwin et al., 1992; Kuck et al., 1995; Grabert, 2014). Levels of preventive and curative activity can vary among them and even among pathosystems for a single fungicide (Mueller et al., 2004; Wong and Wilcox, 2001).

Triazoles are sterol synthesis inhibitors and many have good activity against rust diseases (Kuck et al., 1995). However, triazoles generally have little effect on fungal spore germination, due to the delayed timing of new sterol biosynthesis and precursor accumulation post germination (Pontzen and Scheinpflug, 1989).

Strobilurins interfere with the respiratory chain and generally are very effective against rust diseases (Godwin et al., 1992). Strobilurins are strongly inhibitory to germination of fungal spores (Buck and Williams-Woodward, 2003; Leroux, 1996) and were shown to be toxic to urediniospores of several rust pathogens after very short exposure times (Mueller et al., 2005). Azoxystrobin was shown to exhibit a significant protective effect (as much as 15 days prior to inoculation) and curative activity (seven days post inoculation) (Mueller et al., 2004). In addition, strobilurins can show antisporant activity on lesions in some pathosystems when applied post infection or post symptom development (Wong and Wilcox, 2001; Ypema and Gold, 1999).

Carboxamide fungicides were discovered more than 40 years ago, but had limited application at that time. Their target enzyme is succinate dehydrogenase (SDH, complex II in the mitochondrial respiration chain), which is a functional part of the tricarboxylic cycle and linked to the mitochondrial electron transport chain (Keon et al., 1991). SDHIs with increased spectrum and potency were launched starting in 2003 and new ones continue to be launched today. Many of these display both preventive and curative action combined with improved acropetal movement. However, the risk of development of resistance to this mode of action is considered to be high and resistance management is required (Sierotzki and Scalliet, 2013).

The latest trend in chemical rust control is the application of mixtures of two or three complementary and non-cross-resistant modes of action from among the strobilurins, triazoles and carboxamides. Such mixtures are designed to give some protection against the development of resistance to any one of the modes of action (Bosch et al., 2015). They also give extended periods of control, generally up to 28 days between applications and can be applied before symptom development at a time when conditions become suitable for the initiation of infection.

McFarlane et al. (2006) found that it was uneconomical to apply fungicides, partly due to the severity of rust outbreaks being difficult to predict. However, it is possible to predict the advent of rust infections, given that leaf wetness duration and temperature are often accurate predictors of rust incidence (Webb and Nutter, 1997; Barrera, 2010; Bebber et al., 2016). Leaf wetness duration itself can be predicted using time, temperature, relative humidity, wind speed, solar radiation and precipitation data provided by an automated weather station (Chtioui et al., 1999).

Further enhancing the utility of fungicides, it has been found that those most extensively used against rusts also exhibit plant physiological effects. Plants treated with triazoles often use less water than untreated plants and have improved antioxidant and osmoprotectant status (Fuller and Zajicek, 1995; Rabert et al., 2014). Morpholines and triazoles can induce the accumulation of abnormal sterols in plant tissue and can change the ratio among normal phytosterols. This may explain increased frost hardiness (Burden et al., 1987) and altered insect pest survival and development (Bodnaryk et al., 1997) that have been observed in SBI fungicide–treated plants.

Plant growth stimulation in plants by strobilurin fungicides has been related to increased nitrate uptake and assimilation. Strobilurins can also slow leaf senescence by increasing endogenous cytokinin and reducing the content of 1-aminocyclopropane-1-carboxylic acid, the precursor for the biosynthesis of ethylene. This may be mediated via stimulating the production of nitric oxide, a key messenger in plants that can signal increased antioxidant activity, reduced ethylene biosynthesis and activation of plant defence responses (Glaab and Kaiser, 1999; Venancio et al., 2003; Kanungo and Joshi, 2014).

Carboxamides are also able to exert positive physiological effects on plants. For example, sedaxane increased the efficiency of photosystem II, reduced non-photochemical quenching and improved photosynthesis and biomass in wheat correlating with systemic changes in the expression of genes involved in defence, chlorophyll synthesis and cell wall modification (Ajigboye et al., 2016).

4.2 Biological

Some natural biological control mechanisms are thought to operate against rusts. The parasitic fungus *Darluca filum* (*Eudarluca caricis*) is known to parasitize *Puccinia kuehnii* (Dube et al., 1978). In a study conducted by Pillay et al. (2005) contaminants of brown rust pustules were tentatively identified as fungi known to be hyperparasitic on various rusts, such as *Cladosporium uredinicola* and *C. tenuissimum* (Ryan and Wilson, 1981; Moricca et al., 2001). Other contaminants showed phylogenetic similarity to *Phoma glomerata*, a species known to be antagonistic or hyperparasitic to other fungal plant pathogens (Ouimet et al., 1997; Sullivan and White, 2000). These results suggest that other fungi often inhabit rust pustules, and several may be important components of a hyperparasitic guild of fungi that naturally restrict the progress of rust epiphytotics. Further study is warranted to evaluate the effectiveness of hyperparasitic control of rust pathogens.

4.3 Plant nutrition

Anderson and Dean (1986) suggested that control of rusts might be possible by modifying the nutritional status of the plant. Their study indicated that N and Ca showed consistent positive and negative associations, respectively, with increasing rust intensity, while P, Cu, Mn and Zn appeared to have lesser influence.

It is also possible that rust intensity influences leaf nutrient status. Cadet et al. (2003) demonstrated that sugarcane brown rust strongly interfered with the nutritional physiology of the plant. The presence of rust appeared to favour the uptake of K, to the detriment of Ca, Mg and Si. This effect could be due to uncontrolled transpiration following leaf cuticle rupture (Grimmer et al., 2012). Soil water depletion and root stress may lead to increased root suberization, which is known to have a significant negative effect on Ca and Mg uptake (Ferguson and Clarkson, 1976).

Cadet et al. (2003) suggested that early application of Ca, before rust infection, could reduce rust intensity by limiting the uptake of K. This is supported by the observed 'rock road effect', where rust severity is lower on field edges adjacent to lime ($CaCO_3$) road spills (Anderson et al., 1990). Reductions in brown rust severity have also been achieved by increasing the Si content of the plant (Camargo et al., 2013; Ramouthar et al., 2016). Calcium silicate slags have been shown to reduce rust intensity.

In South Africa, rust infections are often more severe in well-grown sugarcane (Cadet et al., 2003). Similar observations were reported in sugarcane in Florida, USA, and in other crops (Anderson and Dean, 1986). There is also an apparent increased susceptibility of plant cane versus succeeding ratoon crops (Anderson et al., 1988), which may be due to increased N mineralization from organic matter following tillage. Indeed, sugarcane growers applying fertilizer in excess of crop requirements can expect an increase in the incidence and severity of rust infestations (Johnson et al., 2007).

5 Brown rust

5.1 History

The first serious outbreak of brown rust of sugarcane caused by *Puccinia melanocephala* H & P. Syd. was reported in India in 1907 on *Erianthus ravennae* (Cummins, 1971; Sathe, 1971).

Hosts of *Puccinia melanocephala* are wide-ranging and include *Saccharum officinarum*, *S. spontaneum*, *S. barberi*, *S. robustum* and *Miscanthus* species (Dixon et al., 2010).

In South Africa, the sugarcane cultivar Co301 has been continuously affected by brown rust since 1941, following a presumed accidental introduction, perhaps as a contaminant in cane fuzz (true seed) imported from India in 1938 (Nuss and Brett, 1995). Originally misidentified as *Puccinia kuehnii*, this rust was later confirmed to be *Puccinia melanocephala*. Co301, which is also highly susceptible to smut, was replaced by a series of locally selected cultivars (from imported Indian true seed), most showing a high degree of resistance to *Puccinia melanocephala*. However, the release of the susceptible cultivar N55/805 in 1965 saw a resurgence of the disease in the mid- to late 1970s, by which time, this cultivar had increased to comprise 11% of the crop (and in one mill area 20%) (Bailey, 1979, 1995; Lamusse, 1978). During the 1980s, this cultivar was replaced and brown rust again declined in importance until the release of highly susceptible cultivar N29.

By the end of the 1970s, *Puccinia melanocephala* had spread to most sugar industries throughout the world, becoming known as 'common' rust. It is now known as brown rust following the worldwide spread of orange rust. Brown rust was first reported to be found in the Western Hemisphere in 1978 in the Dominican Republic and it also appeared in Australia in 1978. It was observed soon after in most sugarcane-growing locations of the Americas, including Florida in 1979 (Purdy et al., 1983; Ryan and Egan, 1989). The introduction of *Puccinia melanocephala* into the Americas is believed to have occurred via the trans-oceanic transport of spores from Cameroon in Africa (Purdy et al., 1985). Bernard (1980) speculated that brown rust arrived in the Dominican Republic in a dust cloud (Fig. 4).

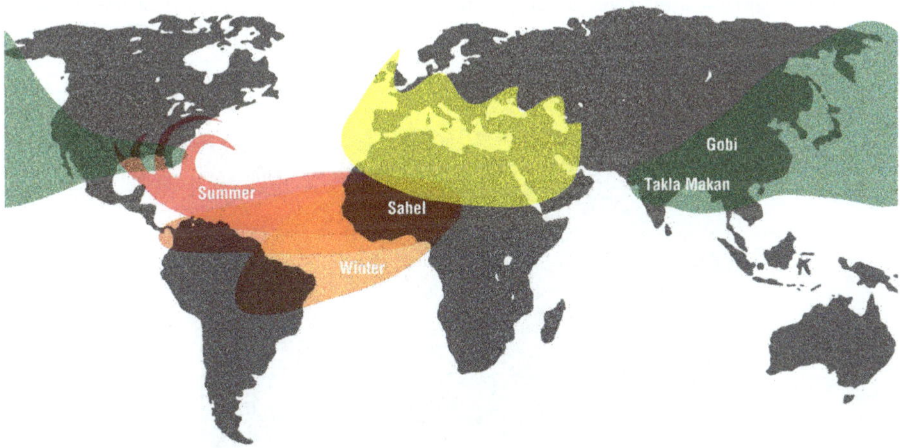

Figure 4 Dust is transported in two major global dust transport systems: (1) from the Sahara and Sahel of Africa to the Americas, Europe and Near East; and (2) from the Takla Makan and Gobi deserts of China, across China, Korea, Japan and the northern Pacific to North America, sometimes exiting over the Atlantic Ocean. Redrawn from Garrison et al. (2003).

More recently, Kelly et al. (2009) reported the first confirmed record of *Puccinia melanocephala* infecting sugarcane in Ethiopia in 2007 despite it having been present in 15 other African countries many years earlier, including the bordering countries of Kenya and Uganda.

5.2 Identification

Brown rust is distinguished by cinnamon-brown to dark-brown elongated pustules that occur on the undersides of the leaf. Urediniospoes are smaller than those of *Puccinia kuenhii* and have more prominent pores and no apical thickening (Fig. 5). Spines are closer and are regularly spaced at 1–1.5 µm apart. This compares to a more irregular spacing of 3–4 µm for *Puccinia kuenhii* (Table 1). Furthermore, the species-specific PCR primers Pm1-F and Pm1-R amplify a 480-bp DNA product from *Puccinia melanocephala* in leaf samples with symptoms of brown rust (Glynn et al., 2010).

5.3 Conditions favouring the disease

Urediniospore germination and appressorium formation usually occur over a fairly wide temperature range from 5°C to 34°C requiring at least 6–8 h of continuous leaf wetness (Purdy et al., 1983; Raid and Comstock, 2000). The optimum temperature was found to be 25°C (Purdy et al., 1983). Hsieh et al. (1977) found that urediniospores germinated over the same range with optimum germination occurring at 22–26°C. Research by Sotomayor et al. (1983) and Raid and Comstock (2000) supported these findings. As with urediniospore germination, appressorium formation occurs over a range of 5–30°C but with an optimum of 15–30°C. Appressorium development is reduced at temperatures of 5–10°C (Purdy et al., 1983; Sotomayor et al., 1983). In addition, Sotomayor et al. (1983) found that urediniospores began to germinate within 1 h of inoculation at 25°C and 30°C and that spores germinated over a temperature range of 5–30°C in 4 h of leaf wetness, but appressorial formation required at least 6 h of leaf wetness (Table 2). The disease is

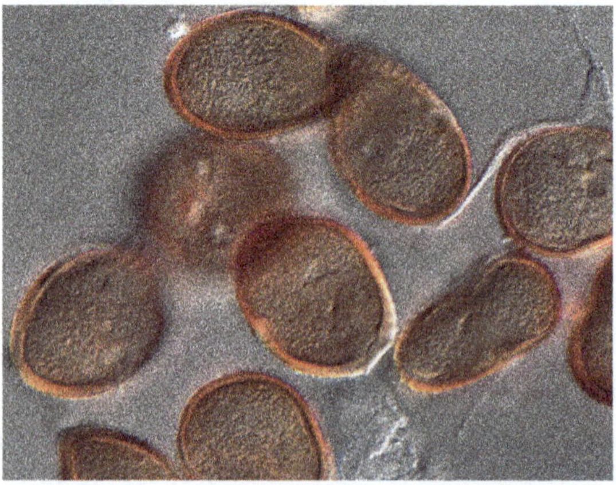

Figure 5 Urediniospores with uniform wall thickness averaging 1.5 µm. Photo courtesy of USDA ARS.

Table 1 Characteristics of sugarcane rusts [adapted from Martin et al. (2017), and references therein]

Characteristic	*Macruropyxis fulva*	*Puccinia kuehnii*	*Puccinia melanocephala*
Lesions	Linear, parallel to mid-rib, usually 2–10 mm	Typically shorter, more oval in shape, occur frequently in groups	Linear, parallel to mid-rib, usually 2–10 mm, up to 20 mm long
Occurrence	Cane of all ages; appears to be favoured by cool, moist weather	Semi-mature to mature cane >6 months; favoured by humid summers, and warm to cool autumns	Young plants, 2–6 months old in spring or autumn, favoured by cool nights and dew
Uredinia			
Colour	Bright orange when fresh, becoming dark reddish brown over time	Orange to reddish brown (Fig. 6)	Cinnamon-brown to dark brown
Size	Up to 3 mm in length	Up to 4 mm in length	Up to 4 mm in length
Position	Primarily hypophyllous, frequently amphigenous, occur more frequently towards leaf tip	Primarily hypophyllous, sometimes amphigenous, and more prevalent in lower half of the leaf	Primarily hypophyllous, sometimes amphigenous, occur more frequently towards leaf tip
Urediniospores			
Colour	Bright orange to orange-red to hyaline	Orange to cinnamon-brown	Cinnamon-brown to dark brown
Size	25–32 × 20–6 μm	33–53 × 21–31 μm	25–39 × 17–28 μm
Shape	Ellipsoid to obovoid, to globose	Obovoid to pyriform or broadly ellipsoidal	Obovoid or ellipsoidal
Wall thickness	1.5–4.9 μm, uniform	1–2.3 μm with apical thickening up to 12 μm	0.8–2.3 μm, uniform
Spines	Abundant, evenly distributed, clustered around spore caps	Moderate, unevenly distributed, not clustered around spore caps	Fine – moderate, evenly distributed, clustered around spore caps (Fig. 7)
Pores	Few observed: 0–3 random distribution	4–5 equatorial	4–5 equatorial
Paraphyses	Unknown	Inconspicuous, sometimes absent	Abundant (Fig. 8)
Teliospores	Unknown	2-celled, rarely 3-celled	2-celled
Spermogonia (pycnia)	Unknown	Unknown	Unknown
Aecia	Unknown	Unknown	Unknown

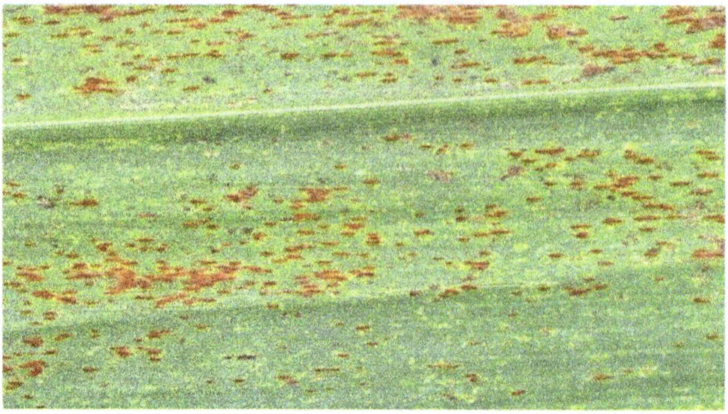

Figure 6 Uredinia are orange to reddish-brown and oval. Photo courtesy of USDA ARS.

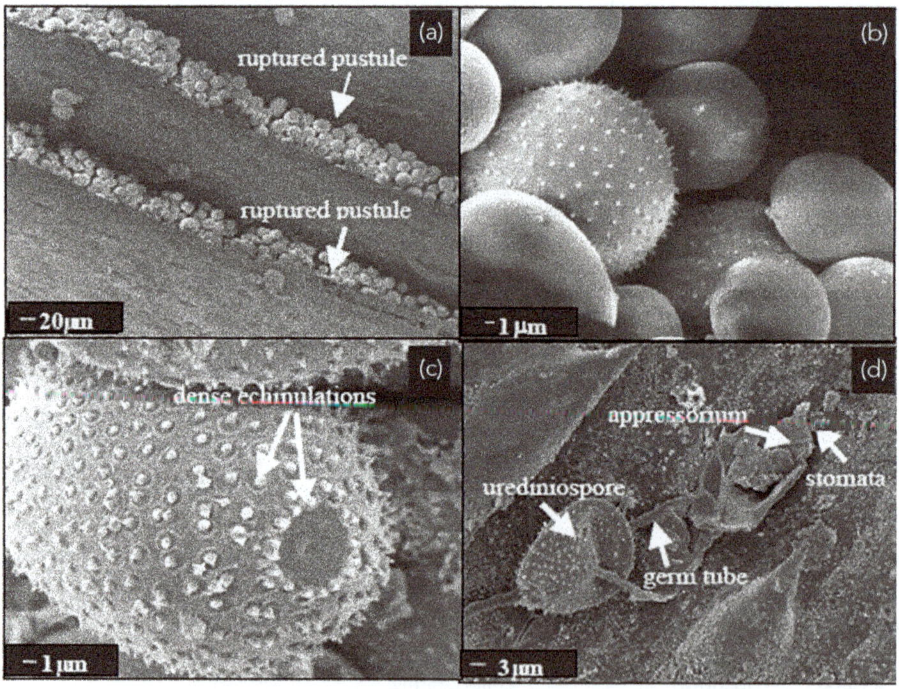

Figure 7 Scanning electron micrographs of (a) ruptured *Puccinia melanocephala* pustules, (b) urediniospores and paraphyses, (c) arrangement of echinulations on the surface of a mature urediniospore and (d) development of germ tubes from a urediniospore. Reproduced from Pillay et al. (2005).

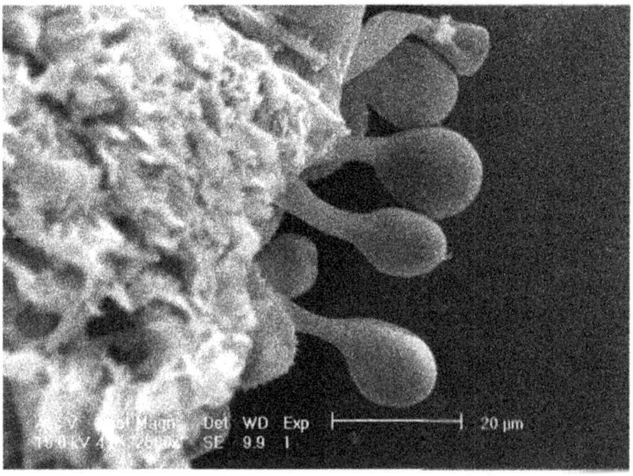

Figure 8 Club-shaped paraphyses contained within uredinia of *Puccinia melanocephala*. Reproduced from Ramouthar (2009).

Table 2 Conditions required for infection by sugarcane rusts

	Temperature (°C)						
Rust	Min	Max	Optimum	Free moisture	Duration (h)	RH (%)	Reference
Brown	5	34	25	Required	6–8	'High'	Purdy et al. (1983)
	5	30	15–30*	Required	≥4 (Opt: ≥6)**		Sotomayor et al. (1983)
	15	30	20–5	Required	≥9 (Opt ≥14)	>98	Ramouthar (2009)
	10	34	18	Required			Hseih and Fang (1983)
Orange	16	25	20			>97	Magarey et al. (2004b)
	18	25					Vasudeva (1958)
	10	34	26				Hseih and Fang (1983)
	10	30	21		Opt 12		Minchio et al. (2011)
Tawny	18	22	19	No, but improved germination in presence of free moisture		100	Martin et al. (2015)

* 15–30°C optimal for appressorium formation but sufficient appressoria developed between 5°C and 10°C to initiate infection.

** urediniospores germinated after 1 h at 25–30°C, 5–30°C after 4 h, but appressorium formation required at least 6 h of free moisture.

more severe in younger plants, with plants between two and six months old being most susceptible (Comstock and Ferreira, 1986; Victoria et al., 1990).

5.4 Yield loss

Yield losses of between 10 and 40% have been recorded in susceptible varieties internationally (Comstock et al, 1992). Following spread into the Caribbean region in 1978, brown rust severely affected the widely grown cultivar B4362 causing yield losses up to 33% (Liu, 1980). By 1982, losses as high as 50% were reported in Mexico (Purdy et al., 1983). In South Africa, losses due to brown rust in the absence of control measures were estimated at 26% of the total yield in the highly susceptible cultivar N29 (McFarlane et al., 2006). Losses result from a combination of a reduction in the number of millable stalks and stalk biomass.

Newly planted sugarcane stands, from 3 to 6 months of age, are more susceptible to brown rust than is ratoon cane (Raid and Comstock, 2006). For plant cane, canopy closure occurs between four and six months and closure can be delayed by severe rust infections. This results in increased weed infestations which incur costs of control and/or result in additional yield losses.

5.5 Resistance

In South Africa, the release of cultivar N29 in 1997 led to a second resurgence of brown rust. N29 was determined during selection to have an intermediate rating for brown rust; however, five years after its commercial release it became evident that N29 was more susceptible than originally determined. This may have been due in part to an increase in the area planted to the cultivar, resulting in a build-up of inoculum levels.

In Florida, brown rust has a history of overcoming disease resistance in cultivars that previously demonstrated resistance (Comstock et al., 2010). In 1979, susceptible cultivars occupied nearly 50% of the state's crop area. By 1986, a newer resistant variety, CP72-1210, occupied 61% of the acreage and ultimately succumbed to the pathogen. Similar shifts from resistance to susceptibility have been reported for several other cultivars, including CP78-1247, which was resistant from 1979 to 1988, (Raid, 1989), CP79-1580 (Dean and Purdy, 1984), CP74-2005 and CL73-239 (Shine et al., 2005). In Louisiana, the cultivar LCP85-384 was so successful that it eventually dominated the production area. In 2000, seven years after release, when LCP85-384 acreage had increased to over 40%, a severe epiphytotic of brown rust occurred in this cultivar that had previously been rated as resistant (Hoy, 2005). Likewise in Argentina resistance in LCP85-384 broke down in 2005, precipitating a phytosanitary crisis (Racedo et al., 2013).

These frequent breakdowns of rust resistance have been linked to the appearance of new pathogenic races (Raid, 1989). Dean and Purdy (1984) reported the occurrence of at least two different races of *Puccinia melanocephala* in Florida, whilst Shine et al. (2005) showed that four pathogenic races were present by the early 1990s. New races develop by overcoming major qualitative resistance genes. In addition, when resistant cultivars transition to become highly susceptible, it suggests that limited underlying non-race-specific quantitative polygenic resistance was present.

Several authors have identified that resistance to brown rust can be quantitative in nature (Hogarth et al., 1993; Hoy et al., 2014; McIntyre et al., 2005; Tai et al., 1981). In addition two large effect resistance genes for brown rust, *Bru1* and *Bru2*, have been reported (Daugrois et al., 1996; Raboin et al., 2006).

The first of these, *Bru1*, was identified on the selfed-progeny used to construct the genetic map of cultivar R570 (Daugrois et al., 1996). The segregation ratio of resistance in the progeny was 3 (resistant):1 (susceptible), indicating that the allele conferring resistance is dominant and that there is one sole copy in the genome of R570. Targeted mapping of *Bru1* identified flanking markers on either side of the gene at 2.2 and 1.9 cM intervals (Asnaghi et al., 2004). This large-effect resistance gene has been shown to be effective against all rust isolates collected from several geographic origins, including the French West Indies and the Mascarene islands (Asnaghi et al., 2001).

Breeding programmes apparently have unknowingly selected for *Bru1* to lesser or greater extents (Barreto et al., 2013; Glynn et al., 2013; Molina et al., 2013; Parco et al., 2014; Racedo et al., 2013; Rutherford and Koch, 2015; Neuber et al., 2017). Two diagnostic molecular markers (R12H16 and 9O20-F4-RsaI) have been developed (Le Cunff et al., 2008; Costet et al., 2012). Recently Wang et al. (2017) have reported an improved *Bru1* diagnostic marker for use with *S. spontaneum*, 9O20-F4-HaeIII, as well as a new dominant *Bru1* diagnostic marker, R12E03-2.

The contribution of *Bru1* towards brown rust resistance within the South African Sugarcane Research Institute's germplasm collection has shown that *Bru1* is present in 172 (56%) of 305 tested genotypes. *Bru1* frequency increased from 46% (in genotypes bred during the period 1975–85) to 68% in those bred during the period 2001–05. This increase is thought to have been driven by increases in disease pressure due to the periodic commercial popularity of susceptible *Bru1*-negative cultivars (N55/805 in the 1970s and; N29 in the early 2000s) (Rutherford and Koch, 2015).

South African *Bru1* prevalence is greater than that found in a similar study in the United States [15% (1975–85), when brown rust first appeared there in 1979, rising to 47% in the current decade (Glynn et al., 2013). Recently, Racedo et al. (2013) reported that 49 of 129 (38%) sugarcane accessions of Argentina were brown rust resistant, but only eight (16%) had *Bru1*, suggesting resistance sources independent of the *Bru1* gene. In China, Li et al. (2018) reported that *Bru1* is the predominant source of genetic resistance and that additional resistance to brown rust is also present.

Since the identification of the *Bru1* resistance gene, breeders have increased the durability of brown rust-resistant cultivars in Florida, with 50% of the clones released between 2011 and 2014 by the CP programme having the *Bru1* gene (Zhao et al., 2015b). *Bru1* appears to be extremely durable, being present in resistant genotypes deployed following the detection of brown rust on cultivar Co301 in South Africa more than 75 years ago. Nevertheless, reliance on one or two resistance genes could prove hazardous. Additional resistance genes independent of *Bru1* and *Bru2* need to be identified and polygenic QR should also be exploited.

Further complicating the breeders task, Zhao et al. (2015b) identified negative relationships between brown and orange rust resistance across families, females and males in the Canal Point (Florida, USA) breeding programme. This suggests that breeding for dual rust resistance will be difficult and that the situation may worsen if and when tawny rust spreads out of Africa.

5.6 Control

Jiang (1985) investigated the efficacy of several fungicides against brown rust and found mancozeb to be the most effective, increasing the yield of variety F176 by 26%. Frequent applications of this non-systemic protectant fungicide were needed and this, coupled with the low net profit obtained after control of the pathogen, limited the use of fungicides in Taiwan.

In the United States, Raid (1992) investigated the use of mancozeb and propiconazole for the control of brown rust on the susceptible variety CP72-1210. Although all fungicide treatments resulted in significant reductions in rust severity, treatments with mancozeb provided better control and higher yields than those with propiconazole only.

Five different fungicides were tested for the control of rust in Mauritius. The best control was achieved with fungicides containing fenpropimorph and mancozeb (Saumtally and Autrey, 1999). In Zimbabwe, trials included the use of triazole fungicides and these were found to significantly reduce brown rust infections (Zvoutete, 2006).

Cost benefit analyses conducted by McFarlane et al. (2006) revealed that it was uneconomical to spray fungicides (mancozeb and propiconazole) due to the severity of rust outbreaks being difficult to predict. Generally, fungicides have been considered to be neither practical nor economical for brown rust control (Ryan and Egan, 1989; Comstock and Raid, 1994). However, this situation could change with the registration of the newer more efficacious fungicides which seem to have collateral plant growth-promoting effects.

6 Orange rust

6.1 History

Orange rust caused by *Puccinia kuenhii* (W. Krüger) E.J. Butler has occurred on sugarcane in the Asia–Oceania region since the 1890s (Ryan and Egan, 1989). This disease was considered minor in Australia up to the end of the 1990s, when it became epiphytotic on the variety Q124, which then constituted around 45% of the Australian sugarcane crop.

It is assumed that this new race subsequently spread around the world. Orange rust was detected for the first time in Florida, Costa Rica, Nicaragua and Guatemala in 2007 (Comstock et al., 2008; Ovalle et al., 2008; Chavarría et al., 2009); Mexico, El Salvador, Panama and Cuba in 2008 (Flores et al., 2009; Pérez-Vicente et al., 2009); Brazil and West Africa in 2009 (Barbasso et al., 2010; Saumtally et al., 2011); Colombia in 2010 (Cadavid et al., 2012); Ecuador in 2011 (Garces et al., 2014) before most recently being reported in Argentina in 2015 (Funes et al., 2016).

6.2 Identification

Observation under a light microscope typically reveals irregularly echinulate urediniospores that are pale in colour with thickened apices and paraphyses that are inconspicuous to absent, as reported by Dixon and Castlebury (2017) (Table 1; Fig. 9).

For molecular confirmation, DNA of urediniospores can be extracted and amplified using Pk1F and Pk1R PCR primers to generate a 527-bp fragment (Glynn et al., 2010). Additionally, the 28S large ribosomal subunit DNA can be sequenced (Aime, 2006), resulting in a PCR fragment and 100% sequence identity with a partial sequence of the

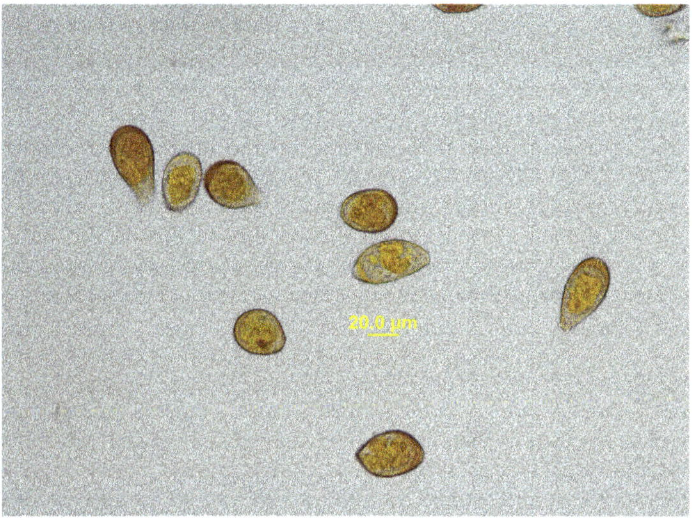

Figure 9 Urediniospores of *Puccinia kuenhii* showing wall thickening in the anterior region. Photo credit: K Braithwaite, Sugar Research Australia.

Puccinia kuehnii 28S ribosomal RNA gene, accession GU058010 (932/932 base pairs, GenBank Accession No. KF202306).

6.3 Conditions favouring the disease

Spore germination experiments suggest that temperatures between 10°C and 34°C allow germination with an optimum between 20°C and 26°C. Germination increases rapidly above 97% RH and free water is not required (Table 2). Hsieh and Fang (1983) found that urediniospores are produced between 10°C and 34°C with an optimum around 26°C.

Climatic conditions are known to exert a significant influence on orange rust (Croft et al., 2000; Staier et al., 2004; Magarey et al., 2004b; Ferrari et al., 2010). For example, Sentelhas et al. (2016) linked orange rust severity data to climatic conditions. Climatic favourability, where air temperature and humidity are conducive for epiphytotic development, is highest in the east of Queensland in Australia, whilst in Brazil, the centre-east region of São Paulo state is most favourable. Such information can help growers to avoid the use of susceptible cultivars in those zones where orange rust poses the greatest risk, and to optimize strategies for chemical control of the disease.

Apan et al. (2004) were able to detect orange rust using the Earth Observing satellite (EO-1) Hyperion instrument for hyperspectral imaging. Several narrow band indices in the near infra-red including a moisture-sensitive band gave the best results. The incorporation of a moisture-sensitive band allowed detection of water loss due to rust rupturing the cuticles of leaves (Croft et al., 2000). There has been consensus that rusts cause an increase in transpiration per unit area of leaf at least in the last stages of disease (Kozlowski, 1978). However, Zhao et al. (2011) found that stomatal conductance and transpiration decreased (along with leaf chlorophyll content and photosynthetic rate) with increasing orange rust sporulation, whilst dark respiration increased.

6.4 Yield loss

The quantitative determination of effects of orange rust on the physiology of individual leaves is the first step towards a broader understanding of crop yield loss (Zhao et al., 2011). Sugarcane yield loss caused by orange rust has been assessed in Australia and the United States. Losses of up to 40% were estimated in individual crops in Australia (Magarey et al., 2001; 2002, 2004a; Staier et al., 2003). In Florida, recoverable sugar yield losses as high as 53% have been recorded on susceptible cultivars (Comstock et al., 2009; Raid et al., 2010), whilst overall industry losses have been estimated at 10% (Rott et al., 2014). Early infections have been shown to reduce stalk population, whilst stalk biomass and sucrose content can be significantly impacted by late infections (Raid et al., 2010; Klosowski et al., 2015).

6.5 Resistance

The emergence of a new race of the pathogen in Australia was suspected to be due to the sudden susceptibility of previously resistant Q124. A previously intermediately resistant cultivar (Q78) was relatively unaffected by orange rust after the outbreak of the disease in Q124, even when grown in proximity. These observations suggest that Q78 possesses quantitative polygenic partial resistance effective against multiple races, whilst Q124 possessed major gene resistance effective against races present before the outbreak. Evidently, QR was largely absent in Q124.

Fortunately, field screening carried out by Magarey and Bull (2009) showed that more than 90% of varieties developed in different breeding programmes from around the world, available for testing in Australia, were rated to be resistant. Likewise in most affected sugarcane industries, there has been more than adequate resistance in commercial cultivars (Atienza and Quimio, 1982; Magarey et al., 2001). Susceptibility in commercial Australian cultivars appears to be rare.

However, in Florida, some sugarcane cultivars only became susceptible 2–3 years after the arrival of orange rust in 2007 (Rott et al., 2014). This shift from resistance to susceptibility could have been due to breakdown of an additional major orange rust resistance gene in the presence of another new race of *Puccinia kuehnii* (Hincapie et al., 2015). Cultivar susceptibility increased from 34% (of acreage) in 2007 to 71% by 2009 (Raid, 2010) and cultivar CP89-2143, which was still resistant in 2009 showed severe symptoms in the 2012 and 2013 crop seasons (Sanjel et al., 2016). Similar to the situation with brown rust, the Florida sugar industry appears to have had low levels of background resistance to orange rust at the time of first disease emergence.

In Brazil, Klosowski et al. (2013) evaluated eight crosses and found that only one showed a pattern of inheritance consistent with the presence of major gene resistance. Their results suggest that resistance to orange rust may be controlled by one major gene and several other minor genes acting quantitatively. Consistent with the results of Sood et al. (2016) in Florida, even in crosses between two highly susceptible genotypes, transgressive segregation was detected as resistant individuals to orange rust were obtained. Progenies of crosses from all resistant and susceptible (RxR, RxS, SxS) parental combinations were composed of higher numbers of resistant clones than expected from a quantitative effect of a major gene plus several minor genes, showing that inheritance of resistance to orange rust is very complex. Therefore, it is important not to eliminate parents susceptible to orange rust that otherwise have high agronomic value because of their ability to produce resistant genotypes in relatively high proportions.

Parental and cultivar resistance to orange rust can be determined by artificial inoculation. Sood et al. (2009) placed 0.5 mL of a *Puccinia kuehnii* spore suspension (10^4 viable spores/mL of water supplemented with 0.002% 1-nonanol) into the leaf whorl of field-grown plants. Four weeks after inoculation, symptoms could be observed in susceptible clones as a band of pustules after they emerged from the whorl. Chapola et al. (2016) compared whorl inoculation with spray inoculation (the latter requiring potted plants, a dew chamber at 22°C for 24 h and glasshouse space) noting that larger severity differences were detected among cultivars with the spray inoculation technique, as measurement of more disease variables was possible. Chapola et al. (2016) concluded that leaf whorl inoculation would be useful at the early stages of breeding, since according to Sood et al. (2009), this method enabled the evaluation of a large number of clones in the field in a short time, with low inoculum use and minimal labour.

Sood et al. (2009) found that *P. kuehnii* spore germination was stimulated by nonanol, whereas that of *Puccinia melanocephala* was not. Spore germination for several rust and smut pathogens of barley, oats, wheat, maize and sorghum have long been known to be stimulated by plant produced nonanol or nonanal and/or 6-methyl-5-hepten-2-one (French et al., 1975). These compounds are known to be present in leaf surface extracts of sugarcane and other Poaceae (Rines et al., 1974; Jiang et al., 2015). Nonanol can overcome self-inhibition by compounds such as 3,4-dimethoxycinnamate where spores in dense suspensions or crowded at a surface (e.g. in pustules) do not germinate (French and Gallimore, 1972; Macko, 1981).

6.6 Control

Given the presence of adequate resistance, the disease has been well controlled with few other measures necessary to maintain high levels of crop production. However, in Australia, a programme was established to screen fungicides for commercial control of the disease until Q124 could be replaced by resistant cultivars. Several fungicides were found to have activity, including tebuconazole and mancozeb, which were registered for commercial use, the first time fungicides had been used for leaf disease control in the Australian sugar industry (Staier et al., 2003).

The Florida sugarcane industry is using fungicides to reduce economic losses from orange rust. At the initiation of the programme, candidate chemistries were selected from three major mode of action classes, the triazoles (FRAC group 3), the carboxamides (FRAC group 7) and the strobilurins (FRAC group 11), alone and in combination and/or alternation. Trial results indicated that the strobilurin fungicides provided the highest level of control, followed by strobilurin/triazole and strobilurin/carboxamide combinations, and finally, the triazole fungicides alone. Spray intervals could be extended up to 28 days, with minimal loss in efficacy. The strobilurin fungicide pyraclostrobin was found to reduce orange rust losses by as much as 40% (Raid and Comstock, 2013). Three to four fungicide applications are needed per crop season to keep orange rust under an economical threshold, whereas only one or two applications are needed to control brown rust, which has a narrower window of susceptibility (Raid et al., 2016).

7 Tawny rust

7.1 History

In 2008, an unknown rust was observed on sugarcane leaves of variety N25 in Swaziland. The uredinia and urediniospores differed from those of *Puccinia kuehnii* and *Puccinia melanocephala* in colour, size and shape (Table 1; Fig. 10). In 2009, this rust was found infecting sugarcane in South Africa. The disease has since spread to most sugarcane-growing regions of South Africa and Swaziland, infecting a wide range of commercial and unreleased South African genotypes, usually in late autumn and spring. Symptoms were also reported in Mozambique and Zimbabwe and were observed for the first time on the grass *Miscanthidium capense* (syn. *Miscanthus ecklonii*) at altitude in the Drakensberg mountains in South Africa in 2016 (Fig. 11) (Martin et al., 2017a).

To date, tawny rust has not been reported outside Southern Africa, but since rust spores are naturally dispersed and spread by wind and atmospheric currents (Purdy et al., 1985), tawny rust is likely to spread further north in Africa and to other continents over time.

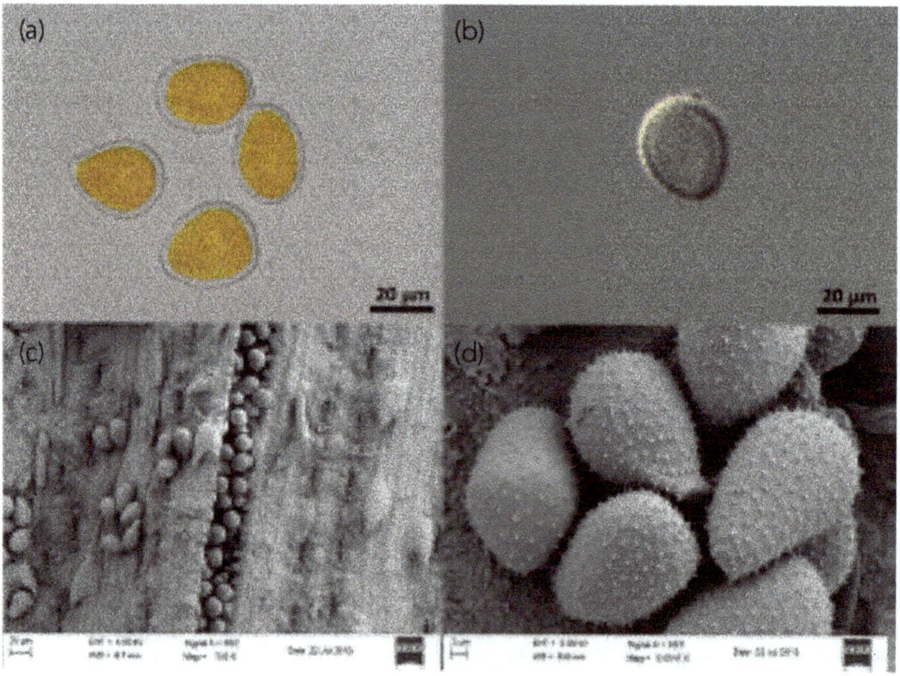

Figure 10 Morphological features of *Macruropyxis fulva*: urediniospores (a, b): urediniospores erumpent from uredinia on the abaxial surface of the sugarcane leaf (SEM) (c): urediniospores (SEM) (d).

Figure 11 Tawny rust (*Macruropyxis fulva*) symptoms on sugarcane (a-d) and *Miscanthidium capense* (e). Uredinia releasing bright orange urediniospores are most common and profuse on the abaxial surface of the sugarcane leaf (b, d) but may also be present on the adaxial surface (c).

7.2 Identification

No identical matches were obtained from the NCBI database for the combined ITS and 28S nuclear large subunit (nLSU) regions for tawny rust first observed in 2008. Phylogenetic analyses based on the same gene region showed that this rust was closely related to *Puccinia sparganioides* Ellis & Barth., which infects several ash species (*Fraxinus* spp.) (Oleaceae) as aecial hosts, and *Macruropyxis fraxini* (Kom.) Azbukina. This group of rusts is phylogenetically distinct from both *Puccinia melanocephala* and *Puccinia kuehnii*. Phylogenetic reconstruction showed that tawny rust falls outside the *Puccinia* I and II clades and groups with the *Macruropyxis* clade (Fig. 12). Hence, the name proposed and accepted for this newly discovered rust species is *Macruropyxis fulva* L.A. Martin, S.A. McFarl. and L.A. Castl., causal agent of tawny rust of sugarcane (Martin et al., 2017a).

A *Macruropyxis fulva*–specific PCR assay has been developed which rapidly distinguishes tawny rust from brown and orange rust (Martin et al., 2017b). This test is particularly useful, given that the symptoms of tawny rust and orange rust can be similar.

7.3 Conditions favouring the disease

Laboratory-based spore germination studies showed that tawny rust spores were able to germinate under both wet and dry conditions at 100% humidity, unlike brown rust spores that require free moisture for germination (Martin et al., 2015) (Table 2). Conditions

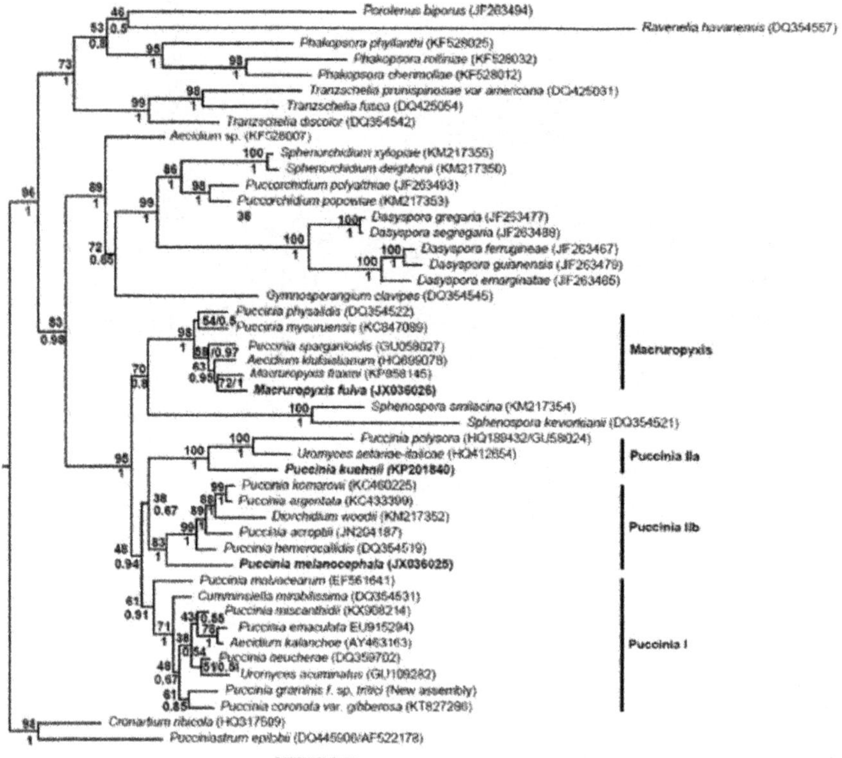

Figure 12 Phylogram indicating the position of *Macruropyxis fulva* relative to related rust fungi within a maximum-likelihood analysis of nuclear rDNA 5.8S, ITS2 and partial LSU sequences. The sequences of *Pucciniastrum epilobii* G. H. Otth and *Cronartium ribicola* J. C. Fisch. were employed as an outgroup. GenBank accession numbers for each rust taxon are given in brackets. Numbers next to nodes show RAxML bootstrap values for 5000 replicates (top) and Bayesian posterior probabilities for two independent runs of 5,000,000 generations (below). Bold vertical bars indicate clades of interest. Taxa representing the three sugarcane rusts are highlighted in bold text. The scale bar (bottom) represents expected numbers of substitutions per site.

for the germination of tawny rust spores were, however, optimal when free moisture was available and at temperatures between 18°C and 22°C (brown rust 20–25°C), with highest germination rates recorded at 19°C. Field observations support these findings, with tawny rust infections tending to begin and end earlier in the spring/summer season than brown rust.

7.4 Resistance

It is thought that *Miscanthidium capense* represents a native original host of *M. fulva* and that since this rust had not previously been described, unlike *Puccinia miscanthidii* Doidge (also on South African *Miscanthidium* – Doidge, 1939), it is likely to have been causing only minor damage to native hosts.

Host specificity normally represents a barrier to rust range expansion, and background levels of plant diversity contain such expansion (Roscher et al., 2007). However, South African *Miscanthidium* is the most closely related genus to *Saccharum* (Evans and Joshi, 2016, 2017) as evidenced by the production of hybrids by Brett (1954). The more related the host and non-host plant species, the greater the probability of successful rust pathogen colonization on the non-host (Bettgenhaeuser et al., 2014). Presumably, elements of basic NHR decrease as host and non-host relatedness increase. In addition, the efficacy of pathogen effectors may also increase with greater relatedness, as the plant molecules targeted by these effectors either become more similar or a greater proportion becomes identical between the host and non-host (Dong et al., 2014). Schulze-Lefert and Panstruga (2011) suggest that changes in pathogen host range, e.g. host range expansions, appear to be driven by variation in pathogen effector repertoires, which subsequently may lead to pathogen speciation. An example of a host range expansion is the spreading of the rice blast disease to wheat in Brazil (Mehta and Baier, 1998).

Also likely is that serial adaptation of the pathogen to the new host has taken place. This phenomenon has been demonstrated artificially in several host–pathogen systems using serial passage experiments. A pathogen is passed for multiple generations through the same host species or genotype specifically testing whether and how rapidly a pathogen is able to adapt to a host that is kept genetically constant (Gilbert and Parker, 2010). Tawny rust was first observed in the commercial variety N25, which was released in 1994 and was still a dominant variety in Swaziland and irrigated regions of South Africa in 2008. The extended longevity of cultivation of commercial sugarcane genotypes in Southern African sugarcane industries could have contributed towards the emergence of tawny rust.

Of the 34 more commonly grown cultivars in the South African sugar industry, 21% are susceptible to tawny rust and are planted over approximately 40 000 ha (16% of the total area under cane). In comparison, 9% of the cultivars grown are susceptible to brown rust, covering approximately 13 000 ha (5% of the industry).

The presence and spread of tawny rust is of major concern to other sugarcane-growing regions worldwide. A parallel might be drawn with *Puccinia psidii* Winter, a native rust of guava (*Psidium guajava*) in South America, which causes only minor damage to the original native host (Coutinho et al., 1998). Introductions of Australian Eucalyptus trees into the Americas provided the pathogen with the opportunity to spread to other members of the Myrtaceae (Glen et al., 2007). In 2010, *Puccinia psidii* was accidentally introduced into Australia where it has become a major plant health concern, affecting 107 new host species (Carnegie and Lidbetter, 2012). Likewise, tawny rust could become a danger to *Saccharum sensu lato* in India and Asia.

8 Conclusion

With their abundant wind dispersed spores, rust fungi are among the most globally mobile plant pathogens. Furthermore, the movement of sugarcane and its influence on the landscape increase the likelihood of evolving new host–rust relationships as has occurred with *Macruropyxis fulva*. The other rust on South African *Miscanthidium*, *Puccinia miscanthidii*, may also pose a risk to sugarcane. Range expansion experiments, as carried out for *Puccinia psidii* on Myrtaceae (Morin et al., 2012), could be applied to *Puccinia miscanthidii* on *Saccharum sensu lato*.

There are many uncertainties when attempting to predict the future for sugarcane–rust interactions particularly in terms of climate change. According to Helfer (2014), even those rusts that are relatively well known do not appear to behave in a generally predictable manner under any of the commonly projected climate change parameters.

The three sugarcane rusts discussed in this chapter differ in terms of optimal conditions for disease progress. However, the following overarching conditions appear to favour rust fungi:

- the availability of high humidity and/or free moisture for several hours favours rust infection,
- host plant uniformity favours rust epiphytotics and
- high nutrient availability and nutrient imbalances favour rust incidence.

Global warming may bring climatic challenges and opportunities for this group of organisms, while increases in carbon dioxide and ozone have been shown to affect host plants and rusts in different ways (Helfer, 2014). To better manage rusts in the future, the implementation of molecular or other surveillance methodologies will be necessary to identify and track new virulent races and to obtain a deeper understanding of rust adaptive capabilities (Ray et al., 2017). Such data could lead to the development of improved models for the prediction of epiphytotics and optimal deployment of durable resistance strategies based on both *R*-genes and QR (Fabre et al., 2015).

Predicting the potential first appearance and subsequent progress of rust diseases is essential for planning optimal fungicide applications. While new modes of action against rusts may appear infrequently (major new fungicides are mainly SDHIs), equally important are new fungicides with enhanced systemicity, curativity and longevity of disease control (Leadbeater, 2015).

A relatively new approach in disease control can consist in the use of compounds able to prime or activate resistance responses in plants. For example, imidacloprid, a neonicotinoid insecticide, has been applied to sugarcane in the furrow at planting for the control of sugarcane thrips (*Fulmekiola serrata* (Kobus) (Thysanoptera: Thripidae)) and yellow sugarcane aphid (*Sipha flava* (Forbes) (Hemiptera: Aphididae)) in South Africa. Observations have indicated that imidacloprid results in reduced brown rust incidence in plant cane and enhanced crop performance under conditions of abiotic stress. Imidacloprid interacts with the salicylate-pathway of plant resistance (Ford et al., 2010) and with poly(ADP-ribose)-polymerase, with the latter affecting energy homeostasis, cell death and stress tolerance (De Block et al., 2005; Thielert, 2006; Vanderauwera et al., 2007). The search is on for compounds of similar structure that are non-insecticidal, but retain the stress and disease-protection effects. It is also possible that synergies could exist between chemical agents and plant breeding for resistance, by selecting crop genotypes more responsive to the chemical priming of resistance responses.

9 References

Agrios, G. N. (2005). *Plant Pathology*, 5th ed. Elsevier, Amsterdam, The Netherlands.

Aime, M. C. (2006). Toward resolving family-level relationships in rust fungi (Uredinales). *Mycosci.* 47: 112–22.

Aime, M. C., Toome, M. and McLaughlin, D. (2014). The Pucciniomycotina, pp. 271–94. In: McLaughlin, D. and Spatafora, J. W. (Eds), *The Mycota VII Part A. Systematics and Evolution*, 2nd ed. Springer-Verlag, Berlin, Germany.

Ajigboye, O. O., Lu, C., Murchie, E. H., Schlatte, C., Swart, G. and Ray, R. V. (2016). Altered gene expression by sedaxane increases PSII efficiency, photosynthesis and growth and improves tolerance to drought in wheat seedlings. *Pestic. Biochem. Physiol.* 137: 49–61.

Ali, S., Gladieux, P., Leconte, M., Gautier, A., Justesen, A., Hovmoller, M. S., Hovmøller, M. S., Enjalbert, J. and de Vallavieille-Pope, C. (2014). Origin, migration routes and worldwide population genetic structure of the wheat yellow rust pathogen *Puccinia striiformis* f.sp. *tritici*. *PLoS Pathol.* 10(1): e1003903. doi:10.1371/journal.ppat.1003903.

Allen, E. A., Hazen, B. E., Hoch, H. C., Kwon, Y., Leinhos, G. M. E., Staples, R. C., Stumpf, M. A. and Terhune, B. T. (1991). Appressorium formation in response to topographical signals by 27 Rust species. *Phytopathol.* 81: 323–31.

Anderson, D. L. and Dean, J. L. (1986). Relationship of rust severity and plant nutrients in sugarcane. *Phytopathol.* 76: 581–5.

Anderson, D. L., Raid, R. N. and Irey, M. (1988). Sugarcane nutrition and rust research: Preliminary results of the 1988 rust research. *Univ. Fla. Sugarcane Grow. Newsl.* 3: 8.

Anderson, D. L., Raid, R. N., Irey, M. and Henderson, L. J. (1990). Association of sugarcane rust severity with soil factors in Florida. *Plant Dis.* 74: 683–6.

Anikster, Y. (1989). Host specificity versus plurivority in barley leaf rusts and their microcyclic relatives. *Mycol. Res.* 93: 175–81.

Apan, A., Held, A., Phinn, S. and Markley, J. (2004). Detecting sugarcane 'orange rust' disease using EO-1 Hyperion hyperspectral imagery. *Int. J. Remote Sens.* 25: 489–98.

Asnaghi, C., D'Hont, A., Glaszmann, J. C. and Rott, P. (2001). Resistance of sugarcane cultivar R570 to *Puccinia melanocephala* isolates from different geographic locations. *Plant Dis.* 85: 282–6.

Asnaghi, C., Roques, D., Ruffel, S., Kaye, C., Hoarau, J. Y., Télismart, H., Girard, J. C., Raboin, L. M., Risterucci, A. M., Grivet, L. and D'Hont, A. (2004). Targeted mapping of sugarcane rust resistance gene (*Bru1*) using bulked segregant analysis and AFLP markers. *Theor. Appl. Genet.* 108: 759–64.

Atienza, C. S. and Quimio, A. J. (1982). Reactions to rust of sugarcane clones and varieties in the Philippines. *Sugarcane Pathol. Newsl.* 29: 3–10.

Bailey, R. A. (1979). Sugarcane rust in South Africa. *Sug. Pathol. Newsl.* 22: 12–13.

Bailey, R. A. (1995). Coping with disease problems: Changing approaches to the challenges presented to the South African sugarcane crops by diseases. *Proc. S. Afr. Sug. Technol. Ass.* 69: 215–16.

Balint-Kurti, P. J. and Johal, G. S. (2009). Maize disease resistance. In: Bennetzen, J. L. and Hake, S. C. (Eds), *Handbook of Maize, Its Biology* (eds). Springer, New York, NY, pp. 229–50.

Barbasso, D., Jordão, H., Maccheroni, W., Boldini, J., Bressiani, J. and Sanguino, A. (2010). First report of *Puccinia kuenhii*, causal agent of orange rust of sugarcane, in Brazil. *Plant Dis.* 94: 1170.

Barrera, W. A. (2010). Effects of environmental variables and crop growth on development of brown rust epidemics in sugarcane. LSU MSc Thesis. 2780. Baton Rouge, LA. http://digitalcommons.lsu.edu/gradschool_theses/2780 (Accessed 20 June 2017).

Barreto, F. Z., Balsalobre, T. W.A., Hoffmann, H. P. and Carneiro, M. S. (2013). Validation of *Bru1* molecular markers linked to resistance to brown rust in a sugarcane germplasm collection. Resumos do 59° Congresso Brasileiro de Genética 112.

Bebber, D. P., Castillo, A. D. and Gurr, S. J. (2016). Modelling coffee leaf rust risk in Colombia with climate reanalysis data. *Phil. Trans. R. Soc. B* 371: 20150458.

Belcher, A. R., Zwonitzer, J. C., Santa Cruz, J., Krakowsky, M. D. , Chung, C. L., Nelson, R., Arellano, C. and Balint-Kurti, P. J. (2012). Analysis of quantitative disease resistance to southern leaf blight and of multiple disease resistance in maize, using near-isogenic lines. *Theor. Appl. Genet.* 124: 433–45.

Bernard, F. A. (1980). Considerations of appearance of sugarcane rust disease in the Dominican Republic. *Proc. Int. Soc. Sugarcane Technol.* 17: 1382–6.

Bettgenhaeuser, J., Gilbert, B., Ayliffe, M. and Moscou, M. J. (2014). Non-host resistance to rust pathogens - a continuation of continua. *Front. Plant Sci.* 5: 664. doi:10.3389/fpls.2014.00664.

Bodnaryk, R. P., Luo, M. and Kudryk, L. (1997). Effects of modifying the phytosterol profile of canola, *Brassica napus* L., on growth, development, and survival of the bertha armyworm, *Mamestra configurata* Walker (Lepidoptera: Noctuidae), the flea beetle, *Phyllotreta cruciferae* (Goeze) (Coleoptera: Chrysomelidae) and the aphids, *Lipaphis erysimi* (Kaltenbach) and *Myzus persicae* (Sulzer) (Homoptera: Aphididae). *Can. J. Plant Sci.* 77: 677–83.

Boller, T. and Felix, G. (2009). A renaissance of elicitors: Perception of microbe-associated molecular patterns and danger signals by pattern-recognition receptors. *Annu. Rev. Plant Biol.* 60: 379–406.

Bosch, F., Paveley, N., Fraaije, B., van den Berg, F. and Oliver, R. (2015). Evidence-based resistance management: A review of existing evidence, pp. 63–76. In: Ishii, H. and Hollomon, D. W. (Eds), *Fungicide Resistance in Plant Pathogens: Principles and a Guide to Practical Management.* Springer, Tokyo, Japan.

Brett, P. G. C. (1954). *Saccharum-Miscanthidium* hybrids. *J. Genet.* 52: 542–6.

Brown, J. (1997). Airborne inoculum. In: Brown, J. F. and Ogle, H. J. (Eds), *Plant Pathogens and Diseases.* Rockvale Publications, Armidale, NSW, Australia, pp. 207–18.

Buck, J. W. and Williams-Woodward, J. L. (2003). The effect of fungicides on urediniospore germination and disease development of daylily rust. *Crop Prot.* 22: 135–40.

Burden, R. S., Clark, T. and Holloway, P. J. (1987). Effects of sterol biosynthesis-inhibiting fungicides and plant growth regulators on the sterol composition of barley plants. *Pestic. Biochem. Physiol.* 27: 289–300.

Butler, E. J. (1914). Notes on some rusts in India. *Ann. Mycol.* 12: 76–82.

Cadavid, M., Angel, J. C. and Victoria, J. I. (2012). First report of orange rust of sugarcane caused by *Puccinia kuenhii* in Colombia. *Plant Dis.* 96: 143.

Cadet, P., McFarlane, S. A. and Meyer, J. H. (2003). Association between nutrients and rust in sugarcane in KwaZulu-Natal. *Proc. S. Afr. Sug. Technol. Ass.* 77: 223–9.

Camargo, de M. S., Amorim, L. and Gomes Junior, A. R. (2013). Silicon fertilisation decreases brown rust incidence in sugarcane. *Crop Prot.* 53: 72–9.

Carmona, E., Vargas, D., Borroto, C. J., Lopez, J., Fernández, A. I., Arencibia, A. and Borrás-Hidalgo, O.(2004). cDNA-AFLP analysis of differential gene expression during the interaction between sugarcane and *Puccinia melanocephala*. *Plant Breed.* 123: 499–501.

Carnegie, A. J. and Lidbetter, J. R. (2012). Rapidly expanding host range for *Puccinia psidii sensu lato* in Australia. *Australas. Plant Pathol.* 41: 13–29.

Carvalho, C. R., Fernandes, R. C., Carvalho, G. M. A., Barreto, R. W. and Evans, H. C. (2011). Cryptosexuality and the genetic diversity paradox in coffee rust, *Hemileia vastatrix*. *PLoS ONE* 6: e26387. doi:10.1371/journal.pone.0026387.

Chakraborty, S., Luck, J., Hollaway, G., Fitzgerald, G. and White, N. (2011). Rust-proofing wheat for a changing climate. *Euphytica* 179: 19–32.

Chapola, R. G., Hoffmann, H. P. and Massola, N. S. (2016). Reaction of sugarcane varieties to orange rust (*Puccinia kuehnii*) and methods for rapid identification of resistant genotypes. *Trop. Plant Pathol.* 41: 139–46.

Chavarría, E., Subirós, F., Vega, J., Ralda, G., Glynn, N. C., Comstock, J. C. and Castlebury, L. A. (2009). First report of orange rust of sugarcane caused by *Puccinia kuenhii* in Costa Rica and Nicaragua. *Plant Dis.* 93: 425.

Chtioui, Y., Francl, L. J. and Panigrahi, S. (1999). Moisture prediction from simple micrometeorological data. *Phytopathol.* 89: 668–72.

Comstock, J. C. and Ferreira, S. A. (1986). Sugarcane rust: Factors affecting infection and symptoms development. *Proc. Int. Soc. Sugar Cane Technol.* 19: 402–10.

Comstock, J. C. and Raid, R. N. (1994). Sugarcane common rust. In: Bhargava, K. S., Rao, G. P., Gillaspie Jr., A. G., Upadhyaya, P. P., Bergamin Filho, A., Agnihotri, V. P. and Chen, C. T. (Eds.) *Current Trends in Sugarcane Pathology*. International Books and Periodicals Supply Service, Delhi, India, pp. 1–10.

Comstock, J. C., Shine Jr., J. M. and Raid, R. N. (1992). Effect of rust on sugarcane growth and biomass. *Plant Dis.* 76: 175–7.

Comstock, J. C., Sood, S. G., Glynn, N. C., Shine, J. M., McKemy, J. M. and Castlebury, L. A. (2008). First report of *Puccinia kuenhii*, causal agent of orange rust of sugarcane, in the United States and Western Hemisphere. *Plant Dis.* 92: 175.

Comstock, J. C., Glynn, N. C. and Davidson, R. W. (2009). Sugarcane brown and orange rusts: Their impact in Florida and the USDA-ARS sugarcane field station response. *J. Amer. Soc. Sugar Cane Technol.* 29: 802.

Comstock, J. C., Glynn, N. C. and Davidson, R. W. (2010). Sugarcane rusts in Florida. *Proc. Int. Soc. Sugar Cane Technol.* 27: 1–9.

Corwin, J. A., Copeland, D., Feusier, J., Subedy, A., Eshbaugh, R., Palmer, C., Maloof, J. and Kliebenstein, D. J. (2016). The quantitative basis of the *Arabidopsis* innate immune system to endemic pathogens depends on pathogen genetics. *PLoS Genet.* 12: e1005789.

Costet, L., Le Cunff, L., Royaert, S., Raboin, L.-M., Hervouet, C., Toubi, L., Telismart, H., Garsmeur, O., Rousselle, Y., Pauquet, J., Nibouche, S., Glaszmann, J.-C., Hoarau, J.-Y. and D'Hont, A. (2012). Haplotype structure around *Bru*1 reveals a narrow genetic basis for brown rust resistance in modern sugarcane cultivars. *Theor. Appl. Genet.* 125: 825–36.

Coutinho, T. A., Wingfield, M. J., Alfenas, A. C. and Crous, P. W. (1998). Eucalyptus rust: A disease with the potential for serious international implications. *Plant Dis.* 82: 819–25.

Croft, B., Magarey, R. and Whittle, P. (2000). Disease management. In: Hogarth, M. and Allsopp, P. (Eds), *Manual of Canegrowing*. Bureau of Sugar Experiment Station, Indooroopilly, pp. 263–89.

Cummins, G. B. (1971). *The Rust Fungi of Cereals, Grasses and Bamboos*. Springer-Verlag, Berlin, Germany.

Cummins, G. B. and Hirastuka, Y. (2003). *Illustrated Genera of Rust Fungi*, 3rd ed. APS Press, St Paul, MN, USA.

Daugrois, J. H., Grivet, L., Roques, D., Hoarau, J. Y., Lombard, H., Glaszmann, J. C. and D'Hont, A. (1996). A putative major gene for rust resistance linked with a RFLP marker in sugarcane cultivar R570. *Theor. Appl. Genet.* 92: 1059–64.

Dean, J. and Purdy, L. (1984). Races of the sugar cane rust fungus, *Puccinia melanocephala*, found in Florida. *Sugar Cane* 1: 15–16.

De Block, M., Verduyn, C., De Brouwer, D. and Cornelissen, M. (2005). Poly(ADP-ribose) polymerase in plants affects energy homeostasis, cell death and stress tolerance. *Plant J.* 41: 95–106.

Dixon, L. and Castlebury, L. 2017. Orange rust of sugarcane - *Puccinia kuehnii*. Systematic mycology and microbiology laboratory, ARS, USDA. Invasive fungi fact sheets, 6 April 2017, Retrieved from: /sbmlweb/fungi/index.cfm.

Dixon, L. J., Castlebury, L. A., Aime, M. C., Glynn, N. C. and Comstock, J. C. (2010). Phylogenetic relationships of sugarcane rust fungi. *Mycol. Prog.* 9: 459–68.

Doidge, E. M. (1939). South African rust fungi III. *Bothalia* 3: 487–512.

Dong, S., Stam, R., Cano, L. M., Song, J., Bozkurt, T. O., Oliva, R., Liu, Z., Tian, M., Win, J., Banfield, M. J., Jones, A. M., van der Hoorn, R. A. and Kamoun, S.(2014). Effector specialization in a lineage of the Irish potato famine pathogen. *Science* 343: 552–5.

Dracatos, P. M., Ayliffe, M., Khatkar, M., Fetch, T., Singh, D. and Park, R. (2014). Inheritance of pre-haustorial resistance to *Puccinia graminis* f. sp. *avenae* in barley (*Hordeum vulgare* L.). *Mol. Plant Microbe Interact.* 27: 1253–62.

Dube, V. P., Charaya, M. U. and Tyagi, S. (1978). The occurrence of fungus *Eudarluca caricis* (Fr.) Eriks. on the rust *Puccinia kuehnii* in India. *Natl. Acad. Sci. Lett.* 1: 387.

Ellison, C. A., Pereira, J. M., Thomas, S. E., Barreto, R. W. and Evans, H. C. (2006). Studies on the rust *Prospodium tuberculatum*, a new classical biological control agent released against the invasive alien weed *Lantana camara* in Australia. 1. Life-cycle and infection parameters. *Australas. Plant Pathol.* 35: 309–19.

Evans, H. C. (1993). Studies on the rust *Maravalia cryptostegiae*, a potential biological control agent of rubber-vine weed *Cryptostegia grandiflora* (Asclepiadaceae: Periplocoideae), in Australia. I: Life cycle. *Mycopath.* 124: 163–74.

Evans, L. D. and Joshi, S. V. (2016). Complete chloroplast genomes of *Saccharum spontaneum, Saccharum officinarum* and *Miscanthus floridulus* (Panicoideae: Andropogoneae) reveal the plastid view on sugarcane origins. *Systemat. Biodiv.* 14(6): 548–71. doi:10.1080/14772000.2016.1197336.

Evans, L. D. and Joshi, S. V. (2017). A new origin of sugarcane: The undiscovered species. *Proc. S. Afr. Sug. Technol. Ass.* (in press).

Fabre, F., Rousseau, E., Mailleret, L. and Moury, B. (2015). Epidemiological and evolutionary management of plant resistance: Optimizing the deployment of cultivar mixtures in time and space in agricultural landscapes. *Evol. Appl.* 8: 919–32.

Ferguson, I. B. and Clarkson, D. T. (1976). Simultaneous uptake and translocation of magnesium and calcium in barley (*Hordeum vulgare* L.) roots. *Planta* 128: 267–9.

Ferrari, J. T., Harakava, R., Domingues, R. J. and Terçariol, I. M. L. (2010). *Ferrugem alaranjada da cana-de-açúcar*. Instituto Biologico, Sao Paulo.

Flores, R. C., Loyo, J. R., Ojeda, R. A. and Rangel, O. C. A. (2009). First report of orange rust of sugarcane caused by *Puccinia kuenhii* in Mexico, El Salvador, and Panama. *Plant Dis.* 93: 1347.

Ford, K. A., Casida, J. E., Chandran, D., Gulevich, A. G., Okrent, R. A., Durkin, K. A., Sarpong, R., Bunnelle, E. M. and Wildermuth, M. C. (2010). Neonicotinoid insecticides induce salicylate-associated plant defense responses. *Proc. Natl. Acad. Sci. USA* 107: 17527–32.

French, R. C. and Gallimore, M. D. (1972). Stimulation of germination of urediniospores of stem rust of wheat in the pustule by n-nonanal and related compounds. *J. Agric. Food Chem.* 20: 421–3.

French, R. C., Gale, A. W., Graham, C. L., Latterell, F. M., Schmitt, C. G., Marchetti, M. A. and Rines, H. W. (1975). Differences in germination response of spores of several species of rust and smut fungi to nonanal, 6-methyl-5-hepten-2-one, and related compounds. *J. Agric. Food Chem.* 23: 766–70.

Fu, D., Uauy, C., Distelfeld, A., Blechi, A., Blechl, A., Epstein, L., Chen, X., Sela, H., Fahima, T. and Dubcovsky, J. (2009). A kinase-START gene confers temperature-dependent resistance to wheat stripe rust. *Science* 323: 1357–60.

Fukuoka, S., Yamanouchi, U., Mizobuchi, R., Yamanouchi, U., Ono, K., Kitazawa, N., Yasuda, N. and Fujita, Y. (2014). Multiple functional polymorphisms in a single disease resistance gene in rice enhance durable resistance to blast. *Sci. Rep.* 4: 4550.

Fukuoka, S., Saka, N., Mizukami, Y., Koga, H., Yamanouchi, U., Yoshioka, Y., Hayashi, N., Ebana, K., Mizobuchi, R. and Yano, M. (2015). Gene pyramiding enhances durable blast disease resistance in rice. *Sci. Rep.* 5: 7773.

Fuller, K. P. and Zajicek, J. M. (1995). Water relations and growth in vinca following chemical growth regulation. *J. Environ. Hort.* 1: 19–21.

Funes, C., Pérez Gómez, S. G., Henriquez, D. D., Di Pauli, V., Bertani, R. P., Fontana, D. P., Rago, A. M., Joya, C. M., Sopena, R. A., González, V., Babi, H., Erazzu, L. E., Cuenya, M. I. and Ploper, L. D. (2016). First report of orange rust of sugarcane caused by *Puccinia kuehnii* in Argentina. *Plant Dis.* 100: 861.

Garcés, F. F., Fiallos, F. F., Silva, E., Martinez, F., Aime, M. C., Comstock, J. C., Glynn, N. C. and Castlebury, L. A. (2014). First report of orange rust of sugarcane caused by *Puccinia kuehnii* in Ecuador. *Plant Dis.* 98: 842.

Garrison, V. H., Shinn, E. A., Foreman, W. T., Griffin, D. W., Holmes, C. W., Kellogg, C. A., Majewski, M. S., Richardson, L. L., Ritchie, K. B. and Smith, G. W. (2003). African and Asian dust: From desert soils to coral reefs. *BioSci.* 53: 469–80.

Garsmeur, O., Charron, C., Bocs, S., Jouffe, V., Samain, S., Couloux, A., Droc, G., Zini, C., Glaszmann, J. C., Van Sluys, M.-A. and D'Hont, A. (2011). High homologous gene conservation despite extreme autopolyploid redundancy in sugarcane. *New Phytol.* 189: 629 642.

Giraud, T., Gladieux, P. and Gavrilets, S. (2010). Linking the emergence of fungal plant diseases with ecological speciation. *Trends Ecol. Evol.* 25: 387–95.

Glaab, J. and Kaiser, W. M. (1999). Increased nitrate reductase activity in leaf tissue after application of the fungicide Kresoxim-methyl. *Planta* 207: 442–8.

Glen, M., Alfenas, A. C., Zauza, E. A., Wingfield, M. J. and Mohammed, C. (2007). *Puccinia psidii*: A threat to the Australian environment and economy – a review. *Australas. Plant Pathol.* 36: 1–16.

Glynn, N. C., Dixon, L. J., Castlebury, L. A., Szabo, L. J. and Comstock, J. C. (2010). PCR assays for the sugarcane rust pathogens *Puccinia kuehnii* and *P. melanocephala* and detection of a SNP associated with geographical distribution in *P. kuehnii*. *Plant Pathol.* 59: 703–11.

Glynn, N. C., Laborde, C., Davidson, R. W., Irey, M. S., Glaz, B., D' Hont, A. and Comstock, J. C. (2013). Utilization of a major brown rust resistance gene in sugarcane breeding. *Mol. Breed.* 31: 323–31.

Godwin, J. R., Anthony, V. M., Clough, J. M. and Godfrey, C. R. A. (1992). ICIA 5504: A novel, broad spectrum, systemic methoxyacrylate fungicide, pp. 435–42. In: *Brighton Crop Prot. Conf.—Pests and Diseases*, Volume 1. Lavenham Press, Lavenham, Suffolk, UK.

Grabert, M. A. (2014). Sugarcane fungicide solutions. http://www.laca1.org/presentations/2014/Sugarcane/Sugarcane%20Fungicide%20Solutions%20-%20Marc%20A%20Grabert,%20BASF.pdf (Accessed 20 July 2017).

Grimmer, M. K., Foulkes, M. J. and Paveley, N. D. (2012). Foliar pathogenesis and plant water relations: A review. *J. Exp. Bot.* 63: 4321–31.

Heath, M. C. (1991). The role of gene-for-gene interactions in the determination of host species specificity. *Phytopathol.* 81: 127–30.

Helfer, S. (2014). Rust fungi and global change. *New Phytol.* 201: 770–80.

Hewitt, H. G. (1998). *Fungicides in Crop Protection.* CAB International, Cambridge, 221p.

Hincapie, M., Lopez Ramos, C., Jameson, A., Raid R. N., Sood, S., Comstock, J. C. and Rott, P. (2015). Detached-leaf assay for assessing variation in pathogenicity of the sugarcane orange rust pathogen (*Puccinia kuehnii*) in Florida. In: *2015 APS Annual Meeting Program*, 1–5 August 2015, Pasadena, CA, abstract, 81p.

Hogarth, D. M., Ryan, C. C. and Taylor, P. W. J. (1993). Quantitative inheritance of rust resistance in sugarcane. *Field Crops Res.* 34: 187–93.

Hoy, J. W. (2005). Impact of rust on LCP 85-384. *Sugar Bull.* 84: 9–13.

Hoy, J. W., Avellaneda, M. C. and Bombecini, J. (2014). Variability in *Puccinia melanocephala* pathogenicity and resistance in sugarcane cultivars. *Plant Dis.* 98: 1728–32.

Hovmøller, M. S., Yahyaoui, A. H., Milus, E. A. and Justesen, A. F. (2008). Rapid global spread of two aggressive strains of a wheat rust fungus. *Mol. Ecol.* 17: 3818–26.

Hovmøller, M. S., Walter, S. and Justesen, A. F. (2010). Escalating threat of wheat rusts. *Science* 329: 369.

Hsieh, W. H. and Fang, J. G. (1983). The uredospore production of *Puccinia melanocephala* and *Puccinia kuehnii* in sugarcanes. *Plant Prot. Bull. Taiwan* 25: 239–44.

Hsieh, W., Lee, C. and Chan, S. (1977). Rust disease of sugarcane in Taiwan: The causal organism *Puccinia melanocephala* Sydow. *Taiwan Sugar* 24: 416–19.

Jiang, D. K. (1985). Chemical control of sugarcane rust (*Puccinia melanocephala* Sydow). *Taiwan Sugar Res. Inst. Annu. Rep.* 108: 25–33.

Jiang, X.-C., Dong, W.-X., Chen, B., Xiao, C., Gui, F.-R., Yan, N.-S., Qian, L. and Li, Z.-Y. (2015). Electrophysiological and oviposition responses of Asian corn borer, *Ostrinia furnacalis* (Lepidoptera: Crambidae), to compounds rinsed from the surfaces of sugarcane and maize leaves. *Eur. J. Entomol.* 112: 295–301.

Johnson, R. M., Grisham, M. P. and Richard Jr., E. P. (2007). Relationship between sugarcane rust severity and soil properties in Louisiana. *Phytopathol.* 97: 748–55.

Jones, J. D. G. and Dangl, J. L. (2006). The plant immune system. *Nature* 444: 323–9.

Kaku, H., Nishizawa, Y., Ishii-Minami, N., Akimoto-Tomiyama, C., Dohmae, N., Takio, K., Minami, E. and Shibuya, N. (2006). Plant cells recognize chitin fragments for defense signaling through a plasma membrane receptor. *Proc. Natl. Acad. Sci. USA* 103: 11086–91.

Kanungo, M. and Joshi, J. (2014). Impact of pyraclostrobin (F-500) on crop plants. *Plant Sci. Today* 1: 174–8.

Kelly, P. L., Reeder, R. and Tafesse, A. (2009). First confirmed report of sugarcane common rust (*Puccinia melanocephala*) in Ethiopia. *Plant Pathol.* 58: 1172.

Keon, J. P. R., White, G. A. and Hargreaves, J. A. (1991). Isolation, characterization and sequence of a gene conferring resistance to the systemic fungicide carboxin from the maize smut pathogen, *Ustilago maydis*. *Curr. Genet.* 19: 475–81.

Kirk, P. M., Cannon, P. F., Minter, D. W. and Stalpers, J. A. (Eds). (2008). *Dictionary of the Fungi*. CABI, Wallingford, UK.

Klosowski, A. C., Bespalhok Filho, J. C., Ruaro, L. and May De Mio L. L. (2013). Inheritance of resistance to orange rust (*Puccinia kuehnii*) in sugarcane families from crosses between parents with different orange rust reactions. *Sugar Tech.* 15: 379–83.

Klosowski, A., Bespalhok, J. C., Ruaro, L., Fragoso, R. B. and De Mio, L. L. M. (2015). Reaction of commercial cultivars and assessment date of Orange rust of sugarcane. *Biosci. J.* 31: 489–98.

Koch, K., Barthlott, W., Koch, S., Hommes, A., Wandelt, K., Mamdouh, W., De-Feyter, S. and Broekmann, P. (2006). Structural analysis of wheat wax (*Triticum aestivum*, c.v. 'Naturastar' L.): From the molecular level to three dimensional crystals. *Planta* 223: 258–70.

Kozlowski, T. T. (1978). *Water Deficits and Plant Growth Vol. V: Water and Plant Disease*. Academic Press, New York, NY, pp. 37–42.

Krattinger, S. G., Lagudah, E. S., Spielmeyer, W., Singh, R. P., Huerta-Espino, J., McFadden, H., Bossolini, E., Selter, L. L. and Keller, B. (2009). A putative ABC transporter confers durable resistance to multiple fungal pathogens in wheat. *Science* 323: 1360–3.

Kuck, K. H., Scheinpflug, H. and Pontzen, R. (1995). DMI fungicides, pp. 205–58. In: Lyr, H. (Ed.), *Modern Selective Fungicides: Properties, Applications, Mechanisms of Action*, 2nd ed. Gustav Fischer Verlag, New York.

Kump, K. L., Bradbury, P. J., Wisser, R. J., Buckler, E. S., Belcher, A. R., Oropeza-Rosas, M. A., Zwonitzer, J. C., Kresovich, S., McMullen, M. D., Ware, D., Balint-Kurti, P. J. and Holland, J. B. (2011). Genome-wide association study of quantitative resistance to southern leaf blight in the maize nested association mapping population. *Nat. Genet.* 43: 163–8.

Lamusse, J. P. (1978). Fifty-third annual review of the milling season in Southern Africa. *Proc. S. Afr. Sug. Technol. Ass.* 53: 1–19.

Leadbeater, A. (2015). Recent developments and challenges in chemical disease control. *Plant Protect. Sci.* 51: 163–69.

Le Cunff, L., Garsmeur, O., Raboin, L. M., Pauquet, J., Telismart, H., Selvi, A., Grivet, L., Philippe, R., Begum, D., Deu, M., Costet, L., Wing, R., Glaszmann, J. C. and D'Hont. A. (2008). Diploid/polyploid syntenic shuttle mapping and haplotype-specific chromosome walking toward a rust resistance gene (Bru1) in highly polyploid sugarcane (2n * 12x * 115). *Genetics* 180: 649–60.

Leme, G. M., Coutinho, I. D., Creste, S., Hojo, O., Carneiro, V. D., Bolzani, S. and Cavalheiro, A. J. (2014). HPLC-DAD method for metabolic fingerprinting of the phenotyping of sugarcane genotypes. *Anal. Methods* 6: 7781–8.

Leonard, K. J. and Szabo, L. J. (2005). Pathogen profile. Stem rust of small grains and grasses caused by *Puccinia graminis*. *Mol. Plant Pathol.* 6: 99–111 .

Leroux, P. (1996). Recent developments in the mode of action of fungicides. *Pestic. Sci.* 47: 191–7.

Li, Z. K., Luo, L. J., Mei, H. W., Paterson, A. H., Zhao, X. H., Zhong, D. B., Wang, Y. P., Yu, X. Q., Zhu, L., Tabien, R., Stansel, J. W. and Ying, C. S. (1999). A 'defeated' rice resistance gene acts as a QTL against a virulent strain of *Xanthomonas oryzae* pv. *oryzae*. *Mol. Genet. Genom.* 261: 58–63.

Li, W. F., Shan, H. L., Zhang, R. Y., Pu, H. C., Wang, X. Y., Cang, X. Y., Yin, J., Luo, Z. M. and Huang, Y. K. (2018). Identification of field resistance and molecular detection of the brown rust resistance gene *Bru1* in new elite sugarcane varieties in China. *Crop Prot.* 103: 46–50.

Liu, L. J. (1980). Observations and considerations on sugarcane rust incidence, varietal reaction and possible occurrence of physiological races. *Sug. Pathol. Newsl.* 25: 5–10.

Macko, V. (1981). Inhibitors and stimulants of spore germination and infection structure formation, pp. 565–83. In: G. Thurian and H. R. Hohl (Eds), *The Fungal Spore: Morphogenetic Controls*. Academic Press, New York.

Magarey, R. C. and Bull, J. I. (2009). Orange rust resistance screening in parent canes at BSES Meringa 2000–2008. *Proc. Aust. Soc. Sugar Cane Technol.* 31: 204–11.

Magarey, R. C., Willcox, T., Croft, B. and Cordingley, A. (2001). Orange rust, a major pathogen affecting crops of Q124 in Queensland in 2000. *Proc. Aust. Soc. Sugar Cane Technol.* 23: 274–80.

Magarey, R. C., Neilsen, W. A. and Bull, J. I. (2004a). The effect of orange rust on sugarcane yield in breeding selection trials in central Queensland: 1999–2001. *Proc. Aust. Soc. Sugar Cane Technol.* 26: 1–7 (CD-ROM).

Magarey, R. C., Neilsen, W. A. and Magnanini, A. J. (2004b). Environmental requirements for spore germination in three sugarcane leaf pathogens. *Proc. Aust. Soc. Sugar Cane Technol.* 26: 1–7 (CD-ROM).

Magarey, R. C., Staier, T. and Willcox, T. G. (2002). Fungicides for control of orange rust in the 2001 Queensland crop. *Proc. Aust. Soc. Sugar Cane Technol.* 24: 1–8 (CD-ROM).

Martin, L. A., Evans, D. L., Rutherford, R. S. and McFarlane, S. A. (2015). Tawny rust: An update on the new species of rust infecting sugarcane in Southern Africa. *Proc. S. Afr. Sug. Technol. Ass.* 88: 309–13.

Martin, L. A., Evans, L. D., Castlebury, L. A., Sifundza, J. T., Comstock, J. C., Rutherford, R. S. and McFarlane, S. A. (2017a). *Macruropyxis fulva* sp. nov., a new rust (Pucciniales) infecting sugarcane in southern Africa. *Australasian Plant Pathol.* 46: 63–74.

Martin, L. A., Rutherford, R. S. and McFarlane, S. A. (2017b). Touchdown PCR assay for the rapid diagnosis of tawny rust caused by *Macruropyxis fulva* on sugarcane. *Australas. Plant Pathol.* 46: 103–5.

McFarlane, K., McFarlane, S. A., Moodley, D. and Rutherford, R. S. (2006). Fungicide trials to determine the effect of brown rust on the yield of sugarcane variety N29. *Proc. S. Afr. Sug. Technol. Ass.* 80: 297–300.

McIntyre, C. L., Casu, R. E., Drenth, J., Knight, D., Whan, V. A., Croft, B. J., Jordan, D. R. and Manners, J. M. (2005). Resistance gene analogues in sorghum and sugarcane and their association with quantitative trait loci for rust resistance. *Genome* 48: 391–400.

Mehta, Y. R. and Baier, A. (1998). Pathogenic variation among *Magnaporthe grisea* isolates attacking triticale and wheat in the state of Paraná. *Summa Phytopathol.* 24: 119–25.

Mills, P., Dehnen-Schmutz, K., Ilbery, B., Jeger, M., Jones, G., Little, R., MacLeod, A., Parker, S., Pautasso, M., Pietravalle, S. and Maye, D. (2011). Integrating natural and social science perspectives on plant disease risk, management and policy formulation. *Phil. Trans. R. Soc. B* 366: 2035–44.

Mims, S. A. and Mims III, F. M. (2003). Fungal spores are transported long distances in smoke from biomass fires. *Atmos. Environ.* 38: 651–5.

Minchio, C. A., Canteri, M. G. and Rocha, J. A. (2011). Germination of uredospores of *Puccinia kuehnii* subjected to different temperatures and incubation times. *Summa Phytopathol.* 37: 211–14.

Molina, L., Queme, J. L. and Rosales, F. (2013). Comparative analysis between phenotype and *Bru1* marker for incidence to brown rust in sugarcane. *Proc. Int. Soc. Sugar Cane Technol.* 28: 738–43.

Moricca, S., Ragazzi, A., Mitchelson, K. R. and Assante, G. (2001). Antagonism of the two- needle pine stem rust fungi *Cronartium flaccidum* and *Peridermium pini* by *Cladosporium tenuissimum* In Vitro and In Planta. *Phytopathol.* 91: 457–68.

Morin, L., Aveyard, R., Lidbetter, J. R. and Wilson, P. G. (2012). Investigating the host-range of the rust fungus *Puccinia psidii* sensu lato across tribes of the family Myrtaceae present in Australia. *PLoS ONE* 7(4): e35434. (doi.org/10.1371/journal.pone.0035434).

Mueller, D. S., Jeffers, S. N. and Buck, J. W. (2004). Effect of timing of fungicide applications on development of rusts on daylily, geranium, and sunflower. *Plant Dis.* 88: 657–61.

Mueller, D. S., Jeffers, S. N. and Buck, J. W. (2005). Toxicity of fungicides to urediniospores of six rust fungi that occur on ornamental crops. *Plant Dis.* 89: 255–61.

Mur, L. A. J., Kenton, P., Lloyd, A. J., Ougham, H. and Prats, E. (2008). The hypersensitive response: The centenary is upon us but how much do we know? *J. Exp. Bot.* 59: 501–20.

Nascimento, J. F., Vida, J. B., Tessmann, D. J., Zambolim, L., Vieira, R. A. and Oliveira, R. R. (2012). Progress of Asian soybean rust and airborne urediniospores of *Phakopsora pachyrhizi* in southern Brazil. *Summa Phytopathol.* 38: 280–7.

Neuber, A. C., Camilo dos Santos, F. R., Costa, J. B., Volpin, M., Xavier, M. A., Perecin, D., Villavicencio Burbano, R. C., de Andrade Landell, M. G., Pinto, L. R. and McIntosh, R. A. (2017). Survey of the *Bru1* gene for brown rust resistance in Brazilian local and basic sugarcane germplasm. *Plant Breed.* 136: 182–8.

Niks, R. E., Qi, X. Q. and Marcel, T. C. (2015). Quantitative resistance to biotrophic filamentous plant pathogens: Concepts, misconceptions, and mechanisms. *Annu. Rev. Phytopathol.* 53: 445–70.

Nirmala, J., Drader, T., Lawrence, P. K., Yin, C., Hulbert, S., Steber, C. M., Steffenson, B. J., Szabo, L. J., von Wettstein, D. and Kleinhofs, A. (2011). Concerted action of two avirulent spore effectors activates reaction to *Puccinia graminis* 1 (*Rpg1*)-mediated cereal stem rust resistance. *Proc. Natl. Acad. Sci. USA* 108: 14676–81.

Noctor, G. and Mhamdi, A. (2017). Climate change, CO_2, and defense: The metabolic, redox, and signaling perspectives. *Trends Plant Sci.* 22: 857–70.

Nurnberger, T. and Lipka, V. (2005). Non-host resistance in plants: New insights into an old phenomenon. *Mol. Plant Pathol.* 6: 335–45.

Nurnberger, T., Brunner, F., Kemmerling, B. and Piater, L. (2004). Innate immunity in plants and animals: Striking similarities and obvious differences. *Immun. Rev.* 198: 249–66.

Nuss, K. J. and Brett, P. G. C. (1995). The release of variety NCo310 in 1945 and its impact on the sugar industry. *Proc. S. Afr. Sug. Technol. Ass.* 69: 3–8.

Oloriz, M. I., Gil, V., Rojas, L., Portal, O., Izquierdo, Y., Jiménez, E. and Höfte, M. (2012). Sugarcane genes differentially expressed in response to *Puccinia melanocephala* infection: Identification and transcript profiling. *Plant Cell Rep.* 31: 955–69.

Ouimet, A., Carisse, O. and Neumann, P. (1997). Environmental and nutritional factors affecting the in vitro inhibition of the vegetative growth of *Venturia inaequalis* by five antagonistic fungi. *Can. J. Bot.* 75: 632–9.

Ovalle, W., Comstock, J. C., Glyn, N. C. and Castlebury, L. A. (2008). First report of *Puccinia kuenhii*, causal agent of orange rust of sugarcane, in Guatemala. *Plant Dis.* 92: 973.

Pachauri, R. K. and Reisinger, A. (2007). *Climate Change 2007: Synthesis Report. Contribution of Working Groups I, II and III to the Fourth Assessment Report of the Intergovernmental Panel on Climate Change*. IPCC, Geneva, Switzerland.

Pangga, I. B., Hanan, J. and Chakraborty, S. (2011). Pathogen dynamics in a crop canopy and their evolution under changing climate. *Plant Pathol.* 60: 70–81.

Papaix, J., Burdon, J. J., Zhan, J. and Thrall, P. H. (2015). Crop pathogen emergence and evolution in agro-ecological landscapes. *Evol. Appl.* 8: 385–402.

Parco, A. S., Avellaneda, M. C., Hale, A. H., Hoy, J. W., Kimbeng, C. A., Pontif, M. J., Gravois, K. A. and Baisakh, N. (2014). Frequency and distribution of the brown rust resistance gene *Bru1* and implications for the Louisiana sugarcane breeding programme. *Plant Breed.* 133: 654–9.

Park, R. F. and Wellings, C. R. (2012). Somatic hybridization in the Uredinales. *Annu. Rev. Phytopathol.* 50: 219–39.

Parlevliet, J. E. (2002). Durability of resistance against fungal, bacterial and viral pathogens: Present situation. *Euphytica* 124: 147–56.

Pérez-Vicente, L., Martín-Triana, E. L., Barroso, F., Martínez-de la Parte, E., Borrás-Hidalgo, O. and Hernández-Estévez, I. (2009). Definitive identification of orange rust of sugarcane caused by *Puccinia kuenhii* in Cuba. *New Disease Rep.* 20: 16.

Pillay, L., McFarlane, S. A. and Rutherford, R. S. (2005). A preliminary report on genetic diversity in populations of sugarcane rust in KwaZulu-Natal. *Proc. S. Afr. Sug. Technol. Ass.* 79: 132–6.

Poland, J. A., Balint-Kurti, P. J., Wisser, R. J., Pratt, R. C. and Nelson, R. J. (2009). Shades of gray: The world of quantitative disease resistance. *Trends Plant Sci.* 14: 21–9.

Pontzen, R. and Scheinpflug, H. (1989). Effects of triazole fungicides on sterol biosynthesis during spore germination of *Botrytis cinerea, Venturia inaequalis*, and *Puccinia graminis* f. sp *tritici*. *Neth. J. Plant Pathol.* 95(Suppl. 1): 151–60.

Purdy, L., Liu, L. J. and Dean, J. (1983). Sugarcane rust, a newly important disease. *Plant Dis.* 67: 1292–6.

Purdy, L. H., Krupa, S. V. and Dean, J. L. (1985). Introduction of sugarcane rust into the Americas and its spread to Florida. *Plant Dis.* 69: 689–93.

Quenouille, J., Paulhiac, E., Moury, B. and Palloix, A. (2014). Quantitative trait loci from the host genetic background modulate the durability of a resistance gene: A rational basis for sustainable resistance breeding in plants. *Heredity* 112: 579–87.

Rabert, G. A., Manivannan, P., Somasundaram, R. and Panneerselvam, R. (2014). Triazole compounds alter the antioxidant and osmoprotectant status in drought stressed *Helianthus annuus* L. plants. *Emir. J. Food Agric.* 26: 265–76.

Raboin, L., Oliveira, K., Le Cunff, L., Telismart, H., Roques, D., Butterfield, M., Hoarau, J. Y. and D'Hont, A. (2006). Genetic mapping in sugarcane, a high polyploidy, using bi-parental progeny: Identification of a gene controlling stalk colour and a new rust resistance gene. *Theor. Appl. Genet.* 112: 1382–91.

Racedo, J., Perera, M. F., Bertani, R., Funes, C., Gonzalez, V., Cuenya, M. I., D'Hont, A., Welin, B. and Castagnaro, A. P. (2013). *Bru1* gene and potential alternative sources of resistance to sugarcane brown rust disease. *Euphytica* 191: 429–36.

Raid, R. N. (1989). Physiological specialization in sugarcane rust (*Puccinia melanocephala*) in Florida. *Plant Dis.* 73: 183.

Raid, R. N. (1992). Fungicidal control of *puccinia melanocephala* on sugar cane. *Sugar Y. Azucar.* 87: 36.

Raid, R. N. (2010). Orange Rust of Sugarcane: Its Importance and Control. Presentation given at the 46th Annual Meeting of the Caribbean Food Crops Society, Dominican Republic. www.cedaf.org.do/eventos/cfcs_2010/presentaciones/03_martes/tstar10p.pdf (Accessed 05 April 2017).

Raid, R. N. and Comstock, J. C. (2000). Common rust. In: Rott, P., Bailey, R. A., Comstock, J. C., Croft B. J. and Saumtally, A. S. (Eds), *A Guide to Sugarcane Diseases*. CIRAD and ISSCT, Montpellier, France, pp. 85–9.

Raid, R. N. and Comstock, J. C. (2006). Sugarcane Rust Disease. EDIS Publication SS-AGR-207. UF/IFAS Extension, Gainesville, FL 32611. http://edis.ifas.ufl.edu/pi207 (Accessed 20 July 2017).

Raid, R. N. and Comstock, J. C. (2013). Evaluation of fungicides for control of orange rust on sugarcane. *Proc. Int. Soc. Sugar Cane Technol.* 28: 1–9.

Raid, R. N., Comstock, J. C. and Glynn, N. C. (2010). A comparison of orange rust epidemics in Florida during 2008 and 2009 and their influence on yield. *Sugar J.* 6: 28.

Raid, R. N., Chaulagain, B., Sanjel, S., Comstock, J. C., Hincapie, M. and Rott, P. (2016). Fungicides as management tools for sugarcane orange rust. In: *Abstracts of the American Phytopathological Society Rust Symposium*, 7–9 March 2016, Pensacola Beach, FL, p. 16.

Ramakrishna, W., Emberton, J., Ogden, M., SanMiguel, P. and Bennetzen, J. L. (2002). Structural analysis of the maize Rp1 complex reveals numerous sites and unexpected mechanisms of local rearrangement. *Plant Cell* 14: 3213–23.

Ramouthar, P. V. (2009). Studies on brown rust (*Puccinia melanocephala*) of sugarcane in South Africa. MSc thesis, University of KwaZulu-Natal, Pietermaritzburg, South Africa. Available from: http://researchspace.ukzn.ac.za/xmlui/bitstream/handle/10413/2625/Ramouthar_Prabashnie_Vengetsamy_2009.pdf?sequence=1 (Accessed 09 April 2017).

Ramouthar, P. V., Caldwell, M. M. and McFarlane, S. A. (2016). Effect of silicon on the severity of brown rust of sugarcane in South Africa. *Eur. J. Plant Pathol.* 145: 53–60.

Ray, M., Ray, A., Dash, S., Mishra, A., Achary, K. G., Nayak, S. and Singh, S. (2017). Fungal disease detection in plants: Traditional assays, novel diagnostic techniques and biosensors. *Biosens. Bioelectron.* 87: 708-23.

Reisige, K., Gorzelanny, C., Daniels, U. and Moerschbacher, B. M. (2006). The C28 aldehyde octacosanal is a morphogenetically active component involved in host plant recognition and infection structure differentiation in the wheat stem rust fungus. *Physiol. Mol. Plant* M68, 33–40.

Rines, H. W., French, R. C. and Daasch, L. W. (1974). Nonanal and 6-methyl-5-hepten-2-one: Endogenous germination stimulators of uredospores of *Puccinia graminis* var. *tritici* and other rusts. *J. Agric. Food Chem.* 22: 96–100.

Ritschel, A. (2005). Monograph of the genus *Hemileia* (Uredinales). *Bibl. Mycol.* 200: 1–132.

Roelfs, A. P. (1982). Effects of barberry eradication on stem rust in the United States. *Plant Dis.* 66: 177–81.

Roscher, C., Schumacher, J., Foitzik, O. and Schulze, E. D. (2007). Resistance to rust fungi in *Lolium perenne* depends on within-species variation and performance of the host species in grasslands of different plant diversity. *Oecologia* 153: 173–83.

Rott, P., Sood, S., Comstock, J. C., Raid, R. N., Gilbert, R. A. and Sandhu, H. S. (2014). Sugarcane orange rust. Publication #SS-AGR-378. EDIS website at http://edis.ifas.ufl.edu/sc099 (Accessed 20 July 2017).

Rouse, M. N., Stoxen, S., Chen, X. M., Szabo, L. J. and Jin, Y. (2009). Diverse stem rust races found in a single field in Washington, USA (Abstr.). *Phytopathol.* 99: S111.

Roux, F., Voisin, D., Badet, T., Balagué, C., Glynn, N. C., Barlet, X., Huard-Chauveau, C., Roby, D. and Raffaele, S. (2014). Resistance to phytopathogens e tutti quanti: Placing plant quantitative disease resistance on the map. *Mol. Plant Pathol.* 15: 427–32.

Rowe, H. C. and Kliebenstein, D. J. (2008). Complex genetics control natural variation in *Arabidopsis thaliana* resistance to *Botrytis cinerea*. *Genetics* 180: 2237–50.

Rutherford, R. S. and van Staden, J. (1996). Towards a rapid near infra-red technique for the prediction of resistance to the sugarcane borer *Eldana saccharina* Walker (Lepidoptera: Pyralidae) using stalk surface wax. *J. Chem. Ecol.* 22: 681–93.

Rutherford, R. S. and Koch, A. C. (2015). Frequency of the *Bru1* gene and potential alternative sources of resistance to sugarcane brown rust. Abstracts of the 49th Congress of the Southern African Society for Plant Pathol. Poster 37.

Ryan, C. C. and Wilson, J. A. (1981). A possible hyperparasite of sugarcane rust: *Cladosporium uredinicola* Speg. *Sug. Pathol. Newsl.* 27: 31–2.

Ryan, C. C. and Egan, B. T. (1989). Rust, pp. 189–210. In: Ricaud, C., Egan, B. T. and Gillaspie Jr., A. G. and Hughes, C. G. (Eds), *Diseases of Sugarcane*. Elsevier, Amsterdam, the Netherlands.

Sanjel, S., Hincapie, M., Sood, S., Comstock, J. C., Raid, R. N. and Rott, P. (2016). Evidence for variation in virulence of the sugarcane orange rust pathogen in Florida. 46th American Society of Sugar Cane Technologists annual meeting, St Pete Beach, FL. *Sugar J.* 79: 29–30.

Sathe, A. V. (1971). Nomenclatural revision of the common rust fungus affecting sugarcane. *Curr. Sci.* 40: 42–3.

Saumtally, S. and Autrey, L. J. C. (1999). Common rust of sugarcane: Progress and prospects. In: Rao G. P., Bergamin Fihlo A., Magarey R. C. and Autrey L. J. C. (Eds), *Sugarcane Pathology Volume 1: Fungal Diseases*. Science Publishers Inc., USA.

Saumtally, A. S., Viremouneix, T. R., Ahondokpê, B., Girard, J.-C. R., Castlebury, L. A., Dixon, L., Glynn, N. C. and Comstock, J. C. (2011). First report of orange rust of sugarcane caused by *Puccinia kuehnii* in Ivory Coast and Cameroon. *Plant Dis.* 95: 357.

Schulze-Lefert, P. and Panstruga, R. (2011). A molecular evolutionary concept connecting nonhost resistance, pathogen host range, and pathogen speciation. *Trends Plant Sci.* 16: 117–25.

Sentelhas, P. C., dos Santos, D. L., Monteiro, L. A., Soares-Colletti, A. R., Pallone Filho, W. J., Donzelli, J. L. and Arrigoni, E. B. (2016). Agro-climatic favorability zones for sugarcane orange rust as a tool for cultivar choice and disease management. *Crop Prot.* 84: 88–97.

Seymour, V. (1985). Physical, chemical and spectral properties of leaf surfaces of *Berberis aquifolium* (Pursh.). *Plant Cell Environ.* 8: 631–8.

Shelton, J. (2012). Epicuticular wax chemistry, morphology, and physiology in sand bluestem, *Andropogon gerardii* ssp. *hallii*, and big bluestem, *Andropogon gerardii* ssp. *gerardii*. M.S. Thesis, Division of Biology, Kansas State University, Manhattan, KS.

Shine Jr., J. M., Comstock, J. C. and Dean, J. L. (2005). Comparison of five isolates of sugarcane brown rust and differential reaction of six sugarcane clones. *Sugar Cane Int.* 23: 24–9.

Sierotzki, H. and Scalliet, G. (2013). A review of current knowledge of resistance aspects for the next-generation succinate dehydrogenase inhibitor fungicides. *Phytopathol.* 103: 880–7.

Singh, R. P, Hodson, D. P., Huerta-Espino, J., Jin, Y., Bhavani, S., Njau, P., Herrera-Foessel, S., Singh, P. K., Singh, S. and Govindan, V. (2011). The emergence of Ug99 races of the stem rust fungus is a threat to world wheat production. *Annu. Rev. Phytopathol.* 49: 465–81.

Sood, S. G., Comstock, J. C. and Glynn, N. C. (2009). Leaf whorl inoculation method for screening sugarcane rust resistance. *Plant Dis.* 93: 1335–40.

Sood, S., Glynn, N. C., Yang, X., Wang, J., McCorkle, K., Rott, P. C. and Comstock, J. C. (2016). Inheritance of resistance to orange rust in sugarcane. *Proc. Int. Soc. Sugar Cane Technol.* 29: 1614–19.

Sotomayor, I. A., Purdy, L. H. and Trese, A. T. (1983). Infection of sugarcane leaves by *Puccinia melanocephala. Phytopathol.* 73: 695–9.

Speth, E. B., Lee, Y. N. and He, S. Y. (2007). Pathogen virulence factors as molecular probes of basic plant cellular functions. *Curr. Opin. Plant Biol.* 10: 580–6.

St Clair, D. A. (2010). Quantitative disease resistance and quantitative resistance loci in breeding. *Annu. Rev. Phytopathol.* 48: 247–68.

Staier, T., Magarey, R. and Willcox, T. G. (2003). Control of orange rust in sugarcane with fungicides. *Proc. Aust. Soc. Sugar Cane Technol.* 25: 1–14 (CD-ROM).

Staier, T. N., Magarey, R. C. and Finlayson, W. A. (2004). Meteorological data collecting, analysis and sugarcane disease forecasting for orange rust. *Proc. Aust. Soc. Sugar Cane Technol.* 26: 1–6 (CD-ROM).

Statler, G. D. and Nordgaard, J. T. (1980). Leaf wettability of wheat in relation to infection by *Puccinia recondita* f.sp. *tritici. Phytopathol.* 70: 641–3.

Stukenbrock, E. H. and McDonald, B. A. (2008). The origins of plant pathogens in agroecosystems. *Annu. Rev. Phytopathol.* 46: 75–100.

Sullivan, R. F. and White Jr., J. F. (2000). Phoma glomerata as a mycoparasite of powdery mildew. *Appl. Environ. Microbiol.* 66: 425–7.

Tai, P. Y. P., Miller, J. D. and Dean, J. L. (1981). Inheritance of resistance to rust in sugarcane. *Field Crops Res.* 4: 261–8.

Thielert, W. (2006). A Unique Product: The story of the imidacloprid stress shield. *Pflanzenschutz-Nachrichten Bayer* 59: 73–86. Bayer AG, Agriculture Centre, Monheim Germany. (ISSN 0340-1723).

Tulloch, A. P., Baum, B. R. and Hoffman L. L. (1980). A survey of epicuticular waxes among genera of Triticeae. 2. Chemistry. *Can. J. Bot.* 58: 2602–15.

Van den Berg, F., Lannou, C., Gilligan, C. A. and van de Bosch, F. (2014). Quantitative resistance can lead to evolutionary changes in traits not targeted by the resistance QTLs. *Evol. Appl.* 7: 370–80.

Vanderauwera, S., De Block, M., Van de Steene, N., van de Cotte, B., Metzlaff, M. and Van Breusegem, F. (2007). Silencing of poly(ADP-ribose) polymerase in plants alters abiotic stress signal transduction. *Proc. Natl. Acad. Sci. USA* 104: 15150–55.

Vasudeva, R. S. (1958). Report of the Division of Mycology and Plant Pathol. Sci. Rept. Agr. Res. Inst. New Delhi, 1956–57, pp. 86–100. From Purdy LH (1985). Sugarcane Rusts. In: Roelfs, A. P. and Bushnell, W. R. (Eds), *The Cereal Rusts Volume II. Diseases, Distribution, Epidemiology, and Control.* Academic Press, London, UK, pp. 237–54.

Venancio, W. S., Rodrigues, M. A. T., Begliomini, E. and Souza, N. L. D. (2003). Physiological effects of strobilurin fungicides on plants. *Cienc. Exatas Terra Cienc. Agr. Eng.* 9: 59–68.

Vetter, M. M., Kronholm, I., He, F., Haweker, H., Reymond, M., Bergelson, J., Robatzek, S. and de Meaux, J. (2012). Flagellin perception varies quantitatively in Arabidopsis thaliana and its relatives. *Mol. Biol. Evol.* 29: 1655–67.

Victoria, J., Moreno, C. and Cassalett, C. (1990). Genotype-environment interaction and its effect on sugar cane rust incidence. *Sugar Cane* 4: 13–17.

Wallace, J. G., Larsson, S. J. and Buckler, E. S. (2014). Entering the second century of maize quantitative genetics. *Heredity* 112: 30–8.

Wang, H., Yang, X. B. and Ma, Z. (2010). Long-distance spore transport of wheat stripe rust pathogen from Sichuan, Yunnan, and Guizhou in Southwestern China. *Plant Dis.* 94: 873–80.

Wang, H., Shi, H. and Wang, Y. (2015). The wetting of leaf surfaces and its ecological significances. Chapter 11. In: Aliofkhazraei, M. (ed.), *Wetting and Wettability*, pp. 295–321 (Accessed 03 June 2017 from: http://www.intechopen.com/books/wetting-and-wettability). ISBN: 978-953-51-2215-9.

Wang, H. B., Chen, P. H., Yang, Y. Q., D'Hont, A. and Lu, Y. H. (2017). Molecular insights into the origin of the brown rust resistance gene *Bru1* among *Saccharum* species. *Theor. Appl. Genet.* 130: 2431–43. (doi:10.1007/s00122-017-2968-3.

Webb, D. H. and Nutter Jr., F. W. (1997). Effects of leaf wetness duration and temperature on infection efficiency, latent period, and rate of pustule appearance of rust in alfalfa. *Phytopathol.* 87: 946–50.

Wellings, C. R. (2007). *Puccinia striiformis* in Australia: A review of the incursion, evolution and adaptation of stripe rust in the period 1979–2006. *Aust. J. Agric. Res.* 58: 567–75.

Wong, F. P. and Wilcox, W. F. (2001). Comparative physical modes of action of azoxystrobilurin, mancozeb, and metalaxyl against Plasmopara viticola (grapevine downy mildew). *Plant Dis.* 85: 649–56.

Ypema, H. L. and Gold, R. E. (1999). Kresoxim-methyl: Modification of a naturally occurring compound to produce a new fungicide. *Plant Dis.* 83: 4–19.

Zhao, D. and Li, Y-R. (2015). Climate change and sugarcane production: Potential impact and mitigation strategies. *Int. J. Agron.*, 2015, 10 .

Zhao, D., Glynn, N. C., Glaz, B., Comstock, J. C. and Sood, S. (2011). Orange rust effects on leaf photosynthesis and related characters of sugarcane. *Plant Dis.* 95: 640–7.

Zhao, D., Davidson, W., Baltzar, M. and Comstock, J. C. (2015a). Field evaluation of sugarcane orange rust for first clonal stage of the CP cultivar development program. *Am. J. Agric. Biol. Sci.* 10: 1–11.

Zhao, D., Davidson, W., Baltzar, M., Comstock, J. C., McCord, P. H. and Sood, S. G. (2015b). Screening for sugarcane brown rust in first clonal stage of the Canal Point sugarcane breeding program. *Agronomy* 5: 341–62.

Zipfel, C. (2009). Early molecular events in PAMP-triggered immunity. *Curr. Opin. Plant Biol.* 12: 414–20.

Zvoutete, P. (2006). Fungicide sprays to control brown rust (*Puccinia melanocephala*) gave variable cane and sugar yield responses in the south-east lowveld of Zimbabwe. *Proc. S. Afr. Sug. Technol. Ass.* 80: 301–6.

Chapter 12

Progress in understanding viruses affecting sugarcane

Kathy Braithwaite, Sugar Research Australia, Australia

1 Introduction

The late twentieth century saw a proliferation of new virus or virus-like discoveries. By 1996, ten previously unidentified sugarcane pathogens or diseases of significance to sugarcane quarantine were reported, with eight of those being known or suspected viruses (Croft, 1996). By 2000, seven viral diseases of sugarcane were officially recognized: Fiji disease (now Fiji leaf gall), mild mosaic, mosaic (caused by two viruses), red leaf mottle, streak, striate mosaic and the leaf fleck caused by bacilliform virus (Rott et al., 2000). *Sugarcane streak mosaic virus* (SCSMV) was only briefly alluded to, and Ramu stunt and sugarcane yellow leaf were not listed as viral diseases. Braithwaite (2001) discussed four new viruses in detail: *peanut clump virus*, *sugarcane bacilliform virus* (SCBV), *sugarcane mild mosaic virus* (SCMMV) and *sugarcane yellow leaf virus* (SCYLV), as they were considered to be the major emerging threats at the time. It is interesting to look at how sugarcane virology has progressed in the intervening years. While the pace of new discoveries has slowed, the range of molecular tools now available to researchers has expanded. This chapter will focus on the most researched sugarcane viruses in the past two decades and will describe the progress that has contributed to our understanding of those viruses.

http://dx.doi.org/10.19103/AS.2017.0035.22

The single biggest change in recent years has been access to low-cost sequencing technologies, termed next-generation sequencing (NGS). The range of platforms available, the reduced costs and the small genome sizes of viruses means that it is now routine to sequence whole viral genomes. This has revolutionized virus discovery and understanding of strain variation in viral populations. But understanding the viral genome sequence alone is not enough to control the disease. Viral diseases represent a significant production constraint for sugarcane. An understanding of vector interactions, spread potential, environmental influences, quarantine implications and control options is essential for managing viral diseases. This chapter will address the broad themes of virus discovery, genetic variation, diagnostic technologies and improving control through biotechnology, using specific viruses as examples. In addition, two case studies will be presented describing the effects that viral diseases can have on sugar industries in developing nations and how their impact can be reduced through research and development. While not all sugarcane viruses will be covered in detail in this chapter, it is hoped that the theme examples and case studies will showcase sugarcane virology in the twenty-first century.

2 Progress in viral discovery

Traditional methods of viral discovery, such as double-stranded RNA analysis, electron microscopy and sap inoculation of indicator hosts, have clearly served sugarcane pathologists well over the years since most of the currently recognized viral diseases were characterized before NGS technologies became common. But how many more viruses are there? New viral diseases arise all the time, for example, when viruses jump hosts, when they increase their geographic range, or through genetic mutation or recombination. New diseases also arise when existing pathogens interact.

NGS is a powerful and generic technique to screen for new viruses. It is non-targeted, requiring no *a priori* knowledge of the pathogen, and hence is able to detect existing strains, new variants and even completely new viruses (Adams et al., 2009; Roossinck et al., 2015). Even techniques considered revolutionary in their day such as enzyme-linked immunosorbent assay (ELISA) and generic polymerase chain reaction (PCR) are species- or genus-specific. The success of NGS lies in bioinformatic analyses, allowing the viral genome to be detected against the background of host DNA by BLAST (basic local alignment search tool) searching for similarities to known viruses or to viral motifs. Sequencing pipelines have even been developed to assist researchers with virus discovery (Ho and Tzanetakis, 2014).

The NGS approach was successfully used to unravel an apparently new and damaging disease of maize observed in Kenya in 2001 (Adams et al., 2013). ELISA and transmission electron microscopy proved negative but sap inoculations indicated the presence of a virus. RNA sequencing, followed by database searching, led to the identification of two viruses, *maize chlorotic mottle virus* (MCMV) and *sugarcane mosaic virus* (SCMV). The combination of MCMV with any of the cereal viruses in the *Potyviridae* family of viruses, including SCMV, is known as maize lethal necrosis disease. The Kenyan viruses had their closest homology to isolates from China (Xie et al., 2011; Gao et al., 2011). Since then, the disease has been confirmed in Rwanda and the Democratic Republic of Congo, and similar symptoms have been reported in Tanzania, Uganda, South Sudan and Ethiopia (Mahuku et al., 2015). It has become a serious issue in sub-Saharan Africa, where maize is a staple food for smallholder farming communities. Recent diversity studies showed that the MCMV sequences were similar, but the SCMV sequences had significant diversity, with at least two groups of SCMV in Kenya (Mahuku et al., 2015). Although this example relates to a maize disease, it is still a cautionary tale of how existing viruses can act synergistically to cause new diseases. MCMV has already been detected in sugarcane growing close to maize in China (Wang et al., 2014). The plants displayed chlorotic and mosaic viral symptoms and were also infected with SCMV. This shows the potential for lethal necrosis to occur in other regions where sugarcane and maize are grown together.

Adams et al. (2013) employed two extraction methods for preferential extraction of double- and single-stranded RNA. The first method targeted replicating viral RNA and the second, total RNA, including plant RNA. Despite this, more MCMV sequences were detected in the total RNA sequencing pool. Both extraction methods revealed a significant background population of bacterial and fungal microbes, but close examination of their taxa and prevalence did not reveal any likely pathogens. A further refinement to viral enrichment of the sequencing template is to target small interfering RNAs (siRNA) and virion-associated nucleic acids (VANA; Candresse et al., 2014). Random sequencing of siRNAs is not only a powerful technique for virus discovery but also provides information on the interaction of the plant antiviral silencing machinery and the virus. Targeting all viral nucleic acids allows for detection of DNA and RNA viruses. By combining reverse transcriptase priming and a Klenow fragment step, both cDNAs and DNAs can be sequenced in a multiplex format. Roossinck et al. (2015) reviews the benefits and drawbacks of each type of viral nucleic acid available as sequencing template.

A VANA-targeted approach was successfully used to detect *sugarcane streak virus* variants in sugarcane held in a quarantine facility by CIRAD (Candresse et al., 2014). Three forms of sugarcane infecting streak viruses are recognized: *sugarcane streak virus* (SSV), *sugarcane streak Réunion virus* (SSRV) and *sugarcane streak Egypt virus* (SSEV) (Shepherd et al., 2008). Once considered strains of *maize streak virus* (Genus *Mastrevirus*, Family *Geminiviridae*), they were shown by genome sequencing to be distinct species of mastreviruses. Candresse et al. (2014) found that two plants known to be infected with SSEV were also infected with a new virus, named *sugarcane white streak virus* (SWSV). The new virus found after BLAST searches of the assembled reads revealed contigs that were homologous with, but only distantly related to, known mastrevirus sequences on GenBank. By combining both siRNA and VANA data, the genome of the novel mastrevirus could be assembled. The symptoms of this new disease are white spots on the base of the leaf blade in the dewlap region becoming fused laterally and appearing as chlorotic stripes. The mild symptoms seen on some infected plants could explain why infection had gone undetected.

This example shows how NGS can be incorporated in quarantine detection procedures. Candresse et al. (2014) proposes a hierarchical approach with the first step being the sequence-independent NGS, followed by secondary sequence-dependant assays. The first step would be carried out on a single plant, possibly displaying unusual symptoms. Once a new virus is confirmed and characterized, a specific assay would be designed to confirm the novel virus in the original host and to screen a wide range of sugarcane germplasm. Sequence data would immediately be available for diagnostic test development. For example, Adams et al. (2013) developed a TaqMan assay to detect MCMV and SCMV in Kenyan maize and barley plants and Candresse et al. (2014) detected SWSV in five plants when screening material originating from Sudan.

3 Progress in detecting strain variation

One benefit arising from the decreased cost of sequencing is that assessing genetic variation within viral populations is now much easier. Previously, genetic variation was observed through host range studies, serological techniques or restriction fragment length polymorphism (RFLP) analysis of PCR amplicons. Now, phylogenetically informative regions of the genome, and even whole genomes, can be routinely sequenced and subjected to phylogenetic analyses. In some cases, a picture of immense genetic diversity results. SCBV will be used to illustrate this.

In the 1990s, SCBV (in the genus *Badnavirus*) was considered a major disease threat due to its worldwide distribution, unreliable symptoms and the question of integrated forms able to cause episomal infections. At that time, badnaviruses were seen as important emerging pathogens in a number of tropical agricultural crops (Borah et al., 2013). Badnaviruses are pararetroviruses with a double-stranded DNA (dsDNA) genome that replicates via a virus-encoded reverse transcriptase (RT). They resemble the caulimoviruses and, like *cauliflower mosaic virus*, they had the potential to be exploited for their promoters to drive transgene expression in transgenic plants, for example, by Tzafrir et al. (1998) and Braithwaite et al. (2004).

Some badnaviruses have been shown to integrate into the host genome, termed endogenous pararetroviruses (EPRVs). The best-studied EPRV is *banana streak virus* (BSV) which integrates into the host banana genome and in certain cultivars is activated by stress, giving rise to episomal infections (Ndowora et al., 1999). The abiotic stresses that trigger activation include micropropagation by *in vitro* culture processes, temperature differences and water stress (Côte et al., 2010). Other proposed EPRVs occur in taro, fig and yam (Yang et al., 2003; Laney et al., 2012; Umber et al., 2014).

A consistent theme with badnaviruses is their high genome variability, even between isolates of the same virus. SCBV is closely related to BSV and both were referred to as strains of the same virus in early literature (Lockhart and Autrey, 1988). Initially, two SCBV species, designated *sugarcane bacilliform MO virus* from Morocco (SCBMOV, Bouhida et al., 1993) and *sugarcane bacilliform IM virus* from Australia (SCBIMV, Geijskes et al., 2002), were officially recognized by the International Committee on Taxonomy of Viruses. Later Muller et al. (2011) added two new species after obtaining full-length sequences from three Guadeloupe isolates. The species were named *sugarcane bacilliform Guadeloupe A virus* (SCBGAV) and *sugarcane bacilliform Guadeloupe D virus* (SCBGDV), representing phylogenetic groups A and D of the seven groups (A to G) within the three major badnavirus clades. Badnaviruses fall into three major clades with SCBV and BSV distributed across clades 1 and 3, while clade 2 contains other badnaviruses from a range of host plants. Phylogenetic analyses also showed no clear separation between SCBV and BSV in Guadeloupe, with both falling into clade 1 (Muller et al., 2011). Since then, Karuppaiah et al. (2013) obtained full-length sequences from five Indian isolates, with three considered unique enough to be proposed as new species: sugarcane bacilliform BB virus, sugarcane bacilliform BO virus and sugarcane bacilliform BR (Black Réunion) virus. By then, badnavirus phylogroups A to K had been recognized.

Further insight into the extent of genetic variation has come from sequencing the partial RT/RNase H genomic region. This conserved region of the badnavirus genome has been consistently targeted for diagnostic primer design and phylogenetic analysis, for example, by Braithwaite et al. (1995), Yang et al. (2003) and Wu et al. (2016). By targeting a shorter region with conserved primers, rather than whole-genome sequencing, a larger number of isolates can be compared. The results show a global picture of immense genetic diversity and numerous tentative novel species, confirming the complex structure of the SCBV population.

A survey by Rao et al. (2014) found SCBV in 28 sugarcane cultivars from the eight major sugarcane-growing states of India. The RT/RNase H region from eight of these isolates was sequenced and found to vary by up to 27%. Five isolates showed highest sequence identity of 86–88% with SCBIMV, while the other three isolates showed highest sequence identity of 75–79% with SCBMOV, but were still distinct enough to be segregated into two, at the time, new badnavirus subclades (SCBV-I and -H). One isolate was identified as a recombinant, having two other Indian SCBV isolates as parents. A more extensive sequence study was performed by Wu et al. (2016). As with Rao et al. (2014), the study began with a survey, initially involving six provinces in China. SCBV was found to be widespread across the major sugarcane-growing regions, with 87 out of 280 leaf samples (31%) collected from plants suspected of virus infection testing positive. An extensive sequence analysis of 94 Chinese and international isolates followed. The results revealed considerable diversity within the SCBV population. While the three major monophyletic badnavirus clades are still recognized, they now encompass 18 subgroups, including five new subgroups SCBV-N to -R. Again, recombination between isolates was demonstrated. These analyses demonstrate that the genetic structure of sugarcane and banana infecting badnaviruses is highly complex and variable.

4 Progress in viral diagnostics

SCYLV came to the sugarcane world's attention in the 1990s when it was shown to be one of the causative agents of yellow leaf syndrome. It was subsequently shown to have a worldwide distribution and considered a major sugarcane pathogen and the subject of a great deal of research. One feature of this virus is the poor association between the presence of the virus and reliable symptoms. Consequently, diagnostic testing is the only accurate means of detecting the virus and this has inspired research into the development and advancement of virus detection methods, which ultimately benefits the detection of other sugarcane viral diseases. SCYLV will be used to illustrate progress in viral diagnostics.

Yellow leaf syndrome was first recognized as a sugarcane disorder in Hawaii in 1989 (Schenck, 2001). Subsequently, a luteovirus was found to be associated with the syndrome and successful transmission from infected plants to healthy plants by aphid vectors was demonstrated (Scagliusi and Lockhart, 2000). Yellow leaf disease was proposed as the name of the viral disease to distinguish it from leaf yellows, a phytoplasma disease also associated with yellow leaf syndrome. Genome sequence comparisons clearly established that SCYLV is a member of the *Luteoviridae* and considered as an emerging virus that has evolved by recombination between luteoviral and poleroviral ancestors (Moonan et al., 2000; Smith et al., 2000).

A range of serological-based tests have been available since the 1990s. Scagliusi and Lockhart (2000) developed a polyclonal antiserum raised against SCYLV for use in ELISA and tissue blot immunoassay (TBIA). TBIA has been widely used due to its simplicity and speed, for example, by Schenck et al. (1997); Comstock et al. (1998); Rassaby et al. (1999)

and Chatenet et al. (2001). Improvements have seen the use of monoclonal antibodies produced from a recombinant read-through protein (Korimbocus et al., 2002b).

The first RT-PCR test was based on primers specific to luteoviruses (Lockhart et al., 1996). By sequencing an amplicon generated from a Florida isolate, the SCYLV-specific primers YLS111 and YLS462 (M. Irey, unpublished) were designed. The primer set was developed into a diagnostic protocol (Comstock et al., 1998) and widely used in many sugar industries, for example, by Aljanabi et al. (2001), Chatenet et al. (2001) and Rassaby et al. (2004). The widespread screening led to two significant conclusions, firstly that SCYLV occurs in most sugarcane-producing countries, and secondly that most sugarcane varieties infected by SCYLV do not exhibit reliable disease symptoms (Schenck, 2001; Zhu et al., 2010). This would explain how the virus spread undetected around the world, even through quarantine (Chatenet et al., 2001).

Soon after the viral genome was sequenced and characterized, it became apparent that two groups of the virus existed: one group contained only isolates from Colombia and the second group was formed by the isolates from other North, South and Central American countries (Moonan and Mirkov, 2002). But further investigations led to the identification of more strains. Abu Ahmad et al. (2006a and b) identified four different genotypes (BRA, PER, REU and CUB) named after the geographical location where they were first detected: Brazil, Peru, Réunion and Cuba, respectively. Specific RT-PCR primers were developed to detect the SCYLV genotypes (Abu Ahmad et al., 2006b). As genotypes BRA and PER are closely related and the RT-PCR test does not distinguish them, they were combined and called genotype BRA-PER. A new generic set of primers were designed to allow detection of all known strains (Girard et al., 2010). Other groups have developed alternative sets of SCYLV primers and modified them for real-time RT-PCR assays (Korimbocus et al., 2002a; Zhu et al., 2010). Xie et al. (2009) developed a one-step quadruplex RT-PCR method for the simultaneous detection of SCYLV, SCMV, SCSMV and *sorghum mosaic virus* (SrMV).

Over the next few years, further genotypes were identified and added to the list. Viswanathan et al. (2008) identified an Indian genotype (IND) and Wang and Zhou (2010) added a Chinese genotype, CHN1. A Hawaiian genotype (HAW) was recognized by ElSayed et al. (2011). Since Abu Ahmad et al. (2007) observed that viral infectivity and pathogenicity differ among the SCYLV genotypes, it is important to have a means of rapidly identifying them. Although sequence analysis is the most accurate method of identifying virus genotypes, a faster and less expensive technique for large numbers of samples would be ideal. An RT-PCR-RFLP assay was developed by Wang et al. (2012). Conserved primers were used to generate a 1.3kb amplicon which was analysed by a triple digest to differentiate the genotypes. With this method, BRA and PER were able to be distinguished, and a new genotype, CHN2, was recognized. A third Chinese genotype has since been identified (CHN3; Lin et al., 2014), making nine genotypes identified so far.

Several novel techniques have become available for SCYLV detection. A method including isothermal nucleic acid sequence-based amplifications (NASBA) and molecular beacon probes was proposed by Gonçalves et al. (2002). The method is highly sensitive, not subject to 'carry-over' contamination problems and does not require gel-based electrophoresis. Despite these advantages, the method, has not become widely accepted. Another non-gel-based method is loop-mediated isothermal amplification (LAMP) which has the potential for simple field-based diagnosis. The detection of LAMP amplification products is achieved by a simple visual evaluation of solution colour change within microcentrifuge tubes. LAMP has already been used successfully for detecting SCMV and SrMV (Keizerweerd et al., 2015). Amata et al. (2016) applied the technique to SCYLV diagnosis in Kenya. When compared with the conventional SCYLV RT-PCR assay, RT-LAMP was similar in sensitivity when detecting the SCYLV-CUB genotype and more sensitive when detecting the SCYLV-BRA genotype; however, several samples escaped detection.

No doubt further genotypes of SCYLV will be detected, including those arising through recombination (Lin et al., 2014). New diagnostic methods will be developed to account for the genetic variation with improved speed and sensitivity. One novel method is high-resolution hyperspectral remote sensing (Grisham et al., 2010) with the potential to remotely detect SCYLV infections in the field away from the laboratory.

5 Progress in controlling viruses

In most sugar industries, diseases are controlled by an integrated approach involving resistant cultivars, disease-free planting material, farm management practices and strict quarantine (Lakshmanan et al., 2005). While there is resistance to different viral pathogens in the *Saccharum* gene pool and sugarcane breeding programmes have successfully exploited these sources, elite high-yielding varieties are often unable to be released commercially due to susceptibility to specific diseases. Generally, management practices targeting insect vectors have not been very effective. Also, the practice of roguing is generally not considered feasible if the infection level exceeds 5%.

For some viral diseases, clean seed cane can be generated through hot water treatment and tissue culture, especially meristem tip culture, with or without the inclusion of antiviral agents. These approaches are not new; serial hot water treatment for SCMV elimination was promoted in the 1970s (for example, see Benda, 1971). Both SCMV and SCSMV can be eliminated through meristem tip culturing (Mishra et al., 2010; Subba Reddy and Sreenivasulu, 2011). Higher rates of elimination occurred when tip culture and chemotherapy were combined (Balamuralikrishnan et al., 2002). The antiviral ribavirin at 50 ppm helped to achieve 95% elimination of SCMV compared to 62% elimination using meristem tip culture alone. SCYLV is also successfully eliminated through meristem tip culture (Fitch et al., 2001; Chatenet et al., 2001) and callus culture (Parmessur et al., 2002), but not through hot water treatment. Yield trials of SCYLV-free lines obtained by meristem culture showed that the absence of SCYLV in a commercial cultivar had higher yield than the line that had not been freed of virus infection, at least in a one-year crop cycle (Lehrer et al., 2009). The benefits of eliminating viruses through tissue culture are transitory in the field because reinfection will occur if aphid numbers are high. But a tissue culture approach is still worthwhile because it ensures virus-free plants can be released from quarantine, maintaining confidence in the germplasm exchange process. It also provides material for epidemiology and yield loss studies (Chatenet et al., 2001).

Introducing specific genetic improvements, in this case virus resistance, directly into elite sugarcane varieties could offer another solution to managing viral diseases. Sugarcane has a complex genome, being highly polyploid and aneuploid. The long period required for breeding new cultivars and low fertility makes traditional breeding laborious. These factors limit the opportunities for crop improvement through conventional breeding and make sugarcane a prime candidate for improvement through genetic engineering (Lakshmanan et al., 2005). Unlike many other members of the Poaceae, most sugarcane cultivars yield regenerable cell lines and several sugarcane transformation methods and selection systems are available (reviewed by Brumbley et al., 2008). However, the variability in the agronomic performance of the transformed clones due to somaclonal variation caused by tissue culture and the transformation procedure means that thorough field evaluation is essential (Gilbert et al., 2005, 2009; Mitchell, 2011). Ideally, the transgenic plants would then be used as parents in breeding programmes.

The most successful approach to develop resistance against viruses is pathogen-derived resistance (PDR), whereby pathogens themselves provide the genes for engineered resistance (Lomonossoff, 1995). For viral PDR, target genes can include full-length, truncated or mutated viral coat proteins; antisense sequences to sections of the viral genome; viral replicases; satellite RNA; and virus movement proteins (Lakshmanan et al., 2005). The transgenes suppress expression of viral RNAs by a process known as post-transcriptional gene silencing (PTGS). PTGS has proven to be an effective resistance mechanism in several plant–virus systems.

The first successful demonstration of PDR in sugarcane was against mosaic disease: Joyce et al. (1998) used the coat protein (CP) gene of SCMV-A, the only strain present in Australia. Ingelbrecht et al. (1999) used an untranslatable form of the SrMV strain SCH CP gene. In both cases, the transformed plants showed a range of phenotypes including immune, resistant, recovery and susceptible. Ingelbrecht et al. (1999) reported that most SrMV-resistant sugarcane lines contained undetectable or reduced CP transcript levels compared with susceptible plants, which is consistent with a PTGS-related resistance mechanism. Although PDR showed potential for increasing yields and disease resistance, researchers began to document the undesirable effects that tissue culture and the transformation process could have on transgenic sugarcane. This led Gilbert et al. (2005) to thoroughly evaluate the agronomic characteristics and field resistance of sugarcane transformed for resistance to SCMV strain E and found variability in both yield and disease resistance.

A transgenic approach has also been trailed for the control of *Fiji disease virus*. This virus (in the genus *Fijivirus*, family *Reoviridae*) has a complex genome consisting of 10 linear molecules of dsRNA. It causes the serious disease Fiji leaf gall throughout the Asia-Pacific region. An epidemic of this disease in Australia during the 1970s was brought under control through an intensive breeding programme but with the loss of valuable germplasm. Transgenic resistance was demonstrated using an untranslatable form of the segment 9 ORF1 (McQualter et al., 2004). One transgenic line showed significantly enhanced resistance to Fiji leaf gall compared with the parent line. However, the molecular phenotypes of the transgenic plants at both the DNA and RNA levels were not entirely consistent with a resistance mechanism based on PTGS, suggesting that PDR is difficult to achieve with dsRNA viruses.

Groups in Florida and Hawaii have attempted to transform sugarcane for improved SCYLV resistance. SCYLV is considered a significant pathogen in Florida and yet the opportunity for traditional breeding for SCYLV resistance is limited. The level of SCYLV resistance in the Canal Point sugarcane population appears to be low, based on the high incidence of the virus in parental clones (Comstock et al., 1998). Thus, for Florida, the effort to generate transgenic lines for SCYLV resistance was considered worthwhile, providing the agronomic performance of the transgenic lines was assessed in detail in the field. Gilbert et al. (2009) used an untranslatable SCYLV CP fragment in the antisense orientation. Molecular characterization was also performed to assess the genetic differences in the transgenic lines compared with the parent line using simple sequence repeat genotyping. In Hawaii, Zhu et al. (2011) used particle bombardment to transform a SCYLV-susceptible cultivar with an untranslatable SCYLV CP gene in a sense orientation. Six out of nine transgenic lines had at least 100-fold lower virus titre than the parent line.

This resistance level, as measured by virus titre and symptom development, was similar to that of a resistant cultivar.

An alternative approach to PTGS is gene silencing through RNA interference (RNAi). Homologous RNA is degraded after the introduction of a dsRNA form, which becomes the target for gene silencing. Guo et al. (2015) used highly conserved sequences of the SrMV CP gene as a silencing target, and prepared an RNAi expression vector with hairpin structures for introduction into sugarcane via *Agrobacterium*-mediated transformation. After herbicide screening, PCR molecular identification and artificial inoculation, anti-SrMV-positive transgenic lines were successfully obtained with an 87.5% success rate.

Despite these published successes, no genetically modified sugarcane has been approved for commercial release anywhere in the world to date (Mitchell, 2011). Although direct genetic engineering can provide advantages over conventional breeding for the development of novel, varied and durable resistance strategies, the technology has some barriers to adoption. Technical issues, such as the large copy number resulting from particle bombardment and the inclusion of antibiotic resistance genes, can be circumvented through alternative systems, including *Agrobacterium*-mediated transformation and minimal DNA cassettes (Joyce et al., 2014). However, there are still issues regarding the lengthy and costly risk assessments to be carried out to ensure that genetically engineered sugarcane is safe. Public perception issues of genetically engineered foods can be unfavourable and the foods are banned in many countries. The new technology, CRISPR/Cas9 (clustered regularly interspaced short palindromic repeats), has been successfully used to deliver potyvirus resistance in cucumber (Chandrasekaran et al., 2016). This approach, based on precise genome editing in plants, does not introduce foreign transgenes, so may lead to increased viral resistance in crop plants without the burden of extensive regulation, making it more publicly acceptable.

6 Case studies

The impact of two viral diseases on sugar industries in developing nations will be presented as case studies. Both diseases are high-risk quarantine threats to the Australian sugar industry and Australian sugarcane pathologists and entomologists have been involved in collaborative research programmes for many years. For both diseases, progress in viral genome sequencing, strain characterization and development of diagnostic methods, in combination with traditional pathology approaches, has led to a greater understanding of the diseases and improved management options.

6.1 Ramu stunt disease in Papua New Guinea

Papua New Guinea (PNG) is considered the home of sugarcane. *Saccharum officinarum* (noble or chewing cane) is a 'native domesticated' sugarcane and is traditionally cultivated alongside *Saccharum edule* ('pit-pit') and other subsistence food plants such as sweet potato, yam, papaya, cassava and banana in garden plots. Garden size can range from a very small plot in a backyard to a larger plot of around 0.4 ha and tended by several families (Fig. 1; Anderson et al., 2009). *Saccharum spontaneum* and *Saccharum robustum* are considered to be 'wild' and are common, especially along roadsides and riparian areas throughout PNG. Commercial sugarcane (*Saccharum* spp. hybrids) is grown only in the Ramu Valley.

Figure 1 A typical village in Papua New Guinea where noble canes (*Saccharum officinarum*) and pit-pit (*Saccharum edule*) are grown with other subsistence food plants such as sweet potato, yam, papaya, cassava and banana in family garden plots.

In the late 1970s, the government of PNG commissioned a study into the technical and economic feasibility of developing a domestic sugar industry in PNG. The study recommended the development of a sugar plantation in the Ramu Valley centred on Gusap. Ramu Sugar, now Ramu Agri-Industries Limited (RAIL), commenced operations in the early 1980s. RAIL's activities comprise oil palm, sugarcane for domestic sugar consumption and ethanol, and beef cattle. Currently, the sugar operations involve approximately 7500 ha, of which approximately 1000 ha is managed by small-scale outgrowers.

Ramu stunt was unknown until 1985–6 when poor yields, stool death and severe stunting became evident in commercial crops at Ramu Sugar, almost destroying the fledgling commercial sugarcane industry. Three of the four commercial sugarcane varieties (Ragnar, Yassawa and Q90) proved susceptible to the disease (Waller et al. 1987; Eastwood 1990). Ragnar, highly susceptible to the disease, was the major variety on the estate, comprising 85–90% of the commercial crop. The fourth variety, Cadmus, was found to be resistant and was used to replant the industry, allowing the company to continue. The disease is characterized by stunting (Fig. 2), failure to ratoon and various ill-defined leaf symptoms which are believed to vary between varieties. The disease poses a major quarantine risk to other sugarcane-producing countries, especially Australia, and restricts the safe movement of sugarcane germplasm from PNG. Australian sugarcane pathologists have had a long association with Ramu stunt: being jointly involved with RAIL in diagnostic test development; performing disease surveys, resistance screening for both Australian and PNG sugarcane varieties; and elucidating disease transmission mechanisms.

Figure 2 Stunting seen in individual stools of Ramu stunt-infected commercial sugarcane on the Ramu estate. This variety is PN97-54.

At the time of the epidemic, the causal agent was unknown but was suspected to be a virus or a phytoplasma. Subsequent research has confirmed that the causal agent is a virus (Braithwaite et al., 2007). Cloning experiments identified a number of sequences with homology to known or proposed members of the viral genus *Tenuivirus*. A preliminary diagnostic RT-PCR test was developed based on a sequence with high homology to tenuivirus RNA polymerase. Since then, various improvements have been made, with the diagnostic test now targeting the disease-specific protein coding region (Braithwaite et al., 2012).

Validation of the diagnostic test used a Ramu stunt resistance screening trial planted on the Ramu estate. Screening trials include standard varieties of known disease ratings (Kuniata et al., 2010) allowing symptoms, disease rating and diagnostic result to be correlated. The next phase of validation was to carry out a survey of commercial canes on the Ramu estate, particularly any showing unusual symptoms. The diagnostic test was an important breakthrough because previously there had been confusion over the symptoms (Waller et al., 1987). The definitive leaf symptom (Fig. 3) is a leaf stripe somewhat similar to that caused by the tenuivirus *maize stripe virus*, although plants showing other symptoms may still be infected. Screening results from the estate survey found that the majority of plants thought to be infected because of leaf yellowing, particularly R570, were not. Plants which showed only yellowing, without any other kind of symptom, were not infected with Ramu stunt.

Figure 3 The definitive leaf stripe symptom seen in the Ramu stunt-infected variety Ragnar.

Since 2001 several surveys through parts of PNG have been undertaken with sampling for Ramu stunt. The surveys targeted *S. officinarum* and *S. edule* in traditional style gardens, and *S. robustum*, *S. spontaneum*, *Miscanthus* and grassy weeds along roadsides and creek and river banks. While a 2001 survey was an extensive survey throughout PNG for sugarcane pests and diseases (Magarey et al., 2002), later surveys were smaller: focussing only on the Ramu Valley, Kassam Pass, Goroka, Madang and Alotau regions (Braithwaite et al., 2014). Ramu stunt infected material was only detected in the Ramu Valley and Alotau and all host plants were *S. officinarum* from traditional gardens. No Ramu stunt was detected in *S. edule*, wild canes or any of the grass weeds tested (blady grass, elephant grass, guinea grass or itch grass). At this stage it appears that Ramu stunt is not common or widespread away from commercial crops.

NGS techniques were used to sequence the genome of the Ramu stunt virus (Mollov et al. 2016). Total RNA was extracted from the variety Ragnar and whole transcriptome shotgun sequencing was performed on an Illumina platform. Six RNAs were identified including the 8902 bp RNA polymerase and five shorter fragments ranging from 1201 to 1675 bp. RNA fragments 2 and 3 had an ambisense arrangement of two open reading frames. Tenuiviruses have large genomes with four to six segments, the largest being around 9 kb, and the smaller ones having an ambisense arrangement (Falk and Tsai, 1998). All six RNA fragments sequenced by Mollov et al. (2016) showed the complementary 5′ and 3′ terminal 10 nucleotide sequences typical of tenuiviruses (Falk and Tsai, 1998). That the RNA5 protein had highest homology to the genus *Phlebovirus* (family *Bunyaviridae*) nucleocapsid protein is also typical for tenuiviruses.

Mollov et al. (2016) has attempted to place the Ramu stunt virus in a phylogenetic context. BLAST analyses showed that the RNA polymerase protein had the highest identity of 38%

to the tenuivirus *rice stripe virus*, followed by *rice grassy stunt virus*. Phylogenetic analysis of the nucleoprotein amino acid sequence (from RNA2) indicated that, while most closely related to tenuiviruses, it was more intermediate between tenuiviruses and phleboviruses but distinct from tospoviruses, the only plant virus genus in the family *Bunyaviridae*. The four hypothetical proteins coded for by RNAs 3, 4 and 6 had highest homology to hypothetical proteins identified by Mahmoud et al. (2007) from *maize yellow stripe virus*, a proposed tenuivirus. Along with the genomic characteristics, various other findings, namely that the vector is a planthopper in the family Delphacidae and the symptoms of leaf stripes of varying width are similar to that seen with some other tenuiviruses, all support the conclusion that the causal agent of Ramu stunt disease is a tenuivirus.

The vector of Ramu stunt disease is the island sugarcane planthopper *Eumetopina flavipes* Muir (Hemiptera) (Kuniata et al., 1994). While the distribution of Ramu stunt appears to be limited to PNG (Magarey et al., 2002), *E. flavipes* extends from PNG throughout islands in the Torres Strait and onto the Northern Peninsula Area (NPA) of Cape York on mainland Australia (Grimshaw and Donaldson, 2007). The planthopper is also present in parts of Southeast Asia including Indonesia, Borneo, Sarawak, Solomon Islands, the Philippines and New Caledonia (Anon, 2014). *E. flavipes* has been recorded from *S. spontaneum*, *S. robustum*, *S. officinarum* and commercial sugarcane (Kuniata et al., 1994).

E. flavipes provides a model system for how insect and human behaviours can influence plant disease invasions. The Torres Strait and NPA represent a high-risk pathway for pest and disease incursions into Australia. Traditionally, noble sugarcanes were traded and gifted within and between cultural groups, including throughout the Torres Strait islands, either as single stalks or bundles, with the leaves attached. *E. flavipes* adults and nymphs were shown to survive for up to six days on cut stalks, suggesting that the social customs of exchanging cut stalks of noble sugarcanes could play an important role in the dispersal of *E. flavipes* (Anderson et al., 2007). Wind-assisted long-distance migration from PNG was demonstrated through modelling (Anderson et al., 2010), but ultimately it was distribution and abundance of host plants that was important in regulating the establishment, growth and population persistence of *E. flavipes* (Anderson et al., 2009).

The use of resistant varieties is the most successful method of controlling Ramu stunt disease in commercial fields. Resistance screening trials have been conducted at RAIL since the late 1980s (Kuniata et al., 2010). Fortunately, the percentage of canes with symptoms in these trials is around 25%, making it relatively easy to select resistant varieties. This means that most clones selected on yield characteristics in the plant improvement programme will be resistant to the disease; and selection for Ramu stunt resistance can occur late in the breeding programme. Yield losses at Ramu Sugar have been minimal since 1986 because of the care taken to cultivate varieties with sufficient resistance (Rauka et al., 2005).

6.2 Sugarcane streak mosaic in Indonesia

Sugarcane is a major crop in Indonesia, contributing over Rupiah 11,500 billion to the economy (Magarey et al., 2010). There are 58 sugarcane factories processing 30 million tonnes of sugarcane from 450 000 ha of land. Farm sizes are small (average around 1 ha) with limited mechanical cultivation or harvesting. The Indonesian sugarcane industry traditionally was centred on the island of Java where over three-quarters of the sugarcane production occurs but in more recent years has expanded into Sulawesi, Sumatra, Sumba and West Papua. Pests and diseases are major contributors to a production decline that has been occurring over the last 40 years. Moth borers are the major pest problem, and mosaic is the major disease.

In recent years, sugar production has declined even more (Putra et al., 2014). At the same time outbreaks of mosaic disease have increased, even infecting varieties known to be resistant to SCMV. A new strain of SCMV or new virus was suspected and SCSMV was confirmed (Damayanti and Putra, 2011). The extent of mosaic disease throughout the Javan industry was not appreciated until an extensive pest and disease survey in 2008–9 where over 930 individual crops were inspected (Magarey et al., 2010; Putra et al., 2014). The survey highlighted the presence of sugarcane mosaic-type symptoms in a large number of crops (Fig. 4). SCSMV was detected in 32% of the 931 fields and in >90% of the crops showing mosaic symptoms. The high incidence of the virus was believed to be due to widespread planting of PS 864, a high-yielding variety with high sugar and good tillering. PS 864 has been widely planted both on Java and outside Java in recent years.

Figure 4 Mosaic-infected plant observed during a survey in West Java. This plant was later confirmed positive for *sugarcane streak mosaic virus*.

SCSMV is one of the most prevalent sugarcane diseases in Asia (Chatenet et al., 2005) occurring in Bangladesh, India, Pakistan, Sri Lanka, Thailand and Vietnam. The disease has very high incidence in India (Rao et al., 2006) and Thailand (Kasemsin et al., 2016), often higher than SCMV. SCSMV is thought to be spread by insect vectors but little is known about the vectors of this disease. SCSMV leaf symptoms include a pale green mosaic pattern (Fig. 5) similar to that caused by SCMV, and so to distinguish them, the disease is referred to as streak mosaic.

Mosaic in sugarcane can be caused by three different pathogens: SCMV, SrMV and SCSMV. SCMV occurs in nearly every sugar industry and was responsible for major losses in the early twentieth century. A number of viral variants had been identified previous to the 1990s (Grisham, 2000). They were classified as strains (A to N) based on their virulence patterns on differential hosts and serological properties. With time, the strains were further researched, described and sorted into pathogens of different viral types, but were all members of the sugarcane mosaic virus complex in the genus *Potyvirus*, family *Potyviridae*, which also includes *Johnson grass mosaic virus*, *maize dwarf mosaic virus*, and *zea mosaic virus* (Chatenet et al., 2005). Only SCMV and SrMV naturally infect sugarcane, each comprising numerous strains, and mixed infections can occur.

Figure 5 The pale green mosaic symptom associated with *sugarcane streak mosaic virus*.

The virus now known as SCSMV was first detected in the United States in quarantined sugarcane from Pakistan and classified as strain SCMV-F (Gillaspie et al., 1978). However, later molecular and serological analyses confirmed that, while it is a member of the *Potyviridae* (Hall et al., 1998; Hema et al., 1999) it is not in the *Potyvirus* genera. Phylogenetic comparisons based on various genomic regions initially suggested that SCSMV could belong in the *Whesmovirus* (Hall et al., 1998), *Tritimovirus* (Hema et al., 1999) or *Susmovirus* genera (Xu et al., 2010), until finally the name *Poacevirus* was accepted (Li et al., 2011). The only other *Poacevirus* is *triticum mosaic virus*.

Since the discovery of SCSMV there has been numerous strains described including from Pakistan, SCSMV-PAK (Hall et al., 1998); southern India, SCSMV-AP (Hema et al., 1999); northern India, SCSMV-UP (Gaur et al., 2006); Japan, SCSMV-JP1 and -JP2; and Indonesia, SCSMV-ID (Li et al., 2011). The last three were detected in quarantine in China, where SCSMV has not yet been observed in the field. Damayanti and Putra (2011) sequenced the CP gene of another Indonesian strain of SCSMV. Their strain shared 92.1% homology to SCSMV-PAK whereas the CP sequence of SCSMV-ID obtained by Li et al. (2011) shared only 85.9% homology. This difference suggests the presence of different strains of SCSMV in Indonesia (Putra and Damayanti, 2012).

An active research programme is now underway in Indonesia to identify possible vectors, alternative hosts, yield effects and establish control strategies (Putra et al., 2014; 2015). A diagnostic test targeting the CP is available to monitor research trials (Damayanti and Putra, 2011) but simple field-based methods, such as LAMP, are now being developed (N. Thompson, unpublished). Transmission trials demonstrated that SCSMV could be transmitted mechanically through wounds made by carborundum and abrasive rubbing pads. Mechanical transmission by knives was also demonstrated (Putra et al., 2015). If the disease is vegetatively transmitted through cane cuttings and spread by cane knives, it would be easy to see how the disease became so widespread. While various aphids are known around the world as mosaic vectors, the vector for SCSMV has not yet been clearly determined from transmission trials (Fig. 6). The virus could not be transmitted by the corn aphid (*Rhopalosiphum maidis*), sugarcane woolly aphid (*Ceratovacuna lanigera*) and sugarcane aphid (*Melanaphis sacchari*), all common species in sugarcane fields in Indonesia, but the brown aphid (*Toxoptera citricida*) was suspected to be a vector (Magarey et al., 2012).

Host range tests showed that SCSMV infects sorghum, maize, *S. officinarum, S. spontaneum, S. robustum, Saccharum sinense, Saccharum barberi* and *Miscanthus.* Several weed grasses, namely *Brachiaria moniliformis, Panicum repens, Paspalum conjugatum* and *Rottboellia exaltata,* were also infected by the virus. Varietal resistance screening showed that some resistant varieties are available, including the high-yielding variety VMC 76-16. Yield loss assessments on variety PS 864 showed that an infection level of 50% or higher resulted in 16–17% and 19–21% reduction in cane tonnage and sugar, respectively. Hot water treatment was trialled and found to only delay appearance and severity of symptoms, but not eradicate the virus.

Recommended management approaches include (1) use of clean planting material, (2) use of resistant varieties, (3) development of a rapid detection method for routine screening and monitoring, (4) avoidance of planting maize in close proximity to sugarcane and (5) use of disinfectants such as Lysol to disinfect cutting tools during planting.

The widespread incidence of the disease makes it very difficult to locate clean planting material, not only for sugarcane crops in Java but also for the establishment and commercial viability of new plantations in other provinces. The disease has been reported from the Indonesian province of West Papua on the island of New Guinea (Putra et al., 2014). This is

Figure 6 Insect vector transmission trial at the Indonesian Sugar Research Institute at Pasuruan.

a cause for concern as New Guinea is the centre of diversity for *S. officinarum*, *S. edule* and *S. robustum*. The spread of the disease into wild canes could result in a loss of valuable germplasm.

The disease is also a major biosecurity threat to neighbouring Asian countries such as China and Australia which are still free of the disease. As the virus is apparently transmitted through knives used to cut canes, entry of SCSMV on illegally imported cuttings of sugarcane is a significant risk. The risk of virus transmission via an unknown insect vector is also high. Monitoring and surveillance are of limited use until the vectors are clearly identified. In Australia, only the mild strain A of SCMV is present. Sugarcane is common in gardens throughout the Torres Strait islands and other parts of northern Australia, and it is not known if the aphids present there can transmit the virus. For this reason, a collaborative research programme is underway to enable the Australian sugar industry to prepare for possible incursions from Indonesia (Magarey et al., 2010 and 2012).

7 Conclusion and future trends

This chapter provides an overview of current research on sugarcane viruses. While only the most intensely studied viral diseases of the last few decades were presented in detail, the themes apply to all sugarcane viral diseases. One major trend emerging from the discussion is that ready access to sequencing technologies has allowed us to appreciate the immense genetic variation between strains and genotypes. This, in turn, affects how diagnostic and management strategies must be adapted to account for the variation. Future trends should see multiplexing become commonplace to save time and money; the routine screening of germplasm in quarantine for new strains, and even for new viruses; and improved methods for in-field screening and the application of novel control options.

8 Where to look for further information

For general virology:

- *International Committee on Taxonomy of Viruses (ICTV)*: 2015 Release
  http://www.ictvonline.org/index.asp
- Plant Viruses Online: Descriptions and Lists from the VIDE Database
  http://sdb.im.ac.cn/vide/refs.htm
- All the Virology on the WWW:
  http://www.virology.net/garryfavwebindex.html

For lists of common and pathogen names for sugarcane diseases:

- *International Society of Plant Pathology ISPPWeb*
  http://www.isppweb.org/names_sugarcane_pathogen.asp#vir and
  http://www.isppweb.org/names_sugarcane_common.asp

Textbooks:

- *Plant Virology 5th Edition*
  R. Hull (2014). Academic Press/Elsevier, London, UK
  ISBN: 978-0-12-384871-0
- *Diseases of Sugarcane: Major Diseases*
  Eds: C. Ricaud, B.T. Egan, A.G. Gillaspie Jr, C.G. Hughes (1989). Elsevier Science Publishers, Amsterdam, The Netherlands
  ISBN: 0-444-42797-X
- *A Guide to Sugarcane Diseases*
  Eds: P. Rott, R.A. Bailey, J.C. Comstock, B.J. Croft and A. S. Saumtally (2000). Cirad/ISSCT, Montpellier, France
  ISBN: 2-87614-386-0
- *Sugarcane Pathology Volume II: Virus and Phytoplasma Diseases*
  Eds: G. P. Rao, R. E. Ford, M. Tosic and D. S. Teakle (2001). Science Publishers Inc., Enfield, USA
  ISBN 1-57808-128-9

For images of sugarcane diseases:

- *The American Phytopathological Society*
  http://www.apsnet.org/publications/imageresources/Pages/ThumbnailsSugarcane.aspx
- *Diseases of Australian Sugarcane – Field Guide*
  R. Magarey (2013). Sugar Research Australia Limited
- *Sugarcane Pests and Diseases – Field Guide - Indonesian version*
  E. Achadian, A. Kristini, R. Magarey, N. Sallam, P. Samson, F.-R. Goebel and K. Lonie (2012). Indonesian Sugar Research Institute, BSES Limited and Australian Centre for International Research

9 Acknowledgements

The two case studies were chosen because of the close ties between SRA pathologists and entomologists and staff at RAIL, Papua New Guinea, and the Indonesian Sugar Research Institute, Indonesian Sweetener and Fibre Crop Research Institute and Bogor Agricultural University, Indonesia. Collaborative research programmes on Ramu stunt and SCSMV have been in place for several years, with funding provided by SRA, Bureau of Sugar Experiment Stations (BSES), Sugar Research and Development Council, Australian Centre for International Agricultural Research (ACIAR) and Queensland Department of Agriculture and Fisheries. The author especially thanks Nicole Thompson (SRA) for helping with the section on SCSMV and for providing the Indonesian photos.

10 References

Abu Ahmad, Y., Rassaby, L., Royer, M., Borg, Z., Braithwaite, K. S., Mirkov, T. E., Irey, M. S., Perrier, X., Smith, G. R. and Rott, P. (2006a). Yellow leaf of sugarcane is caused by at least three different genotypes of sugarcane yellow leaf virus, one of which predominates on the Island of Réunion. *Archives of Virology*, 151: 1355–71. https://doi.org/10.1007/s00705-005-0712-9.

Abu Ahmad, Y., Royer, M., Daugrois, J-H., Costet, L., Lett, J-M., Victoria, J. I., Girard, J-C. and Rott, P. (2006b). Geographical distribution of four *Sugarcane yellow leaf virus* genotypes. *Plant Disease*, 90: 1156–60. https://doi.org/10.1094/pd-90-1156.

Abu Ahmad, Y., Costet, L., Daugrois, J. H., Nibouche, S., Letourmy, P., Girard, J-C. and Rott, P. (2007). Variation in infection capacity and in virulence exists between genotypes of *Sugarcane yellow leaf virus*. *Plant Disease*, 91: 253–9.

Adams, I. P., Glover, R. H., Monger, W. A., Mumford, R., Jackeviciene, E., Navalinskiene, M., Samuitiene, M. and Boonham, N. (2009). Next-generation sequencing and metagenomic analysis: a universal diagnostic tool in plant virology. *Molecular Plant Pathology*, 10: 537–45.

Adams, I. P., Miano, D. W., Kinyua, Z. M., Wangai, A., Kimani, E., Phiri, N., Reeder, R., Harju, V., Glover, R., Hany, U., Souza-Richards, R., Deb Nath, P., Nixon, T., Fox, A., Barnes, A., Smith, J., Skelton, A., Thwaites, R., Mumford, R. and Boonham, N. (2013). Use of next-generation sequencing for the identification and characterization of *Maize chlorotic mottle virus* and *Sugarcane mosaic virus* causing maize lethal necrosis in Kenya. *Plant Pathology*, 62: 741–9.

Aljanabi, S. M., Parmessur, Y., Moutia, Y., Saumtally, S. and Dookun, A. (2001). Further evidence of the association of a phytoplasma and a virus with yellow leaf syndrome in sugarcane. *Plant Pathology*, 50: 628–36.

Amata, R. L., Fernandez, E., Filloux, D., Martin, D. P., Rott, P. and Roumagnac, P. (2016). Prevalence of *Sugarcane yellow leaf virus* in sugarcane-producing regions in Kenya revealed by reverse-transcription loop-mediated isothermal amplification method. *Plant Disease*, 100: 260–8. https://doi.org/10.1094/pdis-05-15-0602-re.

Anderson, K. L., Sallam, M. and Congdon, B. C. (2007). Long-distance dispersal by *Eumetopina flavipes* (Hemiptera: Delphacidae), vector of Ramu stunt: is culture contributing? *Proceedings of the Australian Society of Sugarcane Technologists*, 29: 226–34.

Anderson, K. L., Sallam, N. and Congdon, B. C. (2009). The effect of host structure on the distribution and abundance of the island sugarcane planthopper, *Eumetopina flavipes* Muir, vector of Ramu stunt disease of sugarcane. *Virus Research*, 141: 247–57.

Anderson, K. L., Deveson, T. E., Sallam, N. and Congdon, B. C. (2010). Wind assisted migration potential of the island sugarcane planthopper, *Eumetopina flavipes* (Hemiptera: Delphacidae): implications for managing incursions across an Australian quarantine frontline. *Journal of Applied Ecology*, 47: 1310–19.

Anon. (2014). Dossier on *Eumetopina flavipes* as a pest of sugarcane. http://www.sugarresearch.com.au/icms_docs/179339_Eumetopina_flavipes_dossier.pdf (accessed 25 July 2016).

Balamuralikrishnan, M., Doraisamy, S., Ganapathy, T. and Viswanathan, R. (2002). Combined effect of chemotherapy and meristem culture on sugarcane mosaic virus elimination in sugarcane. *Sugar Tech*, 4: 19–25.

Benda, G. T. A. (1971). Control of sugarcane mosaic by serial heat treatment. *Proceedings of the International Society of Sugar Cane Technologists*, XIV: 955–60.

Borah, B. K., Sharma, S., Kant, R., Johnson, A. M. A., Saigopal, D. V. R. and Dasgupta, I. (2013). Bacilliform DNA-containing plant viruses in the tropics: commonalities within a genetically diverse group. *Molecular Plant Pathology*, 14: 759–71.

Bouhida, M., Lockhart, B. E. L. and Olszewski, N. E. (1993). An analysis of the complete sequence of a Sugarcane bacilliform virus genome infectious to banana and rice. *Journal of General Virology*, 74: 15–22.

Braithwaite, K. S., Egeskov, N. M. and Smith, G. R. (1995). Detection of Sugarcane bacilliform virus using the polymerase chain reaction. *Plant Disease*, 79: 792–6.

Braithwaite, K. S. (2001). New virus and virus-like diseases of sugarcane – an overview. In *Sugarcane Pathology Vol II: Virus and Phytoplasma Diseases*, Eds. G. P. Rao, R. E. Ford, M. Tosic and D. S. Teakle, Science Publishers, Enfield USA, 3–24.

Braithwaite, K. S., Geijskes, R. J. and Smith, G. R. (2004). A variable region of the *Sugarcane bacilliform virus* (SCBV) genome can be used to generate promoters for transgene expression in sugarcane. *Plant Cell Reports*, 23: 319–26.

Braithwaite, K. S., Croft, B. J. and Magarey, R. C. (2007). Progress in identifying the cause of Ramu stunt disease of sugarcane. *Proceedings of the Australian Society of Sugar Cane Technologists*, 29: 235–41.

Braithwaite, K. S., Kombukon, R., Kuniata, L. S. and Magarey, R. C. (2012). Ramu stunt: resistance screening and validation of the diagnostic test. *Proceedings of the Australian Society of Sugar Cane Technologists*, 34 (electronic format).

Braithwaite, K. S., Kuniata, L. S. and Magarey, R. C. (2014). The importance of disease surveys for understanding quarantine pathogens. *Proceedings of the Australian Society of Sugar Cane Technologists*, 36 (electronic format).

Brumbley, S. M., Snyman, S. J., Gnanasambandam, A., Joyce, P., Hermann, S. R., da Silva, J. A., McQualter, R. B., Wang, M-L., Egan, B. T., Paterson, A. H., Albert, H. H. and Moore, P. H. (2009). Sugarcane. *Compendium of Transgenic Crop Plants: Transgenic Sugar, Tuber and Fiber Crops*, 7: 1–58.

Candresse, T., Filloux, D., Muhire, B., Julian, C., Galzi, S., Fort, G., Bernardo, P., Daugrois, J-H., Fernandez, E., Martin, D. P., Varsani, A. and Roumagnac, P. (2014). Appearances can be deceptive: revealing a hidden viral infection with deep sequencing in a plant quarantine context. *PLoS ONE*, 9: e102945.

Chandrasekaran, J., Brumin, M., Wolf, D., Leibman, D., Klap, C., Pearlsman, M., Sherman, A., Arazi, T. and Gal-On, A. (2016). Development of broad virus resistance in non-transgenic cucumber using CRISPR/Cas9 technology. *Molecular Plant Pathology*, 17: 1140–53.

Chatenet, M., Delage, C., Ripolles, M., Irey, M., Lockhart, B. E. L. and Rott. P. (2001). Detection of *Sugarcane yellow leaf virus* in quarantine and production of virus-free sugarcane by apical meristem culture. *Plant Disease*, 85: 1177–80.

Chatenet, M., Mazarin, C., Girard, J.-C., Fernandez, E., Gargani, D., Rao, G. P., Royer, M., Lockhart, B. E. L. and Rott, P. (2005). Detection of *Sugarcane streak mosaic virus* in sugarcane from several Asian countries. *Sugar Cane International*, 23: 12–15.

Comstock, J. C., Irey, M. S., Lockhart, B. E. L. and Wang, Z. K. (1998). Incidence of yellow leaf syndrome in CP cultivars based on polymerase chain reaction and serological techniques. *Sugar Cane*, 4: 21–4.

Côte, F. X., Galzi, S., Folliot, M., Lamagnére, Y., Teycheney, P-Y. and Iskra-Caruana, M-L. (2010). Micropropagation by tissue culture triggers differential expression of infectious endogenous *Banana streak virus* sequences (eBSV) present in the B genome of natural and synthetic interspecific banana plantains. *Molecular Plant Pathology*, 11: 137–44. https://doi.org/10.1111/j.1364-3703.2009.00583.x.

Croft, B. J. (1996). Review of restrictions to free access to germplasm exchange facing Australian and other international sugar industries. In *Sugarcane Germplasm Conservation and Exchange*, Eds B. J. Croft, C. M. Piggin, E. S. Wallis and D. M. Hogarth. ACIAR Proceedings No. 67, 6–9.

Damayanti, T. A. and Putra, L. K. (2011). First occurrence of *Sugarcane streak mosaic* virus infecting sugarcane in Indonesia. *Journal of General Plant Pathology*, 77: 72–4.

Eastwood, D. (1990). Ramu stunt disease: development and consequence at Ramu Sugar Ltd. *Sugar Cane* 2: 15–16.

ElSayed, A. I., Weig, A. R. and Komor, E. (2011). Molecular characterization of Hawaiian *Sugarcane yellow leaf virus* genotypes and their phylogenetic relationship to strains from other sugarcane-growing countries. *European Journal of Plant Pathology*, 129: 399–412.

Falk, B. W. and Tsai, J. H. (1998). Biology and molecular biology of viruses in the genus *Tenuivirus*. *Annual Review of Phytopathology*, 36, 139–63.

Fitch, M. M. M., Lehrer, A. T., Komor, E. and Moore, P. H. (2001). Elimination of *Sugarcane yellow leaf virus* from infected sugarcane plants by meristem tip culture visualised by tissue blot immunoassay. *Plant Pathology*, 50: 676–80.

Gao, B., Cui, X-W., Li, X-D., Zhang, C-Q. and Miao, H-Q. (2011). Complete genomic sequence analysis of a highly virulent isolate revealed a novel strain of *Sugarcane mosaic virus*. *Virus Genes*, 43: 390–7.

Gaur, R. K., Singh, M., Singh, A., Mishra, L. C. and Rao, G. P. (2006). Identification and molecular characterization of sugarcane streak mosaic virus in north India. *Sugar Cane International*, 24: 13–17.

Geijskes, R. J., Braithwaite, K. S., Dale, J. L., Harding, R. M. and Smith, G. R. (2002). Sequence analysis of an Australian isolate of sugarcane bacilliform badnavirus. *Archives of Virology*, 147: 2393–404.

Gilbert, R. A., Gallo-Meagher, M., Comstock, J. C., Miller, J. D., Jain, M. and Abouzid, A. (2005). Agronomic evaluation of sugarcane lines transformed for resistance to *Sugarcane mosaic virus* strain E. *Crop Science*, 45: 2060–7.

Gilbert, R. A., Glynn, N. C., Comstock, J. C. and Davis, M. J. (2009). Agronomic performance and genetic characterization of sugarcane transformed for resistance to Sugarcane yellow leaf virus. *Field Crops Research*, 111: 39–46.

Gillaspie, A. G. Jr., Mock, R. G. and Smith, F. F. (1978). Identification of sugarcane mosaic virus and characterization of strains of the virus from Pakistan, Iran, and Cameroon. *Proceedings of the International Society of Sugar Cane Technologists*, 16: 347–55.

Girard, J-C., Fernandez, E., Daugrois, J.-H., Roques, D., Roumagnac, P. and Rott, P. (2010). Genetic diversity of *Sugarcane yellow leaf virus* in a sugarcane selection plot in Guadeloupe (FWI). *Proceedings of the International Society of Sugar Cane Technologists*, Vol. 27, Veracruz, Mexico.

Gonçalves, M. C., Klerks, M. M., Verbeek, M., Vega, J. and Van den Heuvel, J. F. J. M. (2002). The use of molecular beacons combined with NASBA for the sensitive detection of *Sugarcane yellow leaf virus*. *European Journal of Plant Pathology*, 108: 401–7.

Grimshaw, J. F. and Donaldson, J. F. (2007). Records of two sugarcane pests *Eumetopina flavipes* Muir (Hemiptera: Delphacidae) and *Chilo terrenellus* Pagenstecher (Lepidoptera: Pyralidae) from the Torres Strait and far north Queensland. *Australian Journal of Entomology*, 46: 35–9.

Grisham, M. P. (2000). Mosaic. In *A Guide to Sugarcane Diseases*, Eds. P. Rott, R. A. Bailey, J. C. Comstock, B. J. Croft and A. S. Saumtally. ISSCT-CIRAD: Montpellier, 249–54.

Grisham, M. P., Johnson, R. M. and Zimba, P. V. (2010). Detecting *Sugarcane yellow leaf virus* infection in asymptomatic leaves with hyperspectral remote sensing and associated leaf pigment changes. *Journal of Virological Methods*, 167: 140–5.

Guo, J., Gao, S., Lin, Q., Wang, H., Que, Y. and Xu, L. (2015). Transgenic sugarcane resistant to *Sorghum mosaic virus* based on coat protein gene silencing by RNA interference. *BioMed Research International*, Article ID 861907, 9p. https://doi.org/10.1155/2015/861907.

Hall, J. S., Adams, B., Parsons, T. J., French, R., Lane, L. C. and Jensen, S. G. (1998). Molecular cloning, sequencing, and phylogenetic relationships of a new potyvirus: Sugarcane streak mosaic virus, and a reevaluation of the classification of the *Potyviridae*. *Molecular Phylogenetics and Evolution*, 10: 323–32.

Hema, M., Joseph, J., Gopinath, K., Sreenivasulu, P. and Savithri, H. S. (1999). Molecular characterization and interviral relationships of a flexuous filamentous virus causing mosaic disease of sugarcane (*Saccharum officinarum* L.) in India. *Archives of Virology*, 144: 479–90.

Ho, T. and Tzanetakis, I. E. (2014). Development of a virus detection and discovery pipeline using next generation sequencing. *Virology*, 471–73: 54–60. https://doi.org/10.1016/j.virol.2014.09.019.

Ingelbrecht, I. L., Irvine, J. E. and Mirkov, T. E. (1999). Posttranscriptional gene silencing in transgenic sugarcane. Dissection of homology-dependent virus resistance in a monocot that has a complex polyploid genome. *Plant Physiology*, 119: 1187–98.

Joyce, P. A., McQualter, R. B., Handley, J. A., Dale, J. L., Harding, R. M. and Smith, G. R. (1998). Transgenic sugarcane resistant to sugarcane mosaic virus. *Proceedings of the Australian Society of Sugar Cane Technologists*, 20: 204–10.

Joyce, P., Hermann, S., O'Connell, A., Dinh, Q., Shumbe, L. and Lakshmanan, P. (2014). Field performance of transgenic sugarcane produced using Agrobacterium and biolistics methods. *Plant Biotechnology Journal*, 12: 411–24.

Karuppaiah, R., Viswanathan, R. and Kumar, V. G. (2013). Genetic diversity of *Sugarcane bacilliform virus* isolates infecting *Saccharum* spp. in India. *Virus Genes*, 46: 505–16.

Kasemsin, P., Chiemsombat, P. and Hongprayoon, R. (2016). Characterization and genetic variation of *Sugarcane Streak Mosaic Virus*, a *Poacevirus* infecting sugarcane in Thailand. *Modern Applied Science*, 10: 137–49. doi:10.5539/mas.v10n4p137.

Keizerweerd, A. T., Chandra, A. and Grisham, M. P. (2015). Development of a reverse transcription loop-mediated isothermal amplification (RT-LAMP) assay for the detection of *Sugarcane mosaic virus* and *Sorghum mosaic virus* in sugarcane. *Journal of Virologicial Methods*, 212: 23–9.

Korimbocus, J., Coates, D., Barker, I. and Boonham, N. (2002a). Improved detection of *Sugarcane yellow leaf virus* using a real-time fluorescent (TaqMan) RT-PCR assay. *Journal of Virological Methods*, 103: 109–20.

Korimbocus, J., Preston, S., Danks, C., Barker, I., Coates, D. and Boonham, N. (2002b). Production of monoclonal antibodies to Sugarcane yellow leaf virus using recombinant readthrough protein. *Journal of Phytopathology*, 150: 488–94.

Kuniata, L. S., Young, G. R., Pais, E., Jones, P. and Nagaraja, H. (1994). Preliminary observations on *Eumetopina* sp. (Hemiptera: Delphacidae) as a vector of Ramu stunt disease of sugarcane in Papua New Guinea. *Journal of the Australian Entomological Society*, 33: 185–6.

Kuniata, L. S., Magarey, R. C., Rauka, G. R., Suma, S. and Bull, J. I. (2010). Screening for Ramu stunt resistance at Ramu Agri-Industries, Gusap, PNG 1986–2008. *Proceedings of the Australian Society of Sugar Cane Technologists*, 32: 312–21.

Lakshmanan, P., Geijskes, R. J., Aitken, K. S., Grof, C. L. P., Bonnett, G. D. and Smith, G. R. (2005). Sugarcane biotechnology: The challenges and opportunities. In *Vitro Cellular and Developmental Biology – Plant*, 41: 345–63.

Laney, A. G., Hassan, M. and Tzanetakis, I. E. (2012). An integrated Badnavirus is prevalent in fig germplasm. *Phytopathology*, 102: 1182–9.

Lehrer, A. T., Wu, K. K. and Komor, E. (2009). Impact of *Sugarcane yellow leaf virus* on growth and sugar yield of sugarcane. *Journal of General Plant Pathology*, 75: 288.

Li, W., He, Z., Li, S., Huang, Y., Zhang, Z., Jiang, D., Wang, X. and Luo, Z. (2011). Molecular characterization of a new strain of sugarcane streak mosaic virus (SCSMV). *Archives of Virology*, 156: 2101–4.

Lin, Y. H., Gao, S. J., Damaj, M. B., Fu, H. Y., Chen, R. K. and Mirkov, T. E. (2014). Genome characterization of *Sugarcane yellow leaf virus* from China reveals a novel recombinant genotype. *Archives of Virology*, 159: 1421–9.

Lockhart, B. E. L. and Autrey, L. J. C. (1988). Occurrence in sugarcane of a bacilliform virus related serologically to banana streak virus. *Plant Disease*, 72: 230–3.

Lockhart, B. E. L., Irey, M. J. and Comstock, J. C. (1996). Sugarcane bacilliform virus, sugarcane mild mosaic virus and sugarcane yellow leaf syndrome. In *Sugarcane Germplasm Conservation and Exchange*. Eds B. J. Croft, C. M. Piggin, E. S. Wallis and D. M. Hogarth, ACIAR Proceedings No. 67, 108–12.

Lomonossoff, G. P. (1995). Pathogen-derived resistance to plant viruses. *Annual Review of Phytopathology*, 33: 323–43.

Magarey, R. C., Suma, S., Irawan, Kuniata, L. S. and Allsopp, P. G. (2002). Sik na binatang bilong suka – diseases and pests encountered during a survey of *Saccharum* germplasm 'in the wild' in Papua New Guinea. *Proceedings of the Australian Society of Sugar Cane Technologists*, 24 (CD-ROM).

Magarey, R. C., Kristini, A., Sallam, N., Samson, P. R., Achadian, E., McGuire, P. J., Goebel, R. and Lonie, K. J. (2010). IPM strategies for pest and disease control in Indonesia: Project overview and outcomes from recent ACIAR-funded research. *Proceedings of the Australian Society of Sugar Cane Technologists*, 32 (electronic format).

Magarey, R. C., Kristini, A., Sallam, N., Achadian, E., Samson, P. R., Goebel, R., Thompson, N. P., McGuire, P. J. and Lonie, K. J. (2012). Preparations to enhance Australia's biosecurity: Part 1– Review of IPM for moth borers and Sugarcane streak mosaic in the Javan sugarcane industry. *Proceedings of the Australian Society of Sugar Cane Technologists*, 34 (electronic format).

Mahmoud, A., Royer, M., Granier, M., Ammar, E-D., Thouvenel, J-C. and Peterschmitt, M. (2007). Evidence for a segmented genome and partial nucleotide sequences of maize yellow stripe virus, a proposed new tenuivirus. *Archives of Virology*, 152: 1757–62.

Mahuku, G., Lockhart, B. E., Wanjala, B., Jones, M. W., Kimunye, J. N., Stewart, L. R., Cassone, B. J., Sevgan, S., Nyasani, J. O., Kusia, E., Kumar, P. L., Niblett, C. L., Kiggundu, A., Asea, G., Pappu, H. R., Wangai. A., Prasanna, B. M. and Redinbaugh, M. G. (2015). Maize lethal necrosis (MLN), an emerging threat to maize-based food security in sub-Saharan Africa. *Phytopathology*, 105: 956–65.

McQualter, R. B., Dale, J. L., Harding, R. M., McMahon, J. A. and Smith, G. R. (2004). Production and evaluation of transgenic sugarcane containing a *Fiji disease virus* (FDV) genome segment S9-derived synthetic resistance gene. *Australian Journal of Agricultural Research*, 55: 139–45.

Mishra, S., Singh, D., Tiwari, A. K., Lal, M. and Rao, G. P. (2010). Elimination of *Sugarcane mosaic virus* and *Sugarcane streak mosaic virus* by tissue culture. *Sugar Cane International*, 28: 119–22.

Mitchell, H. J. (2011). Regulation of genetically modified (GM) sugarcane in Australia. *Proceedings of the Australian Society of Sugar Cane Technologists*, 33 (electronic format).

Mollov, D., Maroon-Lango, C. and Kuniata, L. (2016). Detection by next generation sequencing of a multi-segmented viral genome from sugarcane associated with Ramu stunt disease. *Virus Genes*, 52: 152–5.

Moonan, F., Molina, J. and Mirkov, T. E. (2000). *Sugarcane yellow leaf virus*: an emerging virus that has evolved by recombination between luteoviral and poleroviral ancestors. *Virology*, 269: 156–71.

Moonan, F. and Mirkov, T. E. (2002). Analyses of the genotypic diversity among North, South, and Central American isolates of *Sugarcane yellow leaf virus:* Evidence for Colombian origins and for intraspecific spatial phylogenetic variation. *Journal of Virology*, 76: 1339–48.

Muller, E., Dupuy, V., Blondin, L., Bauffe, F., Daugrois, J. H., Laboureau, N. and Iskra-Caruana, M.-L. (2011). High molecular variability of sugarcane bacilliform viruses in Guadeloupe implying the existence of at least three new species. *Virus Research*, 160: 414–19.

Ndowora, T., Dahal, G., LaFleur, D., Harper, G., Hull, R., Olszewski, N. E. and Lockhart, B. (1999). Evidence that badnavirus infection in *Musa* can originate from integrated pararetroviral sequences. *Virology*, 255: 214–20.

Parmessur, Y., Aljanabi, S., Saumtally, S. and Dookun-Saumtally, A. (2002). *Sugarcane yellow leaf virus* and sugarcane yellows phytoplasma: elimination by tissue culture. *Plant Pathology*, 51: 561–6.

Putra, L. K. and Damayanti, T. A. (2012). Major diseases affecting sugarcane production in Indonesia. *Functional Plant Science and Biotechnology*, 6 (Special Issue 2): 124–9.

Putra, L. K., Kristini, A., Achadian, E. M. and Damayanti, T. A. (2014). *Sugarcane streak mosaic virus* in Indonesia: distribution, characterisation, yield losses and management approaches. *Sugar Tech*, 16: 392–9.

Putra, L. K., Astono, T. H., Syamsidi, S. R. C. and Djauhari, S. (2015). Investigation on transmission modes and host range of *Sugarcane streak mosaic virus* in sugarcane (*Saccharum officinarum* L.) in Indonesia. *Journal of Agricultural and Crop Research*, 3: 59–66.

Rao, G. P., Chatenet, M., Girard, J-G. and Rott, P. (2006). Distribution of *Sugarcane mosaic* and *Sugarcane streak mosaic virus* in India. *Sugar Tech*, 8: 79–81. https://doi.org/10.1007/bf02943747.

Rao, G. P., Sharma, S. K., Singh, D., Arya, M., Singh, P. and Baranwal, V. K. (2014). Genetically diverse variants of *Sugarcane bacilliform virus* infecting sugarcane in India and evidence of a novel recombinant *Badnavirus* variant. *Journal of Phytopathology*, 162: 779–87.

Rassaby, L., Girard, J. C., Irey, M. S., Lockhart, B. E. L. and Rott, P. (1999). Survey of sugarcane yellow leaf syndrome in Réunion Island. *Sugar Cane*, 10: 16–18.

Rassaby, L., Girard, J. C., Lemaire, O., Costet, L., Irey, M. S., Kodja, H., Lockhart, B. E. L. and Rott, P. (2004). Spread of *Sugarcane yellow leaf virus* in sugarcane plants and fields on the Island of Réunion. *Plant Pathology*, 53: 117–25.

Rauka, G. R., Magarey, R. C. and Kuniata, L. S. (2005). The current disease situation on the commercial estate of Ramu Sugar, Papua New Guinea. *Proceedings of the Australian Society of Sugar Cane Technologists*, 27: 266–72.

Roossinck, M. J., Martin, D. P. and Roumagnac, P. (2015). Plant virus metagenomics: advances in virus discovery. *Phytopathology*, 105: 716–27.

Rott, P., Bailey, R. A., Comstock, J. C., Croft, B. J. and Saumtally, A. S. (2000). *A Guide to Sugarcane Diseases*. ISSCT-CIRAD: Montpelier.

Scagliusi, S. M. and Lockhart, B. E. L. (2000). Transmission, characterisation, and serology of a luteovirus associated with yellow leaf syndrome of sugarcane. *Phytopathology*, 90: 120–4.

Schenck, S., Hu, J. S., Lockhart, B. E. L. (1997). Use of a tissue blot immunoassay to determine the distribution of *Sugarcane yellowleaf virus* in Hawaii. *Sugar Cane*, 4: 5–8.

Schenck, S. (2001). Sugarcane yellow leaf syndrome: history and current concepts. In *Sugarcane Pathology Vol II: Virus and Phytoplasma Diseases*, Eds. G. P. Rao, R. E. Ford, M. Tosic and D. S. Teakle, Science Publishers, Enfield USA, 25–35.

Shepherd, D. N., Varsani, A., Windram, O. P., Lefeuvre, P., Monjane, A. L., Owor, B. E. and Martin, D. P. (2008). Novel sugarcane streak and sugarcane steak Reunion mastreviruses from southern Africa and La Réunion. *Archives of Virology*, 153: 606–9. https://doi.org/10.1007/s00705-007-0016-3.

Smith, G. R., Borg, Z., Lockhart, B. E., Braithwaite, K. S. and Gibbs, M. J. (2000). Sugarcane yellow leaf virus: a novel member of the *Luteoviridae* that probably arose by inter-species recombination. *Journal of General Virology*, 81: 1865–9.

Subba Reddy, C. V. and Sreenivasulu, P. (2011). Generation of *Sugarcane streak mosaic virus*-free sugarcane (*Saccharum* spp. hybrid) from infected plants by in vitro meristem tip culture. *European Journal of Plant Pathology*, 130: 597–604.

Tzafrir, I., Torbert, K. A., Lockhart, B. E. L., Somers, D. A. and Olszewski, N. E. (1998). The sugarcane bacilliform badnavirus promoter is active in both monocots and dicots. *Plant Molecular Biology*, 38: 347–56.

Umber, M., Filloux, D., Muller, E., Laboureau, N., Galzi, S., Roumagnac, P., Iskra-Caruana, M.-L., Pavis, C., Teycheney, P.-Y. and Seal, S. E. (2014). The genome of African yam (*Dioscorea cayenensis-rotundata* complex) hosts endogenous sequences from four distinct badnaviurs species. *Molecular Plant Pathology*, 15: 790–801.

Viswanathan, R., Balamuralikrishan, M., Karuppaiah, R. (2008). Identification of three genotypes of *Sugarcane yellow leaf virus* causing yellow leaf disease from India and their molecular characterization. *Virus Genes*, 37: 368–79.

Waller, J. M., Egan, B. T. and Eastwood, D. (1987). Ramu stunt, an important new sugarcane disease in Papua New Guinea. *Tropical Pest Management*, 33: 347–9.

Wang, M. Q. and Zhou, G. H. (2010). A near-complete genome sequence of a distinct isolate of *Sugarcane yellow leaf virus* from China, representing a sixth new genotype. *Virus Genes*, 41: 268–72.

Wang, M. Q., Xu, D. L., Li, R. and Zhou, G. H. (2012). Genotype identification and genetic diversity of *Sugarcane yellow leaf virus* in China. *Plant Pathology*, 61: 986–93.

Wang, Q., Zhou, X. P., and Wu, J. X. (2014). First report of *Maize chlorotic mottle virus* infecting sugarcane (*Saccharum officinarum*). *Plant Disease*, 98: 572. https://doi.org/10.1094/pdis-07-13-0727-pdn.

Wu, X.-B., Alabi, O. J., Damaj, M. B., Sun, S.-R., Mirkov, T. E., Fu, H.-Y., Chen, R.–K. and Gao, S.-J. (2016). Prevalence and RT/RNase H genealogy of *Sugarcane bacilliform virus* isolates from China. *Journal of Phytopathology*, 164: 595–607. https://doi.org/10.1111/jph.12483.

Xie, L., Zhang, J., Wang, Q., Meng, C., Hong, J. and Zhou, X. (2011). Characterization of *Maize chlorotic mottle virus* associated with maize lethal necrosis disease in China. *Journal of Phytopathology*, 159: 191–3.

Xie, Y., Wang, M., Xu, D., Li, R. and Zhou, G. (2009). Simultaneous detection and identification of four sugarcane viruses by one-step RT-PCR. *Journal of Virological Methods*, 162: 64–8.

Xu, D.-L., Zhou, G.-H., Xie, Y.-J., Mock, R. and Li, R. (2010). Complete nucleotide sequence and taxonomy of *Sugarcane streak mosaic virus*, member of a novel genus in the family *Potyviridae*. *Virus Genes*, 40: 432–9.

Yang, I. C., Hafner, G. J., Revill, P. A., Dale, J. L. and Harding, R. M. (2003). Sequence diversity of South Pacific isolates of Taro bacilliform virus and the development of a PCR-based diagnostic test. *Archives of Virology*, 148: 1957–68.

Zhu, Y. J., Lim, S. T. S., Schenck, S., Arcinas, A. and Komor, E. (2010). RT-PCR and quantitative real-time RT-PCR detection of *Sugarcane Yellow Leaf Virus* (SCYLV) in symptomatic and asymptomatic plants of Hawaiian sugarcane cultivars and the correlation of SCYLV titre to yield. *European Journal of Plant Pathology*, 127: 263–73.

Zhu, Y. J., McCafferty, H., Osterman, G., Lim, S., Agbayani, R., Lehrer, A., Schenck, S. and Komor, E. (2011). Genetic transformation with untranslatable coat protein gene of *Sugarcane yellow leaf virus* reduces virus titers in sugarcane. *Transgenic Research*, 20: 503–12.

Chapter 13

Recent progress in understanding three major bacterial diseases of sugarcane: gumming, leaf scald and ratoon stunting

Monique Royer, Isabelle Pieretti, and Stéphane Cociancich, CIRAD, France; and Philippe Rott, University of Florida, USA

1 Introduction

All known agronomically important bacterial diseases of sugarcane have been thoroughly described in a book published in 2000 (Rott, 2000a), which provides for each of these diseases an overview of their geographical distribution, diagnosis, transmission, host range, epidemiology and control. The most important progress since the publication of this book is the development of new diagnostic methods based on the detection of bacteria by amplifying DNA (PCR or polymerase chain reaction). These new methods are especially useful for slow-growing or fastidious bacteria for which culture methods are complex.

The PCR technique is performed with synthetic oligonucleotides (called primers) that are specific to a particular bacterial pathogen. Recent partial or full sequencing of the genome of several bacterial pathogens causing diseases of sugarcane allows an easy design of such specific PCR primers. Meanwhile, a new DNA amplification method, the loop-mediated isothermal amplification (LAMP), allows a faster and more specific diagnosis of bacterial infection (Notomi, 2000 and 2015). In contrast to PCR in which the reaction is

http://dx.doi.org/10.19103/AS.2017.0035.23

carried out with a series of alternating temperature steps or cycles, LAMP is carried out at a constant temperature, and does not require a thermal cycler. Because of its isothermal reaction, LAMP is easier to implement than PCR. Moreover, the amplification efficiency of LAMP is higher than PCR because there is no time loss for thermal change based on its isothermal reaction. LAMP is highly specific for the target sequence since its recognition is strictly performed by six independent primers in the initial stage and by four independent primers during the later stages of the LAMP reaction (Notomi, 2000 and 2015). The LAMP method was, for example, successfully applied for the detection of the fastidious bacterial pathogen *Leifsonia xyli* subsp. *xyli* in sugarcane (Liu, 2013; Ghai, 2014). Real-time quantitative PCR (qPCR) protocols specific for *L. xyli* subsp. *xyli* were also recently developed (Fu, 2016; Carvalho, 2016). The qPCR method monitors the amplification of the targeted DNA molecule during the PCR, that is, in real-time, and not at its end, as in conventional PCR or LAMP. This method, which is more difficult to implement than LAMP, offers a powerful tool to detect *L. xyli* subsp. *xyli* because of its higher sensitivity.

Since sugarcane is vegetatively propagated, the use of healthy or resistant planting material is the most effective measure to control bacterial diseases of sugarcane. Hot-water-treated stalk cuttings or healthy material obtained by tissue culture can be used to produce seed-cane in specific disease-free nurseries. In these nurseries, all harvesting instruments must be disinfected either by heating in a flame or by dipping in a chemical solution. Additionally, special care must be taken during germplasm movement between sugarcane-growing locations to avoid long-distance spread of these bacterial pathogens. Quarantine units are specifically designed to detect and eliminate pathogens causing diseases in sugarcane. At Visacane, the sugarcane quarantine service of CIRAD (French Agricultural Research Centre for International Development; http://visacane.cirad.fr), DNA-amplification-based diagnostic methods are routinely used to ensure that sugarcane stalks are disease-free before shipment to a new geographical location. Chapter 19 of this book focuses on biosecurity of sugarcane cultivation and lists other worldwide sugarcane quarantine services.

The present chapter aims at summarizing recent insights provided by both genomic and microscopy analyses of sugarcane pathogens. To date, such data are available only for the causal agents of gumming, leaf scald and ratoon stunting. Although these basic scientific research data do not provide immediate solutions to control these diseases, they do offer the opportunity to formulate hypotheses regarding the origin of pathogens, their adaptation to sugarcane or their ability to spread and cause diseases in sugarcane. Gumming and leaf scald are sugarcane diseases caused by '*Xanthomonas vasicola* pv. *vasculorum*' and *X. albilineans*, respectively. The genus *Xanthomonas* includes many gram-negative plant pathogenic bacteria that collectively infect a broad range of crops and wild plant species. However, individual *Xanthomonas* species or pathovars usually cause disease in only a few plant species and are highly adapted to their hosts. A recent review summarized the current knowledge of the molecular basis of this host specificity, with a particular focus on the ecology, physiology, and pathogenicity of *Xanthomonas* strains (Jacques, 2016). Interestingly, genomic data discussed in the present chapter indicate that, although belonging to the same genus, '*X. vasicola* pv. *vasculorum*' and *X. albilineans* use very different strategies to induce disease in sugarcane. Ratoon stunting is the sugarcane disease caused by *L. xyli* subsp. *xyli*. The genus *Leifsonia* includes gram-positive bacteria isolated from plants but also from soil, Himalayan glaciers, cyanobacterial mats or water samples. The present chapter focuses on genomic and microscopy data indicating that, although belonging to distinct bacterial phyla (Actinobacteria and Proteobacteria, respectively), *L. xyli* subsp. *xyli* and *X. albilineans* use similar strategies to spread in sugarcane.

This chapter also describes genomic comparative data, including data of bacterial strains closely related to '*X. vasicola* pv. *vasculorum*', *X. albilineans* or *L. xyli* subsp. *xyli* that were isolated from plants other than sugarcane. Genome comparisons are expected to help researchers to elaborate hypotheses regarding the origin of the respective ancestors of these three pathogens and their adaptation to sugarcane.

2 Bacterial pathogens of sugarcane

'*X. vasicola* pv. *vasculorum*' is the causal agent of gumming of sugarcane. This vascular pathogen is transmitted mainly by mechanical means during either preparation of cuttings or harvest, and by wind-blow rains (especially under cyclonic conditions characterized by high humidity and warm temperatures, giving the bacteria the opportunity to exude from the infected leaves) (Saumtally, 2000). Foliar symptoms vary from yellow stripes to partial chlorosis, usually followed by necrosis. Red foliar stripes also sometimes develop, and gum exudates can be observed in stalk sections (Ricaud, 1989; see Fig. 1). These symptoms are very similar to those observed for another sugarcane disease called false red stripe.

The taxonomy of '*X. vasicola* pv. *vasculorum*' has been confusing for a long time: some isolates causing sugarcane gumming have been listed as *X. axonopodis* pv. *vasculorum* (Ivey, 2010) and others as *X. campestris* pv. *vasculorum* (Qhobela, 1992). This confusion partly arose from the similarity of the symptoms induced in sugarcane by '*X. vasicola* pv. *vasculorum*', the causal agent of gumming, and *X. axonopodis* pv. *vasculorum*, the causal agent of false red stripe. Based on genomic DNA hybridization (Vauterin, 1995) and cellular fatty acid profiling studies (Dookun, 2000), isolates causing gumming were clearly attributed to '*X. vasicola* pv. *vasculorum*', and this classification was later on confirmed

Figure 1 Symptoms of gumming disease of sugarcane (photo courtesy of Salem Saumtally, Mauritius Sugar Industry Research Institute). (a) Leaves of sugarcane exhibiting yellow stripes caused by '*Xanthomonas vasicola* pv. *vasculorum*'. (b) sugarcane stalk section showing yellow bacterial exsudates of '*X. vasicola* pv. *vasculorum*'.

using PCR-based genetic markers (Rademaker, 2005). The type strain of *X. campestris* pv. *vasculorum* was shown to belong to *X. axonopodis pv. vasculorum* but, to date, no type strain has been defined for '*X. vasicola* pv. *vasculorum*' in the international code of nomenclature of bacteria. Based on the ribosomal DNA 16S/23S sequences, strains causing false red stripe were confirmed to belong to *X. axonopodis* pv. *vasculorum* (Royer, unpublished data obtained from strains isolated in Brazil by Giglioti, 1999). No genome sequence has been published yet for *X. axonopodis* pv. *vasculorum*, explaining why this pathogen has not been included in this chapter.

X. albilineans is the causal agent of leaf scald of sugarcane. *X. albilineans* spreads in the xylem vessels and was recovered from roots, stalks and leaves (Klett, 1994). Disease symptoms vary from a single, white, narrow, sharply defined stripe to complete wilting and necrosis of infected sugarcane leaves, leading to plant death and large yield losses in the field (Rott, 2000b; see Fig. 2). Spread of *X. albilineans* occurs mainly mechanically through the use of contaminated harvesting tools and by distribution and planting of infected seed-cane. However, aerial transmission and epiphytic survival have also been reported for this pathogen (Autrey, 1992; Daugrois, 2003; Champoiseau, 2009). Organized disturbed weather such as tropical depressions, storms, or hurricanes is also considered as responsible for long-distance dissemination of the pathogen that invades the sugarcane leaf canopy (Daugrois, 2003). The sugarcane leaf scald outbreak that occurred in Louisiana in November 1992 was related to the passage of a severe hurricane over the Gulf of Mexico in August 1992 (Hoy, 1994).

X. albilineans produces albicidin, a secreted phytotoxin with antibiotic properties (Birch, 2001). Albicidin is a potent inhibitor of DNA gyrase, an essential bacterial enzyme that catalyses the ATP-dependent negative super-coiling of double-stranded closed-circular DNA (Hashimi, 2007). This broad-spectrum antibiotic may help *X. albilineans* to compete with other bacteria spreading in sugarcane. Albicidin also targets the chloroplast DNA gyrase of sugarcane, inhibits proplastid DNA replication and blocks chloroplast differentiation, resulting in appearance of white foliar stripes (Birch, 2001; Hashimi, 2007). The structure of albicidin was recently elucidated and a total chemical synthesis route was established, providing the basis for the assessment of albicidin as a lead structure for antibacterial drug development and use as medicine (Cociancich, 2015; Kretz, 2015). *X. albilineans* exhibits intra-species genotypic variability, as illustrated by the description of three serotypes (Rott, 1986) and eight genetic groups identified by pulsed-field gel electrophoresis (PFGE) (Davis, 1997).

L. xyli subsp. *xyli* is the causal agent of sugarcane ratoon stunting (or sugarcane ratoon stunt). This pathogen, which spreads in the xylem vessels, does not induce any reliable or characteristic external symptoms. On a worldwide basis, and from an economical point of view, ratoon stunting is the most important bacterial disease of sugarcane. Annual yield losses vary drastically, from negligible up to more than 30% (Bailey, 1999; Davis, 2000; Dean, 1989; Grisham, 2009). Yield reduction is caused by the slower growth of diseased crops with the concomitant production of thinner and shorter stalks and, for the most severe attacks of the pathogen, a reduction in the number of stalks (Davis, 2000; see Fig. 3). An internal red discoloration of vascular bundles at the lower portion of the nodes can be observed in diseased stalks of some sugarcane varieties. Because of the lack of reliable diagnostic symptoms, diagnosis is most often based on the detection of *L. xyli* subsp. *xyli* using laboratory techniques. These techniques include mainly PCR and LAMP assays because *L. xyli* subsp. *xyli* is a fastidious bacterium that cannot be grown easily on culture media in the laboratory (Liu, 2013; Ghai, 2014). *L. xyli* subsp. *xyli*

Figure 2 Symptoms of leaf scald disease of sugarcane (photo courtesy of Philippe Rott). (a) Progress of severity of leaf symptoms, from white lines to necrosis. From left to right: Healthy sugarcane leaf; leaf with white lines; leaf with white lines and chlorosis; leaf with early stage of necrosis; leaf with advanced necrosis; leaf with white lines, chlorosis, and pronounced necrosis. (b) Leaves of sugarcane exhibiting severe bleaching, chlorosis, and necrosis caused by *Xanthomonas albilineans*. (c) Mature sugarcane stalk with leaf scald symptoms caused by *X. albilineans*: numerous side shoots along the stalk and leaves exhibiting bleaching, chlorosis, and necrosis.

is transmitted mechanically through the use of contaminated harvesting tools and by distribution and planting of infected seed-cane. No aerial transmission has been reported for this pathogen. A recent publication argued that a wild sugarcane species, *Saccharum spontaneum*, could be the original host of *L. xyli* subsp. *xyli* (Young, 2016). Consequently, presence of *L. xyli* subsp. *xyli* in commercial sugarcane varieties would result from the acquisition of the pathogen during the interspecific hybridization work undertaken in Java during the 1920s.

Figure 3 Symptoms of ratoon stunting disease of sugarcane (photo courtesy of Philippe Rott). Increased stunting evident in sugarcane infected by *Leifsonia xyli* subsp. *xyli* (two sugarcane rows on the left of the RSD sign) compared to disease-free sugarcane (two sugarcane rows on the right of the RSD sign).

Formerly classified as *Clavibacter xyli* subsp. *xyli* (Davis, 1984), *L. xyli* subsp. *xyli* was removed from the genus *Clavibacter* to create, in 2000, the genus *Leifsonia* together with *L. poae* found in *Poa annua* root galls, and several other *Leifsonia* species isolated from environmental samples such as water, snow or soil (Evtushenko, 2000; Young, 2016). *L. xyli* subsp. *xyli* is remarkably genetically uniform: analysis of isolates of *L. xyli* subsp. *xyli* from nine countries revealed no variation in the 16S ribosomal DNA sequence, nor any genomic re-arrangement based on DNA fingerprinting profiles (Young, 2006).

3 Genomic features of bacterial pathogens of sugarcane

The advent of genome sequencing has transformed the field of plant pathology, providing more tools and resources for understanding diversity and taxonomy of pathogens and for studying the molecular basis of plant bacterial diseases. According to the methods used for sequencing, available bacterial genome sequences are finished genome sequences corresponding to the complete sequence of a linear or circular chromosome or draft genome sequences corresponding to unassembled parts of the sequence of a chromosome. Genome sequences also include the sequence of plasmids which are segments of DNA independent of the main chromosome and capable of replication. Several methods are available to build genomic-based phylogenetic trees and to estimate the genetic distances between bacteria (Zhi, 2012). The average nucleotide identity (ANI), calculated from pairwise comparisons of all genome sequences shared between two given strains, has been proposed as a new metrics for prokaryotic species definition and classification. An ANI value of 95–96% was proposed as a recommended standard for delineating prokaryotic species (Goris, 2007; Richter, 2009).

The process of annotating prokaryotic genomes includes prediction of protein-coding genes, as well as other functional genome units such as structural RNAs, tRNAs, small RNAs, pseudogenes, control regions, direct and inverted repeats, insertion sequences, transposons, and other mobile elements. Several platforms provide prokaryotic genome annotation services such as the platform MicroScope and the pipeline EuGene-PP (Vallenet, 2009; Sallet, 2014). Comparative analyses generally result in identification of candidate genes potentially involved in pathogenicity. These analyses also contribute to determine the 'core genome' containing genes present in all strains of a species, the 'dispensable genome' containing genes present in two or more strains, and finally 'unique genes' specific to single strains of a pathogen.

A draft genome sequence of '*X. vasicola* pv. *vasculorum*' strain NCPPB 702, isolated from sugarcane in Zimbabwe in 1959, was published in 2010 (Studholme, 2010). This genome sequence is 5.5 Mb in size, and it encodes the complete Hrp (hypersensitive response and pathogenicity) type III secretion system (T3SS). Most of plant pathogenic *Xanthomonas* require an Hrp T3SS to secrete and translocate effector proteins into plant cells, and to cause disease (for review, see Büttner, 2010). At least 21 homologues of the T3SS effectors described in other *Xanthomonas* species were identified in the genome of '*X. vasicola* pv. *vasculorum*' NCPPB 702 (Studholme, 2010). Because T3SS effectors participate in host defence suppression, they are regarded as major host-specificity determinants of *Xanthomonas* strains (Jacques, 2016). T3SS effectors are also known to promote pathogen multiplication and dispersion (Jacques, 2016). For these reasons, and despite the lack of experimental data, the T3SS effectors encoded by the genome of '*X. vasicola* pv. *vasculorum*' are expected to play a key role in the strategies used by this pathogen to spread in sugarcane.

The genome sequence of '*X. vasicola* pv. *vasculorum*' also encodes the complete gum locus that is involved in biosynthesis of xanthan, an extracellular polysaccharide (EPS) which contributes to the virulence of most *Xanthomonas* strains (Aslam, 2008; Li, 2011). The production of xanthan is likely linked to the symptom of gumming induced by '*X. vasicola* pv. *vasculorum*' in sugarcane (See Fig. 1). Draft genome sequences of four additional '*X. vasicola* pv. *vasculorum*' strains isolated from sugarcane were published and compared in 2014, revealing intra-species variability, especially in the number and nature of T3SS effectors encoded by each strain (Wasukira, 2014). However, because of the lack of experimental data, much remains unknown about the virulence determinants of '*X. vasicola* pv. *vasculorum*'.

A finished genome sequence of *X. albilineans* strain GPE PC73, isolated from sugarcane in Guadeloupe (French West Indies) in 2003, was published in 2009

(Pieretti, 2009). This strain was previously identified as highly pathogenic to sugarcane (Champoiseau, 2006). The genetic information of *X. albilineans* strain GPE PC73 is encoded on a 3.8-Mb circular chromosome, much smaller than those described to date in other sequenced *Xanthomonas* (approximately 5 Mb), and on three plasmids of 25, 27 and 32 kb. Analysis of this genome sequence showed that *X. albilineans* experienced a genome reduction and that this bacterial species exhibits distinctive genomic features compared to other sequenced *Xanthomonas* species (Pieretti, 2009, 2012, 2015b). For example, the genome of *X. albilineans* lacks the Hrp T3SS and the xanthan biosynthesis loci. In contrast, the genome of *X. albilineans* contains specific loci encoding a SPI-1 (*Salmonella* pathogenicity island-1) T3SS and several non-ribosomal peptide synthetases (NRPS). The SPI-1 T3SS is required in animal or insect-associated bacteria to secrete and translocate effector proteins into cells of their host (Cornelis, 2006).

Inoculation of sugarcane with knockout mutants showed that the SPI-1 T3SS is not required by *X. albilineans* to spread within sugarcane and to cause symptoms (Marguerettaz, 2011). This SPI-1 T3SS is expected to be required for the interaction of *X. albilineans* with an unknown animal or insect host (Marguerettaz, 2011). NRPS enable bacteria to produce small molecules encompassing a broad variety of structures as well as a wide range of biological activities. NRPS loci of *X. albilineans* include the locus directing the biosynthesis of albicidin. Biosynthesis products of other NRPS loci of *X. albilineans* remain unknown (Royer, 2013). Draft genome sequences of *X. albilineans* strains belonging to each of the eight genetic groups identified by PFGE were published in GenBank in 2015. These sequences revealed that most of specific genomic features identified in strain GPE PC73 are conserved in all *X. albilineans* strains (Pieretti, unpublished data). Occurrence of intra-species genetic variability was also confirmed, and ANI was 97.8% for the most distant strains.

A finished genome sequence of *L. xyli* subsp. *xyli* strain CTCB07 was published in 2004 (Monteiro-Vitorello, 2004). Strain CTCB07 was chosen for sequencing because of its routine use in screening for resistance to ratoon stunting in a Brazilian sugarcane-breeding programme. *L. xyli* subsp. *xyli* possesses a 2.6-Mb circular chromosome. Thirteen per cent of the 2,351 genes encoded by this chromosome were considered as pseudogenes (Monteiro-Vitorello, 2004), indicating that *L. xyli* subsp. *xyli* experienced a genome erosion. It was also hypothesized that the ancestor of *L. xyli* subsp. *xyli* was a free-living bacterium, a hypothesis that is supported by other features of the genome of *L. xyli* subsp. *xyli* (Monteiro-Vitorello, 2004 and 2009). Interestingly, genome mining did not result in identification of any pathogenicity gene specific to *L. xyli* subsp. *xyli* (Monteiro-Vitorello, 2004). Furthermore, the genome of *L. xyli* subsp. *xyli* does not encode any cytolysin-mediated translocation (CMT) system that is regarded as an equivalent of the T3SS in gram-positive bacteria. The absence of any CMT system and the low number of putative pathogenicity genes in the genome of *L. xyli* subsp. *xyli* were proposed to consider *L. xyli* subsp. *xyli* as a stealth pathogen that induces only subtle symptoms in sugarcane (Monteiro-Vitorello, 2004 and 2009). A draft genome of strain *L. xyli* subsp. *xyli* GXBZ01 isolated from diseased sugarcane in China in 2014 was published in GenBank in 2015. Comparison of the genome sequences of *L. xyli* subsp. *xyli* strains CTCB07 and GXBZ01 confirmed that the bacterial species causing sugarcane ratoon stunting is genetically uniform (ANI = 99.8%; Pieretti, unpublished data).

4 Transposon mutagenesis studies to identify genes for pathogenicity

Transposons are useful laboratory tools facilitating genome-scale studies of gene function. They are used to obtain libraries of thousands of bacterial mutants. These mutants correspond to independent mutations (independent insertion events of the transposon). The higher the number of mutants, the higher is the probability that the library contains mutants in each gene of the bacterial genome. The screening of these mutants for loss of pathogenicity often leads in identification of genes required for disease expression. PCR amplification and sequencing of the transposon insertion site of the mutants affected in pathogenicity result in identification of the mutated gene. Availability of the genome sequence of the bacterium facilitates this identification. The first limit of this method is that, in some bacteria, transposition does not occur completely at random and distribution of the insertion sites of the transposon is limited to some regions of the genome. The second limit is that the phenotype observed for a given mutant can result from a mutation differing from the transposon insertion. Two strategies can be used to confirm that the phenotype of a mutant is caused by the mutation of the gene in which the transposon inserted. The first one is site-directed mutagenesis to specifically mutate this gene in the wild-type strain. The second strategy (called complementation) consists in restoration of the wild-type phenotype by introduction into the mutant of a functional copy of the mutated gene.

A library of 712 transposon mutants of *L. xyli* subsp. *xyli* strain QPF110 has been screened for the ability to spread in sugarcane (Brumbley, 2004). This time-consuming work consisted in inoculation of sugarcane with each of the 712 mutants. Approximately three months post inoculation, plants were harvested and checked for the presence of *L. xyli* subsp. *xyli* (Brumbley, 2004). Only one mutant failed to colonize sugarcane in repeated experiments. The transposon insertion site of this mutant was located in the protein-coding gene annotated as Lxx12290 in the genome sequence of *L. xyli* subsp. *xyli* strain CTCB07. The protein encoded by this gene is predicted to be an integral membrane protein. Its amino acid sequence shares similarities with Zn-dependent proteases with chaperone function. The link between this mutated gene and the loss of pathogenicity was not confirmed by site-directed mutagenesis, nor by complementation with a functional form of the wild-type gene. Without any further experimental data, the role of this gene remains unknown. This gene is conserved in the species *Leifsonia aquatica*, which is not associated with plants. It is indeed present in many bacteria commonly found in soil, and it is absent in *X. albilineans*.

A library of 1216 transposon mutants of *X. albilineans* strain XaFL07-1 has been produced and screened for mutants affected in pathogenicity (Rott, 2011). Mutants exhibiting reduced capacity in producing disease symptoms or in colonizing the sugarcane stalk were further analysed and several pathogenicity-related genes, so far unknown to occur in *Xanthomonas* spp., were identified. These genes included loci encoding hypothetical proteins, transport proteins and an OmpA family outer-membrane protein (Rott, 2011). The importance of protein OmpA for pathogenicity of *X. albilineans* was confirmed by site-directed mutagenesis and complementation studies (Fleites, 2013). This latter publication also provided new insights into the requirement of protein OmpA for integrity of the bacterial outer membrane. To date, no complementation studies were performed for the other candidate pathogenicity factors identified by Rott (2011). The putative role of the transport candidate proteins in pathogenicity of *X. albilineans* is discussed below.

Although highly interesting and informative, the above-mentioned studies remain incomplete. The major reason is the low number of mutants that were screened. In order to cover the entire genome (i.e. to make sure that each gene was mutated via transposon insertion), at least 6000 and 10 000 mutants should be produced and screened for *L. xyli* subsp. *xyli* and *X. albilineans*, respectively. The major limitation relies in the difficulty to inoculate a high number of mutants into sugarcane plants, a process that is time and space consuming. Additionally, transposition does not appear to occur completely at random in the genome of *L. xyli* subsp. *xyli* strain QPF110 (Brumbley, 2004).

5 Genomic features indicating spreading strategies

As proposed above, '*X. vasicola* pv. *vasculorum*' most likely uses T3SS effectors to suppress sugarcane defences and to promote its multiplication and spread in sugarcane. In contrast, with the exception of the genes coding for toxin albicidin and hypothetical proteins, no genes potentially involved in offensive or counter-defensive mechanisms were identified in the genomes of *X. albilineans* and *L. xyli* subsp. *xyli*. The bacterial pathogen *Xylella fastidiosa*, like *X. albilineans* and *L. xyli* subsp. *xyli*, does not possess any T3SS either, nor uses T3SS effectors to cause diseases in many plant hosts (Simpson, 2000). *X. fastidiosa* is vector-transmitted by various xylem sap-feeding insects and is able to colonize many plant species (citrus, wine grape, coffee, alfalfa, peach, plum, almond, elm, maple, pear, etc.) (reviewed in Chatterjee, 2008). *X. fastidiosa* and *X. albilineans* are phylogenetically close and both belong to the family *Xanthomonadaceae* (Pieretti, 2009). *X. fastidiosa*, *X. albilineans* and *L. xyli* subsp. *xyli* live and multiply essentially in dead xylem cells or tracheary elements. In this rather cloistered environmental niche, these pathogens may have largely avoided surveillance by general and specific plant defence systems and succeed in spreading without T3SS effectors. However, like the rice vascular pathogen *Xanthomonas oryzae* pv. *oryzae*, they may interact with living xylem parenchyma cells (Hilaire, 2001). If they do, they do not use T3SS effectors but other effectors that remain to be identified.

X. fastidiosa, *X. albilineans* and *L. xyli* subsp. *xyli*, which all experienced a reductive genome evolution, most likely use strategies to avoid or limit detection by plant defence systems. Microbe-associated molecular patterns (also called MAMPs) are molecules that can be referred to as small molecular motifs conserved in a class of microbes, for example bacteria. The detection of MAMPs by plant defence systems results in an oxidative burst and callose deposition. The genome erosion experienced by *X. fastidiosa*, *X. albilineans* and *L. xyli* subsp. *xyli* may have resulted in the loss of genes encoding MAMPs, favouring adaptation of these pathogens to the life within their respective hosts. For example, *X. fastidiosa* and *L. xyli* subsp. *xyli* have lost genes required for assembly of the flagellum. Flagellin, the principal component of the bacterial flagellum, is known to activate defence mechanisms in several plant species. Type strain CFBP 2523 of *X. albilineans* has lost genes required for assembly of the flagellum (Darrasse, 2013). However, the genome of *X. albilineans* strain GPE PC73 contains a complete flagellum biosynthesis and assembly locus (Pieretti, 2009) and most strains of *X. albilineans* carry a polar flagellum (Birch, 2001): Rott (1984) has observed a polar flagellum for all 28 analysed strains spanning the genetic diversity of *X. albilineans*. The flagellar cap protein FliD of strain GPE PC73 possesses a polyserine linker (PSL) of 12 serine residues that is absent in the corresponding protein encoded by other *Xanthomonas* species. This linker is conserved in all *X. albilineans* strains for which a draft sequence is available in GenBank (Pieretti, unpublished data). It was proposed that the presence of this PSL may modify the tertiary structure of the protein FliD when compared to other *Xanthomonas* species and, as a consequence, may impair its detection by the sugarcane defence system (Pieretti, 2012). Further experimental studies are required to confirm that the protein flagellin or the protein FliD without any PSL are recognized as MAMPs by the sugarcane defence system.

Because of the carbon-limiting content of the xylem sap, xylem vessels are a stressful environment compared to other niches within a plant, such as the parenchyma. The adaptation of *X. fastidiosa*, *X. albilineans* and *L. xyli* subsp. *xyli* to a xylem-associated lifestyle can be illustrated by their enzymes adapted to the use of plant cell breakdown products as carbon sources. The genome of *X. albilineans* strain GPE PC73 contains 19 genes encoding putative cell-wall degrading enzymes (CWDE; e.g. cellulases, polygalacturonases, rhamnogalacturonases, beta-glucosidases and xylanases), all of which are predicted to be secreted by a type II secretion system (Pieretti, 2012). One of these enzymes, the endoglucanase EngXCA, also encoded by the genome of *X. fastidiosa* and *L. xyli* subsp. *xyli*, has a cellulose-binding domain (CBD) at its C-terminal extremity that is separated from the endoglucanase domain by a linker region consisting of simple or repetitive sequences rich in proline, threonine, serine or glycine. The longer the linker region, the longer will be the physical distance between the endoglucanase domain and the CBD. Long linker regions are known to enhance substrate accessibility (Boraston, 2004). The length of the linker region is higher in *X. fastidiosa*, *X. albilineans* and *L. xyli* subsp. *xyli* in comparison with other plant pathogenic bacteria (see Table 1), suggesting that EngXCA of these pathogens are particularly well adapted to the degradation of carbohydrate substrates. This is even more striking for *X. albilineans* and *X. fastidiosa*, which possess additional cellulases exhibiting a CBD and a long linker region (Pieretti, 2012).

X. fastidiosa, *X. albilineans* and *L. xyli* subsp. *xyli* are also very well equipped to spread in a nutrient-limiting environment. In gram-negative bacteria, outer membrane TonB-dependent transporters (TBDTs) are involved in the active transport and uptake of plant nutrients, such as iron-siderophore complexes, vitamin B12, nickel and plant carbohydrates (Blanvillain, 2007; Schauer, 2008; Déjean, 2013). Therefore, the 15 and 35 TBDT genes found on the chromosomes of *X. fastidiosa* and *X. albilineans*, respectively, could play a key role in the active transport of plant nutrients. Some of these TBDT genes may be adapted for the transport of carbohydrate products provided by CWDE. Interestingly, two TBDT genes of *X. albilineans* are expected to contribute to disease severity and extent of stalk colonization of sugarcane (Rott, 2011). *L. xyli* subsp. *xyli*, like all gram-positive bacteria, does not possess any TBDT gene. However, *L. xyli* subsp. *xyli* has a strikingly large repertoire of ATP-binding cassette transporters (ABC transporters) and a complete phosphoenolpyruvate-dependent phosphotransferase system which could allow this pathogen to efficiently uptake plant nutrients (Monteiro-Vitorello, 2004). ABC transporters utilize the energy of adenosine triphosphate (ATP) to energize the translocation of various substrates across membranes, either for uptake or for export of the substrate. The genome of *L. xyli* subsp. *xyli* encodes 42 ABC transporters whereas the genomes of *X. fastidiosa* and *X. albilineans* encode less of these transporters (26 and 32, respectively). Another gene shared by *X. albilineans* and *L. xyli* subsp. *xyli* could be important for spread in a nutrient-poor environment. This gene encodes an alanine dehydrogenase which was shown, in other bacteria, to permit use of alanine as a nitrogen source (Feng, 2002). It was mistakenly proposed by Pieretti (2012) to encode an oxidoreductase involved in

Table 1 Comparison of the endoglucanase EngXCA encoded by different plant pathogenic bacteria. EngXCA contains a cellulose-binding domain (CBD) in all strains listed in the table. The length of the linker separating the CBD from the endoglucanase part varies between the bacterial species and strains. Two species, *X. sacchari* (strain NCPPB 4393) and *X. albilineans* (strain GPE PC73), encode two EngXCA. Adapted from Pieretti (2012)

Bacterial species and strain number	Host plant	Accession number	Length of the linker
Xylella fastidiosa Temecula 1	Grapevine	PD1851	157 Aa[4]
Xylella fastidiosa 9a5c	Citrus	XF0818	132 Aa
Xanthomonas albilineans GPE PC73	Sugarcane	XALc_2969	96 Aa
		XALc_2967	26 Aa
Leifsonia xyli subsp. *xyli* CTCB07	Sugarcane	Lxx22675[1]	51 Aa
Xanthomonas sacchari NCPPB 4393	Sugarcane	AGDB00000000[2]	44 Aa
			28 Aa
Leifsonia xyli subsp. *cynodontis* DSM46306	Bermuda grass	NC_022438[2]	37 Aa
Xanthomonas oryzae pv. *oryzae* MAFF 311018	Rice	XOO_3789	33 Aa
Xanthomonas translucens DSM 18974	Barley	NZ_CAPJ01000000[2]	29 Aa
Xanthomonas campestris pv. *campestris* ATCC 33913	Cabbage	XCC3521	29 Aa
Clavibacter michiganensis subsp. *michiganensis* NCPPB 382	Tomato	CAA44467[1,3]	27 Aa
'*Xanthomonas vasicola* pv. *vasculorum*' NCPPB 702	Sugarcane	ACHS01000051[2]	24 Aa
Xanthomonas translucens pv. *translucens* UPB787	Barley	NZ_JTEM00000000[2]	23 Aa
Xanthomonas axonopodis pv. *vesicatoria* 85–10	Pepper	XCV0670	21 Aa
Xanthomonas axonopodis pv. *citri* 306	Citrus	XAC0612	19 Aa
Xanthomonas oryzae pv. *oryzicola* BLS256	Rice	XOC_0671	9 Aa

[1]Annotated as CelA in Jahr (2000) and Monteiro-Vitorello (2004).
[2]For non-annotated draft genome sequences, the accession of the complete genome is provided.
[3]In strain NCPPB 382 of *Clavibacter michiganensis* subsp. *michiganensis*, the endoglucanase contains an additional peptidoglycan-binding domain which is located downstream from the CBD.
[4]Amino acid.

protecting from the oxidative burst. This alanine dehydrogenase is absent in *X. fastidiosa*, but it is present in *Xanthomonas* species belonging to the same phylogenetic clade as *X. albilineans*: *X. sacchari*, *X. translucens*, and *X. hyacinthi*.

As shown in other plant–microbe interactions, *X. fastidiosa*, *X. albilineans* and *L. xyli* subsp. *xyli* most likely use various strategies to protect themselves against plant defence systems. ABC transporters may confer resistance against antibacterial toxins produced by plants by exporting the toxin outside of the bacterial cell. Two ABC transporter genes appear to be required for efficient stalk colonization of sugarcane by *X. albilineans* (Rott, 2011). One of these genes confers resistance to novobiocin, confirming that it could confer resistance to a toxin or another antibacterial compound produced by sugarcane (Rott, 2011). Three other ABC transporter genes from *X. albilineans* are also present in the genomes of *L. xyli* subsp. *xyli*, *Acidovorax avenae* subsp. *avenae*, and *X. sacchari*, three bacterial species which are all able to spread in sugarcane (Pieretti, 2012). These three ABC transporters may confer resistance against antibacterial toxins specifically produced by sugarcane. Two putative catalase genes (*srpA* and *katE*) and a bifunctional catalase/peroxidase (*katG*) are present in the genome of *X. albilineans*. These genes may play a role by protecting *X. albilineans* from an oxidative burst resulting from its recognition by the sugarcane plant. A recent study showed that *katG* is required for pathogenicity of *X. albilineans* in sugarcane (Jain, 2016). Out of these three genes, only *katE* is shared by *X. albilineans* and *L. xyli* subsp. *xyli*.

In silico analysis of the genome of *L. xyli* subsp. *xyli* and *X. albilineans* suggested that both pathogens are able to spread within sugarcane xylem vessels with less virulent traits compared to '*X. vasicola* pv. *vasculorum*' that possesses a Hrp T3SS. However, *L. xyli* subsp. *xyli* and *X. albilineans* may use specific and unknown strategies to suppress or weaken sugarcane plant defences. Secretion systems identified in the genomes of *L. xyli* subsp. *xyli* and *X. albilineans* might be involved in the secretion of yet-to-be identified effectors that could be secreted in the xylem and diffuse to the parenchyma target cells (Pieretti, 2012; Monteiro-Vitorello, 2004). Hypothetical protein genes specific to *L. xyli* subsp. *xyli* or *X. albilineans* may correspond to one or more of these effectors. However, based on current genomic data, pathogenicity of both *L. xyli* subsp. *xyli* and *X. albilineans* seems to result mostly from an efficient use of the nutrients available in the xylem vessels of sugarcane rather than a sophisticated dialogue with the sugarcane plant.

6 Bacterial invasion of xylem vessels and non-vascular tissues of sugarcane

Until recently, *X. albilineans* and *L. xyli* subsp. *xyli* were both considered as strictly limited to the xylem vessels of sugarcane. In order to better understand the movement of *X. albilineans* and *L. xyli* subsp. *xyli* inside sugarcane, new tools were developed to monitor these pathogens during the process of their respective colonization of sugarcane. Strains of *X. albilineans* and *L. xyli* subsp. *xyli* were tagged with a non-disruptive fluorescent marker, the green fluorescent protein (GFP). Confocal laser scanning microscopy and regular fluorescence microscopy were used to investigate the localization of the tagged strains of *X. albilineans* and *L. xyli* subsp. *xyli* in sugarcane, respectively. These microscopy methods have the advantage of allowing observations of intact plant tissues. Surprisingly, GFP-tagged strains of both *X. albilineans* and *L. xyli* subsp. *xyli* were observed not only in the xylem vessels but also in non-vascular tissues (Mensi, 2014; Quecine, 2015).

During the early stages of infection (i.e. 5 to 15 days post-inoculation), *X. albilineans* was observed only in the xylem vessels (protoxylem and metaxylem), but at later stages of infection (i.e. 20–60 days post-inoculation), *X. albilineans* was also found in tissues other than the xylem, either in leaves or in stalks (Mensi, 2014). In leaves, *X. albilineans* was observed in the phloem, parenchyma and bulliform cells (see Fig. 4). In stalks, *X. albilineans* was observed in sucrose storage cells and in intercellular spaces between these cells (see Fig. 4). Most of these observations of *X. albilineans* made by confocal microscopy were confirmed by immunocytochemistry and transmission electron microscopy (see Fig. 4, Mensi, 2014). These recently published data are consistent with older observations depicting the presence of *X. albilineans* not only in xylem vessels but also in non-vascular tissues (Martin, 1932; Legaz, 2011).

The GFP-tagged strain of *L. xyli* subsp. *xyli* was observed only in leaves of sugarcane, and no fluorescent signal could be detected in stalks (Quecine, 2015). This latter result was linked to difficulties in finding *L. xyli* subsp. *xyli* in stalks of sugarcane. In leaves, the GFP-tagged *L. xyli* subsp. *xyli* strain was observed in xylem vessels and also in the mesophyll and in the bundle sheath cells surrounding the vascular system (Quecine, 2015). The pathogen was also observed in non-vascular foliar tissues of sugarcane at both early and later stages of infection (Quecine, 2015). These results are in agreement with older observations depicting the presence of *L. xyli* subsp. *xyli* in tracheids and xylem parenchyma (Kao, 1978).

The GFP labelling has also been successfully used to monitor *in planta Clavibacter michiganensis* subsp. *michiganensis*. This gram-positive bacterium is closely related to *L. xyli* subsp. *xyli* and also lacks any secretion system delivering effectors directly into the plant host cells. *C. michiganensis* subsp. *michiganensis* is the causal agent of bacterial wilt and canker, a highly damaging disease of tomato. The GFP-tagged strain of *C. michiganensis* subsp. *michiganensis* was shown to spread efficiently in the xylem vessels of tomato, but the bacterium was also observed in the parenchymatous tissue and in the intracellular spaces (Lelis, 2014).

Like *X. albilineans* and *L. xyli* subsp. *xyli*, *C. michiganensis* subsp. *michiganensis* possesses an endoglucanase EngXCA harbouring a CBD domain and a linker region (see Table 1). The endoglucanase EngXCA of *C. michiganensis* subsp. *michiganensis* is a pathogenicity determinant required for induction of the disease in tomato (Jahr, 2000). Therefore, endoglucanase EngXCA, by allowing an efficient degradation of cell walls, may play a key role in the ability of *X. albilineans*, *L. xyli* subsp. *xyli* and *C. michiganensis* subsp. *michiganensis* to exit the vessels of the xylem and to colonize other tissues. However, *X. fastidiosa*, that also possesses an endoglucanase EngXCA with a long linker region, appears unable to spread in non-vascular tissues since no fluorescent signal has been observed outside of the xylem by monitoring a tagged strain of *X. fastidiosa* in grapevine (Newman, 2003) and citrus (Niza, 2015).

Nevertheless, the work in grapevine presented evidence for movement of *X. fastidiosa* through bordered pits that separate xylem vessels from each other (Newman, 2003). The pit membrane, which lies in the centre of each pit, allows water and nutrients to pass between xylem conduits but limits the spread of vascular pathogens. A mutant of *X. fastidiosa* disrupted in the gene encoding the endopolygalacturonase PglA, involved in degradation of the cell wall, was restricted to the point of inoculation and unable to move systemically in grapevine, indicating that PglA plays a major role in vessel-to-vessel movement (Roper, 2007). Moreover, an *in vitro* experimental study with purified enzymes showed that a combination of both the endoglucanase EngXCA and the endopolygalacturonase PglA of

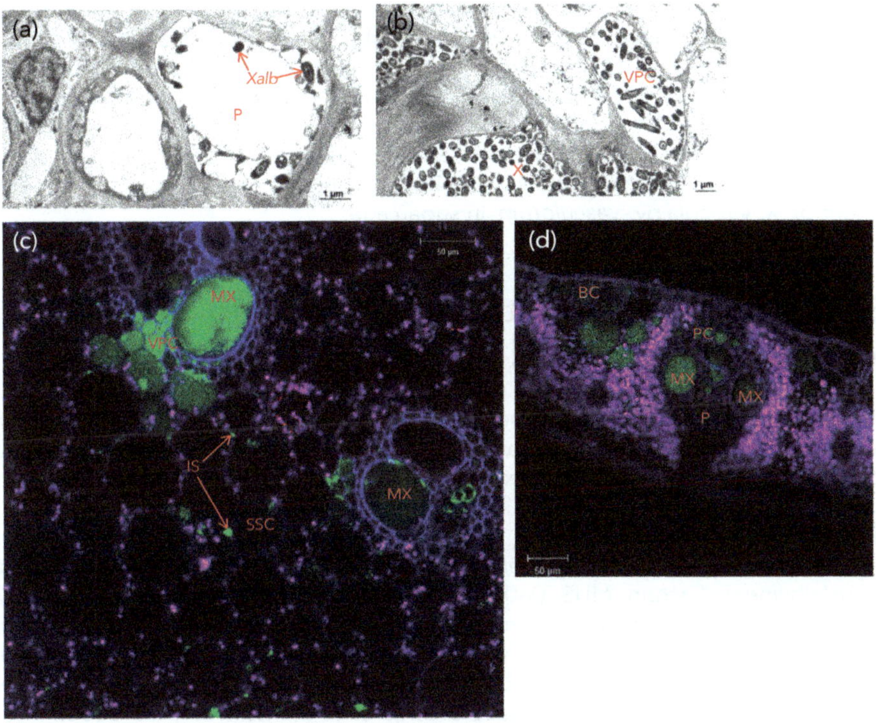

Figure 4 Localization of *Xanthomonas albilineans* in tissues of infected sugarcane plants (photo courtesy of Daniel Gargani and Imène Mensi, CIRAD). (a) and (b) Transmission electron microscopy images of cross sections of sugarcane stalks. *X. albilineans* (*Xalb*) cells correspond to the black round or oval structures or to the black rods. At the right of image (a), the cell containing bacteria corresponds to the phloem (P). At the bottom of image (b), the cell which is delimited by a thickened wall corresponds to the xylem (X). At the right of the image (b), the cell which is delimited by a thin wall corresponds to a vascular parenchyma cell (VPC). (c) and (d) Confocal microscopy images of a cross section of a sugarcane stalk (c) and a sugarcane leaf (d). In confocal microscopy images, the plant cell wall appears in blue, plasts in pink and green fluorescent protein-tagged bacteria in green. One of the two metaxylem vessels (MX) of the vascular bundle observed at the right of image (c) is highly invaded by the bacteria. Only one of the two metaxylem vessels of the vascular bundle observed at the top of image (c) is visible. Surrounding this metaxylem, cells containing plasts correspond to the living vascular parenchyma cells (VPC). This metaxylem and some of the vascular parenchyma cells are invaded by bacteria. Bacteria are also present in the intercellular spaces (IS) between the sucrose storage cells (SSC) that start to be invaded. In the foliar vascular bundle observed in (d), the two metaxylem vessels (MX) are invaded by bacteria. Parenchyma cells (PC) surrounding these metaxylem vessels are also invaded by bacteria. Bacteria containing cells that are located under epidermal cells, outside of the vascular bundle, correspond to bulliform cells (BC). Bacteria are also seen in the phloem (P) cells below the two metaxylem vessels. Scale bars correspond to 1 μm in (a) and (b), and to 50 μm in (c) and (d).

X. fastidiosa is able to open the pores of the pit membranes in grapevine (Pérez-Donoso, 2010), confirming that these two CWDE play a key role in vessel-to-vessel movement of *X. fastidiosa*. Such a vessel-to-vessel movement through bordered pits could probably also occur during systemic infection of sugarcane by *X. albilineans* or *L. xyli* subsp. *xyli*, but this remains to be shown experimentally. Notably, gene *pglA* is also conserved in *X. albilineans*, *L. xyli* subsp. *xyli* and *C. michiganensis* subsp. *michiganensis*. The genome of *X. albilineans* strain GPE PC73 contains two genes encoding a PglA as well as two genes encoding an EngXCA, confirming the importance of these CWDE in pathogenicity of *X. albilineans*. Like most other species of *Xanthomonas*, the genome of '*X. vasicola* pv. *vasculorum*' encodes a PglA and an EngXCA. To date, no data are available regarding the movement of '*X. vasicola* pv. *vasculorum*' in sugarcane.

7 Genomics-based hypotheses on the origin and adaptation of bacterial pathogens of sugarcane

Recent sequencing of the genome of bacteria isolated from plants other than sugarcane allowed us to propose hypotheses regarding the adaptation to sugarcane of '*X. vasicola* pv. *vasculorum*', *X. albilineans* and *L. xyli* subsp. *xyli* and the origin of their respective ancestors.

Full genome sequence analysis of two strains initially classified as *X. albilineans* revealed a novel species named '*Xanthomonas pseudalbilineans*' (Pieretti, 2015a). '*X. pseudalbilineans*' strain MUS 060 was isolated from the wild grass *Paspalum dilatatum* Poir. in 1940 in Mauritius (Orian, 1962). This plant, naturally infected by MUS 060, exhibited symptoms similar to those of leaf scald disease of sugarcane (Orian, 1962). '*X. pseudalbilineans*' strain GPE 39 was isolated in 1997 in Guadeloupe from dew droplets collected on the leaf surface of sugarcane plants (Daugrois, 2003). These plants were grown in a banana plantation area distant from any other sugarcane field, and were obtained from disease-free tissue-cultured sugarcane, suggesting that the main host of strain GPE 39 was not sugarcane (Daugrois, 2003). When inoculated on sugarcane, both '*X. pseudalbilineans*' strains MUS 060 and GPE 39 were unable to spread systemically in the plant (Orian, 1962; Daugrois, 2003). ANI values between strains MUS 060 and GPE 39 *versus X. albilineans* reach 91% at most, showing that they do not belong to the *X. albilineans* species. The ANI value between strain MUS 060 *versus* strain GPE 39 is 96%, indicating that these two strains belong to the same species (Pieretti, 2015a). The latter conclusions were confirmed by a phylogenetic analysis which grouped strains MUS 060 and GPE 39 in a clade distinct from the one containing *X. albilineans* strains (Pieretti, 2015a).

Genomic comparative analyses indicated that genome erosion occurred before the divergence of the species *X. albilineans* and '*X. pseudalbilineans*'. Interestingly, the locus directing biosynthesis of the toxin albicidin is conserved in strain MUS 060, indicating that this locus was present in a common ancestor of *X. albilineans* and '*X. pseudalbilineans*'. Similarly, '*X. pseudalbilineans*' strains MUS 060 and GPE 39 possess two genes encoding the endoglucanase EngXCA and one gene encoding the endopolygalacturonase PglA, indicating that these genes were present in the common ancestor of *X. albilineans* and '*X. pseudalbilineans*'. The second PglA-encoding gene from *X. albilineans* strain GPE PC73 is not present in the two '*X. pseudalbilineans*' strains. The T3SS Hrp and the xanthan biosynthesis loci are both absent in '*X. pseudalbilineans*' strains MUS 060 and GPE 39. This indicates that strategies used by both *X. albilineans* and '*X. pseudalbilineans*' species to spread in plants differ from the ones used by most other species of *Xanthomonas*.

The SPI-1 T3SS is not present either in the two '*X. pseudalbilineans*' strains MUS 060 and GPE 39. The genome of these strains also lacks a locus required for biosynthesis of glycogen which is conserved in *X. albilineans* and other sequenced species of *Xanthomonas*. Notably, this glycogen locus is also absent in *X. fastidiosa*. Although the glycogen pathway is not an essential component for bacterial growth, the inability to synthesize glycogen may reduce the survival time of bacteria when placed in an environment lacking an exogenous carbon source. Orthology comparisons identified a total of 2415 genes shared by *X. albilineans* and '*X. pseudalbilineans*' species, 206 genes specific to *X. albilineans* (absent in '*X. pseudalbilineans*') and 128 genes specific to '*X. pseudalbilineans*' (absent in *X. albilineans*). The putative contribution of some of these genes to the fitness and specific ecological niche of *X. albilineans* and '*X. pseudalbilineans*' is discussed in Pieretti (2015a).

A finished genome sequence of *L. xyli* subsp. *cynodontis* strain DSM46306, which was isolated from Bermuda grass (*Cynodon dactylon*), has been available since 2013 (Monteiro-Vitorello, 2013). This strain causes stunting in Bermuda grass. It possesses a 2.7 Mb circular chromosome, indicating that genome erosion occurred in a common ancestor of *L. xyli* subsp. *xyli* and *L. xyli* subsp. *cynodontis* and that these two pathogens derived from a same free-living bacterium according to the hypothesis proposed for *L. xyli* subsp. *xyli* (Monteiro-Vitorello, 2004). Orthology analysis revealed a total of 627 genes specific to *L. xyli* subsp. *cynodontis* (absent in *L. xyli* subsp. *xyli*) and 750 genes specific to *L. xyli* subsp. *xyli* (absent in *L. xyli* subsp. *cynodontis*) (Monteiro-Vitorello, 2013). The CWDE PglA and EngXCA are both conserved in *L. xyli* subsp. *cynodontis* strain DSM46306, confirming that these enzymes play a key role in the ability of *L. xyli* subsp. *xyli* and *L. xyli* subsp. *cynodontis* to spread in plants (Royer, unpublished Tblastn analyses). Genes encoding these CWDE are not present in the genome of other sequenced species of *Leifsonia*, which are not associated with plants, suggesting that their acquisition played a key role in the adaptation to a plant-associated life style of the common ancestor of *L. xyli* subsp. *xyli* and *L. xyli* subsp. *cynodontis*. It is interesting to note that the linker of the endoglucanase EngXCA encoded by *L. xyli* subsp. *cynodontis* strain DSM46306 is 14 amino acids shorter than that of the EngXCA encoded by *L. xyli* subsp. *xyli* strain CTCB07 (See Table 1).

The putative pathogenicity gene Lxx12290 of *L. xyli* subsp. *xyli* strain CTCB07, which was identified by transposon mutagenesis (Brumbley, 2004), is conserved in *L. xyli* subsp. *cynodontis* strain DSM46306 (Royer, unpublished data). The ANI value between genome sequences of *L. xyli* subsp. *xyli* strain CTCB07 *versus* *L. xyli* subsp. *cynodontis* strain DSM46306 is 92%, indicating that *L. xyli* subsp. *xyli* and *L. xyli* subsp. *cynodontis* could be considered as belonging to two different species (Pieretti, unpublished data). Unfortunately, no detailed study comparing the genomes of *L. xyli* subsp. *xyli* and *L. xyli* subsp. *cynodontis* has been published yet. Nevertheless, strains closely related to *L. xyli* subsp. *cynodontis* were isolated in eastern Australia on several grasses: Rhodes grass (*Chloris gayana*), couch grass (*Elymus repens*) and panic grass (*Panicum maximum*) (Mills, 2001). Sequencing of the genome of these strains will also help to better understand the adaptation of *L. xyli* subsp. *xyli* to sugarcane.

'*X. vasicola* pv. *vasculorum*' is also the causal agent of bacterial leaf streak of maize or corn. This disease is of relatively minor importance in South Africa, but has emerged in 2016 in the United States (http://cropwatch.unl.edu/2016/bacterial-leaf-streak-corn-confirmed-nebraska). '*X. vasicola* pv. *musacearum*' is the causal agent of *Xanthomonas* wilt of banana and plantain (*Musa* spp.), an emergent disease that was first reported in the early 2000s in East Africa (Tushemereirwe, 2004). Draft genome sequences of five strains of '*X. vasicola* pv. *vasculorum*' isolated from sugarcane and maize in Africa before 1962 were compared with draft genome sequences of two strains of '*X. vasicola* pv. *musacearum*' isolated from banana in Africa in 1967 and 2007 (Wasukira, 2014). Additional draft genome sequences are now available in GenBank for *X. vasicola* strains isolated from sorghum, *Holcus*, and *Tripsacum*. The ANI values between genome sequences of strains isolated from banana, sugarcane, maize, sorghum, *Holcus*, and *Tripsacum* are above 98%, confirming that all these strains belong to the species *X. vasicola* (Pieretti, unpublished data). The species *X. vasicola* is therefore able to spread in various plant species, in contrast to *X. albilineans* and *L. xyli* subsp. *xyli* that are limited to a single or very few host plants.

The genomic-based phylogenetic tree published by Wasukira (2014) indicated that the strains of '*X. vasicola* pv. *vasculorum*' from sugarcane and maize form a monophyletic clade distinct from the clade formed by the strains of '*X. vasicola* pv. *musacearum*'. A genomic-based phylogenetic tree including the recently sequenced strains as well as strains isolated from sugarcane and maize is available at the National Center for Biotechnology Information, or NCBI (https://www.ncbi.nlm.nih.gov/) (See Fig. 5). Unfortunately, this tree does not include the two sequenced strains of '*X. vasicola* pv. *musacearum*'. However, this tree clearly shows that the monophyletic clade described by Wasukira (2014) contains strains isolated not only from sugarcane and maize but also from sorghum and *Holcus*. These data could indicate that all strains of this monophyletic clade are able to cause gumming disease in sugarcane. This hypothesis should be further experimentally explored to evaluate the risk of emergence of sugarcane gumming due to transmission from other plants infected by *X. vasicola* strains.

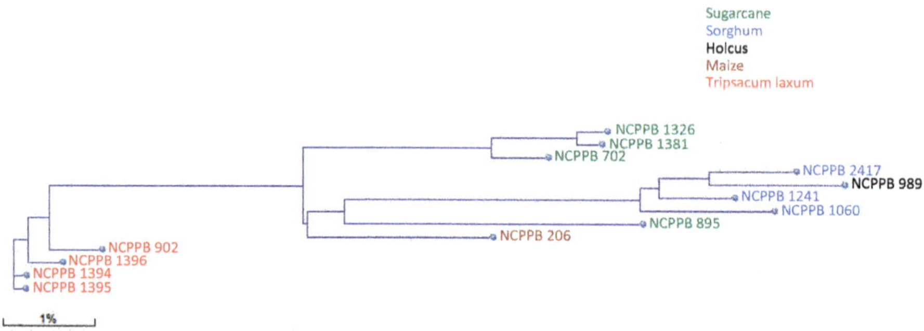

Figure 5 Dendrogram of *Xanthomonas vasicola* strains based on genomic BLAST alignment. Strains written in green (NCPPB 1326, NCPPB 1381, NCPPB 702 and NCPPB 895) were isolated from sugarcane. Strains written in blue (NCPPB 2417, NCPPB 1241 and NCPPB 1060) were isolated from sorghum. Strains written in red (NCPPB 902, NCPPB 1396, NCPPB 1394 and NCPPB 1395) were isolated from *Tripsacum laxum*. The strain written in brown (NCPPB 206) was isolated from maize (*Zea mays*). The strain written in black (NCPPB 989) was isolated from *Holcus* and was previously classified as *Xanthomonas holcicola*. Adapted from the dendrogram available at NCBI (https://www.ncbi.nlm.nih.gov/) on 28 September 2016. This dendrogram is using blast average score as distance. The scale is indicated by the bar at the bottom left of the figure.

8 Conclusion and future trends

Recent insights provided by both genomic and microscopy analyses indicated that, although belonging to two distinct bacterial phyla, *L. xyli* subsp. *xyli* and *X. albilineans* use similar strategies to spread in sugarcane. To date, under natural conditions of infection, these two pathogens were almost exclusively isolated from sugarcane. In very few instances, *X. albilineans* has also been found in maize in Brazil (Sordi, 1986) and in Mauritius (Autrey, 1992). However, the taxonomy of these latter isolates remains questionable because, as it was the case for the former *X. albilineans* isolates now renamed '*X. pseudalbilineans*', classical criteria are not sufficient to confirm that a strain belongs to the species *X. albilineans*. Genome sequencing of these strains is therefore required for a more precise taxonomy.

Based on ANI data, strains phylogenetically close to *L. xyli* subsp. *xyli* and *X. albilineans* that were isolated from other plants most probably belong to other species than *L. xyli* subsp. *xyli* or *X. albilineans*. Meanwhile, the sugarcane pathogen '*X. vasicola* pv. *vasculorum*' appears to use a strategy to infect plants that is different from the one(s) used by *L. xyli* subsp. *xyli* and *X. albilineans*. '*X. vasicola* pv. *vasculorum*' appears also to be more easily transmitted to sugarcane from other *Poaceae*. This difference between *L. xyli* subsp. *xyli* and *X. albilineans*, on the one hand, and '*X. vasicola* pv. *vasculorum*', on the other hand, may indicate that the association with sugarcane is much older for *L. xyli* subsp. *xyli* and *X. albilineans* than for '*X. vasicola* pv. *vasculorum*'. Furthermore, *L. xyli* subsp. *xyli* and *X. albilineans* may be so well adapted to spread in the sugarcane plant that they are unable to spread in other plant species. Additionally, because of their reduced genome, the adaptation of *L. xyli* subsp. *xyli* and *X. albilineans* to a new plant may be more difficult and may require more time than for '*X. vasicola* pv. *vasculorum*'.

Recent genomic insights also open up new prospects regarding further studies to better understand the pathogenicity of *L. xyli* subsp. *xyli*, *X. albilineans* and '*X. vasicola* pv. *vasculorum*'. As mentioned above, several candidate genes potentially required for pathogenicity have been identified, such as the CWDE EngXCA and PglA. To better understand how *L. xyli* subsp. *xyli*, *X. albilineans* and '*X. vasicola* pv. *vasculorum*' succeed in their natural habitat, that is, the sugarcane, genomic data should be completed by transcriptomic data to identify bacterial genes specifically expressed during pathogenesis. Such studies would require extraction of bacterial RNA directly from infected sugarcane samples or alternatively from bacteria incubated with sugarcane leaf homogenates to simulate *in planta* conditions. A similar approach was used for *C. michiganensis* subsp. *michiganensis* (Flügel, 2012). The functional studies based on transposon mutagenesis performed with *X. albilineans* (Rott, 2011) and *L. xyli* subsp. *xyli* (Brumbley, 2004) demonstrated that studies with new and sophisticated molecular tools are feasible in sugarcane, like in other plants that are easier to handle. Scientists should be given the means to develop such ambitious experimental projects on bacterial diseases of sugarcane.

The primary goal of research on plant pathogenic bacteria is to contribute to a better control of plant diseases. As mentioned in the introduction, availability of genome sequences has increased the efficiency of the detection of bacteria, thereby contributing to a better control by limiting the spread of diseases via contaminated planting material. Genomics also facilitates taxonomic assignments and provides powerful tools to develop concrete standards for speciation. However, a major question remains: How can a better understanding of the molecular basis of pathogenicity of bacteria help to formulate novel strategies for disease control? Transgenic plants that are propagated vegetatively are the most desirable because they circumvent genetic variations. For this reason, the transgenic approach appears particularly promising for sugarcane.

Leaf scald has been the first disease for which genetic modification for increased resistance against a systemic sugarcane pathogen has been developed, using a novel mechanism of disease resistance (Zhang, 1999). In this study, the enzyme AlbD, which is able to detoxify and inactivate albicidin, has been successfully used to produce transgenic sugarcanes resistant to leaf scald. This 'antipathogenesis' approach is based on the identification of weaknesses in pathogenesis strategies that can be used as targets for the development of disease control measures (Birch, 2001). A prospective target for the development of novel control methods of leaf scald or ratoon stunting could be the endoglucanase EngXCA. New molecular tools like peptide aptamers could be used to block the activity of this cellulase. Peptide aptamers are small combinatorial proteins that are selected to bind to specific sites on their target molecules (Reverdatto, 2015). The development of such new sophisticated tools and a better understanding of pathogenesis will clear the path toward new strategies to control bacterial diseases of sugarcane. The recent outbreak of *X. fastidiosa* in the European Union, associated with a destructive disease of the olive tree (Cariddi, 2014), has revived studies on this pathogen. New 'antipathogenesis' approaches to control *X. fastidiosa* could arise from these intensive studies. These approaches may also be applied to control sugarcane leaf scald and ratoon stunting, two diseases that are caused by pathogens sharing features with *X. fastidiosa*.

9 Where to look for further information

Kado, C. I. (Ed) (2010), *Plant Bacteriology*, St. Paul, American Phytopathological Society

Lopes, C. A. and Boiteux L. S. (2012), Breeding for Resistance to Bacterial Diseases. In *Plant Breeding for Biotic Stress Resistance*. Fritsche-Neto, R. (Ed), Berlin, Heidelberg. Springer Berlin Heidelberg, pp. 37–55.

van der Wolf, J. and De Boer, S. H. (2015), Phytopathogenic Bacteria. In *Principles of Plant-Microbe Interactions: Microbes for Sustainable Agriculture*. Lugtenberg, B. (Ed), Cham, Springer International Publishing, pp. 65–77.

10 References

Aslam, S. N., Newman, M. A., Erbs, G., Morrissey, K. L., Chinchilla, D., Boller, T., Jensen, T. T., De Castro, C., Ierano, T., Molinaro, A., Jackson, R. W., Knight, M. R. and Cooper, R. M. (2008), 'Bacterial polysaccharides suppress induced innate immunity by calcium chelation'. *Curr. Biol.* 18, 1078–83.

Autrey, L. J. C., Saumtally, S., Dookun, A., Sullivan, S. and Dhayan S. (1992), 'Aerial transmission of the leaf scald pathogen, *Xanthomonas albilineans*'. *Proc. Int. Soc. Sugarcane Technol.* 21, 508–26.

Bailey, R. A. and McFarlane, S. A. (1999), 'The status of ratoon stunting disease of sugarcane in southern and central Africa'. *Proc. Int. Soc. Sugarcane Technol.* 23, 338–46.

Birch, R. G. (2001), '*Xanthomonas albilineans* and the antipathogenesis approach to disease control'. *Mol. Plant Pathol.* 2, 1–11.

Blanvillain, S., Meyer, D., Boulanger, A., Lautier, M., Guynet, C., Denancé, N., Vasse, J., Lauber, E. and Arlat, M. (2007), 'Plant carbohydrate scavenging through TonB-dependent receptors: a feature shared by phytopathogenic and aquatic bacteria'. *PLoS ONE* 2, e224.

Boraston, A. B., Bolam, D. N., Gilbert, H. J. and Davies, G. J. (2004), 'Carbohydrate-binding modules: fine-tuning polysaccharide recognition'. *Biochem. J.* 382, 769–81.

Brumbley, S. M., Petrasovits, L. A., Murphy, R. M., Nagel, R. J., Candy, J. M. and Hermann, S. R. (2004, 'Establishment of a functional genomics platform for *Leifsonia xyli* subsp. *xyli*'. *Mol. Plant Microbe Interact.* 17, 175–83.

Büttner, D. and Bonas, U. (2010), 'Regulation and secretion of *Xanthomonas* virulence factors'. *FEMS Microbiol. Rev.* 34, 107–33.

Cariddi, C., Saponari, M., Boscia, D., De Stradis, A., Loconsole, G., Nigro, F., Porcelli, F., Potere, O. and Martelli, G. P. (2014), 'Isolation of a *Xylella fastidiosa* strain infecting olive and oleander in Apulia, Italy'. *J. Plant Pathol.* 96, 1–5.

Carvalho, G., da Silva, T. G. E. R., Munhoz, A. T., Monteiro-Vitorello, C. B., Azevedo, R. A., Melotto, M. and Camargo, L. E. A. (2016), 'Development of a qPCR for *Leifsonia xyli* subsp. *xyli* and quantification of the effects of heat treatment of sugarcane cuttings on Lxx'. *Crop Prot.* 80, 51–55.

Champoiseau, P., Daugrois, J. H., Pieretti, I., Cociancich, S., Royer, M. and Rott, P. (2006), 'High variation in pathogenicity of genetically closely related strains of *Xanthomonas albilineans*, the sugarcane leaf scald pathogen, in Guadeloupe'. *Phytopathology* 96, 1081–91.

Champoiseau, P., Rott, P. and Daugrois, J. H. (2009), 'Epiphytic populations of *Xanthomonas albilineans* and subsequent sugarcane stalk infection are linked to rainfall in Guadeloupe'. *Plant Dis.* 93, 339–46.

Chatterjee, S., Almeida, R. P. P. and Lindow, S. (2008), 'Living in two worlds: the plant and insect lifestyles of *Xylella fastidiosa*'. *Annu. Rev. Phytopathol.* 46, 243–71.

Cociancich, S., Pesic, A., Petras, D., Uhlmann, S., Kretz, J., Schubert, V., Vieweg, L., Duplan, S., Marguerettaz, M., Noëll, J., Pieretti, I., Hügelland, M., Kemper, S., Mainz, A., Rott, P., Royer, M. and Süssmuth, R. D. (2015), 'The gyrase inhibitor albicidin consists of *p*-aminobenzoic acids and cyanoalanine'. *Nat. Chem. Biol.* 11, 195–7.

Cornelis, G. R. (2006), 'The type III secretion injectisome'. *Nat. Rev. Microbiol.* 4, 811–25.

Darrasse, A., Carrère, S., Barbe, V., Boureau, T., Arrieta-Ortiz, M. L., Bonneau, S., Briand, M., Brin, C., Cociancich, S., Durand, K., Fouteau, S., Gagnevin, L., Guérin, F., Guy, E., Indiana, A., Koebnik, R., Lauber, E., Munoz, A., Noël, L. D., Pieretti, I., Poussier, S., Pruvost, O., Robène-Soustrade, I., Rott, P., Royer, M., Serres-Giardi, L., Szurek, B., van Sluys, M.-A., Verdier, V., Vernière, C., Arlat, M., Manceau C., and Jacques, M.-A. (2013), 'Genome sequence of *Xanthomonas fuscans* subsp. *fuscans* strain 4834-R reveals that flagellar motility is not a general feature of xanthomonads'. *BMC Genomics* 14, 761.

Daugrois, J. H., Dumont, V., Champoiseau, P., Costet, L., Boisne-Noc, R. and Rott, P. (2003), 'Aerial contamination of sugarcane in Guadeloupe by two strains of *Xanthomonas albilineans*'. *Eur. J. Plant Pathol.* 109, 445–58.

Davis, M. J., Gillaspie, A. G., Vidaver, A. K. and Harris, R. W. (1984), '*Clavibacter*: a new genus containing some phytopathogenic coryneform bacteria, including *Clavibacter xyli* subsp. *xyli* sp. nov., subsp. nov. and *Clavibacter xyli* subsp. *cynodontis* subsp. nov., pathogens that cause ratoon stunting disease of sugarcane and bermudagrass stunting disease'. *Int. J. Syst. Bacteriol.* 34, 107–17.

Davis, M. J., Rott, P., Warmuth, C. J., Chatenet, M. and Baudin, P. (1997), 'Intraspecific genomic variation within *Xanthomonas albilineans*, the sugarcane leaf scald pathogen'. *Phytopathology* 87, 316–24.

Davis, M. J. and Bailey, R. A. (2000), 'Ratoon stunting. In *A Guide to Sugarcane Diseases*. Rott, P., Bailey, R. A., Comstock, J. C., Croft, B. J. and Saumtally, A. S. (Eds). Montpellier, France, Cirad-ISSCT, pp. 49–54.

Dean, J. L. and Davis, M. J. (1989), 'Yield losses caused by ratoon stunting disease of sugarcane in Florida'. *J. Am. Soc. Sugarcane Technol.* 10, 66–72.

Déjean, G., Blanvillain-Baufumé, S., Boulanger, A., Darrasse, A., Dugé de Bernonville, T., Girard, A. L., Carrére, S., Jamet, S., Zischek, C., Lautier, M., Solé, M., Büttner, D., Jacques, M. A.,

Lauber, E. and Arlat, M. (2013), 'The xylan utilization system of the plant pathogen *Xanthomonas campestris* pv. *campestris* controls epiphytic life and reveals common features with oligotrophic bacteria and animal gut symbionts'. *New Phytol.* 198, 899–915.

Dookun, A., Stead, D. E. and Autrey, L. J. (2000), 'Variation among strains of *Xanthomonas campestris* pv. *vasculorum* from Mauritius and other countries based on fatty acid analysis'. *Syst. Appl. Microbiol.* 23, 148–55.

Evtushenko, L. I., Dorofeeva, L. V., Subbotin, S. A., Cole, J. R. and Tiedje, J. M. (2000), '*Leifsonia poae* gen. nov., sp. nov., isolated from nematode galls on *Poa annua*, and reclassification of "*Corynebacterium aquaticum*" Leifson 1962 as *Leifsonia aquatica* (ex Leifson 1962) gen. nov., nom. rev., comb. nov. and *Clavibacter xyli* Davis et al. 1984 with two subspecies as *Leifsonia xyli* (Davis et al. 1984) gen. nov., comb. Nov'. *Int. J. Syst. Evol. Microbiol.* 1, 371–80.

Feng, Z., Cáceres, N. E., Sarath, G. and Barletta, R. G. (2002), '*Mycobacterium smegmatis* L-alanine dehydrogenase (Ald) is required for proficient utilization of alanine as a sole nitrogen source and sustained anaerobic growth'. *J. Bacteriol.* 184, 5001–10.

Fleites, L. A., Mensi, I., Gargani, D., Zhang, S., Rott, P. and Gabriel, D. W. (2013), '*Xanthomonas albilineans* OmpA1 appears to be functionally modular and both the OMC and C-like domains are necessary for leaf scald disease of sugarcane. *Mol. Plant Microbe Interact.* 26, 1200–10.

Flügel, M, Becker, A., Gartemann, K. H. and Eichenlaub R. (2012), 'Analysis of the interaction of *Clavibacter michiganensis* subsp. *michiganensis* with its host plant tomato by genome-wide expression profiling'. *J. Biotechnol.* 160, 42–54.https://doi.org/10.1016/j.jbiotec.2012.01.023

Fu, H. Y., Sun, S. R., Wang, J. D., Ahmad, K., Wang, H. B., Chen, R. K. and Gao, S. J. (2016), 'Rapid and quantitative detection of *Leifsonia xyli* subsp. *xyli* in sugarcane stalk juice using a real-time fluorescent (TaqMan) PCR assay'. *Biomed. Res. Int.* 2016, Article ID 2681816, 8 pages.

Ghai, M., Singh, V., Martin, L. A., McFarlane, S. A., van Antwerpen, T. and Rutherford, R. S. (2014), 'A rapid and visual loop-mediated isothermal amplification assay to detect *Leifsonia xyli* subsp. *xyli* targeting a transposase gene'. *Lett. Appl. Microbiol.* 59, 648–57.

Giglioti, E. A., Almeida, I. M. G., Clerc, F., Rott, P., Chatenet, M., Beriam, L. O., Carvalho, M. L. V., Masuda, Y. and Matsuoka, S. (1999), 'False red stripe: A new disease of sugarcane in Brazil'. *Proc. Int. Soc. Sugarcane Technol.* 23, 410–11.

Goris, J., Konstantinidis, K. T., Klappenbach, J. A., Coenye, T., Vandamme, P. and Tiedje, J. M. (2007), 'DNA-DNA hybridization values and their relationship to whole-genome sequence similarities'. *Int. J. Syst. Evol. Microbiol.* 57, 81–91.

Grisham, M. P., Johnson, R. M. and Viator, R. V. (2009), 'Effect of ratoon stunting disease on yield of recently released sugarcane cultivars in Louisiana'. *J. Am. Soc. Sugarcane Technol.* 29, 119–27.

Hashimi, S. M., Wall, M. K., Smith, A. B., Maxwell, A. and Birch, R. G. (2007), 'The phytotoxin albicidin is a novel inhibitor of DNA gyrase'. *Antimicrob. Agents Chemother.* 51, 181–7.

Hilaire, E., Young, S. A., Willard, L. H., McGee, J. D., Sweat, T., Chittoor, J. M., Guikema, J. A. and Leach, J. E. (2001), 'Vascular defense responses in rice: Peroxidase accumulation in xylem parenchyma cells and xylem wall thickening'. *Mol. Plant Microbe Interact.* 14, 1411–19.

Hoy, J. W. and Grisham, M. P. (1994), 'sugarcane leaf scald distribution, symptomatology, and effect on yield in Louisiana'. *Plant Dis.* 78, 1086–7.

Ivey, M. L. L., Tusiime, G. and Miller, S. A. (2010), 'A polymerase chain reaction assay for the detection of *Xanthomonas campestris* pv. *musacearum* in banana'. *Plant Dis.* 94, 109–14.

Jacques, M. A., Arlat, M., Boulanger, A., Boureau, T., Carrère, S., Cesbron, S., Chen, N. W., Cociancich, S., Darrasse, A., Denancé, N., Fischer-Le Saux, M., Gagnevin, L., Koebnik, R., Lauber, E., Noël, L. D., Pieretti, I., Portier, P., Pruvost, O., Rieux, A., Robène, I., Royer, M., Szurek, B., Verdier, V. and Vernière, C. (2016), 'Using ecology, physiology, and genomics to understand host specificity in *Xanthomonas*'. *Annu. Rev. Phytopathol.* 54, 163–87.

Jahr, H., Dreier, J., Meletzus, D., Bahro, R. and Eichenlaub, R. (2000), 'The endo-beta-1,4-glucanase CelA of *Clavibacter michiganensis* subsp. *michiganensis* is a pathogenicity determinant required for induction of bacterial wilt of tomato'. *Mol. Plant Microbe Interact.* 13, 703–14.

Jain, M., Fleites, L., Zhang, S. and Gabriel D. (2016), 'The bifunctional catalase/peroxidase KatG is required for pathogenicity of *Xanthomonas albilineans* on sugarcane'. Poster 718-P. 2016 APS annual meeting. 30 July to 3 August 2016, Tampa, FL, USA.
Kao, J. and Damann, K. E. Jr. (1978), 'Microcolonies of the bacterium associated with ratoon stunting disease found in sugarcane xylem matrix'. *Phytopathology* 68, 545–51.
Klett, P. and Rott, P. (1994), 'Inoculum sources for the spread of leaf scald disease of sugarcane caused by *Xanthomonas albilineans* in Guadeloupe'. *J. Phytopathol.* 142, 283–91.
Kretz, J., Kerwat, D., Schubert, V., Grätz, S., Pesic, A., Semsary, S., Cociancich, S., Royer, M. and Süssmuth, R. D. (2015), 'Total synthesis of albicidin: A lead structure from *Xanthomonas albilineans* for potent antibacterial gyrase inhibitors'. *Angew. Chem. Int. Ed. Engl.* 54, 1969–73.
Legaz, M. E., Blanch, M., Piñón, D., Santiago, R., Fontaniella, B., Blanco, Y., Solas, M. T. and Vicente, C. (2011), 'sugarcane glycoproteins may act as signals for the production of xanthan in the plant-associated bacterium *Xanthomonas albilineans*'. *Plant Signal Behav.* 6, 1132–9.
Lelis, F. V. V., Czajkowski, R., de Souza, R. M., Ribeiro, D. H. and van der Wolf, J. M. (2014), 'Studies on the colonization of axenically grown tomato plants by a GFP-tagged strain of *Clavibacter michiganensis* subsp michiganensis'.. *Eur. J. Plant Pathol.* 139, 53–66.https://doi.org/10.1007/s10658-013-0362-9
Li, J. and Wang, N. (2011), 'Genome-wide mutagenesis of *Xanthomonas axonopodis* pv. *citri* reveals novel genetic determinants and regulation mechanisms of biofilm formation'. *PLoS ONE* 6, e21804.
Liu, J., Xu, L., Guo, J., Chen, R., Grisham, M. P. and Que, Y. (2013), 'Development of loop-mediated isothermal amplification for detection of *Leifsonia xyli* subsp. *xyli* in sugarcane'. *BioMed Res. Inter.* 2013, Article ID 357692, 8 pages.
Marguerettaz, M., Pieretti, I., Gayral, P., Puig, J., Brin, C., Cociancich, S., Poussier, S., Rott, P. and Royer, M. (2011), 'Genomic and evolutionary features of the SPI-1 type III secretion system that is present in *Xanthomonas albilineans* but is not essential for xylem colonization and symptom development of sugarcane leaf scald'. *Mol. Plant Microbe Interact.* 24, 246–59.
Martin, J. P., Carpenter, C. W. and Weller, D. M. (1932), 'Leaf scald disease of sugarcane in Hawaii'. *Planters' Record* 36, 145–96.
Mensi, I., Vernerey, M. S., Gargani, D., Nicole, M. and Rott, P. (2014), 'Breaking dogmas: the plant vascular pathogen *Xanthomonas albilineans* is able to invade non-vascular tissues despite its reduced genome'. *Open Biol.* 4, 130116.
Mills, L., Leaman, T. M., Taghavi, S. M., Shackel, L., Dominiak, B. C., Taylor, P. W. J., Fegan, M. and Teakle, D. S. (2001), '*Leifsonia xyli*-like bacteria are endophytes of grasses in eastern Australia'. *Australas. Plant Pathol.* 30, 145–51.
Monteiro-Vitorello, C. B., Camargo, L. E., Van Sluys, M. A., Kitajima, J. P., Truffi, D., do Amaral, A. M., Harakava, R., de Oliveira, J. C., Wood, D., de Oliveira, M. C., Miyaki, C., Takita, M. A., da Silva, A. C., Furlan, L. R., Carraro, D. M., Camarotte, G., Almeida, N. F. Jr., Carrer, H., Coutinho, L. L., El-Dorry, H. A., Ferro, M. I., Gagliardi, P. R., Giglioti, E., Goldman, M. H., Goldman, G. H., Kimura, E. T., Ferro, E. S., Kuramae, E. E., Lemos, E. G., Lemos, M. V., Mauro, S. M., Machado, M. A., Marino, C. L., Menck, C. F., Nunes, L. R., Oliveira, R. C., Pereira, G. G., Siqueira, W., de Souza, A. A., Tsai, S. M., Zanca, A. S., Simpson, A. J., Brumbley, S. M. and Setúbal, J. C. (2004), 'The genome sequence of the gram-positive sugarcane pathogen *Leifsonia xyli* subsp. *xyli*'. *Mol. Plant Microbe Interact.* 17, 827–36.
Monteiro-Vitorello, C. B., Zerillo, M. M., Van Sluys, M. A. and Aranha Camargo, L. E. (2009), 'Genome sequence-based insights into the biology of the sugarcane pathogen *Leifsonia xyli* subsp. *xyli*'. In *Plant pathogenic bacteria Genomics and Molecular Biology*. Jackson, R. W. (ed.), Norfolk, UK, Caister Academic Press, pp. 135–46.
Monteiro-Vitorello, C. B., Zerillo, M. M., Van Sluys, M. A., Camargo, L. E. and Kitajima J. P. (2013), 'Complete genome sequence of *Leifsonia xyli* subsp. *cynodontis* strain DSM46306, a Gram-positive bacterial pathogen of grasses'. *Genome Announc.* 1, e00915–13.https://doi.org/10.1128/genomea.00915-13

Newman, K. L., Almeida, R. P., Purcell, A. H. and Lindow, S. E. (2003), 'Use of a green fluorescent strain for analysis of *Xylella fastidiosa* colonization of *Vitis vinifera*'. *Appl. Environ. Microbiol.* 69, 7319–27.

Niza, B., Coletta-Filho, H. D., Merfa, M. V., Takita, M. A. and de Souza, A. A. (2015), 'Differential colonization patterns of *Xylella fastidiosa* infecting citrus genotypes'. *Plant Pathol.* 64, 1259–69.

Notomi, T., Okayama, H., Masubuchi, H., Yonekawa, T., Watanabe, K., Amino, N. and Hase, T. (2000), 'Loop-mediated isothermal amplification of DNA'. *Nucleic Acids Res.* 28, e63.

Notomi, T., Mori, Y., Tomita, N. and Kanda, H. (2015), 'Loop-mediated isothermal amplification (LAMP): principle, features, and future prospects'. *J. Microbiol.* 53, 1–5.

Orian, G. (1962), 'A disease of *Paspalum dilatatum* in Mauritius caused by a species of bacterium closely resembling *Xanthomonas albilineans* (Ashby) Dowson'. *Rev. Agric. Sucr. Ile Maurice* 41, 7–20.

Pérez-Donoso, A. G., Sun, Q., Roper, M. C., Greve, L. C., Kirkpatrick, B. and Labavitch J. M. (2010), 'Cell wall-degrading enzymes enlarge the pore size of intervessel pit membranes in healthy and *Xylella fastidiosa*-infected grapevines'. *Plant Physiol.* 152, 1748–59.https://doi.org/10.1104/pp.109.148791

Pieretti, I., Royer, M., Barbe, V., Carrere, S., Koebnik, R., Cociancich, S., Couloux, A., Darrasse, A., Gouzy, J., Jacques, M. A., Lauber, E., Manceau, C., Mangenot, S., Poussier, S., Segurens, B., Szurek, B., Verdier, V., Arlat, M. and Rott, P. (2009), 'The complete genome sequence of *Xanthomonas albilineans* provides new insights into the reductive genome evolution of the xylem-limited *Xanthomonadaceae*'. *BMC Genomics* 10, 616.

Pieretti, I., Royer, M., Barbe, V., Carrere, S., Koebnik, R., Couloux, A., Darrasse, A., Gouzy, J., Jacques, M. A., Lauber, E., Manceau, C., Mangenot, S., Poussier, S., Segurens, B., Szurek, B., Verdier, V., Arlat, M., Gabriel, D. W., Rott, P. and Cociancich, S. (2012), 'Genomic insights into strategies used by *Xanthomonas albilineans* with its reduced artillery to spread within sugarcane xylem vessels'. *BMC Genomics* 13, 658.

Pieretti, I., Cociancich, S., Bolot, S., Carrère, S., Morisset, A., Rott, P. and Royer, M. (2015a), 'Full genome sequence analysis of two isolates reveals a novel *Xanthomonas* species close to the sugarcane pathogen *Xanthomonas albilineans*'. *Genes (Basel)* 6, 714–33.https://doi.org/10.3390/genes6030714

Pieretti, I., Pesic, A., Petras, D., Royer, M., Süssmuth, R. D. and Cociancich, S. (2015b), 'What makes *Xanthomonas albilineans* unique amongst xanthomonads'. *Front. Plant Sci.* 6, Article 289.

Qhobela, M. and Claflin, L. E. (1992), 'Eastern and southern African strains of *Xanthomonas campestris* pv. *vasculorum* are distinguishable by restriction fragment length polymorphism of DNA and polyacrylamide gel electrophoresis of membrane proteins'. *Plant Pathol.* 41, 113–21.

Quecine, M. C., Silva, T. M., Carvalho, G., Saito, S., Mondin, M., Teixeira-Silva, N. S., Camargo, L. E. A. and Monteiro-Vitorello, C. B. (2015), 'A stable *Leifsonia xyli* subsp. *xyli* GFP-tagged strain reveals a new colonization niche in sugarcane tissues'. *Plant Pathol.* 65, 154–62.

Rademaker, J. L. W., Louws, F. J., Schultz, M. H., Rossbach, U., Vauterin, L., Swings, J. and de Bruijn, F. J. (2005), 'A comprehensive species to strain taxonomic framework for *Xanthomonas*'. *Phytopathology* 95, 1098–1111.

Reverdatto, S., Burz, D. S. and Shekhtman, A. (2015), 'Peptide aptamers: Development and applications'. *Curr. Top. Med. Chem.* 15, 1082–1101.

Ricaud, C. and Autrey, L. J. C. (1989), 'Gumming disease'. In *Diseases of Sugarcane. Major Diseases*. Ricaud, C., Egan, B. T., Gillapsie, A. G. Jr. and Hughes, C. G. (Eds), Amsterdam, The Netherlands, Elsevier Science Publishers BV, pp. 21–38.

Richter, M. and Rossello-Mora, R. (2009), 'Shifting the genomic gold standard for the prokaryotic species definition'. *Proc. Natl. Acad. Sci. USA* 106, 19126–31.

Roper, M. C., Greve, L. C., Warren, J. G. Labavitch, J. M. and Kirkpatrick, B. C. (2007), '*Xylella fastidiosa* requires polygalacturonase for colonization and pathogenicity in *Vitis vinifera* grapevines'. *Mol. Plant Microbe Interact.* 20, 411–19.

Rott, P. (1984). 'Apport des cultures *in vitro* à l'étude de l'échaudure des feuilles de canne à sucre (*Saccharum* sp.) causée par *Xanthomonas albilineans* (Ashby) Dowson'. Thèse de doctorat de 3e cycle, Orsay, France, Université de Paris-Sud.

Rott, P., Arnaud, M. and Baudin, P. (1986), 'Serological and lysotypical variability of *Xanthomonas albilineans* (Ashby) Dowson, causal agent of sugarcane leaf scald disease'. *J. Phytopathol.* 116, 201–11.

Rott, P., Bailey, R. A. Comstock, J. C., Croft, B. J. and Saumtally, A. S. (Eds) (2000a), *A Guide to Sugarcane Diseases*. Montpellier, France, Cirad–ISSCT.

Rott, P. and Davis, M. J. (2000b), 'Leaf scald'. In *A Guide to Sugarcane Diseases*. Rott, P., Bailey, R. A., Comstock, J. C., Croft, B. J. and Saumtally, A. S. (Eds), Montpellier, France, Cirad-ISSCT, pp. 38–44.

Rott, P., Fleites, L., Marlow, G., Royer, M. and Gabriel, D. W. (2011), 'Identification of new candidate pathogenicity factors in the xylem-invading pathogen *Xanthomonas albilineans* by transposon mutagenesis'. *Mol. Plant Microbe Interact.* 24, 594–605.

Royer, M., Koebnik, R., Marguerettaz, M., Barbe, V., Robin, G. P., Brin, C., Carrere, S., Gomez, C., Hügelland, M., Völler, G. H., Noëll, J., Pieretti, I., Rausch, S., Verdier, V., Poussier, S., Rott, P., Süssmuth, R. D. and Cociancich, S. (2013), 'Genome mining reveals the genus *Xanthomonas* to be a promising reservoir for new bioactive non-ribosomally synthesized peptides'. *BMC Genomics* 14, 658.

Sallet, E., Gouzy, J. and Schiex, T. (2014). 'EuGene-PP: a next generation automated annotation pipeline for prokaryotic genomes'. *Bioinformatics* 30, 2659–61.

Saumtally, A. S. and Dookun, A. (2000), 'Gumming'. In *A Guide to Sugarcane Diseases*. Rott, P., Bailey, R. A., Comstock, J. C., Croft, B. J. and Saumtally, A. S. (Eds), Montpellier, France, Cirad-ISSCT, pp. 32–7.

Schauer, K., Rodionov, D. A. and de Reuse, H. (2008), 'New substrates for TonB-dependent transport: do we only see the "'tip of the iceberg"?'. *Trends in Biochem. Sci.* 33, 330–8.

Simpson, A. J., Reinach, F. C., Arruda, P., Abreu, F. A., Acencio, M., Alvarenga, R., Alves, L. M., Araya, J. E., Baia, G. S., Baptista, C. S., Barros, M. H., Bonaccorsi, E. D., Bordin, S., Bové, J. M., Briones, M. R., Bueno, M. R., Camargo, A. A., Camargo, L. E., Carraro, D. M., Carrer, H., Colauto, N. B., Colombo, C., Costa, F. F., Costa, M. C., Costa-Neto, C. M., Coutinho, L. L., Cristofani, M., Dias-Neto, E., Docena, C., El-Dorry, H., Facincani, A. P., Ferreira, A. J., Ferreira, V. C., Ferro, J. A., Fraga, J. S., França, S. C., Franco, M. C., Frohme, M., Furlan, L. R., Garnier, M., Goldman, G. H., Goldman, M. H., Gomes, S. L., Gruber, A., Ho, P. L., Hoheisel, J. D., Junqueira, M. L., Kemper, E. L., Kitajima, J. P., Krieger, J. E., Kuramae, E. E., Laigret, F., Lambais, M. R., Leite, L. C., Lemos, E. G., Lemos, M. V., Lopes, S. A., Lopes, C. R., Machado, J. A., Machado, M. A., Madeira, A. M., Madeira, H. M., Marino, C. L., Marques, M. V., Martins, E. A., Martins, E. M., Matsukuma, A. Y., Menck, C. F., Miracca, E. C., Miyaki, C. Y., Monteiro-Vitorello, C. B., Moon, D. H., Nagai, M. A., Nascimento, A. L., Netto, L. E., Nhani, A. Jr., Nobrega, F. G., Nunes, L. R., Oliveira, M. A., de Oliveira, M. C., de Oliveira, R. C., Palmieri, D. A., Paris, A., Peixoto, B. R., Pereira, G. A., Pereira, H. A. Jr., Pesquero, J. B., Quaggio, R. B., Roberto, P. G., Rodrigues, V., de M Rosa, A. J., de Rosa, V. E. Jr., de Sá, R. G., Santelli, R. V., Sawasaki, H. E., da Silva, A. C., da Silva, A. M., da Silva, F. R., da Silva, W. A. Jr., da Silveira, J. F., Silvestri, M. L., Siqueira, W. J., de Souza, A. A., de Souza, A. P., Terenzi, M. F., Truffi, D., Tsai, S. M., Tsuhako, M. H., Vallada, H., Van Sluys, M. A., Verjovski-Almeida, S., Vettore, A. L., Zago, M. A., Zatz, M., Meidanis, J. and Setubal, J. C. (2000), 'The genome sequence of the plant pathogen *Xylella fastidiosa*'. *Nature* 406, 151–9.

Sordi, R. A. and Tokeshi, H. (1986), 'Presence of *Xanthomonas albilineans* in guttation droplets of sugarcane and sweet corn leaves showing leaf scald disease symptoms'. Sociedade dos Técnicos Açucareiros do Brasil, 4 July to August, 60–3.

Studholme, D. J., Kemen, E., MacLean, D., Schornack, S., Aritua, V., Thwaites, R., Grant, M., Smith, J. and Jones, J. D. (2010), 'Genome-wide sequencing data reveals virulence factors implicated in banana *Xanthomonas* wilt'. *FEMS Microbiol. Lett.* 310, 182–92.

Tushemereirwe, W., Kangire, A., Ssekiwoko, F., Offord, L. C., Crozier, J., Boa, E., Rutherford, M. and Smith J. J. (2004), 'First report of *Xanthomonas campestris* pv. *musacearum* on banana in Uganda'. *Plant Pathol.* 53, 802.

Vallenet, D., Belda, E., Calteau, A., Cruveiller, S., Engelen, S., Lajus, A., Le Fèvre, F., Longin, C., Mornico, D., Roche, D., Rouy, Z., Salvignol, G., Scarpelli, C., Thil Smith, A. A., Weiman, M. and Médigue, C. (2013), 'MicroScope – an integrated microbial resource for the curation and comparative analysis of genomic and metabolic data'. *Nucleic Acids Res.* 41(Database issue), D636–47.

Vauterin, L., Hoste, B., Kersters, K. and Swings, J. (1995), 'Reclassification of *Xanthomonas*'. *Inter. J. Syst. Bacteriol.* 45, 472–89.

Wasukira, A., Coulter, M., Al-Sowayeh, N., Thwaites, R., Paszkiewicz, K., Kubiriba, J., Smith, J., Grant, M. and Studholme, D. J. (2014), 'Genome sequencing of *Xanthomonas vasicola* pathovar *vasculorum* reveals variation in plasmids and genes encoding lipopolysaccharide synthesis, type-IV pilus and type-III secretion effectors'. *Pathogens* 3, 211–37.

Young, A. J., Petrasovits, L. A., Croft, B. J., Gillings, M. and Brumbley, S. M. (2006), 'Genetic uniformity of international isolates of *Leifsonia xyli* subsp. *xyli*, causal agent of ratoon stunting disease of sugarcane (*Saccharum* interspecific hybrids)'. *Austr. Plant Pathol.* 35, 503–11.

Young, A. J. (2016), 'Possible origin of ratoon stunting disease following interspecific hybridization of *Saccharum* species'. *Plant Pathol.* 65, 1403–10.

Zhang, L., Xu, J. and Birch, R. G. (1999), 'Engineered detoxification confers resistance against a pathogenic bacterium'. *Nat. Biotechnol.* 17, 1021–4.

Zhi, X. Y., Zhao, W., Li, W. J. and Zhao, G. P. (2012), 'Prokaryotic systematics in the genomics era'. *Antonie Van Leeuwenhoek* 101, 21–34.

Chapter 14

Managing the soil biological community to improve soil health and reduce losses from nematode pests

Graham R. Stirling, Biological Crop Protection Pty Ltd, Australia

1 Introduction

Plant-parasitic nematodes are important pests of sugarcane. A survey of nematologists from more than 70 countries (Sasser and Freckman, 1987) indicated that crop losses due to nematodes were higher in sugarcane (annual average losses of 15.3%) than in any other crop except banana, coconut and a few vegetable crops. Sugarcane also has more pest nematodes than most other cultivated crops, with Cadet and Spaull (2005) indicating that 310 species in 48 genera had been recorded from the roots and/or rhizosphere of sugarcane.

Given that several comprehensive reviews of the ecology, pathogenicity and control of nematodes in sugarcane are available (Holtzmann, 1964; Prasad, 1972; Birchfield, 1984;

http://dx.doi.org/10.19103/AS.2017.0035.24

Spaull and Cadet, 1991; Cadet and Spaull, 2005), the initial part of this chapter will provide no more than a brief overview of these pests. It largely focuses on the management practices that have been used to reduce losses from plant-parasitic nematodes, with particular emphasis on progress made in the last 15 years.

The traditional way agricultural industries have dealt with pest nematodes is to identify the key pests and then devise management practices to control them (e.g. cultivar resistance, crop rotation and nematicides). However, this approach has not been successful in sugarcane for several reasons. First, sugarcane has a greater range of potentially damaging nematodes than any other agricultural crop, and it is almost impossible to find rotation crops that are effective against all of them, or develop sugarcane cultivars with resistance to all species. Second, the crop is perennial and so crop rotation only has an impact on nematode populations early in the cropping cycle. Third, the use of nematicides is limited by economics and the potential for off-site environmental impacts. Perhaps, more importantly, such approaches do little to overcome the broader soil health problems that limit the productivity of sugarcane. Consequently, the latter part of this chapter takes a more holistic view of the problem and considers how nematode pests might be managed in future. It looks at the nematode community as a whole (including the free-living nematode community) and argues that it is part of a soil food web that provides many important ecosystem services. It then considers how this interacting group of organisms might be managed to improve soil and plant health. Key outcomes of such an approach are an improvement in soil structural properties, greater levels of nutrient mineralization from organic matter and an increase in the suppressive services that keep nematode pests under control. Consequently, roots begin to function more effectively and plants are able to tolerate higher populations of plant-parasitic nematodes.

2 Nematode pests of sugarcane

The most important plant-parasitic nematode genera associated with sugarcane, together with some of the most widely distributed species, are listed in Table 1. However, it must be recognized that the list does not provide detailed information on nematode occurrence in particular cane-growing regions. The genera in the table are certainly relatively common on sugarcane, but species other than those listed will almost certainly occur in some regions. For example, *Pratylenchus zeae* is the most frequently encountered species of *Pratylenchus*, but 19 other species have been reported on sugarcane (Spaull and Cadet, 1991). In the same way, *Meloidogyne incognita* and *Meloidogyne javanica* are widespread on sugarcane, but five other *Meloidogyne* species have been identified (Cadet and Spaull, 2005). Another important point to recognize is that the abundance of individual nematode species is influenced by factors such as soil texture, pH, soil carbon and the nitrogen and phosphorus content of the soil (Godefroid et al., 2013), and this means that nematode distribution will vary considerably between and within regions.

What is also clear from the literature is that in some sugarcane-growing countries, information on nematode distribution is based on a relatively limited data set. In such situations, it is important that results from comprehensive surveys are used to determine research priorities and decide whether nematode management programmes should be implemented. Since nematode taxonomy at a species level has sometimes been inadequate in the past, it is also important to ensure that nematodes are identified

Table 1 Plant-parasitic nematodes commonly associated with sugarcane: distribution, relative pathogenicity and symptoms produced on roots

Common name	Genera	Common species	Feeding group	Geographic distribution	Pathogenicity	Symptoms on roots
Root-knot	*Meloidogyne*	*javanica, incognita*	Sedentary endoparasite	Widespread, but mainly restricted to sandy and well-structured loamy soils	+++	Swelling and galling of root tips, reduced root elongation
Root-lesion	*Pratylenchus*	*brachyurus, zeae*	Migratory endoparasite	Present in almost all sugarcane soils	+++	Reddish-purple, elliptically shaped lesions on young roots (later turning brown to black), necrosis, destruction of fine roots, reduced root mass
Dagger	*Xiphinema*	*elongatum*	Ectoparasite	Typically occur in about one-third of cane fields	++	Swelling and malformation of root tips and stunting of roots
Needle	*Paralongidorus*		Ectoparasite	Not widely distributed but relatively common in some countries (e.g. South Africa)	++	Swelling and malformation of root tips
Stubby root	*Paratrichodorus*	*minor*	Ectoparasite	Common in sands and loamy sands. Less frequent in fine-textured soils	++	Roots are typically stunted and have a stubby appearance. Fine roots are lacking
Stunt	*Tylenchorhynchus*	*annulatus*	Ectoparasite	Widespread. Generally found in 60–80% of cane fields	++	Root systems are sparse, with signs of necrosis and moderate to severe stunting of lateral roots
Lance	*Hoplolaimus*		Migratory endoparasite	Widespread in some cane-growing regions	++	Purple-red necrotic lesions and root girdling. Restricted growth and development of roots
Ring	Various genera in the family Criconematidae		Ectoparasite	Relatively common but not readily detected with extraction methods that rely on motility	++	General debilitation of the root system and loss of fine roots
Spiral	*Helicotylenchus*	*dihystera*	Ectoparasite or semi-endoparasite	Widespread. Often found in more than 75% of cane fields	+	Reduced root growth, discolouration and lesions. *Helicotylenchus dihystera*, the most common species, causes no pronounced symptoms and has little impact on growth
Reniform	*Rotylenchulus*	*parvus, reniformis*	Sedentary endoparasite	Relatively common in some cane-growing regions	+	*Rotylenchulus parvus*, the most common species, causes little damage

correctly. In some cases, this may mean using modern molecular techniques to confirm identifications based on morphology. Perhaps the best example is the recent discovery of *Pratylenchus parazeae* on sugarcane in China (Wang et al., 2015). This new species was found in fields that were also infested with *P. zeae* and both morphological and molecular tests were needed to clearly distinguish the two species.

2.1 Key pests

Although many different nematodes are capable of feeding on sugarcane roots, there seems to be a consensus amongst sugarcane nematologists that based on their distribution and pathogenicity, there are two key nematode pests.

Root-knot nematode (*Meloidogyne* spp.) is the most commonly recognized nematode pest of sugarcane for two reasons. First, it produces galls and elongated swellings on root tips, and although these protuberances may be small and difficult to detect, they are visible to the naked eye and differ from the symptoms produced by any other root pathogen. Second, the damage caused to the root system results in symptoms that are usually apparent above-ground. Root-knot nematode is most commonly found in coarse-textured sandy soils where nutrients and water are limiting, and in such situations, leaves begin to wilt and curl and the number and length of stalks is reduced when root proliferation and functionality is impaired. Seven species of *Meloidogyne* have been reported on sugarcane, but only two species (*M. javanica* and *M. incognita*) are widespread.

The damage caused by root-lesion nematode (*Pratylenchus* spp.) is usually not as obvious as root-knot nematode because the symptoms it produces (lesioning, root discolouration and a lack of fine roots) are not distinctive. Similar below-ground symptoms can be caused by fungal pathogens, root-feeding arthropods and abiotic factors such as compaction and poor aeration. Also, above-ground symptoms are relatively subtle (yellowing of leaves, shorter stalks and reduced stalk mass) and may only be apparent when nematode-free plots are available for comparison. *P. zeae* is by far the most widespread species of *Pratylenchus* and is probably the most important nematode pest of sugarcane worldwide. It not only invades roots and destroys the cells on which it feeds, but also provides avenues of entrance for fungi and bacteria that only incite disease when damaged root cells provide them with an entry point.

2.2 Second-tier pests

Sugarcane is unusual amongst agricultural crops in having a wide range of nematode pests. In addition to the key pests discussed above, many other plant-parasitic nematodes can cause damage, including stunt nematode (*Tylenchorhynchus*), dagger nematode (*Xiphinema*), needle nematode (*Paralongidorus*), lance nematode (*Hoplolaimus*), stubby root nematode (*Paratrichodorus*) and ring nematode (various genera in the Criconematidae). Most of these nematodes are ectoparasites and when present at high population densities they damage root tips, limit root elongation and destroy fine roots.

2.3 Other plant-feeding nematodes

A third group of plant-feeding nematodes includes spiral nematode (*Helicotylenchus*, *Scutellonema*, *Rotylenchus*), reniform nematode (*Rotylenchulus*), sheath nematode (*Hemicycliophora*), pin nematode (*Paratylenchus*) and burrowing nematode (*Achlysiella*).

One of the spiral nematodes (*Helicotylenchus dihystera*) is perhaps the most common species on sugarcane worldwide, and although it is pathogenic to sugarcane in pots (Apt and Koike, 1982; Rao and Swarup, 1975), field studies suggest that it causes little damage and is often associated with high-yielding areas of the field (Cadet et al., 2002; Cadet and Spaull, 2005). The other genera are either relatively non-pathogenic or their importance has not been determined. Several genera in the Tylenchidae (e.g. *Tylenchus*, *Psilenchus* and *Filenchus*) should also be mentioned here, as they occur in most sugarcane soils. They mainly feed on root epidermal cells and root hairs and cause little damage, but some genera (e.g. *Filenchus*) have been cultured on fungi and may be fungivores (Yeates et al., 1993; Okada and Kadota, 2003).

3 Economic importance of plant-parasitic nematodes

Results from pathogenicity experiments in pots and the responses obtained when non-volatile nematicides are applied in the field clearly demonstrate that plant-parasitic nematodes are economically important pests of sugarcane (Birchfield, 1984; Spaull and Cadet, 1990; Cadet and Spaull, 2005). Yield increases of 10–20% are commonly obtained when nematode populations are reduced with nematicides, but the extent of the response is determined by many factors, with soil texture perhaps being the most important. Nematodes restrict the growth of the root system, but the resultant effects on water uptake will be much greater in coarse-textured soils with low water-holding capacities. In heavy-textured soils, crops are better able to cope with loss of roots because water is more readily available. In the same way, nematodes cause fewer losses in situations where crops are well irrigated or where nutrient supplies are adequate.

Although nematicides have proved a useful tool for identifying situations where nematodes are causing crop losses, they also have their limitations. The main problem is that non-volatile nematicides such as aldicarb, fenamiphos and carbofuran have a limited half-life in soil, and the single application usually applied in field trials protects roots for only a few weeks after planting. In an attempt to reduce nematode populations for a longer period, Blair and Stirling (2007) applied nematicides 3–4 times per year and found that root-lesion nematode populations were sometimes reduced by only 60–80%. Control of root-knot nematode was often inconsistent, particularly in ratoons. This suggests that responses from nematicides may underestimate the losses caused by nematodes. On the other hand, organophosphate and carbamate nematicides are also insecticidal, and so it is possible that growth and yield increases may be partly due to their effects on soil invertebrates other than nematodes.

4 Nematode control measures

4.1 Crop rotation

Rotational cropping is a standard practice in most forms of agriculture due to the many benefits it provides: provision of nutrients for the next crop, reduction in levels of deleterious pests and diversification of farm income. However, it was rarely used in sugarcane until research in Australia demonstrated that yields improved by 15–30% following legume crops such as soy bean or peanut and that these yield improvements carried through to subsequent ratoons. In fact, the yield forgone by leaving land out of sugarcane for 12 months was more than recovered over the whole crop cycle (Garside and Bell, 2001; 2011a; Garside et al., 2005). Breaking the monoculture with pasture or a bare fallow produced similar yield responses, but regardless of the type of break, longer-duration breaks produced the greatest yield responses. Since responses to fumigating the soil with methyl bromide were obtained in the same experiments (Garside and Bell, 2011a,b), it was assumed that the crop, bare fallow and pasture breaks were reducing populations of the fungal and nematode pathogens that are always associated with the roots of monocultured sugarcane.

Subsequent studies have shown that rotation crops almost always provide some control of root-lesion and root-knot nematodes, the two most important nematode pests of sugarcane. Populations of *P. zeae* were reduced by consecutive legume crops and remained lower than monocultured sugarcane for the first 6 months after replanting (Stirling et al., 2001), while observations in commercial fields indicated that fallow legumes reduced numbers of the pest by an average of 84% (Stirling et al., 2007). Pot studies have shown that many potentially useful rotation crops host lower numbers of *P. zeae* than sugarcane (Sundararaj and Mehta, 1993a; Stirling et al., 2010a; Berry et al., 2011; Obici et al., 2011; Santana et al., 2012), while in the field, an 18-month-old crop of sunn hemp and velvet bean reduced populations of *P. zeae* to very low levels (Berry et al., 2011). Similar host effects have been observed for root-knot nematode, with velvet bean, peanut and some varieties of soy bean and cowpea reducing nematode populations to very low levels (Stirling et al., 2006a,b). Forage sorghum is also a poor host of root-knot nematode, but its susceptibility to *P. zeae* limits its usefulness as a rotation crop for sugarcane (Stirling et al., 2010a).

In the absence of an appropriate rotation crop, populations of nematode pests will increase rapidly when the land is replanted to sugarcane. High numbers of *P. zeae* (>2000 nematodes/g root) can be found in roots within 5–10 weeks of planting, and at these infestation levels, root biomass and the length of the sett root system is reduced and primary shoot growth is affected (Pankhurst et al., 2001). Thus, the early growth of sugarcane can be improved by growing a non-host crop for 6–12 months. However, as nematode populations have usually returned to very high levels by the end of the plant crop (Stirling et al., 2001; Pankhurst et al., 2005c; Stirling 2008), rotation crops have relatively little impact on losses caused by nematodes.

4.2 Resistance and tolerance

Modern sugarcane hybrids are crosses between *Saccharum officinarum* and *Saccharum spontaneum* and since they have been derived from a relatively narrow ancestral base, they generally react in much the same way to nematodes. Although nematode resistance has been reported in sugarcane (Cadet and Spaull, 1990; Spaull and Cadet 2005), the commercial varieties grown in most countries are highly susceptible. Consequently, scientists have looked elsewhere for sources of resistance and found that clones of *S. spontaneum* and *Erianthus arundinaceus*, and some backcross clones derived from these wild canes, are resistant or moderately resistant to *M. javanica* and *P. zeae* (Stirling et al., 2012; Bhuiyan et al., 2016a,b).

Although plant breeders are now attempting to transfer resistance genes from wild species into commercially acceptable sugarcane varieties, it is questionable whether this approach will ultimately be successful. Sugarcane is attacked by a diverse community of plant-parasitic nematodes, and if varieties that limit the reproduction of key species are developed, it is almost certain that the ecological niche they occupy will be filled by other plant-feeding nematodes.

Given that there is only a remote chance of developing varieties with resistance to the wide spectrum of nematodes capable of attacking sugarcane, future breeding programmes should focus on selecting for tolerance (i.e. varieties that grow well in spite of the damage caused by plant-feeding nematodes). Since new varieties are always tested in soils infested with plant-parasitic nematodes, traditional selection programmes tend to select material with some tolerance. However, the rate of progress could be improved by focusing selection programmes on sites that are heavily infested with key nematode pests. Another option is to compare yields in infested and relatively nematode-free plots (established using a combination of rotation crops and nematicides) and select varieties that do not suffer major yield losses in nematode-infested soil.

What is ultimately required in sugarcane is a combination of resistance and tolerance. Sources of resistance to key pests may be identified and incorporated into commercial varieties, but those varieties must also have the capacity to cope with the damage caused by the many other plant-feeding nematodes that are found in sugarcane soils.

4.3 Nematicides

During the period from 1960 to 1990, nematode control in sugarcane focused on non-volatile nematicides and the results of that work are summarized by Birchfield (1984) and Cadet and Spaull (1991). Since then, field trials in various countries have continued to show that sugarcane yields can be increased by applying a nematicide.

- United States: Aldicarb increased yield by 15%, but responses occurred only at sites with sandy soils (Bond et al., 2000).
- South Africa: Aldicarb applied annually to 10 sugarcane cultivars at two sites increased long-term sugarcane production by a factor of three at one site and five at the other site (Cadet and Spaull, 2003).
- Australia: Fenamiphos and aldicarb applied 3–4 times/year increased the yield of plant and first ratoon crops by an average of 15.3 and 11.6%, respectively (Blair and Stirling, 2007).

Although organophosphate and carbamate nematicides have been a vital tool for measuring the effects of nematodes on sugarcane, they are relatively expensive and have never been widely used in the sugar industry worldwide. Nematicides are sometimes seen as an economically viable option in sandy soils, but because responses tend to be small and erratic in heavier soils, most of the soils in which sugarcane is grown have never been treated. Concerns about the toxicity and environmental impact of chemicals such as aldicarb and fenamiphos (Thomason, 1987) have also limited their use and led to their withdrawal or deregistration in many countries. New nematicides such as abamectin, fluensulfone and fluopyram are less toxic and have been registered for use on some crops in some countries, but it remains to be seen whether they are sufficiently cost-effective to be used on sugarcane.

4.4 Conclusion

Given the outcomes of the work discussed above, it is clear that the traditional approaches used to control plant-parasitic nematodes in sugarcane have their limitations: rotation crops reduce nematode populations for relatively short periods; varieties with resistance to one or two nematode species are unlikely to perform because most sugarcane fields are infested with five or six damaging species, and nematicides provide only temporary control and are limited by cost and concerns about their environmental impact. Better management strategies are obviously required and an approach which focuses on managing the whole biological community rather than the pest nematode component is discussed in the remainder of this chapter.

5 Nematode management from a systems perspective

In the past, those attempting to develop solutions to soil-related problems in agriculture have failed to view the soil ecosystem in its entirety. Instead, they subdivided the soil into its parts (physics, chemistry and biology), focused on the first two components and dealt with the biological component as if pests and pathogens were the only organisms that lived in soil. They then tried to manage the system without recognizing that interactions between various components of the system are always occurring. Thus, both scientists and practitioners failed to acknowledge that a practice introduced to influence one component of the system has 'flow-on' effects that impinge on other parts of the system.

In the same way, those involved in agriculture have been good at defining specific problems such as soil compaction, low levels of organic matter, unhealthy root systems, poor nutrient-use efficiency, heavy infestations of soilborne pathogens and severe sub-soil sodicity. Steps have then been taken to find solutions, but in the process there has been a lack of recognition that the problems are interrelated. Having broken the system into fragments for many years, we are ill-prepared to deal with the whole. We are only aware of some parts of the system and since we do not comprehend its complexity, we cannot deal with problems in a holistic manner.

Once we accept that a systems approach is necessary when dealing with problems caused by plant-parasitic nematodes, we can no longer focus only on the pest. Instead, we need to think holistically about the soil/root environment, consider all the organisms in the soil biological community and recognize the following:

- The various nematodes that feed on roots may have different life histories and feeding habits but they are all using the same resource. Thus, they interact with each other. Since they also compete for that resource with fungal and oomycete pathogens, the presence of these competitors will not only increase root damage but also influence nematode population densities.
- Plant-parasitic nematodes are only one part of a nematode community that also contains bacterivores, fungivores, omnivores and predators. All these nematodes help mineralize nutrients from organic matter and are food sources for natural enemies. Some also prey on plant parasites.
- Nematodes are just one component of a complex and diverse soil food web. Bacteria, fungi, microarthropods, macrofauna and many other soil organisms occupy the same environment and they influence each other through mechanisms such as competition, antibiotic production, parasitism and predation.
- Plant-feeding nematodes will damage plant roots, but their effects on plant growth will be influenced by environmental factors. Thus, in situations where water and nutrients are readily available, crops can produce relatively high yields despite a poorly functioning root system.

- Soil physical and chemical factors influence nematode damage thresholds (the number of nematodes required to cause economic loss). Thus, in situations where root growth is restricted by impediments such as compaction, acidity, waterlogging, sodicity and salinity, nematode populations do not have to be very high before additional crop losses begin to occur.

6 Nematode pests: only one component of the soil biological community

In the same way that agricultural scientists working on soil-related problems have failed to view the soil ecosystem as a whole, the biologists who work with sugarcane have tended to be inward-looking. Entomologists, plant pathologists and nematologists have almost always focused on individual pests and pathogens and have rarely considered how they interact with other organisms in the rhizosphere. They have also failed to recognize that control measures developed for one pest will inevitably have an impact on other pests or their natural enemies. Thus, from an ecological perspective, those attempting to resolve nematode problems on sugarcane must appreciate that the target pest is only one component of a much wider soil biological community.

6.1 Plant-parasitic nematodes: a community of pest species

Many different plant-parasitic nematodes coexist in sugarcane fields, and since they occupy similar ecological niches, their population densities will be influenced by the interactions that occur between them. The results of a survey in India support this contention as the population densities of *P. zeae* and *Hoplolaimus indicus* were negatively associated, presumably because the two endoparasites were competing for similar feeding sites. In contrast, two nematodes with different feeding habits (*Tylenchorhynchus annulatus* and *H. dihystera*) had a positive association (Sundararaj and Mehta, 1993b). Interactions between nematodes were also observed in a Brazilian study in pots, because when *P. zeae* and *M. incognita* were inoculated together, an increase in the initial population of one species caused a reduction in the population of the other (Fontana et. al., 2015).

Since the nematode species that feed on sugarcane roots are always part of a community of pest species, the traditional approach of assessing pathogenicity in single-species experiments is questionable. It is more logical to consider the effects of the nematode community as a whole. Bond et al. (2004a) took this approach in Louisiana and found that three relatively common ectoparasites (*Mesocriconema xenoplax*, *Paratrichodorus minor*, *T. annulatus*) acted collectively to reduce shoot biomass and the number of tillers per plant. The important message from such studies is that problems caused by multiple pest species can be resolved only through community management (Cadet and Spaull, 2005).

6.2 Free-living nematodes

Although plant-parasitic nematodes are the dominant group of nematodes in most sugarcane soils, a diverse population of free-living nematodes also occur, particularly in healthy soils. These nematodes are often characterized by their feeding habits, with four major groups (bacterivores, fungivores, omnivores and predators) commonly recognized. From an ecological perspective, free-living nematodes play an important role in soil as they prey on microbes and other soil organisms, thereby encouraging microbial turnover and nutrient cycling (Ingham et al., 1986; Yeates and Wardle, 1996; Ferris et al., 1998; Chen and Ferris, 1999). They also help regulate populations of other soil organisms, including nematodes (Stirling, 2014). The fact that plant parasites and bacterivores dominate sugarcane soils and numbers of omnivores and predators are usually relatively low (Stirling et al., 2007; Mondino et al., 2010; Rodrigues et al., 2011) is an indication that many important biological functions are impaired.

6.3 Fungal and oomycete pathogens of roots

Sugarcane roots often have an unhealthy appearance as they are usually stunted, brown to black in colour and lack root hairs. A range of oomycete pathogens have been isolated from roots showing such symptoms, but the most important are various species of *Pythium* (Hoy, 2000) and *Pachymetra chaunorhiza*, the cause of Pachymetra root rot (Croft and Magarey, 2000). Fungi are also commonly associated with unhealthy sugarcane roots, with 49 species from 24 genera having been isolated in Australian studies (Magarey et al., 1995; 2005a; Magarey 1996). A wide range of dark sterile (dematiaceous) and white sterile fungi were also found in those studies and most of them caused reductions in shoot or root weight and darkening of roots in pot experiments (Magarey et al., 1995; 2005a). In many sugarcane soils, fungal pathogens are more important than nematode pests, and in such situations, it is important to recognize that the crop may not respond unless both groups of pathogens are controlled (Pankhurst et al., 2001; 2005b; Garside and Bell, 2011b).

Studies by Bond et al. (2004b) provide yet another example of interactions between nematodes and other root pathogens. Plant-parasitic nematodes and the oomycete *Pythium arrhenomanes* both reduced shoot and root growth when inoculated separately onto sugarcane. However, interactions between the pathogens were observed when they were inoculated together. The oomycete reduced populations of plant-parasitic nematodes and this resulted in a less than additive response for plant growth suppression.

6.4 Beneficial species

Soil is the most complex and diverse ecosystem on Earth. One gram of healthy soil will contain more than 10^{10} bacterial cells and hundreds of metres of fungal hyphae together with a vast array of microfauna (protozoa, nematodes, rotifers and tardigrades), mesofauna (springtails, mites, symphyla and enchytraeids) and macrofauna (molluscs, millipedes, isopods and earthworms). The role of the microflora (bacteria and fungi) is to break down plant-derived detritus, while the soil fauna feed on microbes and on each other and are, therefore, profoundly important in governing microbial processes (Wardle, 2002; Coleman and Crossley, 2003; Bardgett, 2005). The vast majority of these organisms are beneficial to the plant as they redistribute organic matter within the soil profile; improve soil structure, aeration and drainage; fix nitrogen from the atmosphere; mineralize nutrients; solubilize phosphorus; produce plant hormones; degrade toxic compounds; enhance nutrient uptake; parasitize or prey on pests and pathogens; and enhance disease resistance. Given their vital role in maintaining soil and plant health, these organisms must be considered when nematode management programmes are being developed.

7 Integrated soil biology management

The main reason plant-parasitic nematodes are important pests of sugarcane is that cane-growing soils have been exploited for many years and are physically, chemically and biologically degraded. The regulatory mechanisms that normally suppress nematode populations no longer operate effectively, while a suboptimal physical and chemical environment means that plants are unable to tolerate the damage that pest nematodes cause to their root systems. Thus, inputs of water, fertilizer and pesticides are required to maintain production, further weakening an already-depleted soil biological community. Given that carbon levels in sugarcane soils are usually more than 50% lower than undisturbed soils in their natural state; that soil organic matter plays a central role in improving a soil's physical, chemical and biological properties; and that improvements in soil health result from interactions between occupants of the detritus-based food web; the ultimate solution to nematode problems is to focus on increasing carbon inputs, minimizing carbon losses and reducing management impacts on soil organisms, thereby building an active and diverse biological community. The advantage of this approach is that it targets the primary cause of the nematode problem (management-induced diminution of the soil biota and shortcomings in the farming system) rather than the secondary effect (the presence of a nematode community dominated by plant parasites). Improved soil health and enhanced sustainability will be the most important outcomes of such an approach, but biological mechanisms that suppress nematodes will also be activated. Other outcomes are likely to be broad-spectrum suppression of other root pathogens and an increased capacity of crops to tolerate the effects of these pathogens.

The important point in the preceding paragraph is that the management practices required to enhance soil biological activity and diversity are the foundation on which effective systems of minimizing losses from plant-parasitic nematodes can be built. This argument is based on the premise that once crop rotation and cover cropping become integral components of the farming system, cropping frequency is increased; bare fallows are eliminated; tillage is minimized; crop residues are retained on the soil surface; levels of soil organic matter are raised; compaction and other suboptimal soil constraints are removed; inputs of nutrients are optimized and pesticides are used judiciously, biological control mechanisms will begin to operate effectively and provide enhanced levels of nematode suppression.

The advantage of the above approach is that it shifts the focus from the pest. Instead, the aim is to foster the biological resources that must be in place if an agroecosystem is to provide a full range of ecosystem services. It is, therefore, a wider concept than integrated pest management, an approach that is commonly used to minimize damage caused by insect pests, weeds, plant pathogens and nematodes. Termed 'integrated soil biology management' by Stirling (2014), it aims to build an active and diverse soil biological community that will not only suppress nematode pests but also improve the soil's physical properties, enhance biological nutrient cycling and provide the many other services provided by the soil biota.

8 Improving soil health and reducing losses from nematode pests

The effects of agricultural management on soil biology and soil health are well recognized (Lehman et al. 2015; Stirling et al. 2016) and are summarized in Table 2. Examples from sugarcane can be found in Pankhurst et al. (2003), Stirling (2008), Stirling et al. (2010b), Cardoso et al. (2015) and Franco et al. (2015; 2016). Sugar producers around the world must, therefore, face the challenge of integrating a range of tactics that promote soil health into the sugarcane farming system. Evidence presented in the final sections of this chapter indicates that, when this is done, the production system will become more profitable and sustainable and losses from nematode pests will be reduced.

Table 2 Generalized effects of agricultural management practices on soil health

Tactics that tend to reduce soil health	Tactics that tend to promote soil health
Frequent tillage, aggressive tillage	No till, minimum till, strategic tillage
Bare fallows	Cover crops
Monoculture	Diverse crop rotations, intercropping, a pasture phase in the rotation
Random traffic from farm machinery	Controlled traffic
Removal of crop residues	Retention of crop residues as mulch, amending soil with organic matter
High inputs of inorganic fertilizers at a single time	Banded fertilizers, split fertilizer applications, controlled-release formulations, organic fertilizers, manures
Weeds controlled primarily with herbicides	Weeds controlled primarily by mulching and/or cultural methods
Pests and diseases controlled with pesticides	Integrated pest management

9 Sugarcane farming systems to improve soil health and enhance sustainability

A sugarcane farming system that incorporated many of the soil-health-promoting practices listed in Table 2 (e.g. retention of crop residues, minimum tillage, legume rotation crops and controlled traffic using GPS guidance) was first tested in Australia at the beginning of this century (Bell et al., 2003; Garside et al., 2004). Results of that research (summarized by Garside et al. (2005), Bell et al. (2007), Stirling (2008) and Stirling et al. (2016)) clearly show that a range of soil health benefits were obtained:

- A much smaller area of the field is compacted to the point where root growth is reduced.
- Total and labile carbon levels increase, particularly near the soil surface.

- Rainfall infiltration rates are higher due to more earthworms and greater macroporosity.
- Surface crusting is reduced and aggregate stability improves.
- The water-holding capacity of the soil increases.
- Soil losses to erosion are reduced.
- The cation-exchange capacity increases.
- Pathogen loads are temporarily reduced (due to rotation with crops that do not host sugarcane pathogens).

Given the soil health benefits obtained with the new farming system, it was not surprising that the land was more productive. Although there was a 12-month delay in replanting sugarcane (because of the need to accommodate a rotation crop), higher yields were obtained in the plant crop and subsequent ratoons and so, over the whole crop cycle, the yield that was forgone was more than recovered. The new system was also more profitable and sustainable because other benefits were obtained, including additional income from crops such as soy bean and peanut, reduced fuel costs, greatly reduced need for nitrogen fertilizer (due to nitrogen fixation by the legumes), improved timeliness of operations, lower labour requirements and fewer pesticide inputs.

10 The impact of improved farming systems on nematode pests and their natural enemies

In the 15 years since the new farming system was first tested in Australia, it has become increasingly apparent that it provides a range of benefits that combine to reduce losses from plant-parasitic nematodes. Evidence to support that contention is summarized below.

- The rotation component of the farming system reduces populations of *P. zeae* and *M. javanica* to relatively low levels and these low populations are usually maintained for 3–6 months after sugarcane is planted (Stirling et al., 2001; 2007).
- When sugarcane is harvested green and crop residues are retained, there is a gradual increase in soil organic carbon levels during the crop cycle. However, when the soil is tilled prior to planting the next crop, the benefits of residue retention are negated and soil carbon returns to the levels observed at the beginning of the cycle (Bell et al., 2001). Thus, when tillage is discontinued, soil carbon levels tend to increase and eventually stabilize at a higher level. Ultimately, this has additional effects that impact on plant-parasitic nematodes (see below).
- Organic amendments with a high C:N ratio (e.g. sugarcane residue, sawdust and grass hay) enhance suppressiveness to plant-parasitic nematodes in pots (Stirling et al., 2003). A subsequent experiment in the field confirmed the beneficial effects of sugarcane residue as an amendment, as roots growing in the amended soil contained 95% fewer *P. zeae* than roots in the unamended control (Stirling et al., 2005). Since mulching soil with sugarcane residue also reduced populations of *P. zeae* (Stirling et al., 2011a), these results suggested that losses caused by nematode pests should decline when biological activity is stimulated by carbon inputs.

- Bioassays of soil from sugarcane fields have shown that the soil under a blanket of crop residues is often highly suppressive to plant-parasitic nematodes. In a bioassay of soil from one site, inoculated nematodes were readily recovered from soil where the soil biota had been eliminated by heat, but 84% fewer nematodes were recovered from untreated soil. The carbon level in this highly suppressive soil was 38.2 g C/kg (Stirling et al., 2011b).
- In a field trial in which sugarcane was planted following conventional tillage or by direct drilling, tillage initially reduced populations of *P. zeae*. However, the nematode resurged strongly in the conventionally tilled plots and so populations of root-lesion nematode tended to be higher than in the direct-drill plots at the end of the plant crop. Even greater tillage effects were observed with *M. javanica* and *Xiphinema elongatum*, as populations of both nematodes were significantly higher at the end of the season following conventional tillage. Furthermore, conventional tillage was detrimental to omnivorous and predatory nematodes (Stirling et al., 2011c).
- Measurements in the field and results from pot experiments have shown that there is an inverse relationship between soil C levels and the number of *P. zeae*/g root (Stirling, 2017). Data from a range of soils showed that increasing the soil C content by 0.5% usually reduced *Pratylenchus* populations in roots by about 35%, but sometimes the reduction was as high as 80%, Studies in South African sugarcane soils have also shown that populations of *P. zeae* are negatively correlated with the organic matter content (Rimé et al., 2003).
- Populations of plant-parasitic nematodes tend to decline once the new farming system has been in place for one or two crop cycles. Another benefit from a crop loss perspective is that damage thresholds will have increased because the physical and chemical health of the soil will have improved. Improvements in the capacity of the soil to hold water and nutrients and a reduction in soil strength mean that roots grow more readily and so plants can tolerate higher nematode populations.

What is clear from the points made above is that enhanced suppression of root pests and pathogens is one of the biological benefits from the new farming system. This type of suppression is due to the combined effects of many antagonists (Hoitink and Boehm, 1999; Stone et al., 2004; Stirling 2014; Larkin 2015) and, because it is enhanced by inputs of organic matter, is sometimes referred to as organic matter-mediated suppression. It occurs to some extent in all sugarcane soils, but its effects are most apparent in the surface soil under the blanket of cane residue, where soil carbon levels are relatively high. This type of suppression results from the activity of many organisms and they act collectively against pests and pathogens thorough mechanisms such as parasitism, predation, competition and antibiosis.

The agents responsible for suppressing various pests and pathogens are not always known, but many bacteria and fungi are antagonistic to *Pythium* (Whipps and Lumsden, 1991; Hoitink and Boehm, 1999; Nelson et al. 2004) and there are situations where this widely distributed pathogen is suppressed in sugarcane soils (Pankhurst et al., 2005a). Suppression of *P. chaunorhiza* has also been detected (Magarey et al., 2004; Pankhurst et al., 2005a), but research is required to identify the main antagonists. The natural enemies responsible for suppressing nematode pests have not been fully determined, but nematode-trapping fungi and predatory nematodes are almost certainly involved. In one highly suppressive soil, 14 nematode-trapping species in the genera *Arthrobotrys*,

Dactylella, *Dactylellina*, *Drechslerella* and *Gamysella* were detected and very high population densities of predatory nematodes (>3 nematodes/g soil) were also present (Stirling et al., 2011b). Soil microarthropods are also likely to be contributing, as more than 30 species of nematophagous mites were found in well-managed sugarcane soils (Stirling and Walter, 2016). Initial work with one mesostigmatid mite (*Protogamasellus* sp.) showed that it fed voraciously on nematodes in the laboratory and markedly reduced populations of root-lesion, stunt and microbivorous nematodes in pots.

Although these Australian studies suggest that soils will become more suppressive to plant-parasitic nematodes when the new sugarcane farming system is adopted, it is also clear that improvements do not occur immediately. Most sugarcane soils have been cultivated aggressively for more than 100 years, and soil carbon levels are much lower than they could be. In a study that used perennial grass pasture as a standard reference, the average carbon level in 11 conventionally cropped sugarcane soils was 14 g C/kg compared with 27 g C/kg in soils from paired sites under pasture (Stirling et al., 2010b). Soil carbon levels had not increased after 5–7 years of the new sugarcane farming system (although there had been an improvement in some biological properties), indicating that it will take time to rectify the damage caused by years of conventional farming. Nevertheless, the results of the above survey are encouraging, as unpublished data from the study showed that at five of the eleven survey sites soil carbon levels under perennial grass pasture were greater than 30 g C/kg soil (range 30.5 to 43.8 g C/kg soil), levels that are comparable to the highly suppressive soil identified by Stirling et al. (2011b). Since sugarcane is somewhat similar to an undisturbed perennial grass pasture when grown using the improved farming system, it is not unreasonable to expect that organic matter levels will eventually increase to the point where well-managed soils are at least partially suppressive to plant-parasitic nematodes, particularly in the upper 10 cm of the profile.

The bacterial genus *Pasteuria* is another antagonist that plays a role in regulating nematode populations in sugarcane soils. First observed attacking root-knot nematodes in Mauritius (Williams, 1960), it has since been found on sugarcane in many countries, including South Africa, the United States, Papua New Guinea and Australia (Spaull, 1984; Birchfield, 1984; Bridge, 1986; Stirling 2009). *Pasteuria* is known to parasitize hundreds of nematode species and once the host is infected it usually does not reproduce (Chen and Dickson, 1998; 2004; Stirling, 2014). A survey by Spaull (1984) in South Africa showed that *Pasteuria penetrans* (the species which attacks root-knot nematode) occurred in more than half of the sugarcane fields with coarse-textured sandy soils. However, *Meloidogyne* populations were generally higher in these fields than in non-infested fields, suggesting that the parasite was not reducing nematode populations but merely removing surplus individuals.

More recent work in Australia has shown that *Pasteuria* is widely distributed in sugarcane fields and attacks four important plant-parasitic nematodes (Stirling and Walter, 2016). *Pasteuria* was detected in about half of the fields sampled and endospores were seen adhering to *M. javanica*, *P. zeae*, *T. annulatus* and *H. dihystera*. Infestation levels were generally low but there were instances where 20–60% of the nematodes were encumbered with spores and some nematodes had more than 10 spores attached. A pot experiment in which *P. penetrans* spores in soil from one of the heavilyinfested sites were killed by autoclaving showed that when the soil was re-inoculated with *M. javanica*, the nematode population increased to 288 000 nematode eggs/plant after 9 months. In contrast, only 1520 eggs were recovered from the roots of plants in untreated soil (Stirling, unpublished). This shows that *P. penetrans* can reach high population densities in sugarcane fields and is capable of suppressing populations of root-knot nematode to relatively low levels.

Studies of a soil where more than 20% of the root-lesion nematodes (*P. zeae*) were encumbered with endospores of *Pasteuria thornei* also showed that this species can kill significant numbers of its host nematode. When the soil was inoculated with *P. zeae* and the nematodes were retrieved 5 weeks later using a method that can recover spore-filled nematodes (Oostendorp *et al.*, 1991), more than 50% of the nematodes were infected by the parasite (Stirling, unpublished).

The results outlined above raise questions about why a host-specific parasite such as *Pasteuria* has not increased to high levels in sugarcane soils, despite the fact that its nematode hosts have been present in most soils for more than 100 years. The most likely reason is that sugarcane soils have been regularly tilled and this disrupts the intimate relationship between host and parasite, which occurs in microsites occupied by plant roots. It is in such microsites that endospores produced within parasitized nematodes are released into the environment. Since plant roots form channels in the soil during growth and those roots later decay to form open biopores (Cresswell and Kirkegaard, 1995), concentrations of *Pasteuria* endospores are likely to be highest in root channels utilized by previous crops. When a root from the next crop grows into such a channel, there is a high probability that an endospore will adhere to a nematode as it attempts to invade or feed on the root. However, when the endospores are dispersed by tillage, they are much less likely to come into contact with a host nematode. Unpublished data from the two Australian sites with the highest levels of *Pasteuria* on *P. zeae* provide some evidence to support that contention. Perennial grass pastures were growing at these sites until sugarcane was planted 18 and 22 years ago, respectively, and assays of soil from adjacent fields still under pasture showed that both spore-encumbered and infected specimens of *P. zeae* were readily recoverable from these soils. The fact that *P. thornei* had multiplied to high levels in these undisturbed grass pastures suggests that the high infestation levels at the two sugarcane sites were due to carryover from the previous pasture. Since sugarcane is also a perennial grass, the most likely reason *Pasteuria* has not increased to high levels in most sugarcane fields is that the soil is usually tilled when the crop is replanted.

11 Future challenges

11.1 Increasing organic carbon levels in sugarcane soils

There is overwhelming evidence to indicate that levels of soil organic matter in conventionally tilled agricultural soils have declined precipitously over the last 100 years (Reeves, 1997; Paustian et al., 2000; Franzluebbers, 2004). Sugarcane soils are no different, as soil carbon levels in most cane-growing regions are likely to be much lower than when they were first brought into production (Stirling et al., 2010; Franco et al., 2015). Since organic matter is the energy source that fuels the soil food web, sugarcane soils can never be biologically healthy until levels of soil organic matter are increased.

Sugarcane produces more biomass than most other agricultural crops, with between 13 and 20 tonnes of dry crop residues usually retained at harvest. Retention of these residues as mulch (sometimes referred to as a trash blanket) could be expected to increase the amount of carbon sequestered in soil, and results obtained in several countries indicate that this does occur. However, the improvements have often been relatively small and limited to the upper 50 mm of the soil profile (Blair et al., 1998; Graham et al., 2002;

Razafimbelo et al., 2006; Canellas et al., 2010; Cerri et al., 2011; Thorburn et al., 2012). Presumably, the main reason for this is that sugarcane soils are usually cultivated at the end of every crop cycle to ameliorate the soil compaction problems caused by heavy vehicle trafficking. In such situations, green cane trash blanketing has no measureable impact on total organic carbon stocks (Page et al., 2013).

Simulation models suggest that soil carbon levels should more than double after 35 years of green cane trash blanketing (Thorburn et al., 2000), with the extent of the increase depending on the amount of crop residue retained (Robertson and Thorburn, 2000). Consequently, trash blanketing cannot, on its own, be expected to quickly increase soil organic carbon levels. Organic amendments will also be needed to hasten the process. Studies in many different industries and environments have shown that many types of amendments are effective, provided they are applied repeatedly (Diacono and Montemurro, 2010).

Vast quantities of organic residues are generated in the sugar milling process, and these wastes are the obvious materials to use as soil amendments in the sugar industry. Mill mud (filter cake and filter mud), bagasse and mill ash are the most important waste products, with mill mud widely used around the world as an organic fertilizer and soil ameliorant. The beneficial effects of mill mud are largely ascribed to the nitrogen and phosphorus it contains, but mill mud also improves soil physical properties, increases water infiltration rates, improves water retention, increases cation-exchange capacity and reduces sodicity (Georges, 1985; Henry and Rhebergen, 1994; Kingston 1999). Yield responses usually occur when mill mud and other amendments are applied, but in the past these benefits were obtained by incorporating the amendments into soil before the crop was planted. Once the sugar industry moves to minimum till systems, other methods of applying these materials will have to be developed. Also, the application rates and timings that provide the greatest benefits will have to be determined.

From a biological perspective, organic amendments have a long history of use against plant-parasitic nematodes (see reviews by Muller and Gooch, 1982; Akhtar and Malik, 2000; Oka, 2010; Thoden et al., 2011; McSorley, 2011; Renco, 2013). Early work summarized by Spaull and Cadet (1991) and more recent studies (Jonathan et al., 1991; Mehta and Sundararaj, 1995; Albuquerque et al., 2002) have shown that organic amendments are a potentially useful nematode management tactic in sugarcane. Some of these amendments may have produced nematicidal quantities of ammonia (Rodriguez-Kabana, 1986) or various repellents and toxicants (Chitwood, 2002), but there is a large body of evidence to indicate that organic materials also act by providing food and protection for biological control agents and enhancing their activity (Stirling, 2014). Thus, one of the challenges of the future is to better understand the biological changes that take place when different amendments are added to soil. Stirling et al. (2003) added sawdust, sugarcane trash, grass hay, legume hay, feedlot manure, poultry manure, chitin and mill mud to a sugarcane soil and found that, 7 months later, the amendments with a high C:N ratio had enhanced suppressiveness to *M. javanica* and *P. zeae* to a greater extent than the nitrogenous materials. The most suppressive soils had a fungal-dominant biology and high numbers of omnivorous nematodes. Since products such as mill mud stimulate bacteria rather than fungi, this result raises questions about whether they are the most appropriate materials to reduce populations of plant-parasitic nematodes.

Although organic amendments generally have positive effects on the soil biota, we need to be aware that some of them may be detrimental to soil organisms. High soil phosphorus levels decrease mycorrhizal colonization of sugarcane roots (Kelly et al., 2001;

Magarey et al., 2005b) and since 4–5 typical applications of mill mud/ash mixtures can increase soil phosphorus concentrations to excessive levels (Barry et al., 2001; Qureshi et al., 2001), it is likely that this amendment has negative effects on these important fungi. The environmental consequences of using mill by-products as soil amendments are also a concern, as concentrations of heavy metals such as cadmium may increase in soil when they are applied as amendments (Barry et al., 1998). Cadmium is toxic to microorganisms and invertebrates and may increase in soil to concentrations known to reduce microbial biomass and soil microbial activity (Masto et al., 2011; Hassan et al., 2016). Thus, the need to enhance the sustainability of the sugar industry by recycling sugar mill and distillery wastes must be balanced against the need to ensure that the amendments actually improve biological activity and diversity in sugarcane soils.

11.2 Holistic management of the soil biological community

The other major challenge facing the sugar industry is to ensure that research scientists, extension staff, growers and millers understand that the soil biological community provides many important ecosystem services and must be managed holistically. This means that management practices aimed at overcoming physical constraints such as compaction, providing the crop with nutrients and dealing with pest and disease problems do not have unforeseen side effects such as those detailed below:

- Tillage is detrimental to many components of soil-based food webs (Hendrix et al. 1986; Wardle 1995) and this means that an already-depleted soil biological community suffers further degradation when tillage implements are used to repair compacted sugarcane soils. A more sustainable solution is to control traffic and eliminate the need for tillage.
- Sugarcane requires inputs of nitrogen, but it is important to recognize that, when used excessively, inorganic fertilizers are detrimental to some beneficial organisms. For example, high concentrations of nitrogen are toxic to nematode-trapping fungi and omnivorous and predatory nematodes (Tenuta and Ferris, 2004; Stirling et al., 2015). Consequently, research is required to determine whether controlled-release fertilizers, products containing nitrification inhibitors and practices that reduce nitrogen concentrations in soil water (e.g. multiple rather than single applications of fertilizers such as urea) reduce the negative effects of the fertilizer. Another example is the need to prevent over-use of phosphorus fertilizers and amendments such as mill mud, as high phosphorus levels are detrimental to arbuscular mycorrhizal fungi (Magarey et al. 2005b). These fungi must be valued because they improve soil structure, enhance plant nutrient uptake, alleviate drought stress and improve plant resistance to nematodes and other root pathogens (Stirling 2014).
- One of the most common methods of dealing with weeds and soilborne pests and pathogens in sugarcane culture is to apply a pesticide. What is not commonly recognized is that chemically based pest control comes at a price: various non-target effects and multiple environmental impacts that can be detrimental in the long term. An excellent review by Sanchez-Bayo (2011) provides many examples of these effects and some that are relevant to sugarcane are given below.
  - Fungicides are often applied to sugarcane setts before they are planted, but the possibility that they may destroy (at least temporarily) the beneficial mycorrhizal symbioses that improve plant health in many different ways is rarely considered.

When applied to soil, fungicides are likely to have an even greater impact and will almost certainly affect fungi such as *Metarhizium*, which are important biocontrol agents of cane grubs (Milner et al., 2002; Samson et al., 2010). Fungal parasites and predators of nematodes will also be affected.

- As a group, soil-applied insecticides can decimate populations of the mesofauna (e.g. mites and Collembola) and macrofauna (e.g. carabid and staphylinid beetles, earwigs, centipedes and spiders), all of which help regulate populations of other soil organisms, including pests. The insecticide imidacloprid, widely used for cane grub control, has sub-lethal effects on earthworms, which redistribute organic matter within the soil profile and play an important role in improving soil structure, aeration and drainage (Sanchez-Bayo, 2011).
- The nematicide aldicarb not only affects the free-living nematode community but also eliminates predacious mites, which are particularly susceptible to this chemical (Sanchez-Bayo, 2011).
- Herbicides, commonly used for weed control in sugarcane, are generally considered to have minimal impacts on the soil biology. A recent review of the impact of glyphosate on soil microbial biomass and soil microbial respiration, for example, indicated minimal effects at typical application rates (Nguyen 2016). However, herbicides are known to affect the abundance of predatory nematodes and the development and growth rate of earthworms (Correia and Moreira, 2010; Zhao et al., 2013; Garcia-Perez et al., 2016), indicating that they do have an impact on the soil fauna. Herbicides also have many indirect effects on the soil biological community because when ground cover disappears and roots are destroyed, the labile carbon sources that fuel the soil food web are no longer present. In another example of these indirect effects, sugarcane yields in weedy plantations in Louisiana were 19% higher than in weed-free plantations because the presence of broadleaf weeds enhanced populations of beneficial carabids, ants and spiders that control *Diatraea saccharalis*, the sugarcane borer (Ali and Reagan, 1985).

12 Conclusions

To conclude, many of the practices used in modern sugarcane culture have disastrous long-term effects on the terrestrial ecosystem. The end result is that soilborne pests and pathogens dominate the biological community in sugarcane soils. Consequently, ever-increasing inputs of fertilizer, irrigation water, pesticides and fuel are required to achieve yields that may be considered acceptable, but are well below the yield potential for a given environment. The only way forward is to undertake research that provides cane growers with a combination of soil management practices that will improve root function, increase levels of soil organic matter and restore the suppressive services that have been lost from sugarcane soils. This is a challenging task but is achievable in the long term, provided specialists in soil science, agronomy, plant nutrition, and pest and disease management are prepared to 1) take a holistic view of the soil environment, 2) consider how the undesirable side effects of individual management practices on the soil flora and fauna can be minimized and 3) establish educational programmes that improve the sugar industry's understanding of soil biological processes.

13 Where to look for further information

The reasons why integrated soil biology management is the best long-term option for improving the health of agricultural soils and reducing losses from nematode pests are explained in chapters 11 and 13 of Stirling (2014). Supporting evidence can be found in a review by Lehman et al. (2015) which outlines the ecosystem services provided by soil organisms, discusses the effects of agricultural management practices on soil health and explains how the soil biology can be used more effectively to mitigate soil degradation. A book by Stirling et al. (2016) is also useful because it focuses on the tactics that can be used by growers to build an active and diverse soil biological community capable of improving soil structure, enhancing plant nutrient uptake and suppressing root pests and pathogens. A chapter devoted specifically to sugarcane is included.

14 References

Akhtar, M. and Malik, A. (2000), Roles of organic soil amendments and soil organisms in the biological control of plant-parasitic nematodes: a review. *Bioresource Technology* 74, 35–47.

Albuquerque, P. H. S., Pedrosa, E. M. R. and Moura, R. M. (2002), Host nematode relationships in *Meloidogyne* spp. infested soil amended with stillage and filter cake. *Nematologia Brasiliera* 26, 27–34.

Ali, A. D. and Reagan, T. E. (1985), Vegetation manipulation impact on predator and prey populations in Louisiana sugarcane ecosystems. *Journal of Economic Entomology* 78, 1409–14.

Apt, W. J. and Koike, H. (1982), Pathogenicity of *Helicotylenchus nannus* and its relation with *Pythium graminicola* on sugarcane in Hawaii. *Phytopathology* 52, 798–802.

Bardgett, R. D. (2005), *The Biology of Soil: A Community and Ecosystem Approach*. Oxford: Oxford University Press.

Barry, G. A., Price, A. M. and Lynch, P. J. (1998), Some implications of the recycling of sugar industry by-products. *Proceedings Australian Society of Sugarcane Technologists* 20, 52–5.

Barry, G. A., Rayment, G. E., Jeffrey, A. J. and Price, A. M. (2001), Changes in cane soil properties from applications of sugarmill by-products. *Proceedings Australian Society of Sugarcane Technologists* 23, 185–91.

Bell, M. J., Halpin, N. V., Orange, D. N. and Haines, M. (2001), Effect of compaction and trash blanketing on rainfall infiltration in sugarcane soils. *Proceedings Australian Society of Sugarcane Technologists* 23, 161–7.

Bell, M. J., Halpin, N. V., Garside, A. L., Moody, P. W., Stirling, G. R. and Robotham, B. J. (2003), Evaluating combinations of fallow management, controlled traffic and tillage options in prototype sugarcane farming systems at Bundaberg. *Proceedings Australian Society of Sugarcane Technologists* 25 (CD ROM).

Bell, M. J., Stirling, G. R. and Pankhurst, C. E. (2007), Management impacts on health of soils supporting Australian grain and sugarcane industries. *Soil & Tillage Research* 97, 256–71.

Berry, S. D., Rhodes, R., Foster, J., Risede, J-M. and van Antwerpen, R. (2011), The effect of cover crops on plant parasitic-nematodes of sugarcane. *International Journal of Pest Management*, 57, 363–75.

Bhuiyan, S. A., Croft, B. J., Stirling, G. R., Wong, E., Jackson, P. and Cox, M. (2016a), Assessment of resistance to root-lesion and root-knot nematodes in Australian hybrid clones of sugarcane and its wild relatives. *Australasian Plant Pathology* 45, 165–73.

Bhuiyan, S. A., Croft, B. J., Wong, E., Ogden-Brown, J., Turner, M., Parfitt, R., Magarey, R., Bull, J. and Cox, M. (2016b), Effects of Pachymetra root rot and nematodes on some elite sugarcane clones in Australia. *Proceedings Australian Society of Sugarcane Technologists* 38, 54–64.

Birchfield, W. (1984), Nematode parasites of sugarcane. In W. R. Nickle (ed.), *Plant and Insect Nematodes*. Marcel Dekker Inc., 571–88.

Blair, G. J., Chapman, L., Whitbread, A. M., Ball-Coelho, B., Larson, P. and Tiessen, H. (1998), Soil carbon changes resulting from sugarcane trash management at two locations in Queensland Australia and in north-east Brazil. *Australian Journal of Soil Research* 36, 873–81.

Blair, B. L. and Stirling, G. R. (2007), The role of plant-parasitic nematodes in reducing the yield of sugarcane in fine-textured soils in Queensland, Australia. *Australian Journal of Experimental Agriculture* 47, 620–34.

Bond, J. P., McGawley, E. C. and Hoy, J. W. (2000), Distribution of plant-parasitic nematodes on sugarcane in Louisiana and efficacy of nematicides. *Journal of Nematology* 32, 493–501.

Bond, J. P., McGawley, E. C. and Hoy, J. W. (2004a), The impact of nematodes on sugarcane cultivars. *Nematropica* 34, 235–43.

Bond, J. P., McGawley, E. C. and Hoy, J. W. (2004b), Sugarcane growth as influenced by nematodes and *Pythium arrhenomanes*. *Nematropica* 34, 245–56.

Bridge, J. (1986), Plant nematode survey of Ramu Sugar Limited, Ramu Valley, Papua New Guinea, Report. CAB International, Institute of Parasitology, St. Albans.

Cadet, P., Spaull, V. W. and McArthur, D. G. (2002), Role of plant-parasitic nematodes and abiotic soil factors in growth heterogeneity of sugarcane on a sandy soil in South Africa. *Plant and Soil* 246, 259–71.

Cadet, P. and Spaull, V. W. (2003), Effect of nematodes on the sustained production of sugarcane in South Africa. *Field Crops Research* 83, 91–100.

Cadet, P. and Spaull, V. W. (2005), Nematode parasites of sugarcane. In Luc, M., Sikora, R. A. and Bridge, J. (eds), *Plant Parasitic Nematodes in Subtropical and Tropical Agriculture*. CAB International, Wallingford, 645–74.

Canellas, L. P., Busato, J. G., Dobbss, L. B., Baldotto, M. A., Rumjanek, V. M. and Olivares, F. L. (2010), Soil organic matter and nutrient pools under long-term non-burning management of sugarcane. *European Journal of Soil Science* 61, 375–83.

Cardoso, M., De, O., Pedrosa, E. M. R., Ferris, H., Rolim, M. M., Vicente, T. F., Da, S., David, M. F. and De, L. (2015), Comparing sugarcane fields and forest fragments: the effect of disturbance on soil physical properties and nematode assemblages. *Soil Use and Management* 31, 397–407.

Cerri, C. C., Galdos, M. V., Maia, S. M. F., Bernoux, M., Feigl, B. J., Powlson, D. and Cerri, C. E. P. (2011), Effect of sugarcane harvesting systems on soil carbon stocks in Brazil: an examination of existing data. *European Journal of Soil Science* 62, 23–8.

Chen, J. and Ferris, H. (1999), The effects of nematode grazing on nitrogen mineralization during fungal decomposition of organic matter. *Soil Biology & Biochemistry* 31, 1265–79.

Chen, S. and Dickson, D. W. (2004), Biological control of nematodes by bacterial antagonists. In Z. X. Chen, S. Y. Chen and D. W. Dickson (eds), *Nematology Advances and Perspectives, Volume 2, Nematode Management and Utilization*. CAB International, Wallingford, 1041–82.

Chen, Z. X. and Dickson, D. W. (1998), Review of *Pasteuria penetrans*: biology, ecology and biological control potential. *Journal of Nematology* 30, 313–40.

Chitwood, D. J. (2002), Phytochemical based strategies for nematode control. *Annual Review of Phytopathology* 40, 221–49.

Coleman, D. C. and Crossley, D. A. Jr. (2003), *Fundamentals of Soil Ecology*. Cambridge: Academic Press.

Correia, F. V. and Moreira, J. C. (2010), Effects of glyphosate and 2,4-D on earthworms (*Eisenia foetida*) in laboratory tests. *Bulletin of Environmental Contamination and Toxicology* 85, 264–8.

Cresswell, H. P. and Kirkegaard, J. A. (1995), Subsoil amelioration by plant roots- the process and the evidence. *Australian Journal of Soil Research* 33, 221–39.

Croft, B. J. and Magarey, R. C. (2000), Pachymetra root rot. In P. Rott, R. A. Bailey, J. C. Comstock, B. J. Croft and A. S. Saumtally (eds), *A Guide to Sugarcane Diseases*. CIRAD and ISSCT, 126–30.

Diacono, M. and Montemurro, F. (2010), Long-term effects of organic amendments on soil fertility. A review. *Agronomy for Sustainable Development* 30, 401–22.

Ferris, H., Venette, R. C., van der Meulen, H. R. and Lau, S. S. (1998), Nitrogen mineralization by bacterial-feeding nematodes: verification and measurement. *Plant & Soil* 203, 159–71.

Fontana, L. F., Dias-Arieira, C. R., Mattei, D., Severino, J. J., Biela, F. and Arieira, J. O. (2015), Competition between *Pratylenchus zeae* and *Meloidogyne incognita* on sugarcane. *Nematropica* 45, 1–8.

Franco, A. L. C., Cherubin, M. R., Pavinato, P. S., Cerri, C. E. P., Six, J., Davies, C. A. and Cerri, C. C. (2015), Soil carbon, nitrogen and phosphorus changes under sugarcane expansion in Brazil. *Science of the Total Environment* 515–16, 30–8.

Franco, A. L. C., Bartz, M. L. C., Cherubin, M. R., Baretta, D., Cerri, C. E. P., Feigl, B. J., Wall, D. H., Davies, C. A. and Cerri, C. C. (2016), Loss of soil (macro) fauna due to expansion of Brazilian sugarcane acreage. *Science of the Total Environment* 563–4, 160–8.

Franzluebbers, A. J. (2004), Tillage and residue management effects on soil organic matter. In F. Magdoff and R. R. Weil (eds), *Soil Organic Matter in Sustainable Agriculture*. CRC Press, Boston, 227–68.

Garcia-Perez, J. A., Alarcon, E., Hernandez, Y. and Hernandez, C. (2016), Impact of litter contaminated with glyphosate-based herbicide on the performance of *Pontoscolex corethrurus*, soil phosphatase activities and soil pH. *Applied Soil Ecology* 104, 31–41.

Garside, A. L. and Bell, M. J. (2001), Fallow legumes in the Australian sugar industry: review of recent research findings and implications for the sugarcane cropping system. *Proceedings Australian Society of Sugarcane Technologists* 23, 230–45.

Garside, A. L. and Bell, M. J. (2011a), Growth and yield responses to amendments to the sugarcane monoculture: effects of crop, pasture and bare fallow breaks and soil fumigation on plant and ratoon crops. *Crop & Pasture Science* 62, 396–412.

Garside, A. L. and Bell, M. J. (2011b), Growth and yield responses to amendments to the sugarcane monoculture: towards identifying the reasons behind the response breaks. *Crop & Pasture Science* 62, 776–89.

Garside, A. L., Watters, T. S., Berthelsen, J. E., Sing, N. J., Robotham, B. G. and Bell, M. J. (2004), Comparisons between conventional and alternative sugarcane farming systems which incorporate permanent beds, minimum tillage, controlled traffic and legume fallows. *Proceedings Australian Society of Sugarcane Technologists* (CD ROM).

Garside, A. L., Bell, M. J., Robotham, B. G., Magarey, R. C. and Stirling, G. R. (2005), Managing yield decline in sugarcane cropping systems. *International Sugar Journal* 107, 16–26.

Georges, J. E. W., Mohamed, M. S. and Harvey, W. O. (1985), The reduction of soil compaction by sugarcane by-products used as soil amendments. *Proceedings West Indies Sugar Technologists* 2, 301–21.

Godefroid, M., Delaville, L., Marie-Luce, S. and Quénéhervé, P. (2013), Spatial stability of a plant-feeding nematode community in relation to macro-scale soil properties. *Soil Biology & Biochemistry* 57, 173–81.

Graham, M. H., Haynes, R. J. and Meyer, J. H. (2002), Soil organic matter content and quality: effects of fertilizer applications, burning and trash retention on a long-term sugarcane experiment in South Africa. *Soil Biology & Biochemistry* 34, 93–102.

Hassan, W., Bano, R., Bashir, S. and Aslam, Z. (2016), Cadmium toxicity and soil biological index under potato (*Solanum tuberosum* L.) cultivation. *Soil Research* 54, 460–8.

Hendrix, P. F., Parmelee, R. W., Crossley, D. A. Jr., Coleman, D. C., Odum, E. P. and Groffman, P. M. (1986), Detritus food webs in conventional and no-tillage agroecosystems. *BioScience* 36, 374–80.

Henry, P. C. and Rhebergen, W. (1994), A review of the effectiveness of gypsum, filtercake and deep ploughing for ameliorating irrigated duplex soils in Swaziland. *Proceedings South African Sugar Technologists Association* 68, 59–64.

Hoitink, H. A. J. and Boehm, M. J. (1999), Biocontrol within the context of soil microbial communities: a substrate-dependent phenomenon. *Annual Review of Phytopathology* 37, 427–46.

Holtzmann, O. V. (1964), Nematodes and sugarcane. In C. G. Hughes, E. V. Abbott and C. A. Wismer (eds), *Sugar-cane Diseases of the World*. Elsevier, Amsterdam, 318–41.

Hoy, J. W. (2000), Pythium root rot. In P. Rott, R. A. Bailey, J. C. Comstock, B. J. Croft and A. S. Saumtally (eds), *A Guide to Sugarcane Diseases*. CIRAD and ISSCT, 141–6.

Ingham, R. E., Trofymow, J. A., Ingham, E. R. and Coleman, D. C. (1985), Interactions of bacteria, fungi, and their nematode grazers: effects on nutrient cycling and plant growth. *Ecological Monographs* 55, 119–40.

Jonathan, E. I., Krishnamoorthy, S. V., Mandharan, M. L. and Muthukrishnan (1991), Effect of organic amendments on the control of sugarcane nematodes. *Bharatiya Sugar* 16, 37–40.

Kelly, R. M., Edwards, D. G., Thompson, J. P. and Magarey, R. C. (2001), Responses of sugarcane, maize and soybean to phosphorus and vesicular-arbuscular mycorrhizal fungi. *Australian Journal of Agricultural Research* 52, 731–43.

Kingston, G. (1999), A role for silicon, nitrogen and reduced bulk density in yield responses to sugar mill ash and filter mud/ash mixtures. *Proceedings Australian Society of Sugarcane Technologists* 21, 114–21.

Larkin, R. P. (2015), Soil health paradigms and implications for disease management. *Annual Review of Phytopathology* 53, 199–221.

Lehman, R. M., Cambardella, C. A., Stott, D. E., Acosta-Martinez, V., Manter, D. K. et al. (2015) Understanding and enhancing soil biological health: the solution for reversing soil degradation. *Sustainability* 7, 988–1027.

Magarey, R. C. (1996), Microbiological aspects of sugarcane yield decline. *Australian Journal of Agricultural Research* 47, 307–22.

Magarey, R. C., Yip, H. Y., Bull, J. I. and Johnson, E. J. (1995), Recent studies into the soil biology of yield decline. *Proceedings Australian Society of Sugarcane Technologists* 17, 128–33.

Magarey, R. C., Bull, J. I. and Camilleri, J. R. (2004), A suppression assay for assessing the effect of management practices on Pachymetra root rot. *Proceedings Australian Society of Sugarcane Technologists* 26 (CD ROM).

Magarey, R. C., Bull, J. I. and Reghenzani, J. R. (2005b), The influence of vesicular arbuscular mycorrhizae (VAM) on sugarcane growth in the field. *Proceedings Australian Society of Sugarcane Technologists* 27, 282–90.

Magarey, R. C., Yip, H. Y. and Bull, J. I. (2005a), Dematiaceous fungi, a cause of poor root health in sugarcane. *Proceedings Australian Society of Sugarcane Technologists* 27, 344–52.

Masto, R. E., Ahirwar, R., George, J., Ram, L. C. and Selvi, V. A. (2011), Soil biological and biochemical response to cadmium exposure. *Open Journal of Soil Science* 1, 8–15.

McSorley, R. (2011), Overview of organic amendments for management of plant-parasitic nematodes, with case studies from Florida. *Journal of Nematology* 43, 69–81.

Mehta, U. K. and Sundararaj, P. (1995), Efficacy of some new soil amendments for the control of the lesion nematode in sugarcane. *Nematologia Mediterranea* 23, 321–3.

Mondino, E. A., Tavares, O. C. H., Lima, E. and Berbara, R. L. L. (2010), Communities of nematodes in sugarcane under different tillage and harvest systems. *Nematropica* 40, 203–15.

Milner, R. J., Samson, P. R. and Bullard, G. K. (2002), FI 1045: a profile of a commercially useful isolate of *Metarhizium anisopliae* var. *anisopliae*. *Biocontrol Science and Technology* 12, 43–58.

Muller, R. and Gooch, P. S. (1982), Organic amendments in nematode control. An examination of the literature. *Nematropica* 12, 319–26.

Nelson, E., Kageyama, K., Van Dijk, K. and Windstam, S. (2004), Biological control of soilborne diseases: important concepts from a model system. *Acta Horticulturae* 635, 115–30.

Nguyen, D. B., Rose, M. T., Rose, T. J., Morris, S. G. and van Zwieten, L. (2016), Impact of glyphosate on soil microbial biomass and respiration: A meta-analysis. *Soil Biology & Biochemistry* 92, 50–7.

Obici, L. V., Dias-Arieira, C. R., Klosowski, E. S., Fontana, L. F., Cunha, T. P. L., Santana, S. M. and Biela, F. (2011), Effect of leguminous plants on *Pratylenchus zeae* and *Helicotylenchus dihystera* in naturally infested soils. *Nematropica* 41, 215–22.

Oka, Y. (2010), Mechanisms of nematode suppression by organic soil amendments- a review. *Applied Soil Ecology* 44, 101–15.

Okada, H. and Kadota, I. (2003), Host status of 10 fungal isolates for two nematode species, *Filenchus misellus* and *Aphelenchus avenae*. *Soil Biology & Biochemistry* 35, 1601–7.

Oostendorp M., Hewlett T.E., Dickson D.W. and Mitchell D.J. (1991) Specific gravity of spores of Pasteuria penetrans and extraction of spore-filled nematodes from soil. *Journal of Nematology* 23, 729–32.

Page, K. L., Bell, M. and Dalal, R. C. (2013), Changes in total soil organic carbon stocks and carbon fractions in sugarcane systems as affected by tillage and trash management in Queensland, Australia. *Soil Research*, 51, 608–14.

Pankhurst, C. E., Blair, B. L., D'Amato, C. D., Bull, J. I. and Magarey, R. C. (2001), Quantification of the effects of fumigation and nematicide treatments on the early shoot and root growth of sugarcane in a yield decline soil. *Proceedings Australian Society of Sugarcane Technologists* 23, 260–7.

Pankhurst, C. E., Magarey, R. C., Stirling, G. R., Blair, B. L., Bell, M. J. and Garside, A. L. (2003), Management practices to improve soil health and reduce the effects of detrimental soil biota associated with yield decline of sugarcane in Queensland, Australia. *Soil and Tillage Research* 72, 125–37.

Pankhurst, C. E., Blair, B. L., Magarey, R. C., Stirling, G. R., Bell, M. J. and Garside, A. L. (2005a), Effects of rotation breaks and organic matter amendments on the capacity of soils to develop biological suppression towards soil organisms associated with yield decline of sugarcane. *Applied Soil Ecology* 28, 271–82.

Pankhurst, C. E., Blair, B. L., Magarey, R. C., Stirling, G. R. and Garside, A. L. (2005b), Effects of biocides and rotation breaks on soil organisms associated with the poor early growth of sugarcane in continuous monoculture. *Plant and Soil* 268, 255–69.

Pankhurst, C. E., Stirling, G. R., Magarey, R. C., Blair, B. L., Holt, J. A., Bell, M. J. and Garside, A. L. (2005c), Quantification of the effects of rotation breaks on soil biological properties and their impact on yield decline of sugarcane. *Soil Biology and Biochemistry* 37, 1121–30.

Paustian, K., Six, J., Elliott, E. T. and Hunt, H. W. (2000), Management options for reducing CO_2 emissions from agricultural soils. *Biogeochemistry* 48, 147–63.

Prasad, S. K. (1972), Nematode diseases of sugar-cane. In J. M Webster (ed.), *Economic Nematology*, Academic Press, Cambridge, 144–58.

Qureshi, M. E., Wegener, M. K. and Mason, F. M. (2001), Economics of sugarmill by-products as a source of nutrients in the Australian sugar industry in Mackay. *Proceedings Australian Society of Sugarcane Technologists* 23, 192–8.

Rao, V. R. and Swarup, G. (1975), Pathogenicity of the spiral nematode, *Helicotylenchus dihystera*, to sugarcane. *Indian Journal of Nematology* 4, 160–6.

Razafimbelo, T., Barthès, B, Larré-Larrouy, M-C., Luca, E. F. D., Laurent, J-Y., Cerri, C. C. and Feller, C. (2006), Effect of sugarcane residue management (mulching versus burning) on organic matter in a clayey Oxisol from southern Brazil. *Agriculture, Ecosystems & Environment* 115, 285–9.

Reeves, D. W. (1997), The role of organic matter in maintaining soil quality in continuous cropping systems. *Soil & Tillage Research* 43, 131–67.

Renčo, M. (2013), Organic amendments of soil as useful tools of plant parasitic nematodes control. *Helminthologia* 50, 3–14.

Rimé, D., Nazaret, S., Gourbière, F., Cadet, P. and Moënne-Loccoz, Y. (2003), Comparison of sandy soils suppressive and conducive to ectoparasitic nematode damage on sugarcane. *Phytopathology* 93, 1437–44.

Robertson, F. A. and Thorburn, P. J. (2000), Trash management- consequences for soil carbon and nitrogen. *Proceedings Australian Society of Sugarcane Technologists* 22, 225–9.

Rodrigues, C. V. M. A., Pedrosa, E. M. R., Oliveira, A. K. S., Leitão, D. A. H. S. and Oliveira, N. J. V. (2011), Vertical distribution of nematode communities associated with sugarcane. *Nematropica* 41, 5–11.

Rodriguez-Kábana, R., Morgan-Jones, G. and Chet, I. (1987), Biological control of nematodes: soil amendments and microbial antagonists. *Plant & Soil* 100, 237–47.

Samson, P. R., Bade, G. S. and Harris, W. J. (2010), Efficacy of Biocane™ against southern one-year canegrub, *Antitrogus consanguineus*. *Proceedings Australian Society of Sugarcane Technologists* 32, 50–61.

Sánchez-Bayo, F. (2011), Impacts of agricultural pesticides on terrestrial ecosystems. In F. Sánchez-Bayo, P. J. van den Brink and R. M. Mann (eds), *Ecological Impacts of Toxic Chemicals*. Bentham Science Publishers, Sharjah, 63–87.

Santana, S. M., Dias-Arieira, C. R., Biela, F., Cunha, T. P. L., Chiamolera, F. M., Puerari, H. H. and Fontana, L. F. (2012), Root-lesion nematodes management with antagonistic plants in sugarcane growing areas. *Nematropica* 42, 63–71.

Sasser, J. N. and Freckman, D. W. (1987), A world perspective on nematology: the role of the Society. In J. A. Veech and D. W. Dickson (eds), *Vistas on Nematology*. Society of Nematologists, 7–14.

Spaull, V. W. (1984), Observations on *Bacillus penetrans* infecting *Meloidogyne* in sugarcane fields in South Africa. *Revue de Nématologie* 7, 277–82.

Spaull, V. W. and Cadet, P. (1991), Nematode parasites of sugarcane. In M. Luc, R. A. Sikora and J. Bridge (eds), *Plant Parasitic Nematodes in Subtropical and Tropical Agriculture*. CAB International, Wallingford, 461–91.

Stirling, G. R. (2008), The impact of farming systems on soil biology and soilborne diseases: examples from the Australian sugar and vegetable industries – the case for better integration of sugarcane and vegetable production and implications for future research. *Australasian Plant Pathology* 37, 1–18.

Stirling, G. R. (2009), An improved sugarcane farming system enhances suppression of root-knot nematode, but what organisms are responsible? *Program handbook, Fifth Australasian Soilborne Diseases Symposium*, Thredbo, NSW, Australia, 80–2.

Stirling, G. R. (2014), *Biological Control of Plant Parasitic Nematodes. 2nd edition. Soil Ecosystem Management in Sustainable Agriculture*. CAB International, Wallingford.

Stirling, G. R. (2017), Soil carbon, root health and nematode pests in sugarcane soils. 1. Root and soil health and its relationship to soil carbon levels. *Proceedings Australian Society of Sugarcane Technologists* 39, 155–65.

Stirling, G. R. and Walter, D. E. (2016), Improving soil health and enhancing suppressive services in sugarcane soils. *Journal of Nematology* 48, 372–73.

Stirling, G. R., Blair, B. L., Pattemore, J. A., Garside, A. L. and Bell, M. J. (2001), Changes in nematode populations following fallow, fumigation and crop rotation, and implications for the role of nematodes in yield decline. *Australasian Plant Pathology* 30, 323–35.

Stirling, G. R., Wilson, E. J., Stirling, A. M., Pankhurst, C. E., Moody, P. W. and Bell, M. J. (2003), Organic amendments enhance biological suppression of plant-parasitic nematodes in sugarcane soils. *Proceedings Australian Society of Sugarcane Technologists* 25: CD ROM.

Stirling, G. R., Wilson, E. J., Stirling, A. M., Pankhurst, C. E., Moody, P. W., Bell, M. J. and Halpin, N. (2005), Amendments of sugarcane trash induce suppressiveness to plant-parasitic nematodes in sugarcane soils. *Australasian Plant Pathology* 34, 203–11.

Stirling, G. R., Berthelsen, J. E., Garside, A. L. and James, A. T. (2006a), The reaction of soybean and other legume crops to root-knot nematodes (*Meloidogyne* spp.) and implications for crops grown in rotation with sugarcane. *Australasian Plant Pathology* 35, 707–14.

Stirling, G. R., Berthelsen, J. E., James, A. T. and Agnew, J. R. (2006b), The impact of root-knot nematodes (*Meloidogyne* spp.) on legume crops grown in rotation with sugarcane. *Proceedings Australian Society of Sugarcane Technologists* 28, 351–8.

Stirling, G. R., Moody, P. and Stirling, A. M. (2007), The potential of nematodes as an indicator of the biological status of sugarcane soils. *Proceedings of the Australian Society of Sugarcane Technologists* 29, 339–51.

Stirling, G. R., Halpin, N. V., Dougall, A. and Bell, M. J. (2010a), Status of winter cereals, other rotation crops and common weeds as hosts of lesion nematode (*Pratylenchus zeae*). *Proceedings of the Australian Society of Sugarcane Technologists* 32, 62–70.

Stirling, G. R., Moody, P. W. and Stirling, A. M. (2010b), The impact of an improved sugarcane farming system on chemical, biochemical and biological properties associated with soil health. *Applied Soil Ecology* 46, 470–7.

Stirling, G. R., Halpin, N. V. and Bell, M. J. (2011a), A surface mulch of crop residues enhances suppressiveness to plant-parasitic nematodes in sugarcane soils. *Nematropica* 41, 109–21.

Stirling, G. R., Rames, E., Stirling, A. M. and Hamill, S. (2011b), Factors associated with the suppressiveness of sugarcane soils to plant-parasitic nematodes. *Journal of Nematology* 43, 135–48.
Stirling, G. R., Halpin, N. V., Bell, M. J. and Moody, P. W. (2011c), Impact of tillage and residues from rotation crops on the nematode community in soil and surface mulch during the following sugarcane crop. *International Sugar Journal* 113, 56–64.
Stirling, G. R, Cox, M. C. and Ogden-Brown, J. (2012), Resistance to plant-parasitic nematodes (*Pratylenchus zeae* and *Meloidogyne javanica*) in *Erianthus* and crosses between *Erianthus* and sugarcane. *International Sugar Journal* 114, 30–6.
Stirling, G. R., Stirling, A. M., Schmidt, S. and Robinson, N. (2015), Impact of nitrogen inputs to a sugarcane soil on plant-parasitic nematodes and their natural enemies. *Proceedings of the Australian Society of Sugarcane Technologists* 37, 204–11.
Stirling, G. R., Hayden, H., Pattison, T. and Stirling, M. (2016), *Soil Health, Soil Biology, Soilborne Diseases and Sustainable Agriculture*. A guide. CSIRO Publishing.
Stone, A. G., Scheuerell, S. J. and Darby, H. M. (2004), Suppression of soilborne diseases in field agricultural systems: Organic matter management, cover cropping, and other cultural practices. In F. Magdoff and R. R. Weil (eds), *Soil Organic Matter in Sustainable Agriculture*. CRC Press, Boca Raton, 131–77.
Sundararaj, P. and Mehta, U. K. (1993a), Host status of some economic crops to *Pratylenchus zeae* and their influence on subsequent sugarcane crops. *Indian Journal of Nematology* 23, 165–9.
Sundararaj, P. and Mehta, U. K. (1993b), Patterns of interspecific associations in plant parasitic nematodes of sugarcane ecosystems. *Nematologia Mediterranea* 21, 275–7.
Tenuta, M. and Ferris, H. (2004), Sensitivity of nematode life-history groups to ions and osmotic tensions of nitrogenous solutions. *Journal of Nematology* 36, 85–94.
Thoden, T. C., Korthals, G. W. and Termorshuizen, A. J. (2011), Organic amendments and their influences on plant-parasitic and free-living nematodes: a promising method for nematode management? *Nematology* 13, 133–53.
Thomason, I. J. (1987), Challenges facing nematology: environmental risks with nematicides and the need for new approaches. In J. A. Veech and D. W. Dickson (eds), *Vistas on Nematology*. Society of Nematologists, 469–76.
Thorburn, P. J., Keating, B. A., Robertson, F. A. and Wood, A. W. (2000), Long-term changes in soil carbon and nitrogen under trash blanketing. *Proceedings Australian Society of Sugarcane Technologists* 22, 217–24.
Thorburn, P. J., Meier, E. A., Collins, K. and Robertson, F. A. (2012), Changes in soil carbon sequestration, fractionation and soil fertility in response to sugarcane residue retention are site-specific. *Soil & Tillage Research* 120, 99–111.
Wang, H., Zhuo, K., Ye, W. and Liao, J. (2015), Morphological and molecular charaterisation of *Pratylenchus parazeae* n. sp. (Nematoda: Pratylenchidae) parasitizing sugarcane in China. *European Journal of Plant Pathology* 143, 173–91.
Wardle, D. A. (1995), Impacts of disturbance on detritus food webs in agro-ecosystems of contrasting tillage and weed management practices. *Advances in Ecological Research* 26, 105–83.
Wardle, D. A. (2002), *Communities and Ecosystems: Linking the Aboveground and Belowground Components*. Monographs in Population Biology 34, Princeton University Press, Princeton.
Whipps, J. M. and Lumsden, R. D. (1991), Biological control of *Pythium* species. *Biocontrol Science and Technology* 1, 75–90.
Williams, J. R. (1960), Studies on the nematode soil fauna of sugarcane fields in Mauritius. 5. Notes upon a parasite of root-knot nematodes. *Nematologica* 5, 37–42.
Yeates, G. W. and Wardle, D. A. (1996), Nematodes as predators and prey: relationships to biological control and soil processes. *Pedobiologia* 40, 43–50.
Yeates, G. W., Bongers, T., de Goede, R. G. M., Freckman, D. W. and Georgieva, S. S. (1993), Feeding habits in soil nematode families and genera- an outline for soil ecologists. *Journal of Nematology* 25, 315–31.
Zhao, J., Neher, D. A., Fu, S., Li, Z. and Wang, K. (2013), Non-target effects of herbicides on soil nematode assemblages. *Pest Management Science* 69, 679–84.

Chapter 15

Progress in understanding and managing insect pests affecting sugarcane

François-Régis Goebel, CIRAD, France; Julien M. Beuzelin, University of Florida, USA; and Mike J. Way, South African Sugarcane Research Institute (SASRI), South Africa

1 Introduction

Sugarcane, *Saccharum* spp. hybrids, is the source of two-thirds of the world's sugar production. The crop is grown on over 20 million hectares in more than hundred countries, including the European overseas territories of Guadeloupe, Martinique and Réunion; the United States; and South Africa. In many sugarcane-producing regions, the sugarcane industry plays a key role in the economy. For example, one million people are dependent on the sugar industry in South Africa (South African Industry Directory, 2016/2017). Sugarcane has also been identified as a potential major source of natural, renewable and environmentally friendly energy in the form of bioethanol and electricity (Botha, 2009).

Sugarcane is susceptible to a wide range of pests and diseases that increase production costs and decrease yield, thus reducing final revenue. There are approximately 1500 arthropod pests and 80 plant diseases recorded worldwide in sugarcane (Williams, 1931; Box, 1953; Goebel and Sallam, 2011); however, only a few species cause economic loss (Sallam, 2006; Conlong and Rutherford 2009; Allsopp, 2010; Showler et al., 2012, Goebel et al., 2014). Stalk borers (Lepidoptera: Crambidae, Noctuidae and Pyralidae)

http://dx.doi.org/10.19103/AS.2017.0035.25

and white grubs (Coleoptera: Scarabaeidae) are regarded as major pests, whereas other insects such as armyworms (Lepidoptera: Noctuidae), aphids (Hemiptera: Aphididae), planthoppers (Hemiptera: Delphacidae), scales (Hemiptera: Coccoidea) and thrips (Thysanoptera: Thripidae) are occasional and localized pests of more limited importance. Thus, continuous monitoring of arthropod populations feeding on sugarcane is needed to identify detrimental infestations and implement management tactics based on research. Because sugarcane agroecosystems are dynamic environments with constant changes in production practices, regulations, weather and climatic patterns, and under the threat of invasive species, pest management research must respond timely and appropriately to any changing circumstances.

Although different tactics can be implemented to manage specific regional pest situations, traditional tactics such as host plant resistance and biological control have been globally adopted. However, a need has been recognized for integrated multi-tactic approaches and a broadening in scope of management to encompass the situation at landscape levels instead of individual fields. Thus, an area-wide integrated pest management (AW-IPM) approach has now emerged. In addition, technological advances in remote sensing, geographic information systems (GIS) and mathematical modelling now allow the development of improved early detection systems and simulation of population dynamics for pest management decision-making. The development and implementation of biosecurity strategies to lower risks posed by imminent biological invasions has also become necessary because of the increase in global travel and trade, and climate change (Cilas et al., 2016).

Unravelling ecological processes governing pest population dynamics in the context of changing pest management paradigms, new technologies and globalization contributes towards refined management plans. This chapter addresses determination of pest-associated crop losses, biological control using parasitoids and entomopathogens, push-pull strategies with the use of companion plants, agricultural practices and their influence on pest population dynamics, insecticide use and biosecurity plans. The focus is on two groups of major sugarcane insect pests: lepidopteran stalk borers and white grubs. New tools for pest detection and management such as modelling, remote sensing and use of GIS are also discussed.

2 Categories of sugarcane pests

2.1 Stalk feeders

Sugarcane stalks are attacked by numerous species of insects often categorized according to the section of the stalk and the age of the crop on which they feed: top borers, internode or stalk borers, and shoot borers. All borers may cause the stalk to dry, which is referred to as a 'dead-hearts'. Lepidopteran borers are predominant in all sugarcane-growing regions, except in Australia and Fiji (Sallam, 2006). Worldwide, there are approximately 50 species of lepidopteran sugarcane borers in the genera *Chilo, Eldana, Sesamia, Diatraea, Scirpophaga, Elasmopalpus, Eoreuma, Telchin, Tetramoera* and *Acigona* (Long and Hensley, 1972; Kfir et al., 2002). Most species are polyphagous and feed on maize (*Zea mays*), rice (*Oryza sativa*), sorghum (*Sorghum bicolor*), millet (e.g. *Pennisetum glaucum*) and wild grasses (Kfir et al., 2002). Borer larvae burrow into the sugarcane stalk where they are protected from insecticides, and therefore, host plant resistance and biological control are usually the main components of stalk borer management strategies. Several coleopteran species can also be stalk feeders. In Southeast Asia and Thailand, larvae of the longhorn beetle (Coleoptera: Cerambycidae) *Dorysthenes buqueti* burrow in sugarcane stubble and feed at the base of the stalk, causing stalk death and crop failure (Sommartya and Suasa-ard, 2006). Weevil borers (Coleoptera: Curculionidae), *Metamasius hemipterus* and *Rhabdoscelus obscurus*, which are also problematic on palms, are present in many sugarcane-producing countries.

2.2 Root feeders

Below-ground insect pests attack sugarcane roots, causing stools to dry out and stalks to lodge. The majority of root feeders are beetle larvae, along with white grubs and wireworms (Coleoptera: Elateridae). White grubs that attack roots worldwide are species in the subfamilies Dynastinae, Rutelinae and Melolonthinae, and genera include *Hoplochelus*, *Dermolepida*, *Lepidiota*, *Heteronychus*, *Adoretus* and *Anomala*. Chemical granules or entomopathogenic fungus are applied to soil for controlling white grubs (Allsopp, 2010). In Florida, the wireworm *Melanotus communis* is abundant in organic and mineral soils (Cherry and Stansly, 2008). This pest is controlled by prophylactic applications of granular phorate at planting but also by flooding fields prior to planting if possible (Cherry et al., 2015). Populations of the once economically important white grub *Tomarus subtropicus* have dramatically declined since the 1980s, likely associated with increased flooding in subsiding organic soils (Cherry et al., 2017). In sugarcane regions of Central and South America, two spittlebugs (Hemiptera: Cercopidae), *Mahanarva fimbriolata* and *Aeneolamia varia*, are major root feeding pests. Termites, *Macrotermes* spp. (Isoptera: Termitidae), can be devastating to the roots and sometimes stalks of drought-stressed crops in regions of Africa and Asia.

2.3 Leaf feeders

Insect pests causing defoliation by feeding directly on sugarcane leaves include armyworms (Lepidoptera: Noctuidae) and locusts (Orthoptera: Acrididae). Population outbreaks are unpredictable and intermittent (Vreyssen et al., 2007). Armyworm outbreaks have been attributed to extensive mechanical harvesting of sugarcane crops, the presence of trash blankets left in fields that assist with weed control and preservation of soil moisture but provide a refuge for the larvae, or the prevalence of fields burnt at harvest. Damage by armyworms usually occurs four to five weeks after harvest; however, marked reduction in sugarcane growth occurs only with severe injury (Isas et al., 2016). Locusts occasionally damage sugarcane, particularly in sub-Saharan regions of Africa, and insecticides are typically used to combat them, while biological control using parasitoids and entomopathogens are being researched.

Mites can feed on the surface of sugarcane leaves. The sugarcane yellow mite, *Oligonychus sacchari* (Prostigmata: Tetranychidae), is an occasional and minor pest of sugarcane in Iranian sugarcane fields, where infestations generally occur during late May-early August, and the lower leaves of sugarcane are usually colonized initially. However, prolonged heavy infestations accompanied by extensive injury to the middle and upper leaves of young plants reduce plant growth (Singh et al., 2003; Nikpay et al., 2013). The sugarcane rust mite, *Abacarus sacchari* (Prostigmata: Eriophyidae), infests sugarcane in Florida (Nuessly et al., 2015), and the severity of injury and potential yield losses depend on sugarcane variety (G. S. Nuessly, unpublished data).

2.4 Sap feeders

Insects feeding on sugarcane sap include aphids, planthoppers, scale insects, whiteflies (Aleyrodidae) and mealybugs (Pseudococcidae). Sap feeders can be virus vectors. The sugarcane aphid, *Melanaphis sacchari*, is a known vector of sugarcane mosaic virus and sugarcane yellow leaf virus (McAllister et al., 2008). *Perkinsiella saccharicida* (Hemiptera: Delphacidae) can transmit Fiji disease. However, in regions like South Africa, the insect can occur, but the disease is absent. Other occasional and minor pests include the sugarcane whitefly, *Neomaskellia andropogonis* (Hemiptera: Aleyrodidae), which has increased in prevalence and has caused increasing amounts of damage in areas of Asia (Nikpay, 2016). Sugarcane thrips, *Fulmekiola serrata* (Thysanoptera: Thripidae), invaded KwaZulu-Natal Province in South Africa and quickly spread throughout the region during a recent drought (Way et al., 2006).

3 Estimating sugarcane crop losses from pests

Determining whether an insect has an economic impact requires the study of the relationship between insect density, injury and crop losses. Economic injury level (EIL) guides pest management decisions based on crop losses, economic value of the agricultural commodity and cost of management must be determined (Stern et al., 1959; Mumford and Norton, 1984; Pedigo et al., 1986). Crop loss studies are important, and should be done using commercial varieties on a regular basis in different production areas under various climatic environments (Walker, 1987). This is a key research area used to justify any IPM strategy. Different approaches have traditionally been adopted to assess crop losses in terms of cane and sugar yield, including manipulation of pest pressure in the field using insecticide applications or artificial infestations (Walker, 1987). Successful trials have been conducted in Réunion, South Africa and Indonesia to compare naturally infested plots, artificially infested plots (by releasing eggs or larvae reared in laboratory colonies), and plots treated with insecticides or release of parasitoids (Goebel et al., 1999; Goebel and Way, 2003; Goebel et al., 2014). These field trials are time consuming because they involve treatments, observations, destructive sampling techniques, measuring and weighting of cane samples. Loss estimations derived from each trial are site-specific because interacting agronomic, climatic and inherent crop characteristics impact on insect pest ecology and thus final yield. Nevertheless, crop loss trials provide key information about the biology and populations dynamics of the target pest, the injury caused to the crop and the impact on various yield components.

In Réunion, the impact of *Chilo sacchariphagus* (Lepidoptera: Crambidae) injury on yield is generally high at all sites where high infestation levels are observed (Fig. 1) (Goebel et al., 1999). This pest had a clear impact on cane weight with the highest reduction of 40 metric tons/ha (26%) at Saint-Paul (dry area) compared to a maximum yield of 160 metric tons/ha in the plots protected with pyrethroids every 14 days over 8 months. In dedicated trials, losses in metric tons of sugar were associated with reduced stalk weight.

In South Africa, the African stalk borer *Eldana saccharina* (Lepidoptera: Pyralidae), a major pest since the 1970s, has a greater impact on sugar content than on cane weight, with a decrease of 21 to 50% in highly infested plots. In all field trials, insecticide applications every 14 days over 8 months maintained very low levels of injury with corresponding high

Figure 1 *Chilo sacchariphagus* larva (a) and adult (b) (Photo F. R. Goebel).

yields (Goebel and Way, 2003). Comparing fields with a wide range of injury levels allowed yield loss estimation of biomass, tons cane per ha, sugar quality and quantity measured as brix, Pol, sucrose, tons sugar per hectare, and economic loss. *E. saccharina* had a clear impact on most yield components. The same dedicated field trials have been used to demonstrate yield losses of between 18% and 27% (tons cane/ha) and between 16% and 24% (tons sucrose/ha) by the sugarcane thrips *Fulmekiola serrata* in South Africa (Way et al., 2004).

In Indonesia, a crop loss trial with *C. sacchariphagus* corroborated the results obtained in Réunion, demonstrating a clear impact on cane yield (Goebel et al., 2014). In this trial in East Java, varying doses of *Trichogramma chilonis* (T1 and T2), a full insecticide application (T3) and a control without any treatment (T4) were used to quantify the yield losses from *C. sacchariphagus*, as well as *C. auricilius* and *Scirpophaga excerptalis* that also bore into the stalk at the same time (Fig. 2). Tables 1–4 show the efficacy of the egg parasitoid at rates of 100,000 individuals/ha released over 4 months, resulting in significant yield reduction decrease, compared to plots treated with insecticides (Goebel et al., 2014).

Feeding by stalk borer larvae reduces the size and mass of sugarcane stalks in addition to the quantity of juice. It also retards maturation, and infested stalks contain higher levels of gums and non-sugar juice components. Goebel et al. (1999) observed that *C. sacchariphagus* injury influences sugarcane plant physiology in a similar manner to water stress, particularly when internodes become severely dry during the growth phase. Regression analysis confirmed that plant biomass is significantly reduced by *E. saccharina*. These variables are usually not included in crop loss studies, due to the payment system mainly based on sugar extracted (or recoverable) in various sugar industries. Sucrose content, which is affected to a greater extent by several borer species, has historically been the most valuable component of the crop. However, the recent interest in producing ethanol from sugarcane may result in a change in emphasis. The impact on cane weight caused by *C. sacchariphagus*, and to a lesser extent by *E. saccharina*, suggests that these

Figure 2 *Scirpophaga excerptalis* larva (a) and adult (b) (Photo F. R. Goebel).

Table 1 Effect of biological control or insecticidal control treatments on stalk characteristics and yields

Treatment	Stalk height (cm)	Diameter (cm)	Weight (kg/stalk)	Cane (t/ha)	Sucrose (t/ha)
T1-Tricho1	246.0 bc	2.57 a	2.00 a	98.9 b	7.4 b
T2-Tricho2	271.9 ab	2.70 a	2.11 ab	115.9 ab	9.1 ab
T3-insecticide	288.9 a	2.74 a	2.33 ab	133.8 a	10.6 a
T4-control	219.5 c	2.58 a	1.73 b	87.6 b	6.4 c
F	6.25	1.61	4.99	4.21	5.60
P	0.003	0.223	0.008	0.014	0.005

For each variable, the means followed by the same letter are not statistically different ($P > 0.05$, Student-Newmans-Keuls).

Table 2 Effect of treatments T1, T2, T3 and T4 on fibre and sucrose contents

Treatment	Fibre % cane	Brix % cane	Purity	Pol % cane	Sucrose (CCS)
T1-Tricho1	11.1 a	18.2 a	76.4 ab	13.8 a	7.6 a
T2-Tricho2	10.2 ab	18.4 a	76.8 a	14.2 a	7.8 a
T3-insecticide	9.2 b	19.4 a	77.0 a	14.9 a	8.0 a
T4-control	12.1 a	17.8 a	75.8 b	13.5 a	7.3 a
F	6.01	2.50	2.67	1.65	1.79
P	0.004	0.109	0.065	0.213	0.179

For each variable, the means followed by the same letter are not statistically different ($P > 0.05$, Student-Newmans-Keuls).

Table 3 Estimated crop losses relative to the treatment associated with the greatest yield (T3-insecticide)

Treatment	No stalks/ha	Loss (%)	Cane yield (t/ha)	Loss (%)	Sucrose yield (t/ha)	Loss (%)
T3*	57,458	–	133.8	–	10.6	–
T2	54,833	4.6	115.8	13.4	9.1	14.1
T1	49,667	13.6	98.9	26.0	7.4	30.2
T4	51,167	10.9	87.6	34.5	6.4	39.6

* Attainable yield = maximum potential of the crop without any significant impact from the moth borers. Loss % = T3-Tn/T3 × 100

Table 4 Comparison of losses (height/weight) from damaged and undamaged stalks by *S. excerptalis*

Stalk samples	N	Height (cm)	Loss (cm)	Weight (kg/stalk)	Loss (kg/stalk)
Undamaged stalks (T3)	481	288.9	–	2.33	–
Stalks without *S. excerptalis* (T4)	420	245.3	43.6 (15.1%)	1.98	−0.35 (15.0%)
Stalks with *S. excerptalis* (T4)	80	181.5	107.0 (37.0%)	1.38	0.95 (40.8%)

Loss % = T3-Treatment/T3 × 100

stalk borers cause adverse effects on plant growth and physiology, as reflected by a significant reduction in stalk length and diameter (Goebel et al., 1999; Goebel and Way, 2003; Goebel and Way, 2006; Goebel et al., 2014). Reduction in plant biomass appears to depend on the crop stage that the stalk borer attacks, and early infestations usually have a greater impact on harvest biomass compared to later infestations (Goebel and Way, 2006). Therefore, control strategies developed against *C. sacchariphagus* in Réunion and Indonesia focus on the early phase of the crop cycle, thereby reducing significant final yield losses. The same trends are reported for other sugarcane stalk borers (Sampson and Kumar, 1983; Goebel et al., 1999; Goebel and Way, 2003). In these studies a strong correlation was detected between internodes bored and damaged stalks (R2 = 0.82; y = 0.22 × −0.61), thus the damaged stalk index is a reliable measure for routine surveys. In conclusion, yield loss field trials conducted in different countries on key borer species illustrated the importance of considering all aspects of yield loss, including quality (sucrose concentration), sugarcane biomass (including fibre), particularly in the context of a possible high demand for green energy, biofuels and bioproducts in the future.

4 Biological control and habitat management of sugarcane pests

Classical and new association biological control (augmentative release of native or introduced species of insect parasitoids) and entomopathogens (fungi, viruses, nematodes or bacteria in sprayable formulation) are applied in various countries to control sugarcane insect pests, typically as part of an integrated pest management strategy. Since the 1960s, the effectiveness of egg, larval and pupal parasitoids against lepidopteran stalk borers has been demonstrated many times in many agricultural crops. Recently, entomopathogens (e.g. *Metarhizium* and *Beauveria*) are being researched more widely. Although biological control is environmentally friendly, caution is required when a parasitoid or fungus is introduced into tropical islands, such as Réunion, Mauritius or Madagascar, because there might be detrimental effects on the endemic fauna.

4.1 Field releases of hymenoptera parasitoids

One biological control method adopted in sugarcane is the field release of parasitic wasps such as egg parasitoids *Trichogramma* spp. and *Telenomus* spp. or larval parasitoids *Cotesia* spp. to control stalk borers (Hassan, 1993; Li, 1994; Botelho, 1999; Goebel et al., 2010; Para et al., 2011; Nikpay et al., 2014). These parasitoids are used worldwide (Fig. 3). In Brazil, they are mass-produced in dedicated units at the sugar mill and released at densities of 200 000 *Trichogramma* per ha and 6000 *Cotesia* per ha (Parra et al., 2011; Parra, 2014). In Brazil, biological control of *Diatraea saccharalis* (Lepidoptera: Crambidae) using *Cotesia flavipes* encompasses more than 2 million hectares of sugarcane, and an additional 500 000 ha when the use of *Trichogramma galloi* is included (Parra, 2014). *Cotesia* spp. are also released in Thailand and India to successfully control several species of stalk borers. *Telenomus busseolae* (Hymenoptera: Scelionidae) is a biological control agent released against stalk borers in Iran, where it is routinely mass-produced for release across Khuzestan Province (Nikpay et al., 2014). In Southeast Asia and India, *T. chilonis* is a key biological control agent, and its released in 4-month-old sugarcane at 100 000 adults per ha over 8 weeks reduced stalk borer injury by 50%, with occasionally better results under relatively high infestations (Goebel, 2014). Parasitoid release can complement insecticide use, provided production units are cost effective, since mass rearing is a vital factor influencing the success of biological control programmes. A large rearing unit has been developed in South Africa (Conlong, 1994) to mass-rear insect hosts and parasitoids and is currently the subject of research activities.

In India, farmers mass-produce *Trichogramma* parasitoids in collaboration with E.I.D. Parry, a local company, within *Trichogramma* rural entrepreneur production centres (Sankar

Figure 3 (a) *T. chilonis* parasitizing *C. sacchariphagus* eggs, (b) black parasitized egg batch. Photo: (F. R. Goebel)

et al., 2014). These efforts have resulted in a reduction in injury from *C. sacchariphagus indicus*, and the entrepreneur model has been adopted locally and abroad.

4.2 Entomopathogens for scarab management in sugarcane

Biological control of sugarcane pests using entomopathogenic fungi primarily targeting Coleoptera and Hemiptera is growing rapidly worldwide. Such biological control programmes are being developed in South America, Thailand and Australia with the use of *Metarhizium anisopliae* (Fig. 4 b). The biological control product BioCane™, containing spores of *M. anisopliae* isolate FI-1045, is available for soil application against the white grub *Dermolepida albohirtum* (Coleoptera: Scarabaeidae) (Fig. 4 a) in sugarcane in Australia (Samson et al., 2016). This fungus is also used to control the longhorn beetle *D. buqueti* in Thailand (Suasa-ard et al., 2008). In Brazil and other Latin American countries, the spittlebugs *M. fimbriolata* and *A. varia* are controlled mainly using *M. anisopliae*, and the application techniques in the field has constantly evolved (Parra, 2014).

In Réunion, *Hoplochelus marginalis* (Coleoptera: Scarabaeidae) was accidentally introduced in 1973 from Madagascar, its native range. The rapid spread of this pest throughout all sugarcane plantations was attributed mainly to a lack of predators, parasitoids and entomopathogenic fungus, which ensure natural control in Madagascar. Following years of chemical control through the 1980s, *Beauveria hoplochelii* (Ascomycota: Hypocreales) (Fig. 4 c), previously described as *B. brongniartii*, was discovered in Madagascar on another species of Hoplochelus. The fungus was introduced into Réunion, where it successfully controlled the pest in most of the sugarcane regions, leading to discontinuation of insecticide use (Jeuffrault et al., 2004; Robène-Soustrade et al., 2015). *H. marginalis* attacks other crops and wild grasses, and its continued suppression by this fungus remains fragile.

4.3 Push-pull strategy and companion plants in sugarcane agroecosystems

The importance of employing companion plants or 'service' plants in and around crop fields has increased in agroecological pest management strategies (Andow, 1991; Nicholls and Altieri, 2004; Ratnadass et al., 2012). Natural vegetation in the vicinity of the crop can influence pest and natural enemy behaviour. In East Africa the push-pull strategy is used to reduce infestations in maize by *Chilo partellus* (Cook et al., 2007). In South Africa, females of *E. saccharina* prefer to oviposit on indigenous sedges rather than sugarcane,

Figure 4 White grub healthy larva (a), killed by *Metarhizium* (b) and by *Beauveria* (c)

and neonate larvae preferentially feed on indigenous sedges rather than sugarcane. In the field, *E. saccharina* oviposition in sugarcane can be reduced by planting within the sugarcane rows *Melinis minutiflora,* an indigenous grass (Barker et al., 2006). Although many parasitoids attack the stalk borer in indigenous habitats, parasitism levels in sugarcane fields are not observed. Behavioural studies, using olfactometers and gas chromatograph analysis of volatiles emitted from undamaged and damaged sugarcane and indigenous host plants, have been conducted in an attempt to explain this phenomenon. Profiles of volatiles emitted by injured indigenous hosts and sugarcane are different and olfactometer studies indicate that one of the parasitoids in particular responds strongly to those volatiles emitted from injured indigenous hosts, but not to volatiles released from injured sugarcane. These findings have been consolidated into a practical AW-IPM strategy for *E. saccharina* that entails, among other practices, increasing the area of wetland adjacent to sugarcane, and thus the area where this pest can be 'pulled' into habitat with the natural enemies (Webster, 2009). At the same time *M. minutiflora* is planted strategically within sugarcane fields to repel *E. saccharina* viz. 'push' thereby minimizing oviposition on sugarcane (Conlong et al., 2009; Cockburn et al., 2014). In Réunion, experiments with plants already known for their attractiveness or repellence are being conducted to reduce damage by *C. sacchariphagus* on sugarcane farms experiencing high pest pressure. The role of *Erianthus arundinaceus* as trap crop for this species has already shown positive results (Nibouche et al., 2012).

5 Cultural control of sugarcane pests

Cultural practices for pest management aim to modify the environment to disrupt the life cycle of target pests, reduce relative attractiveness and suitability of the crop, and increase natural enemy populations (Pedigo, 2002). The development of these cultural practices, which must be compatible with other agronomic production practices, generally requires detailed knowledge of interactions between pests, their food sources and shelter (crop and alternate hosts), their competitors (other pests), and natural enemies (predators, parasitoids and pathogens) (Tillman et al., 2002).

In sugarcane grown in Louisiana, manipulating cultural practices can be used to improve management of the stalk borer *D. saccharalis*. Sugarcane is planted between August and October in the state, with August plantings allowing minimum overlap between planting and harvesting operations (Garrison et al., 2000). In addition, August plantings are associated with better root establishment and are usually conducive to higher yields although yield response to planting dates can be variety specific (Viator et al., 2005b). However, fields planted in August have historically shown increased *D. saccharalis* infestations (Charpentier and Mathes, 1969). Field experiments comparing the effects of four planting dates between early August and mid-November under current production practices showed that sugarcane planted in early August, being available for an extended period of time before the onset of cold temperatures, sustained greater *D. saccharalis* infestations during the fall than sugarcane planted later (Beuzelin et al., 2011a, Fig. 5). These experiments suggest that early August plantings increase overwintering populations. Thus, early August plantings, particularly if using a susceptible cultivar, should be avoided in areas where *D. saccharalis* has been a recurrent problem.

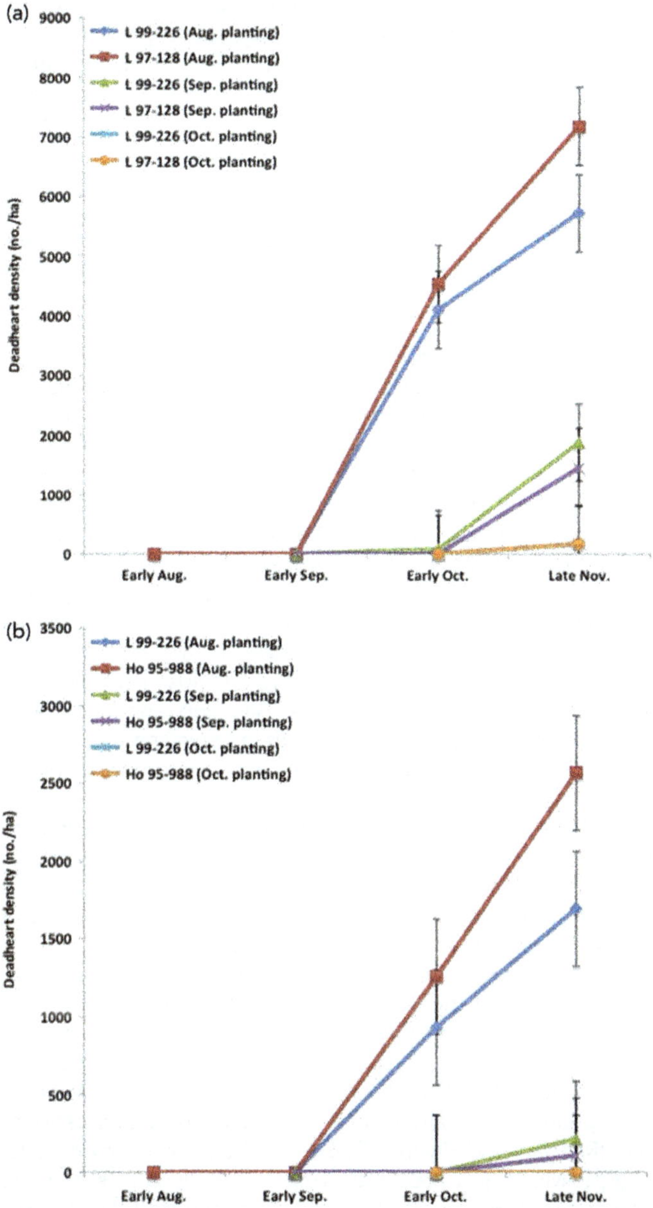

Figure 5 Densities of *D. saccharalis*-caused deadhearts (LS means ± SE) during the fall in sugarcane as affected by planting date in a) Patoutville (2006) and b) Bunkie (2007), Louisiana (Beuzelin et al., 2011a). Planting date impacted deadheart densities in 2006 ($F = 54.2$; d.f. = 2,54; $P < 0.001$) and 2007 ($F = 11.7$; d.f.=2,9; $P = 0.003$).

Harvest residue left in the field over the winter was once considered the most important source of *D. saccharalis* populations infesting new sugarcane growth the following spring (Ingram et al., 1951). However, Kirst and Hensley (1974) showed that although tops of sugarcane stalks left in the field after harvest are initially heavily infested with small larvae, they decay rapidly and do not serve as habitats for overwintering *D. saccharalis* populations. Thus, harvest residue can be left on the soil surface to allow maximum exposure of overwintering *D. saccharalis* larvae to cold temperatures and natural enemies (Kirst and Hensley, 1974). Because underground portions of stubble and newly planted stalks are considered important overwintering habitats, in addition to delaying planting date as possible, planting *D. saccharalis*-free sugarcane seed pieces and prompt ploughing after final harvest are recommended to reduce *D. saccharalis* overwintering populations (Gravois et al., 2014).

The study of sugarcane fields infested with grass weeds, broadleaf weeds, or a mixture of both, showed that weedy fields were associated with a higher abundance and diversity of predators than weed-free sugarcane fields (Ali and Reagan, 1985; Showler et al., 1990; Showler and Reagan, 1991). However, Ali and Reagan (1985) reported that the presence of weeds was not associated with differences in *D. saccharalis* injury and production of adults. Nevertheless, fields with broadleaf weeds infestations did not suffer yield loss and produced an increase in net return to the grower. It was concluded that annual broadleaf weeds, at a subcompetitive level, were beneficial in that they reduced herbicide and cultivation costs while increasing diversity. Showler and Reagan (1991) showed that the presence of annual weeds in sugarcane fields caused at least 25% less injury from *D. saccharalis* compared to weed-free fields. However, the presence of these weeds decreased sugarcane biomass, tiller density and sugar yields. These losses were partially counterbalanced by decreased cultivation costs. Allowing annual broadleaf weeds to grow can therefore assist in managing *D. saccharalis* in Louisiana, but this practice has not been adopted under the current production model, which promotes a high level of weed control.

Maize, sorghum and rice are crop hosts of *D. saccharalis* that can be commercially grown in adjacent fields in Louisiana. Sosa (1990) compared *D. saccharalis* oviposition among four sugarcane clones, a corn cultivar, a sorghum cultivar and a rice cultivar. Despite oviposition preference variations among sugarcane clones, sugarcane attractiveness was always equal to or higher than that of other grass crop hosts. These observations suggest that *D. saccharalis* will readily infest major grass crop hosts, when available in the agroecosystem. In addition, non-Bt corn (i.e. not producing *Bacillus thuringiensis* insecticidal proteins targeting lepidopteran pests) and sorghum are highly suitable hosts that senesce during the summer, producing *D. saccharalis* populations capable of infesting sugarcane (Flynn et al., 1984; Reagan and Flynn, 1986; Castro et al., 2004). Thus, non-Bt corn and sorghum potentially increase *D. saccharalis* populations when grown in sugarcane areas (Reagan and Flynn, 1986) and should be grown as far as possible from sugarcane fields. Modelling area-wide population dynamics of *D. saccharalis* on different sugarcane cultivars, Bessin et al. (1991) demonstrated that the size and spatial arrangement of areas planted to susceptible and resistant cultivars influenced *D. saccharalis* population build-up. Thus, the arrangement in space of hosts with varying levels of suitability for *D. saccharalis* will be expected to impact population build-up on an area-wide basis.

Agricultural farming practices have been integrated into insect pest management in Louisiana sugarcane, primarily targeting *D. saccharalis* (Hensley, 1971; Beuzelin et al.,

2009). However, another stalk borer, *Eoreuma loftini* (Lepidoptera: Crambidae), now threatens the Louisiana sugarcane industry. This invasive species was first detected in southwest Louisiana in 2008 (Hummel et al., 2010), and by early 2016 it was confirmed to be present in nine parishes in south Louisiana (Wilson et al., 2015, 2016b). Cultural practices contributing to *D. saccharalis* management that likely assist in controlling *E. loftini* include reducing the number of overwintering larvae by avoiding August plantings and ploughing as quickly as possible the stubble in fallow fields. In addition, *E. loftini* prefers corn to sugarcane for laying eggs whereas sorghum is equally preferred to sugarcane in the greenhouse (Showler et al., 2011). Non-Bt corn also exhibits more adult exit holes than sugarcane, indicating greater *E. loftini* immature performance on corn, whereas sorghum exhibits equal to or greater than the numbers of exit holes in sugarcane. In commercial fields, *E. loftini* injury to sugarcane increases as non-Bt corn senesces (Showler et al., 2012). Thus, grass crop hosts, and non-Bt corn in particular, should be grown as far as possible from sugarcane fields as recommended for *D. saccharalis* management. However, Bt corn adjacent to sugarcane fields would be expected to attract *E. loftini* adults while causing high larval mortality (Showler et al., 2013; Beuzelin unpublished data), which might be beneficial.

Whereas *D. saccharalis* females lay eggs on both sides of green leaves in the upper canopy (Holloway, 1928; Sosa, 1990), *E. loftini* females oviposit in cracks and folds, mostly on dry leaves of dead shoots or of the lower (<80 cm) portion of live sugarcane stalks (Van Leerdam et al., 1984; 1986). This behaviour has been associated with increased oviposition on drought-stressed sugarcane, because of increased availability of folds and dry leaf material for oviposition. In addition, increased nutritional quality of drought-stressed sugarcane, with increased concentrations of certain free amino acids, might be associated with increased oviposition (Reay-Jones et al., 2007; Showler and Castro, 2010a,b). In a greenhouse study, *E. loftini* females laid on average >10 times more eggs on drought-stressed sugarcane than on well-watered sugarcane (Showler and Castro, 2010a). Thus, irrigation is a cultural practice that can specifically assist in managing *E. loftini* in drought-stressed sugarcane. Irrigation reduced the probability of a bored internode by 60% in a 2-yr field experiment (Reay-Jones et al., 2005). The oviposition behaviour of *E. loftini* might also justify manipulation of leaf residue left on the soil surface after green sugarcane harvesting, which might increase the availability of oviposition substrate and could lead to increased *E. loftini* infestations (A. T. Showler, personal communication). Post-harvest residue, however, is typically burnt during the late winter in Louisiana to prevent excess soil moisture and to increase soil temperatures (Viator et al., 2005a, 2009). This practice can negatively impact some groups of soil-associated predators (White et al., 2011).

In addition to irrigation and harvest residue management, the impact of soil fertilization and the role of non-crop habitat management on *E. loftini* populations and pest severity have been the focus of recent research on cultural practices for *E. loftini* management. Increasing levels of nitrogen (N) fertilization in sugarcane by applying compost at planting, in addition to conventional N fertilization in the spring, increases *E. loftini* injury in plant cane and the first ratoon (Showler, 2015). Increasing conventional N fertilization from 0 to 135 kg N/ha in a 2-yr field study also increased 2.9- and 3.5-fold the percentage of *E. loftini*-bored internodes in high-biomass and sweet sorghum (Van Weelden et al., 2016a). Thus, a balance between increases in yield and *E. loftini* injury associated with N fertilization must be taken into account. These results are consistent with other studies that show an increase in stalk borer infestations with over-application of nitrogen (Lopez et al., 1983; Atkinson and Nuss, 1989; Goebel et al., 2005). The application of silicon

(Si) soil amendments has been associated with increased resistance to stalk borers in sugarcane (Kvedaras et al., 2007; Keeping et al., 2013; White and White, 2014), and is a cultural practice that might counteract the effects of N fertilization on *E. loftini*. This aspect of soil nutrient management has been particularly studied for *E. saccharina*, whose pest severity in South Africa sugarcane is influenced by applications of nitrogen and silicon (Kvedaras et al., 2007). While over-application of nitrogen can increase two- to threefold the level of infestation, silicon in the soil (and taken up by the plant) can significantly reduce this infestation. Silicon in the plant seems to act as a mechanical barrier to the borer larvae whose mandibles were severely worn after penetration in the stalk (Kvedaras and Keeping, 2009). Weed management and the presence of weeds can also influence pest infestation in sugarcane fields. *E. loftini* infests numerous weedy grasses in non-crop habitats adjacent to rice fields in southeast Texas (Beuzelin et al., 2011b). Because johnsongrass and vaseygrass – *Sorghum halepense* and *Paspalum urvillei*, respectively – are ubiquitous perennial weeds thought to play a role in *E. loftini* population overwintering and build-up in Texas and Louisiana (Beuzelin et al., 2011b, 2013), destruction of these weeds along field margins and ditch banks is recommended in the fall and spring to reduce overwintering populations (Beuzelin et al., 2016). However, the contribution of non-crop habitats to *E. loftini* populations infesting sugarcane fields has not been quantified. In addition, the role of these habitats as a reservoir for natural enemies has not been determined.

Cultural practices can assist in managing *D. saccharalis* and *E. loftini* in Louisiana sugarcane, as they generally do for other sugarcane pests. Whereas a better quantification of the impact of certain practices on each stalk borer species is needed, integrating practices for successful management of the two species, as a pest complex, will be needed. Managing these two insects as a stalk borer complex is important because they share comparable life histories but also exhibit different behaviours, suggesting that practices negatively impacting one species might be favourable to the other in some instances. In addition to integrating multiple cultural practices targeting multiple pests, optimizing spatial implementation of those practices will affect their effectiveness because sugarcane stalk borers are mobile and attack numerous host plants. Thus, the impact of cultural practices implemented at the field, farm or regional level should be understood to optimize implementation, which will lead to greater benefits for farmers and encourage further adoption of refined cultural practices.

6 Insecticides

Foliar insecticides are generally not used for control of aboveground sugarcane insect pests worldwide because of detrimental non-target effects, poor efficacy, cost, restrictive regulations, but also the availability of more efficacious and sustainable management tactics (Goebel and Sallam, 2011). Nevertheless, in countries such as Australia, South Africa and the United States, insecticides are routinely used mainly to combat stalk borers and white grubs. In Australia, control of greyback canegrub *Dermolepida albohirtum* and childers grubs (*Antitrogus parvulus, Lepidiota negatoria and Antitrogus consanguineus*) using Suscon® Blue (140 g/Kg chlorpyrifos ethyl) and Suscon® Maxi (50 g/kg imidacloprid) has prevented considerable losses for up to three years (Chandler and Tucker, 2011) to the industry in North Queensland (Allsopp, 2010). In South Africa, FASTAC EC

(Alpha-cypermethrin) is registered for use against *E. saccharina* under specific conditions to target neonates before they bore into the stalk (Leslie et al., 2006). However the use of insecticides against this borer is permitted only under specific conditions such as in aged crops and drought seasons. Furthermore, the impact of these treatments on the arthropod fauna (ants, spiders and other predators abundant in sugarcane fields) is yet to be precisely quantified.

In Louisiana, since the 1950s, insecticides have played a major role in *D. saccharalis* management (Ingram et al., 1951; Hensley, 1971; Posey et al., 2006). In that industry, classical and augmentation biological control attempts have been unsuccessful (White and Reagan, 1999; White et al., 2004) and the use of cultivar resistance and manipulation of cultural practices alone have not maintained populations below economic levels (Bessin et al., 1990; Posey et al., 2006). In the United States, the use of insecticides has been influenced by environmental issues and the development of resistance (Vines et al., 1984, Southwick et al., 1995; Akbar et al., 2008). In the mid-1990s only pyrethroids were registered. Research therefore focused on developing reduced-risk insecticides with narrow ranges to decrease application hazards and negative impact on wildlife, thereby conserving natural enemies (including the red imported fire ant, *Solenopsis invicta* (Hymenoptera: Formicidae)) and preventing aphid resurgence (Reagan and Posey 2001). Tebufenozide, an ecdysone agonist that causes larvae to produce a malformed cuticle (Wing et al., 1988; Dhadialla et al., 1998), was granted a permanent federal registration in 1998 (Reagan and Posey 2001). Advantages of this insecticide include a strong specificity to lepidopteran insects and little to no toxicity to parasitoids and predators in sugarcane fields (Woolwine et al., 1995; Reagan et al., 1997). Concerns about *D. saccharalis* insecticide resistance following the widespread use of tebufenozide (Akbar et al., 2008) led to the study of novaluron, a chitin synthesis inhibitor, which was granted a permanent federal registration in 2009 for use on sugarcane in the United States (Beuzelin et al., 2010). Aerial applications of novaluron in large plot (≥0.4 ha) field studies were not associated with negative effects on major non-target arthropods including *S. invicta*, predaceous beetles, and spiders (Beuzelin et al., 2010). However, potential sub-lethal effects on *D. saccharalis*'s natural enemies remain undetermined. Chlorantraniliprole and flubendiamide, two recently developed ryanodine receptor activators, obtained permanent federal registrations in 2010 and 2011, respectively. These insecticides are highly effective against *D. saccharalis* in sugarcane (Beuzelin et al., 2014; Wilson et al., 2016a, Fig. 6) and exhibit very low mammalian toxicity (Lahm et al., 2009). However, although chlorantraniliprole and flubendiamide are relatively more toxic to lepidopteran insects (Lahm et al., 2009), impacts on non-target arthropods of the Louisiana sugarcane agroecosystem have not been evaluated.

Overall, management of *D. saccharalis* in Louisiana can rely on reduced-risk insecticides with three modes of action with an ecdysone agonist, a chitin synthesis inhibitor, and two ryanodine receptor activators (Gravois et al., 2014). Pyrethroids are also registered and offer a fourth mode of action despite the fact that they are generally not recommended because of short residual activity and potential effects on non-target arthropods (Beuzelin et al., 2010). Thus, foliar applications of the several recommended insecticides when *D. saccharalis* populations reach the threshold of approximately 5% of sugarcane stalks with at least one live larva behind leaf sheaths are expected to be economically and environmentally sustainable. The conventional use of foliar insecticides in Louisiana sugarcane, however, will be challenged by the establishment of *E. loftini*. The invasive stalk borer can be targeted by the same insecticides as those used against *D. saccharalis*

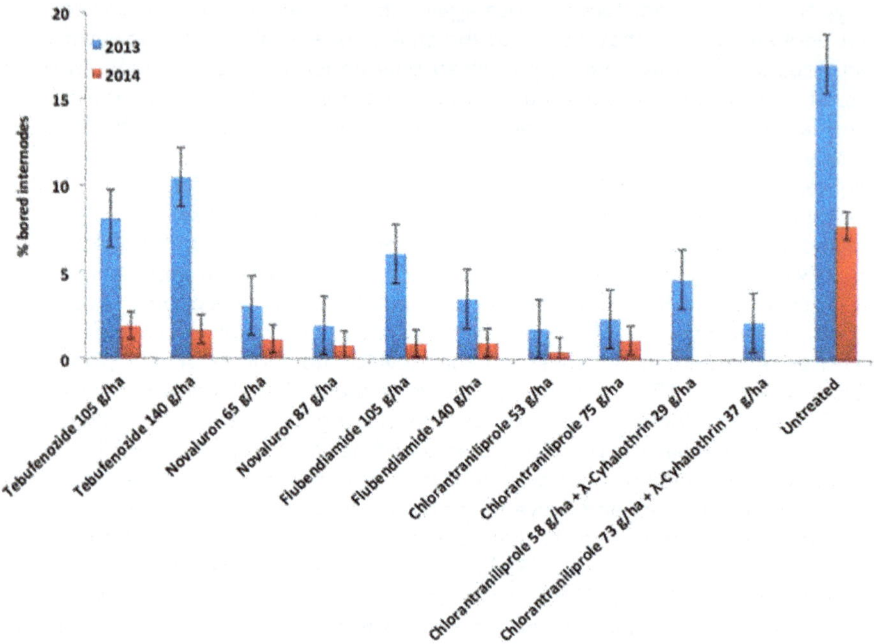

Figure 6 End of season *D. saccharalis* injury (LS means ± SE) in insecticide efficacy evaluations conducted in Patoutville (2013, 2014), Louisiana (Beuzelin et al., 2014; Wilson et al., 2016a). Insecticide applications impacted *D. saccharalis* injury in 2013 ($F = 7.9$; d.f.=10,30; $P < 0.001$) and 2014 ($F = 8.1$; d.f.=8,24; $P < 0.001$).

but yield protection in numerous field studies has been inconsistent (Wilson et al., 2012; Wilson, 2016). Egg laying in sugarcane plant cracks and folds in the lower canopy and quick entry of young *E. loftini* larvae into stalks are likely the major factors reducing insecticide efficacy as insecticides typically target eggs and larvae before they enter sugarcane stalks (Reay-jones et al., 2005; Wilson et al., 2012). In addition, a substantial proportion of *E. loftini* populations are expected to be active year-round in Louisiana sugarcane, resulting in overlapping generations that require thorough scouting for management decisions throughout the growing season (Beuzelin et al., 2016; Wilson, 2016). The use of *E. loftini* pheromone traps, by detecting trends in *E. loftini* population dynamics, might assist in indicating when to increase frequency and intensity of larval scouting. However, pheromone trap-assisted scouting for refined insecticide use has showed inconsistent results (Wilson et al., 2012, 2016).

The future of foliar insecticide use in Louisiana sugarcane will depend on the relative abundance of *D. saccharalis* and *E. loftini*. In situations where *E. loftini* becomes the predominant stalk borer, successful management might only be achieved with improved insecticide timing, more effective active ingredients, or additional tactics such as conventional or transgenic host plant resistance. Genetically engineered sugarcane clones producing Bt proteins, *Galanthus nivalis* agglutinin or soybean proteinase inhibitors express *D. saccharalis* and *E. loftini* resistance traits (Arencibia et al., 1997; Braga et al., 2001; Sétamou et al., 2002a,b; Falco and Silva-Filho, 2003). Insect-resistant transgenic sugarcane varieties might therefore complement insecticide use although interest in the commercial development of such varieties is low in the United States. Management of sap feeders might also impact foliar insecticide use patterns in the future. The West Indian canefly, *Saccharosydne saccharivora* (Hemiptera: Delphacidae), has historically been observed feeding on sugarcane in Louisiana but at population levels not thought to cause economic losses until recently. Unusually high *S. saccharivora* infestations have been reported statewide during the 2012 and 2016 growing seasons, possibly because of preceding mild winters. In one field study in 2012, the absence of insecticidal protection with imidacloprid or a mixture of imidacloprid and lambda-cyhalothrin was associated with reductions in sugarcane and sugar yields of 26 and 34%, respectively (White et al., 2013; W. H. White, unpublished data). Despite the absence of a threshold for *S. saccharivora* management, registered pyrethroids were applied to control the pest. Crisis exemptions also allowed the legal use of imidacloprid in 2012 and of acetamiprid in 2016 over thousands of hectares because of concerns about non-target impacts of pyrethroids. Field evaluations of recently developed insecticides with documented activity against sap feeders showed that sulfoxaflor suppressed *S. saccharivora* populations but provided excellent control of the sugarcane aphid, *Melanaphis sacchari* (Zehntner), another sap feeder of concern in Louisiana (Wilson et al., 2013). Flupyradifurone provided excellent control of *S. saccharivora* (Beuzelin and Wilson, unpublished data). In addition, although not evaluated for *M. sacchari* management in sugarcane, flupyradifurone provides excellent control of the aphid in grain sorghum (Van Weelden et al., 2016b). Thus, there is interest in expanded use of foliar insecticides in Louisiana sugarcane for *S. saccharivora* and *M. sacchari* management. The continuation and possible expansion of foliar insecticide use will have to rely on results of future studies addressing pest status and density-yield loss relationships for sap feeders but also non-target impacts of insecticides to prevent disruption of the agroecosystem.

7 Remote sensing, GIS and population modelling of sugarcane pests

With the growing use of remote sensing and computer modelling in pest management, there is an opportunity to understand temporal and spatial dynamics of insect populations at a depth never possible before. Remote sensing is a powerful tool being developed for pest surveillance in sugarcane to predict areas of potential outbreaks and to target intervention programmes, generally at the landscape level (Lefko et al., 1998). GIS are becoming increasingly more important in pest management programmes because they can be used to create maps and conduct geostatistical analysis of spatial interactions that occur at larger scales (Becker et al., 2005). Relating insect distributions to physiographical elements of the landscape is essential to predict future pest population dynamics (Hunter, 2002; Thies et al., 2003). Field observations on environmental conditions, including vegetation, water and topography, can be combined in a GIS to enhance interpretation of remote sensing data and to facilitate characterization of the landscape in terms of pest movements and resulting infestation (Liebhold et al., 1993).

In South Africa, mathematical models have been developed to simulate the population growth of *E. saccharina* and its dispersal in temporally changing and spatially heterogeneous environments (Potgieter et al., 2011, 2013). These models have been refined into agent-based modelling to better simulate field conditions (Van Vuuren et al., 2015). For monitoring sugarcane thrips *F. serrata*, remote sensing is being researched, and identification of putative markers and development of a rating system for resistance to sugarcane thrips is underway (Abdel-Rahman et al., 2010; Joshi et al., 2013).

In Australia, satellite imagery was also investigated as a method for detecting white grub infestations over large areas. High spatial resolution, multispectral and panchromatic, satellite images were acquired in May-June, corresponding with the months when symptoms of feeding by the greyback canegrubs *D. albohirtum* are most visible (Fig. 7). Images taken over three years were processed using geographic object-based image analysis (GEOBIA). Results indicated that grub damage within sugarcane fields could be detected using very-high-resolution imagery (Fig. 8). However, in some situation, canegrub

Figure 7 Aerial photo of field damaged by *D. albohirtum* (visible holes in sugarcane plots).

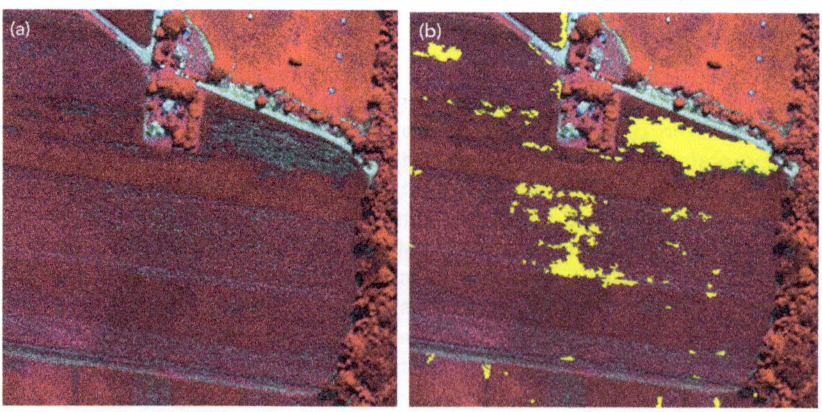

Figure 8 Checking for grub damage by image processing. Texture analysis by infrared (a); resegmentation of potential grub damage object (yellow) (b).

damage was difficult to identify because other problems such as water-logging, feral pig (*Sus scrofa*) damage or weed infestations appeared similar (Sallam et al., 2015).

Australian research using satellite images have generated infestation risk maps for the Mulgrave area in North Queensland (Zellner et al., 2014). A landscape map of the study site's land cover was first produced with seven classes: buildings, bare soil, sugarcane, herbaceous vegetation, isolated woody vegetation, riparian woods and rainforests. Satisfactory results were obtained on the discrimination between vegetation and non-vegetation areas, due to the high spatial resolution of the panchromatic channel and fine scale segmentation (mean object size = 1.5 m^2), which includes single trees in the vegetation domain. The separation of sugarcane and other vegetation at the second level (mean object size = 490 m^2) is a crucial processing step, since all types of non-sugarcane vegetation border the target classes. An overall accuracy of 97% is reached at this level, with a kappa coefficient of 0.95 that represents a stable basis for more detailed classification of non-sugarcane vegetation. Remaining problems are the loss of small individual trees into neighbouring objects, the classification of herbaceous surfaces to forest and sugarcane classes, and the confusion along transitions from sugarcane to woody vegetation. The map derived from the model allowed the development of a risk map for grub damage. Figure 9 shows the spatial distribution of sugarcane fields with three risk levels of damage: low, medium and high. Risk is higher adjacent to the forest and near the palm plantations.

Natural and riparian vegetation, private gardens, and isolated trees have been identified as key landscape elements in the ecology of *D. albohirtum*. The adults (beetles, Fig. 10) have the ability to fly to and from sugarcane fields after a period of mating and feeding on trees called 'feeding and roosting' trees (Fig. 8). A list of these trees is available and can

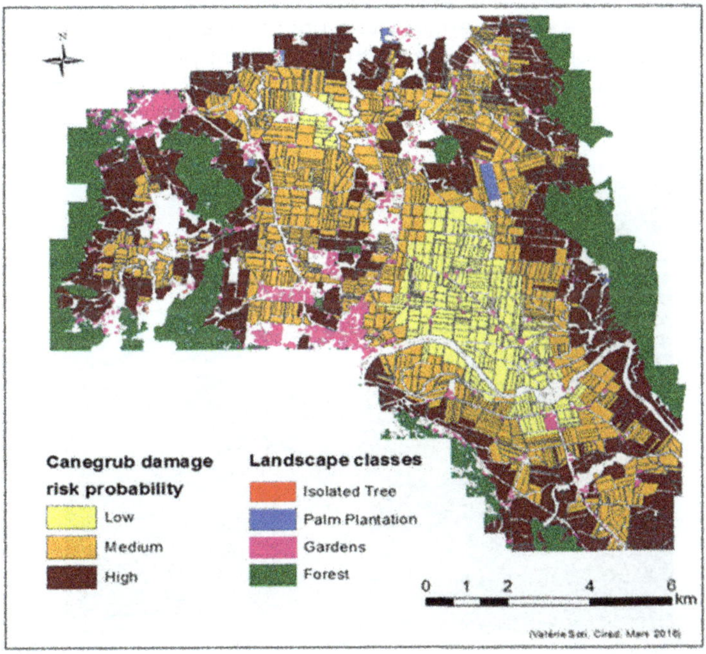

Figure 9 Risk map of canegrub damage in the Mulgrave study area (Zellner et al., 2014).

Figure 10 Couple of *D. albohirtum* adults resting on a fig tree.

be updated as knowledge progresses (Goebel et al., 2010b). Feeding and roosting trees can be used as indicators of local beetle populations.

Because trees are food sources and aggregating sites, they are key elements of vegetation surrounding sugarcane fields and surveys should be conducted on a regular basis to inspect these 'hotspots' to estimate beetle populations. It was discovered that an almond tree (*Calophyllum inophyllum*) growing near sugarcane fields in the Mulgrave area attracts hundreds of beetles each year (Goebel et al., 2010) although there is no leaf injury and the beetles are merely aggregating prior to mating. This type of infestation can serve as a population indicator during the main flight period, and the presence of these trees undoubtedly contributes to increased damage in adjacent sugarcane fields (Fig. 11).

Figure 11 Feeding and roosting trees bordering sugarcane fields (Mulgrave, Queensland).

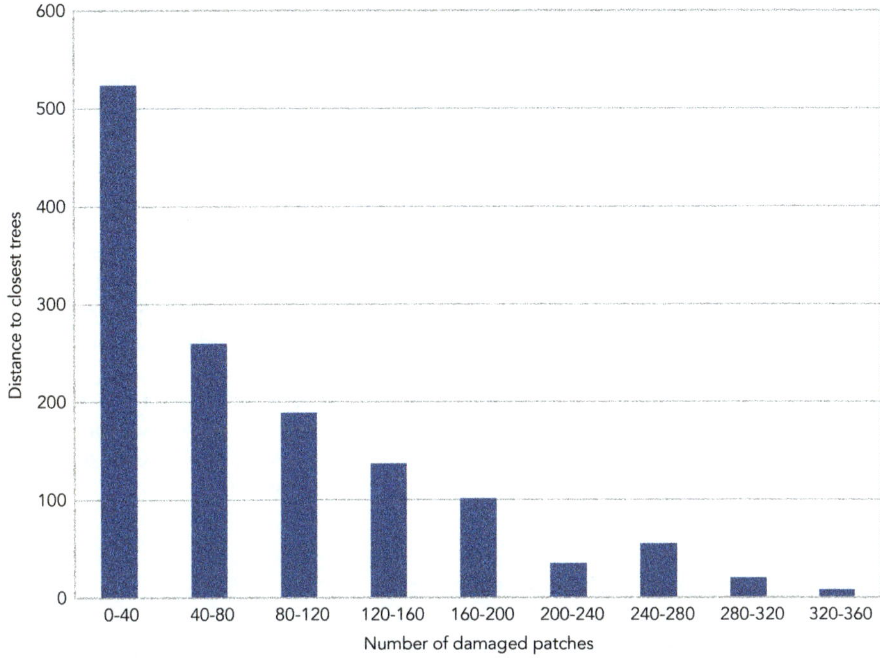

Figure 12 Distance (m) from tree edges to damaged patches in sugarcane fields (2010 data).

Distances beetles are able to fly have been poorly documented but seem relatively short (Goebel et al., 2010) (Fig. 12), which presumably explains that the greatest level of injury is located along river creeks with feeding and roosting trees. There is scope to treat strips of vegetation from the edge to 200 metres inside the field to reduce the damage on a wide scale. Risk maps generated from satellite images are valuable tools to help growers manage their pest problems on targeted areas.

8 Biosecurity

Climate change and growing international trade in plant products have increased the risk of major pests and diseases spreading among sugarcane-growing regions. Biosecurity in all sugarcane-producing regions is an ongoing process that involves threat identification, incursion risk analysis, pathway risk analysis, surveillance detection, diagnostics strategies and industry biosecurity planning. Industry biosecurity planning is adopted for risk prioritization and ensuring rapid response through an Emergency Plant Pest Response Deed process that allows agreement on who will pay for incursion response before the incursion happens based on the scale of the threat. Australian plant health agencies have adopted strict practices to safeguard the country's agricultural industries, including sugarcane. Australian farmers grow sugarcane in tropical Queensland and Western Australia and are threatened by exotic pest incursions from neighbouring countries, such as Papua New Guinea and Indonesia – two sugar-producing countries known for their rich biodiversity. A case history is sugarcane smut *Ustilago scitaminea*, which arrived from Southeast Asia into Australia in 2006 after sugarcane plantations were placed in northern Western Australia into the path of Asian air currents. The disease subsequently spread to Queensland (Sallam and Allsopp, 2005). Australia has developed an advanced biosecurity and quarantine system implemented by the Australian Quarantine and Inspection Service (AQIS). The Australian sugarcane industry so far remains free of any major stalk borer pest problems. However, pathway analysis showed that 22 of 36 exotic stalk borer species have the potential to invade Australia (Sallam and Allsopp, 2005; Sallam, 2006; Sallam et al., 2010). Insect species listed in Table 5 are considered to present the greatest risk for the Australian sugarcane industry with severe consequences for the capacity of this industry to develop a viable biofuel and other bioproduct capacity (Goebel and Sallam, 2011).

In South Africa, biosecurity measures have been developed to avoid the spread of *C. sacchariphagus* from Mozambique (Way et al., 2012). However, there have been recent introductions into this region, with sugarcane thrips *F. serrata* (Way et al., 2006) and yellow sugarcane aphid *Sipha flava* (Hemiptera: Aphididae) (Way et al., 2014).

9 Conclusion and future trends

Studying insect populations at the landscape level demands the involvement of multiple disciplines such as entomology, agronomy, ecology, plant breeding, physiology and mathematical modelling. The key role of the entomologist is to supply scientific information to improve current management strategies and develop new ones. It is therefore a complex and stimulating scientific challenge in a context of strong societal demand. Pest management in sugarcane has evolved rapidly, partly due to the recent development of technologies that have advanced the ability to adopt an area-wide approach that will deal with issues of reduced surrounding natural habitats and concomitant pest range expansion due to climate change. Pesticide use is being integrated with novel tactics that can be incorporated into an IPM strategy. For example, the use of the sterile insect technique (SIT) may become a possibility for *E. saccharina* management, with demonstration of F1 sterility in this stalk borer (Walton, 2011; Walton and Conlong, 2016). Pest status is being considered in the context of the influence of the landscape in and around sugarcane fields. Better knowledge of how insect pests interact with plants and the ecosystem is essential for the development of effective and environmentally sound IPM strategies. In particular, knowledge is warranted of soil fertility in sugarcane agroecosystems which plays a major role in insect-soil interactions and therefore in crop protection (Altieri and Nicholls, 2003). Pests are influenced by agricultural practices, namely fertilizers, irrigation, weed management and pesticides, which can cause high infestations with economic loss. Good crop management is therefore the first step to maintain low levels of pest infestation. Conventional pest management in sugarcane needs to view the challenge at the level of the landscape or even the territory with the global approach of AW-IPM (Vreyssen et al., 2007). Such an approach is knowledge intensive and requires involvement of stakeholders at the technical and political level since this is a global strategy (Fig. 13).

Table 5 Australian sugarcane industry high priority plant pest threat list, 2016 (Sugarcane biosecurity/Plant Health Australia)

Insect pest/common name	Scientific name	Entry potential	Establishment potential	Spread potential	Economic impact	Overall risk
Planthopper (vector of Ramu stunt disease)	*Eumetopina flavipes*	HIGH	HIGH	HIGH	EXTREME	EXTREME
Planthopper (vector Fiji leaf gall disease)	*Perkinsiella vastatrix*	HIGH	HIGH	HIGH	EXTREME	EXTREME
Planthopper (vector Fiji leaf gall disease)	*Perkinsiella vitiensis*	HIGH	HIGH	HIGH	EXTREME	EXTREME
Dark headed rice borer	*Chilo terrenellus*	HIGH	HIGH	HIGH	EXTREME	EXTREME
Pink stalk borer	*Sesamia grisescens*	HIGH	HIGH	HIGH	EXTREME	EXTREME
Yellow top borer	*Chilo infuscatellus*	MEDIUM	HIGH	HIGH	EXTREME	EXTREME
Internode borer	*Chilo sacchariphagus*	MEDIUM	HIGH	HIGH	EXTREME	EXTREME
Spotted stem borer	*Chilo tumidicostalis*	MEDIUM	HIGH	HIGH	EXTREME	EXTREME
African stalk borer	*Eldana saccharina*	LOW	HIGH	HIGH	EXTREME	HIGH
Root borer	*Polyocha depressella*	LOW	HIGH	HIGH	EXTREME	HIGH
Pyrilla	*Pyrilla perpusilla*	MEDIUM	HIGH	HIGH	HIGH	HIGH
Whitefly	*Aleurolobus barodensis*	HIGH	HIGH	HIGH	HIGH	HIGH
Woolly aphid	*Ceratovacuna lanigera*	HIGH	HIGH	HIGH	HIGH	HIGH
Top borer	*Scirpophaga excerptalis*	HIGH	MEDIUM	HIGH	HIGH	HIGH

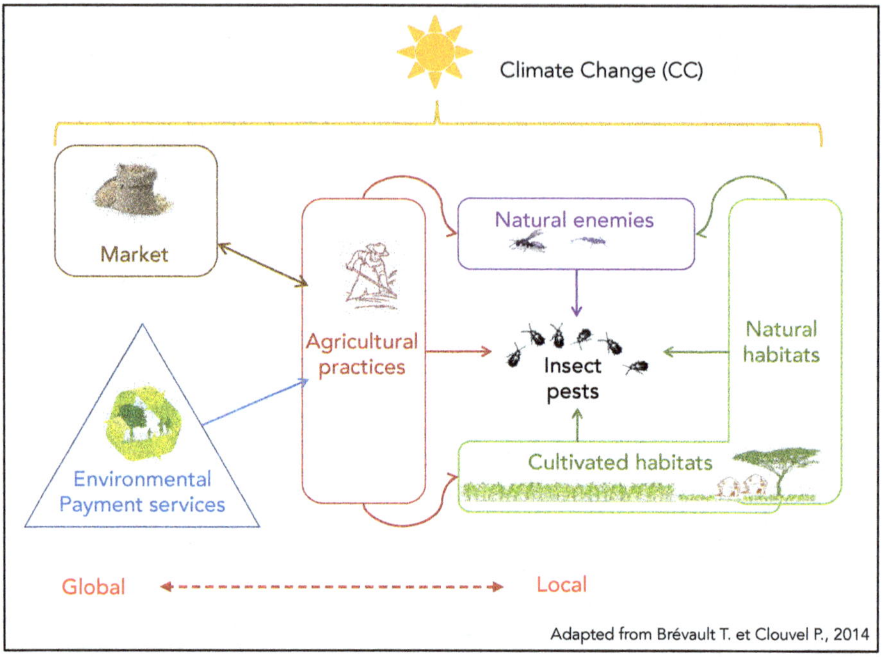

Figure 13 All relationships to consider in an agroecosystem under the pressure of climate change: pests and their natural enemies, link with non-cultivated areas, cropping practices as well as economic and political settings on different scales (markets, especially export markets, public environmental policies), which can impact agricultural practices.

10 Where to look for further information

Information on pests, IPM and future research on this topic can be found in the main research institutions (among others) mentioned in this chapter such as: South African Sugarcane Research Instute (www.sasa.org.za); the French Agricultural Research Centre for International Development CIRAD (www.cirad.fr), Sugar Research Australia Limited (SRA Ltd) (www.sugarresearch.com.au), University of Florida/Institute of Food and Agricultural Sciences Everglades Research and Education Center (www.ufl.edu).

11 References

Abdel-Rahman, E. M., Ahmed, F. B., van den Berg, M. and Way, M. J. (2010), Potential of spectroscopic data sets for sugarcane thrips (*Fulmekiola serrata* Kobus) damage detection. *International Journal of Remote Sensing* 31: 4199–216.

Akbar, W., Ottea, J. A., Beuzelin, J. M., Reagan, T. E. and Huang, F. (2008), Selection and life history traits of tebufenozide-resistant sugarcane borer (Lepidoptera: Crambidae). *Journal of Economic Entomology* 101: 1903–10.

Ali, A. D. and Reagan, T. E. (1985), Vegetation manipulation impact on predator and prey populations in Louisiana sugarcane ecosystems. *Journal of Economic Entomology* 78: 1409–14.

Allsopp, P. (2010), Integrated management of sugarcane whitegrubs in Australia: an evolving success. *Annual Review of Entomology* 55: 329–49.

Altieri, M. A. and Nicholls, C. I. (2003), Soil fertility management and insect pests: harmonizing soil and plant health in agroecosystems. *Soil and Tillage Research* 72: 203–11.

Andow, D. A. (1991), Vegetational diversity and arthropod population response. *Annual Review of Entomology* 36: 561–86.

Arencibia, A., Vázquez, R. I., Prieto, D., Téllez, P., Carmona, E. R., Coego, A., Hernández, L., Gustavo, A. and Selman-Housein, G. (1997), Transgenic sugarcane plants resistant to stem borer attack. *Molecular Breeding* 34: 247–55.

Atkinson, P. R. and Nuss, K. J. (1989), Associations between host-plant nitrogen and infestation of the sugarcane borer, *Eldana saccharina* Walker (Lepidoptera: Pyralidae). *Bulletin of Entomological Research* 79: 489–506.

Barker, A. L., Conlong, D. E. and Byrne, M. J. (2006), Habitat management using *Melinis minutiflora* to decrease the infestation of sugarcane by *Eldana saccharina* (Lepidoptera: Pyralidae). *Proceedings of the South African Sugar Technologists' Association* 80: 226–35.

Botha, F. C. (2009), Energy yield and cost in a sugarcane biomass system. *Proceedings of the Australian Society of Sugar Cane Technologists* 31: 1–10.

Box, H. E. (1953), List of sugar-cane insects: A synonymic catalogue of sugar-cane insects and mites of the world, and their insect parasites and predators, arranged systematically. London: Commonwealth Institute of Entomology.

Becker, A. A., French, B. W. and Chandler, L. D. (2005), Using GIS in areawide pest management: A case study in South Dakota. *Transactions in GIS* 9: 109–27.

Bessin, R. T., Moser, E. B. and Reagan, T. E. (1990), Integration of control tactics for management of the sugarcane borer (Lepidoptera: Pyralidae) in Louisiana sugarcane. *Journal of Economic Entomology* 83: 1563–9.

Bessin, R. T., Stinner, R. E. and Reagan, T. E. (1991), Modeling the areawide impact of sugarcane varieties and predation on sugarcane borer (Lepidoptera: Pyralidae). *Environmental Entomology* 20: 252–7.

Beuzelin, J. M., Reagan, T. E., Akbar, W., Flanagan, J. W., Cormier, H. J. and Blouin D. C. (2009), Impact of Hurricane Rita storm surge on sugarcane borer (Lepidoptera: Crambidae) management in Louisiana. *Journal of Economic Entomology* 102: 1054–61.

Beuzelin, J. M., Akbar, W., Mészáros, A., Reay-Jones, F. P. F. and Reagan, T. E. (2010), Field assessment of novaluron for sugarcane borer, *Diatraea saccharalis* (F.) (Lepidoptera: Crambidae), management in Louisiana sugarcane. *Crop Protection* 29: 1168–76.

Beuzelin, J. M., Mészáros, A., Akbar, W. and Reagan, T. E. (2011a), Sugarcane planting date impact on fall and spring sugarcane borer (Lepidoptera: Crambidae) infestations. *Florida Entomologist* 94: 242–52.

Beuzelin, J. M., Mészáros, A., Reagan, T. E., Wilson, L. T., Way, M. O., Blouin, D. C. and Showler, A. T. (2011b), Seasonal infestations of two stem borers (Lepidoptera: Crambidae) in non-crop grasses of Gulf Coast rice agroecosystems. *Environmental Entomology* 40: 1036–50.

Beuzelin, J. M., Wilson, L. T., Showler, A. T., Mészáros, A., Wilson, B. E., Way, M. O. and Reagan, T. E. (2013), Oviposition and larval development of a stem borer, *Eoreuma loftini*, on rice and non-crop grass hosts. *Entomologia Experimentalis et Applicata* 146: 332–46.

Beuzelin, J. M., Wilson, B. E., VanWeelden, M. T. and May, D. M. (2014), Small plot evaluation of insecticides against the sugarcane borer, 2013. *Arthropod Management Tests* 39: F6.

Beuzelin, J. M., Wilson, B. E., VanWeelden, M. T., Mészáros, A., Way, M. O., Stout, M. J. and Reagan, T. E. (2016), Biology and management of the Mexican rice borer (Lepidoptera: Crambidae) in Rice in the United States. *Journal of Integrated Pest Management* 7: 7.

Botelho, P. S. M., Parra, J. R. P., Das Chagas Neto, J. F. and Oliveira, C. P. B. (1999), Association of the egg parasitoid *Trichogramma galloi* Zuchi (Hymenoptera: Trichogrammatidae) with the larval parasitoid *Cotesia flavipes* (Cam.) (Hymenoptera: Braconidae) to control the sugarcane borer *Diatraea saccharalis* (Fabr.) (Lepidoptera: Crambidae). *Anais do Sociedade Entomologica do Brasil* 28: 491–6.

Braga, D. P. V., Arrigoni, E. D. B., Burnquist, W. L., Silva-Filho, M. C. and Ulian, E. C. (2001), A new approach for control of *Diatraea saccharalis* (Lepidoptera: Crambidae) through the expression of an insecticidal Cryla(b) protein in transgenic sugarcane. *Proceedings of the International Society of Sugar Cane Technologists* 24: 331–6.

Castro, B. A., Leonard, B. R. and Riley, T. J. (2004), Management of feeding damage and survival of southwestern corn borer and sugarcane borer (Lepidoptera: Crambidae) with *Bacillus thuringiensis* transgenic field corn. *Journal of Economic Entomology* 97: 2106–16.

Chandler, K. J. and Tucker, G. R. (2011), SuSCon Maxi and control of childers, negatoria and southern one-year canegrubs in sugarcane. *Proceedings of the Australian Society of Sugar Cane Technologists* 33: 1–10.

Charpentier, L. J. and Mathes R. (1969), Cultural practices in relation to stalk moth borer infestations in sugar cane, in J. R. Williams, J. R. Metcalf, R. W. Mungomery and R. Mathes (eds), *Pests of Sugarcane*. Elsevier publishing Co., New York, pp. 163–74.

Cherry, R. and Stansly, P. (2008), Abundance and spatial distribution of wireworms (Coleoptera: Elateridae) in Florida sugarcane fields on muck versus sandy soils. *Florida Entomologist* 91: 383–7.

Cherry, R. H., Nuessly, G. H. and Sandhu, H. S. (2015), Insect management in sugarcane. UF/IFAS Extension Publication ENY-406, Gainesville, Florida.

Cherry, R., McCray, M. and Sandhu, H. (2017), Changes in the relative abundance of soil insect pests in Florida sugarcane. *Journal of Entomological Science*: in press.

Cilas, C., Goebel, F. R., Babin, R. and Avelino, J. (2016), Tropical crop pests and diseases in a climate change setting – A few examples, in E. Torquebiau (ed.), *Climate Change and Agriculture Worldwide*. Springer, Dordrecht, Netherlands, pp. 73–82.

Cockburn, J., Coetzee, H., van den Berg, J. and Conlong, D. (2014), Large-scale sugarcane farmers' knowledge and perceptions of *Eldana saccharina* Walker (Lepidoptera: Pyralidae), push-pull and integrated pest management. *Crop Protection* 56: 1–9.

Conlong, D. E. (1994), A review and perspectives for the biological control of the African sugarcane stalk borer *Eldana saccharina* Walker (Lepidoptera: Pyralidae). *Agriculture, Ecosystems and Environment* 48: 9–17.

Conlong, D. E. and Rutherford, R. S. (2009), Conventional and New Biological and Habitat Interventions for Integrated Pest Management Systems: Review and Case Studies using *Eldana saccharina* Walker (Lepidoptera: Pyralidae). Chapter 10, Peshin, R. and Dhawan, A. K. (Eds). Vol. 1. Springer. 689pp. ISBN 978-1-4020-8991-6, pp. 241–61.

Cook, S. M, Khan, Z. R. and Pickett, J. A. (2007), The use of push-pull strategies in integrated pest management. *Annual Review of Entomology* 52: 375–400.

Dhadialla, T. S., Carlson, G. R. and Le, D. P. (1998), New insecticides with ecdysteroidal and juvenile hormone activity. *Annual Review of Entomology* 43: 545–69.

Falco, M. C. and Silva-Filho, M. C. (2003), Expression of soybean proteinase inhibitors in transgenic sugarcane plants: effects on natural defense against *Diatraea saccharalis. Plant Physiology and Biochemistry* 41: 761–6.

Flynn, J. L., Reagan, T. E. and Ogunwolu, E. O. (1984), Establishment and damage of the sugarcane borer (Lepidoptera: Pyralidae) in corn as influenced by plant development. *Journal of Economic Entomology* 77: 691–7.

Garrison, D. D., Dufrene, E. O. and Legendre, B. J. (2000), Effect of planting date on yields of sugarcane cultivars grown in Louisiana. *Journal of the American Society of Sugar Cane Technologists* 20: 115.

Goebel, R., Fernandez, E., Tiber, R. and Alauzet, C. (1999), Dégâts et pertes de rendement sur la canne à sucre dus au foreur *Chilo sacchariphagus* (Bojer) à l'île de la Réunion (Lep.: pyralidae). *Annales de la Société Entomologique de France* 35: 476–81.

Goebel, R., Way, M. J. and Gossard, C. (2005), The status of *Eldana saccharina* (Lepidoptera: Pyralidae) in the South African sugar industry based on regular survey data. *Proceedings of the South African Sugar Technologists' Association* 79: 337–46.

Goebel, F. R. and Way, M. (2009), Crop losses due to two sugarcane stem borers in Réunion and South Africa. *Sugar Cane International* 27: 107–11.

Goebel, F. R., Tabone, E., Do Thi Khanh, H., Roux, E., Marquier, M. and Frandon, J. (2010), Biocontrol of *Chilo sacchariphagus* (Lepidoptera: Crambidae) a key pest of sugarcane: lessons from the past and future prospects. *Sugar Cane International* 28: 128–32.

Goebel, F. R., Sallam, N., Samson, P. R. and Chandler, K. (2010), Quantifying spatial movement of the greyback cane beetle in the sugarcane landscape: data available and research needs. *Proceedings of the Australian Society of Sugar Cane Technologists* 32: 71–83.

Goebel, F. R. and Sallam, N. (2011). New pest threats for sugarcane in the new bioeconomy and how to manage them. *Current Opinion in Environmental Sustainability* 3, 81–9.

Gravois, K., Viator, S., Reagan, T. E., Beuzelin, J. M., Griffin, J. L., Tubaña, B. S. and Hoy, J. W. (2014), Sugarcane production handbook. *Louisiana State University Agricultural Center Publication* 2859, Baton Rouge, Louisiana.

Hassan, S. A. (1993), The mass rearing and utilisation of *Trichogramma* to control lepidopterous pests: achievements and outlook. *Pesticide Science* 37: 387–91.

Hensley, S. D. (1971), Management of sugarcane borer populations in Louisiana, a decade of change. *Entomophaga* 16: 133–46.

Holloway, T. E., Halley W. E., Loftin U. C. and Heinrich, C. (1928), The sugar-cane borer in the United States. *U. S. Department of Agriculture Technical Bulletin* 41.

Hummel, N. A., Hardy,T., Reagan, T. E., Pollet, D. K., Carlton, C. E., Stout, M. J., Beuzelin, J. M., Akbar, W. and W. H. White. (2010), Monitoring and first discovery of the Mexican rice borer *Eoreuma loftini* (Lepidoptera: Crambidae) in Louisiana. *Florida Entomologist* 93: 123–4.

Hunter, M. D. (2002), Landscape structure, habitat fragmentation, and the ecology of insects. *Agricultural and Forest Entomology* 4: 159–66.

Ingram, J. W., Bynum, E. K., Mathes, R., Haley, W. E. and Charpentier, L. J. (1951), Pests of sugarcane and their control. *U. S. Department of Agriculture Circular* 878.

Isas, M., Pérez, M. L. P., Salvatore, A., Gastaminza, G., Willink, E., and White, W. (2016), Impacts of crop residue on damage by sugarcane pests during the tillering phase in Argentina. *Florida Entomologist* 99: 1–5.

Jeuffrault, E., Rolet, A., Reynaud, B., Manikom, R., Georger, S., Taye, T., Chiroleu, F., Fouillaud, M. and Vercambre, B. (2004), Vingt ans de lutte contre le ver blanc de la canne à sucre à la Réunion: Un succès, mais il reste des questionnements scientifiques pour confirmer la durabilité de la lutte biologique. *Phytoma-la Défense des Végétaux* 573: 16–19.

Joshi S. V., Zhou M. M., Leslie G. W., Way M. J. and Keeping M. G. (2013), Comparison of methods for determining thrips (*Fulmekiola serrata*) damage and implications for resistance screening. *Proceedings of the South African Sugar Technologists' Association* 86: 291–4.

Keeping, M. G., Meyer, J. H. and Sewpersad, C. (2013), Soil silicon amendments increase resistance of sugarcane to stalk borer *Eldana saccharina* Walker (Lepidoptera: Pyralidae) under field conditions. *Plant and Soil* 363: 297–318.

Kirst, L. D. and Hensley, S. D. (1974), A study of the overwintering populations of *Diatraea saccharalis* (F.) in Louisiana. *Proceedings of the International Society of Sugar Cane Technologists* 15: 475–87.

Kfir, R., Overholt, W. A., Khan, Z. R. and Polaszek, A. (2002), Biology and management of economically important lepidopteran cereal stem borers in Africa. *Annual Review of Entomology* 47: 701–31.

Kvedaras, O. L., Byrne, M. J., Coombs, N. E. and Keeping, M. G. (2009), Influence of plant silicon and sugarcane cultivar on mandibular wear in larval *Eldana saccharina* Walker (Lepidoptera: Pyralidae). *Agricultural and Forest Entomology* 11: 301–6.

Kvedaras, O. L., Keeping M. G., Goebel, F. R. and Byrne, M. (2007), *Eldana saccharina* Walker (Lepidoptera: Pyralidae) larval performance and stalk damage in sugarcane: influence of plant silicon and feeding site. *International Journal of Pest Management* 53: 183–94.

Lahm, G. P., Cordova, D. and Barry, J. D. (2009), New and selective ryanodine receptor activators for insect control. *Bioorganic and Medical Chemistry* 17: 4127–33.

Leslie, G. W., Stranack, R. A., and de Haas, O. (2006), Progress in the use of Fastac (alpha-cypermerthrin) for the control of the sugarcane borer *Eldana saccharina* (Lepidoptera: Pyralidae) and an assessment of its commercial impact. *Proceedings of the South African Sugar Technologists' Association* 80: 236–44.

Lefko, S. A., Pedigo, L. P., Batchelor W. D. and Rice, M. E. (1998), Spatial modeling of preferred wireworm (Coleoptera: Elateridae) Habitat. *Environmental Entomology* 27: 184–90.

Li, L. Y. (1994), Worldwide use of *Trichogramma* for biological control on different crops: a survey, in E. Wajnberg and S. A. Hassan (eds), *Biological Control with Egg Parasitoids*, CAB International, Berkshire, UK, pp. 37–53.

Liebhold, A. M. and Rossi, R. E. (1993), Geostatistics and geographics information systems in applied insect ecology. *Annual Review of Entomology* 38: 303–27.

Long, W. H. and Hensley, S. D. (1972), Insect pests of sugarcane. *Annual Review of Entomology* 17: 149–76.

Lopez, E., Fernandez, C., Lopez, O. (1983), Effect of nitrogen fertilization on *Diatraea saccharalis* (Fbr.) incidence on sugarcane. *Proceedings of the International Society of Sugar Cane Technologists* 18: 910–14.

McAllister, C. D., Hoy, J. W. and Reagan, T. E. (2008), Temporal increase and spatial distribution of sugarcane yellow leaf and infestations of the aphid vector, *Melanaphis sacchari*. *Plant Disease* 92, 607–15.

Mumford, J. D. and Norton, G. A. (1984), Economics of decision making in pest management. *Annual Review of Entomology* 29: 157–74

Nibouche, S., Tibère, R. and Costet, L. (2012), The use of *Erianthus arundinaceus* as a trap crop for the stem borer *Chilo sacchariphagus* reduces yield losses in sugarcane crops: preliminary results. *Crop Protection* 42: 10–15.

Nicholls, C. I. and Altieri M. A. (2004), Agroecological bases of ecological engineering for pest management, in G. M. Gurr, S. D. Wratten and M. A. Altieri (eds), *Ecological Engineering for Pest Management: Advances in Habitat Manipulation for Arthropods*. CSIRO publishing, Collingwood, Australia, pp. 33–54.

Nikpay, A., (2016), Damage assessment of sugarcane whitefly *Neomaskellia andropogonis* Corbett and population dynamics on seven commercial varieties in southwest of Iran. *Sugar Tech*, DOI 10.1007/s12355–016-0451-0.

Nikpay, A., Kord, H., Goebel, R. and Sharafizadeh, P. (2014), Assessment of natural parasitism of sugarcane moth borers *Sesamia* spp by *Telenomus busseolae* in five commercial cultivars. *Sugar Tech*, 16(3): 325–8.

Nuessly, G. H., Cherry, R. H. and Sandhu, H. S. (2015). Miscelaneous insect pests of Florida sugarcane. *UF/IFAS Extension Publication* ENY-667, Gainesville, Florida.

Parra, J. R. P., Botelho, P. S. M. and Pinto, A. S. (2010), Biological control of pests as a key component for sustainable sugarcane production, in L. A. B. Cortez (ed.), *Sugarcane Bioethanol: RandD for Productivity and Sustainability*. Blücher, São Paulo, SP, Brazil, pp. 441–50.

Parra, J. R. (2014), Biological control in Brazil: an overview. *Scientia Agricola* 71: 420–9.

Pedigo, L. P., Hutchins, S. H. and Higley, L. G. (1986), Economic injury levels in theory and practice. *Annual Review of Entomology* 31: 341–68.

Pedigo, L. P. (2002). *Entomology and Pest Management* (4th edition) Prentice Hall, Upper Saddle River, NJ.

Posey, F. R., White, W. H., Reay-Jones, F. P. F., Gravois, K., Salassi, M. E., Leonard, B. R. and Reagan, T. E. (2006), Sugarcane borer (Lepidoptera: Crambidae) management threshold assessment on four sugarcane cultivars. *Journal of Economic Entomology* 99: 966–71.

Potgieter, L., van Vuuren, J. H. and Conlong, D. E. (2011), Modelling the effects of the sterile insect technique applied to *Eldana saccharina* Walker in sugarcane. *ORiON* 28: 59–84.

Potgieter, L., van Vuuren, J. H. and Conlong, D. E. (2013), A reaction-diffusion model for the control of *Eldana saccharina* Walker in sugarcane using the sterile insect technique. *Ecological Modelling* 250: 319–28.

Ratnadass, A., Fernandes, P., Avelino, J. and Habib, R. (2012), Plant species diversity for sustainable management of crop pests and diseases in agroecosystems: a review. *Agronomy for Sustainable Development* 32: 273–303.

Reagan, T. E. and Flynn, J. L. (1986), Insect pest management of sweet sorghum in sugarcane production systems of Louisiana: problems and integration, in W. H. Smith (ed.), *Biomass Energy Development*. Plenum Publishing, Inc., New York, pp. 227–39.

Reagan, T. E. and Posey, F. R. (2001), Development of an insecticide management program that enhances biological control. *Proceedings of the International Society of Sugar Cane Technologists* 24: 370–3.

Reagan, T. E., Rodriguez, L. M., Ostheimer, E. A. and Woolwine, A. E. (1997), Comparison of insecticide effects on a braconid parasitoid, *Cotesia chilonis*, in sugarcane. *Arthropod Management Tests* 22: 321.

Reay-Jones, F. P. F., Showler, A. T., Reagan, T. E., Legendre, B. L., Way, M. O. and Moser, E. B. (2005), Integrated tactics for managing the Mexican rice borer (Lepidoptera: Crambidae) in sugarcane. *Environmental Entomology* 34: 1558–65.

Reay-Jones, F. P. F., Wilson, L. T., Showler, A. T., Reagan, T. E. and Way, M. O. (2007), Role of oviposition preference in an invasive crambid impacting two graminaceous host crops. *Environmental Entomology* 36: 938–51.

Robène-Soustrade, I., Jouen, E., Pastou, D., Payet-Hoarau, M., Goble, T., Linderme, D., Lefeuvre, P., Calmes, C., Reynaud, B., Nibouche and S., Costet, L. (2015), Description and phylogenetic placement of *Beauveria hoplocheli* sp. nov. used in the biological control of the sugarcane white grub, *Hoplochelus marginalis*, on Reunion Island. *Mycologia* 107: 1221–32.

Sallam, M. N. and Allopp, P. G. (2005), Our home is girt by sea – but how well are we prepared in Australia for exotic cane borers? *Proceedings of the Australian Society of Sugar Cane Technologists* 27: 358–66.

Sallam, M. N. (2006), A review of sugarcane stemborers and their natural enemies in Asia and Indian Ocean Islands: an Australian perspective. *Annales de la Société Entomologique de France* 42: 263–83.

Sallam, N., Kristini, A., Achadian, E., Sochib, M. and Adi, H. (2010), Monitoring sugarcane moth borers in Indonesia: towards better preparedness for exotic incursions in Indonesia. *Proceedings of the Australian Society of Sugar Cane Technologists* 32: 181–92.

Sankar, M., Jaigeetha, S. and Manjunatha, S. R. (2014), Exploitation of biocontrol agents, *Trichogramma chilonis* and *Tetrastichus howardi* on yield improvement in sugarcane at EID Parry (India) Ltd., Sugar Mill Command Areas. *International Journal of Innovative Research and Development* 3(8): 314–18.

Sampson, M. A. and Kumar, R. (1983), Borer damage and estimation of loss caused by sugarcane stemborers in Southern Ghana. *Insect Science and its Application* 6: 705–10.

Sampson, P. R., Staier, T. N. and Bull, J. I. (2016), Evaluation of an application procedure for *Metarhizium anisopliae* in sugarcane ratoons for control of the white grub *Dermolepida albohirtum*. *Crop Protection* 25: 741–7.

Sétamou, M., Bernal, J. S., Legaspi, J. C., Mirkov, T. E. and Legaspi, B. C., Jr. (2002), Evaluation of lectin-expressing transgenic sugarcane against stalkborers (Lepidoptera: Pyralidae): effects on life history parameters. *Journal of Economic Entomology* 95: 469–77.

Showler, A. T. (2015), Effects of compost and chicken litter on soil nutrition, and sugarcane physiochemistry, yield, and injury caused by Mexican rice borer, *Eoreuma loftini* (Dyar) (Lepidoptera: Crambidae). *Crop Protection* 71: 1–11.

Showler, A. T. and Castro, B. A. (2010a), Influence of drought stress on Mexican rice borer (Lepidoptera: Crambidae) oviposition preference in sugarcane. *Crop Protection* 29: 415–21.

Showler, A. T. and Castro, B. A. (2010b), Mexican rice borer (Lepidoptera: Crambidae) oviposition site selection stimuli on sugarcane, and potential field applications. *Journal of Economic Entomology* 103: 1180–6.

Showler, A. T. and Reagan, T. E. (1991), Effects of sugarcane borer, weed, and nematode control strategies in Louisiana sugarcane. *Environmental Entomology* 20: 358–70.

Showler, A. T., Reagan T. E. and Knaus, R. M. (1990), Sugar-cane weed community interactions with arthropods and pathogens. *Insect Science and Its Application* 11: 1–11.

Showler, A. T., Beuzelin, J. M. and Reagan, T. E. (2011), Alternate crop and weed host plant oviposition preferences by the Mexican rice borer (Lepidoptera: Crambidae). *Crop Protection* 30: 895–901.

Showler, A. T., Wilson, B. E. and Reagan, T. E. (2012), Mexican rice borer (Lepidoptera: Crambidae) injury to corn greater than to sorghum and sugarcane under field conditions. *Journal of Economic Entomology* 105: 1597–1602.

Showler, A. T., Cook S. C. and Abrigo, V. (2013), Transgenic Bt corn varietal resistance against the Mexican rice borer, *Eoreuma loftini* (V. Dyar) (Lepidoptera: Crambidae), and implications to sugarcane. *Crop Protection* 48: 57–62.

Sosa, O., Jr. (1990), Oviposition preference by the sugarcane borer (Lepidoptera: Pyralidae). *Journal of Economic Entomology* 83: 866–8.

Southwick, L. M., Willis G. H., Reagan T. E. and Rodriguez L. M. (1995), Residues in runoff and on leaves of azinphosmethyl and esfenvalerate applied to sugarcane. *Environmental Entomology* 24: 1013–17.

Stern, V. M., Smith, R. F., van den Bosch, R. and Hagen, K. S. (1959), The integrated control concept. *Hilgardia* 29: 81–101.

Suasa-ard, W., Sommartya, P., Buchatian, P., Puntongcum, A. and Chiangsin, R. (2008), Effect of *Metarhizium anisopliae* on infection of sugarcane stems borer, *Dorysthenes buqueti* Guerin (Coleoptera: Cerambycidae) in laboratory. *Proceedings of the 46th Kasetsart University Annual Conference, Kasetsart, Thailand* 155–60.

Thies, C., Steffan-Dewenter, I. and Tscharntke, T. (2003), Effect of landscape context on herbivory and parasitism at different spatial scales. *Oïkos* 101: 18–25.

Tilman, D., Cassman, K. G., Matson, P. A., Naylor, R. and Polasky, S. (2002), Agricultural sustainability and intensive production practices. *Nature* 418: 671–7.

Van Leerdam, M. B., Johnson, K. J. R. and Smith J. W., Jr. (1984), Effects of physical characteristics and orientation on oviposition by *Eoreuma loftini* (Lepidoptera: Pyralidae). *Environmental Entomology* 13: 800–2.

Van Leerdam, M. B., Johnson, K. J. R. and Smith, J. W., Jr. (1986), Ovipositional sites of *Eoreuma loftini* (Lepidoptera: Pyralidae) in sugarcane. *Environmental Entomology* 15: 75–8.

VanWeelden, M. T., Wilson, B. E., Beuzelin, J. M., Reagan, T. E. and Way, M. O. (2016a), Impact of nitrogen fertilization on Mexican rice borer (Lepidoptera: Crambidae) injury and yield in bioenergy sorghum. *Crop Protection* 84: 37–43.

VanWeelden, M. T., Beuzelin, J. M. and May, D. M. (2016b), Evaluation of foliar insecticides for management of the sugarcane aphid in sorghum, 2016. *Arthropod Management Tests* 41: tsw067.

Van Vuuren, B. J., Potgieter, L. and Van Vuuren, J. H (2015), The design of an agent-based model for simulating the population dynamics of *Eldana saccharina* (Lepidoptera: Pyralidae). *Proceedings of the South African Sugar Technologists' Association* 88: 304–8.

Viator, R. P., Johnson, R. M. and Richard, E. P., Jr. (2005a), Challenges of post-harvest residue management in the Louisiana sugarcane industry. *Sugar Cane International* 23(6): 3–7.

Viator, R. P., Richard E. P., Jr., Garrison D. D., Dufrene E. O., Jr. and Tew T. L. (2005b), Sugarcane cultivar yield response to planting date. *Journal of the American Society of Sugar Cane Technologists* 25: 78–87.

Viator, R. P., Johnson, R. M. and Richard, E. P., Jr. (2009), Effects of mechanical removal and incorporation of post-harvest residue on ratoon sugarcane yields. *Sugar Cane International* 27(4): 149–52.

Vines, R. C., Reagan, T. E., Sparks, T. C. and Pollet, D. K. (1984), Laboratory selection of *Diatraea saccharalis* (F.) (Lepidoptera: Pyralidae) for resistance to fenvalerate and monocrotophos. *Journal of Economic Entomology* 77: 857–63.

Vreyssen, M. J. B., Robinson, A. S. and Hendrichs, J. (2007), *Area-wide Control of Insect Pests: From Research to Field Implementation*. Springer, Dordrecht, The Netherlands.

Walker, P. T. (1987), Quantifying the relationship between insect populations, damage, yield and economic thresholds, in P. S Teng (ed.), *Crop Losses Assessment and Pest Management*, St-Paul, Minnesota, USA: APS, pp. 114–25.

Walton, A. J. (2011), Radiation biology of *Eldana saccharina* Walker (Lepidoptera: Pyralidae). M.Sc. Thesis, Stellenbosch University, Stellenbosch, South Africa.

Walton, A. J. and Conlong, D. E. (2016), General biology of *Eldana saccharina* (Lepidoptera: Pyralidae): a target for sterile insect technique. *Florida Entomologist* 99 (special issue #1): 30–5.
Way, M. J., Stiller, M., Leslie, G. W., Conlong, D. E., Keeping, M. G. and Rutherford, R. S. (2006), *Fulmekiola serrata* (Kobus) (Thysanoptera: Thripidae), a new pest in southern African sugarcane. *African Entomology* 14: 401–3.
Way, M. J., Rutherford, R. S., Sewpersad, C., Leslie, G. W. and Keeping, M. G. (2010), Impact of sugarcane thrips, *Fulmekiola serrata* (Kobus) (Thysanoptera: Thripidae) on sugarcane yield in field trials. *Proceedings of the South African Sugar Technologists' Association* 83: 244–56.
Way, M. J., Conlong, D. E. and Rutherford, R. S. (2012), Biosecurity against invasive alien Insect pests: A case study of *Chilo sacchariphagus* (Lepidoptera: Crambidae) in the southern African region. *International Sugar Journal* 114: 359–63.
Way, M. J., Conlong, D. E., Martin, L. A., Mcfarlane, S. A., Stranack, R., Keeping, M. G. and Rutherford, R. S. (2014), First record of yellow sugarcane aphid, *Sipha flava* (Homoptera: Aphididae) in the South African Sugarcane Industry. *Proceedings of the South African Sugar Technologists' Association* 87: 53–7.
Webster, T. M., Brenchley, P. G. and Conlong, D. E. (2009), Progress in the area-wide integrated pest management plan for *Eldana saccharina* Walker (Lepidoptera: Pyralidae) in the Midlands North region of KwaZulu-Natal. *Proceedings of the South African Sugar Technologists' Association* 82: 471–85.
White, W. H. and Reagan, T. E. (1999), Biological control of the sugarcane borer with introduced parasites in Louisiana. *Sugar Bulletin* 77: 21–4.
White, W. H., Reagan, T. E., Smith, J. W., Jr. and Salazar, J. A. (2004), Refuge releases of *Cotesia flavipes* (Hymenoptera: Braconidae) into the Louisiana sugarcane ecosystem. *Environmental Entomology* 33: 627–32.
White, W. H., Viator R. P. and White, P. M. (2011), Effects of post-harvest residue and methods of residue removal on ground inhabiting arthropod predators in sugarcane. *Journal of the American Society of Sugar Cane Technologists* 31: 39–50.
White, W. H., Dalley C. D., and Viator B. J. (2013), West Indian canefly (Fulgorid) experiences from the 2012 season. *Sugar Bulletin* 91(7): 21–3.
White, W. H., and White P. M. (2014), Sugarcane borer resistance in sugarcane as affected by silicon applications in potting medium. *Journal of the American Society of Sugar Cane Technologists* 33: 38–54.
William, F. X. (1931), The insects and other invertebrates of Hawaiian sugarcane fields. Hawaiian Sugar Planter's Association and Advertiser Publishing Company, Honolulu, Hawaii. 400 p.
Wilson, B. E. (2016), Mexican rice borer (*Eoreuma loftini*) pheromone trap efficacy and role in invasive species monitoring and pest management. Ph.D. dissertation, Louisiana State University, Baton Rouge, Louisiana.
Wilson, B. E., Showler, A. T., Reagan, T. E. and Beuzelin, J. M. (2012), Improved chemical control for the Mexican rice borer (Lepidoptera: Crambidae) in sugarcane: larval exposure, a novel scouting method, and efficacy of a single aerial insecticide application. *Journal of Economic Entomology* 105: 1998–2006.
Wilson, B. E., VanWeelden, M. T., Beuzelin, J. M. and Mészáros A. (2013), Assessment of insecticides against a complex of hemipteran pests in sugarcane, 2012. *Arthropod Management Tests* 38: F73.
Wilson, B. E., Hardy, T. N., Beuzelin, J. M., VanWeelden, M. T., Reagan, T. E., Miller, R., Meaux, J., Stout, M. J. and Carlton C. E. (2015), Expansion of the Mexican rice borer (Lepidoptera: Crambidae) into rice and sugarcane in Louisiana. *Environmental Entomology* 44: 757–66.
Wilson, B. E., Beuzelin, J. M., VanWeelden, M. T., May, D. M. and Reagan, T. E. (2016a), Small plot evaluation of insecticides against the sugarcane borer, 2014. *Arthropod Management Tests* 41: F50.
Wilson, B. E., Stout, M. J., Beuzelin, J. M. and Reagan, T. E. (2016b), Mexican rice borer continues spread in Louisiana. *Louisiana Agriculture* 59(1): 22–3.

Wing, K. D., Slawecki, R. A. and Carlson, G. R. (1988), RH 5849, a nonsteroidal ecdysone agonist: effects on larval Lepidoptera. *Science* 241: 470–2.

Woolwine, A. E., Rodriguez, L. M., Ostheimer, E. A. and Reagan, T. E. (1995), Effects of aerially applied insecticide for SCB on non-target arthropods, 1994. *Arthropod Management Tests* 20: 257.

Zellner, P., Lelong, C., Soti, V., Tran, A., Sallam, N., Robson, A. and Goebel, F. R. (2014), A remote sensing and GIS approach to the relationship between canegrub infestations and natural vegetation in the sugarcane landscape of Queensland, Australia. *IOBC/WPRS bulletin*, 100: 153–8.

Chapter 16

Integrated disease management strategies in sugarcane cultivation

Jack C. Comstock and Sushma G. Sood, USDA-ARS, USA

1 Introduction

Sugarcane is susceptible to a number of diseases that can greatly impact sugarcane production and cause very significant yield losses. More than 100 bacterial, fungal, phytoplasma and viral diseases are present in sugarcane growing areas worldwide (Rott et al., 2000). Some of the most significant diseases are illustrated in Fig. 1.

Elements of the cane production system that influence disease spread and development are listed below. The sugarcane crop is planted using vegetative planting material, sourced from farmer nurseries. The planting operation involves cutting stalks and placement of short sections of the cut stalks into field soil. The crop is grown for approximately 12 months before the above-ground vegetative material is harvested, either machine-cut or hand harvested and the cane sent to the sugar factory for processing. The crop regrows from below-ground stubble and is harvested again roughly 12 months later (ratoon crop). The number of ratoon crops varies with country, variety and the state of the below-ground stubble, which can be affected by the harvesting operations. The crop is grown as a monoculture and in some countries fields have been continuously cropped with sugarcane for 50–100 years. Large proportions of vast regions may grow a single crop (sugarcane) and be dominated by a single variety. Variety yield decline is a phenomenon closely associated with sugarcane cropping. Varieties tend to have a limited commercial production life; for productivity to remain high, this requires the consistent release of new high-yielding varieties.

http://dx.doi.org/10.19103/AS.2017.0035.27

Figure 1 (a–f) Photographs of symptoms of sugarcane diseases. (a) Brown rust with brownish pustules containing brown colored urediniospores on the underside of the leaf (Erick Chavarria, Liga Agricola Industrial de la Caña de Azúcar (LAICA), Costa Rica). (b) Orange rust with pustules containing orange colored urediniospores on the underside of the leaf (Jack Comstock, retired, formerly of the USDA-ARS Sugarcane Field Station, Canal Point, Florida). (c) A sorus (commonly called a whip) of sugarcane smut. The whip developed as the young plant emerged because the seedpiece used for planting was infected (Jack Comstock, retired, formerly of the USDA-ARS Sugarcane Field Station, Canal Point, Florida). (d) Numerous whips of sugarcane smut on the extremely susceptible variety, Q 158, at the USDA-ARS Sugarcane Field Station, Canal Point, Florida due to natural infection (Jack Comstock, retired, formerly of the USDA-ARS Sugarcane Field Station, Canal Point, Florida). (e and f) Internal symptoms of pineapple sett rot of setts. The soil-borne pathogen, *Ceratocystis paradoxa*, invades the cut ends of the setts after planting (Erick Chavarria, Liga Agricola Industrial de la Caña de Azúcar (LAICA), Costa Rica).

Figure 1 (*Continued*) (g–k) Photographs of symptoms of sugarcane diseases. (g, h and i) Symptoms of red stripe: (g) Foliar red stripe symptoms on upper leaves of almost all plants in the field. (h) Red stripe on an individual plant. (i) Internal stalk deterioration (Jack Comstock, retired, formerly of the USDA-ARS Sugarcane Field Station, Canal Point, Florida). (j and k) Pokkah boeng (Jack Comstock, retired, formerly of the USDA-ARS Sugarcane Field Station, Canal Point, Florida).

The cultivation system lends itself to disease development and to rapid disease escalation. Systemic (those that occur within the plant tissues) bacterial, fungal, phytoplasma and viral diseases may infect planting material in nurseries, and this can lead to widespread disease incidence in freshly planted crops. Diseases that invade the freshly cut stalks of the planting material may reduce crop germination and lead to reduced plant stands. A number of diseases are transmitted mechanically, either via planting or harvesting equipment; this leads to extensive spread of disease in plant crops and mounting disease incidence in ratoon crops. Soil-borne diseases are favoured by the monoculture system and the extended length of time for which some farms and regions have been continuously cropped with sugarcane. Leaf diseases are not uncommon in cane, and broad plantings of a single crop across whole regions can lead to poor canopy health, particularly when

Figure 1 (*Continued*) (l–s) Photographs of symptoms of sugarcane diseases. (l, m and n) Downy mildew symptoms: (l) Reddish brown streaks on the spindle leaves. (m) The streaks on the underside of the leaf with some whitish downy appearance. (n) White downy appearance of the pathogen sporulation on the underside of the leaf (Jack Comstock, retired, formerly of the USDA-ARS Sugarcane Field Station, Canal Point, Florida). (o) Dry top rot symptoms showing a desiccated shriveled stalk (Jack Comstock, retired, formerly of the USDA-ARS Sugarcane Field Station, Canal Point, Florida). (p) A plot of ratoon stunt-infected sugarcane on the right and a healthy (taller) plot on the left showing the reduced growth due to the disease (Diego Luzuriaga, Florida Crystals Corp., Pahokee, Florida). (q, r and s) Symptoms of leaf scald. (q) A classic diagnostic 'pencil-line' symptom where a single light yellow vascular bundle extends toward the leaf edge. (r) White-yellowish spindle leaves. (s) Discolored side shoots emerging from the lateral buds of the stalk (Jack Comstock, retired, formerly of the USDA-ARS Sugarcane Field Station, Canal Point, Florida).

Figure 1 (*Continued*) (t–w) Photographs of symptoms of sugarcane diseases. (t) Yellow leaf symptoms showing a yellowing of the underside of mid-rib with yellowing/necrosis extending into the leaf blade (Jack Comstock, retired, formerly of the USDA-ARS Sugarcane Field Station, Canal Point, Florida). (u) Mosaic symptoms. In this case the plant is infected with sugarcane mosaic but similar symptoms occur in sorghum mosaic and sugarcane streak mosaic (Jack Comstock, retired, formerly of the USDA-ARS Sugarcane Field Station, Canal Point, Florida). (v and w) White leaf symptoms: (v) An emerging plant that is entirely white. (w) An emerging yellowish plant (Jack Comstock, retired, formerly of the USDA-ARS Sugarcane Field Station, Canal Point, Florida).

a susceptible variety is grown across vast areas. The extensive nature of the crop means that some forms of disease management may be unprofitable in some circumstances (for instance, chemical control). Some systemic diseases are also spread in the maturing crop, either by windblown spores or via insect vectors. These diseases include some of the most important ones in industries worldwide. Plant nutrition and soil type may also influence the incidence of some diseases.

Use of different disease management strategies, according to when and how diseases enter crops, is essential to reduce yield losses. Disease management is based on six management principles: avoidance, eradication, exclusion, immunization (resistance), protection and therapy (NAS, 1968).

In this chapter, we explore and outline the various disease management strategies used in sugarcane cropping systems around the world and the potential for IDM to minimize losses and maximize profitability. Sources for further information on disease management are also provided.

2 Integrated disease management (IDM)

IDM strategies for sugarcane must address all the major diseases present and also prevent the introduction of new pathogens and new pathogenic strains or races. The management practices must be cost-effective and minimize yield losses. The IDM strategies for an industry must be flexible to allow for changes to address new issues. Since disease is an outcome of interactions between three components, pathogen, host and environment, it is important to use an integrated approach to manage one or all components to achieve sustainable plant disease management. IDM strategies for some sugarcane diseases are outlined in Table 1.

3 Managing diseases by controlling the pathogens

3.1 Exclusion

The best possible management strategy is to ensure a particular disease is excluded from a production area – farmers then do not need to consider the disease. The greatest emphasis should be placed on this strategy, if the disease potentially poses a severe threat to commercial crops. Exclusion of sugarcane pathogens and infected seedcane by national and regional regulations and local government/industry practices in conjunction with strict quarantine is extremely important to limit disease spread. On a world basis, national and/or regional quarantines take steps to prevent the introduction of new pathogens and new races or strains of pathogens. Figure 2a–d show photographs of the National Plant Germplasm Quarantine Center of the United States Department of Agriculture – Animal Plant Health Inspection Service.

Quarantine restricts sugarcane introductions (emphasizing vegetative cuttings, setts) and uses diagnostic assays and therapeutic/eradication practices to prevent an incursion of all vegetatively propagated sugarcane pathogens. Sugarcane varieties are usually exchanged in the form of vegetative cuttings (setts, seedpieces, etc.). The importation of cuttings poses a risk of introducing systemic pathogens. The importation of true-seed (commonly called fuzz) from sexual crosses is also used to transfer genes between breeding programmes with significantly less risk (Maroon-Lango et al., 2013). Introductions of fuzz have also been used to select varieties for commercial production in countries that do not have crossing programmes (Afghan et al., 2013). Several countries import sugarcane fuzz with minimal risk (Maroon-Lango et al., 2013).

Exclusion played an important role in the Hawaiian sugarcane industry. Smut was first observed in Hawaii in 1971 (Comstock et al., 1983). Attempts to locate and destroy infected fields were made but were not successful. However, smut was initially only on the islands of Oahu and Maui and not present on those of Hawaii and Kauai. Therefore, strict industry regulations included only transferring healthy seedcane obtained from healthy plots established in an isolated location (an inter-island quarantine) on the windward side of Oahu. The inter-island quarantine was planted with sugarcane seedcane that was hot water–treated at 52°C for 45 minutes (Byther and Steiner, 1973). This essentially eliminated smut from the seedcane used for planting in the inter-island quarantine. Seedcane cut from this location was given a shorter 30-minute hot water treatment (HWT) prior to shipment. These precautions prevented the introduction of smut to the island of Hawaii for several years.

Table 1 IDM strategies for important diseases of sugarcane

Disease	Pathogen	Distribution	Integrated disease management strategies
Bacterial diseases			
Leaf scald	*Xanthomonas albilineans*	Worldwide	Cold-soak long hot-water treatment, sanitation, clean seedcane, resistance, biological control
Ratoon stunt	*Leifsonia xyli* subsp. *xyli*	Worldwide	Long hot-water treatment, sanitation, clean seedcane, resistance
Red stripe	*Acidovorax avenae* subsp. *avenae*	Central America, Argentina	Short hot-water treatment (SHWT), sanitation, clean seed-cane, resistance
Fungal diseases			
Brown rust	*Puccinia melanocephala*	Worldwide	Resistance, chemical control
Orange rust	*Puccinia kuehnii*	Wide (spreading)	Resistance, chemical control
Pachymetra root rot	*Pachymetra chaunorhiza*	Australia	Resistance, sanitation
Pineapple sett rot	*Ceratocystis paradoxa*	Worldwide	Chemical control, cultural practices
Red rot	*Glomerella tucumanensis*	Worldwide (major importance in India)	Resistance, biological and chemical control, clean seedcane
Smut	*Sporisorium scitamineum*	Worldwide except Fiji	Resistance, SHWT, roguing
Tawny rust	*Macruropyxis fulva*	South Africa, Swaziland, Mozambique	Resistance, chemical control
Other diseases			
Chlorotic streak	*Cercozoa*	Worldwide	SHWT, clean seedcane, drainage
White leaf	Phytoplasma	Southeast Asia	Clean seedcane, fertilization, roguing
Viral diseases			
Fiji leaf gall	*Fiji disease virus*	Australia, Fiji	Clean seedcane, resistance, vector control
Mosaic	*Sugarcane mosaic virus* *Sorghum mosaic virus*	Worldwide	Clean seedcane, resistance
Yellow leaf	*Sugarcane yellow leaf virus*	Worldwide	Clean seedcane, resistance

Figure 2 (a) Entrance sign of the National Plant Germplasm Quarantine Center of the United States Department of Agriculture – Animal Plant Health Inspection Service, Beltsville, Maryland. (b) Imported sugarcane clones grown in enclosed greenhouse during quarantine. (c) Tissue culture room used to grow plants after meristem culturing to eliminate viruses. (d) Tissue cultures of virus-free plants. (Martha Malapi-Wight, USDA-APHIS, Plant Germplasm Quarantine Program, Beltsville, Maryland).

One of the best examples of exclusion is sugarcane smut in Australia (see Fig. 1c and d). Smut was essentially worldwide except in Australia when a smut incursion was identified in 1998 in a small segment of the Australian industry (in Western Australia) (Magarey et al., 2009; Croft et al., 2008). Australia's geographic isolation and its strict national quarantine regulations prevented the introduction of smut for decades (Magarey et al., 2009; Croft et al., 2008). After the smut incursion, internal governmental and industry regulations restricted the movement of vegetative seedcane from Western Australia unless proper HWTs were applied. This regulation prevented smut spread to the major sugarcane production region of Queensland. The windborne teliospores of the pathogen, *Sporisorium scitamineum*, were the likely means for the spread of smut to the Western Hemisphere and probably led to the introduction of smut to Queensland, Australia, in 2006. Exclusion saved the industry many millions of dollars and allowed time to prepare IDM strategies prior to the 2006 incursion (Magarey et al., 2009; Croft et al., 2008). This exclusion policy was merited because the Q varieties of Australia were noted for their smut susceptibility in other countries (see Fig. 1d). Prior to the smut incursion, a major smut screening programme was initiated in Indonesia incorporating Australian Q varieties and clones in the later stages of their breeding programme in an effort to increase smut resistance. The effort provided information on the seriousness of the threat, identified Q varieties that were resistant and provided hands-on experience with the disease prior to its introduction.

Exclusion has a higher probability of success with non-windborne diseases such as Pachymetra root rot, caused by *Pachymetra chaunorhiza*, a soil-borne disease. Pachymetra root rot is only found in Australia where the movement of soil from infested fields is prevented via regulation (Magarey, 1994). In addition, clones in the variety development programme are screened for *Pachymetra chaunorhiza* resistance. The movement of *Pachymetra chaunorhiza* to other continents is unlikely because it is not windborne and can only be transmitted via seedcane contaminated with soil and possibly in soil on implements imported from areas where *Pachymetra chaunorhiza* is present. Standard quarantine procedures would normally detect and prevent its introduction to other countries and regions.

Another soil-borne disease, dry top rot caused by *Ligniera vasculorum* (see Fig. 1o), was first reported in Puerto Rico (Abbott, 1964) in the 1920–30s. There were no reports anywhere until 1994 in Florida when it was observed in a small field at the USDA-ARS Sugarcane Field Station (Comstock et al., 1994). The spread of dry top rot is unlikely because of its soil-borne nature. The pathogen is present in infected stalks that near maturity appear to be dying of drought. Infected stalks are not suitable to use as seedcane. In greenhouse experiments symptomatic stalks planted in flats usually do not germinate or if young plants do develop they usually die or are extremely weakened. The spread of this disease has been prevented by not moving seedcane from the infested field. The incidence of dry top rot has not increased over the years (Comstock, personal observation); the practice of not cutting infected stalks for seedcane has limited its spread.

3.2 Eradication and reduction of pathogen inoculum

Complete eradication of a major sugarcane disease from a large sugarcane industry is extremely difficult and rare. An example is downy mildew in Australia. Although downy mildew (see Fig. 1l,m and n) was documented as a major disease in Australia in the early 1900s, it was eliminated from the country with the last observation of the disease in a commercial field in 1957 (Hughes and Robinson, 1961). Eradication was achieved by planting resistant varieties, roguing diseased plant-infested fields and not planting alternative host crops, such as corn, near sugarcane fields. Conidia of downy mildew are windblown; however, the conidia are fragile and neither survive for long nor travel long distances.

Partial eradication or elimination of pathogens from vegetative planting material by certain HWTs is commonly used for eliminating systemic pathogens. The eradication of bacterial and fungal pathogens and selection of seedcane with lower pathogen incidences reduces pathogen inoculum and minimizes economic losses; disease buildup may also be delayed by several years.

3.2.1 Hot water treatment and tissue culture

HWT is effective in either partially or completely eliminating the pathogens causing chlorotic streak, smut, leaf scald and ratoon stunting disease (often called RSD) (Viswanathan, 2012; Gillaspie and Davis, 1992). Stalks are stripped of all the leaves before HWT. Whole stalk,

Figure 3 (a) Roofed hot water treatment facility in Guatemala, showing blue treatment baskets and a hoist in background (Victor Azañon, La Unión Sugar Mill, Guatemala). (b) Hot water facility at Fatima Sugar Mills Limited in Pakistan showing a boiler and (c) The support structure with hoist and one full basket being brought out of the tank (Dr. Syed Zia Ul Hussnain, Director of Research and Development, Fatima Sugar Mills Limited, Multan, Pakistan). (d) Facility in South Africa showing the submerged baskets with sugarcane stalks being treated and a basket of cane prior to treatment and (e) during treatment and (f) disinfecting cane knives prior to cutting (Sharon McFarlane, South Africa Sugarcane Research Institute (SASRI), Mount Edgecombe, South Africa).

two-bud or one-bud setts can be hot water treated. Long HWT (LHWT) (50°C for 3 h) is used to treat RSD-infected stalks in many countries. In some cases, RSD and leaf scald-infected canes are treated by cold-soak LHWT (canes are soaked in slow percolating tap water for 40 h before LHWT); this treatment is more effective, particularly for leaf scald (Steindl, 1972). Smut and chlorotic streak can be controlled by short HWT (50°C for 30 minutes). The HWT facilities vary in appearance, but the primary requirement is to have accurate temperature control (see Fig. 3a–f). The effectiveness is dependent on the temperature and duration of treatment. HWT is also effective in reducing the incidence of the white leaf disease caused by a phytoplasma (Ling, 1962).

Unfortunately, heat therapy is not effective for viral diseases such as sugarcane mosaic and yellow leaf. Meristem tissue culture is used to culture a very few meristem cells that have less probability of being infected. These meristem cells can develop virus-free plantlets for the development of disease-free seedcane (Fitch et al., 2001). Although this process may not be true eradication it is a process that identifies and selects virus-free cells for initial propagation to establish vegetative planting sources. If the source plant is disease-free, a micropropagation technique can be used to multiply shoots in culture

Figure 4 Micropropagation of plantlets from disease-free seedcane. (a) Cutting a single bud seedpiece from a disease-free stalk. (b) Young shoots emerging from single bud cuttings. (c) Shoots with meristems excised from the germinated seedpiece. (d) Small meristem placed on agar medium. (e) Shoot multiplication. (f) Separation of shoots for further multiplication. (g) Young plantlets in flats. (h) Young plantlet ready for transplanting to the field (Dr. Andrea Maldonado, CENGICAÑA, Guatemala).

and produce plantlets much faster than normal vegetative propagation. Figure 4a–h show the micropropagation starting with a disease-free plant, excising of the apical apex, trimming down to a smaller piece and placing on culture medium, initiating growth and multiplication of shoots that are separated and placed in a rooting medium for plantlet growth and hardening before transplantation to the field. Sugarcane mosaic, yellow leaf and Fiji disease viruses were successfully removed by apical meristem culture (Fitch et al., 2001; Wagih et al., 1995; Leu, 1972). In Thailand, disease-free plants were extensively produced by meristem tissue culture to manage sugarcane white leaf (SCWL) disease (Wongkaew and Fletcher, 2004).

3.2.2 Sanitation

Sanitation is used to prevent the spread of systemic pathogens, especially the bacterial pathogens *Xanthomonas albilineans* and *Leifsonia xyli* subsp. *xyli* that cause leaf scald (see Fig. 1q, r and s) and ratoon stunt (see Fig. 1p), respectively. These pathogens can be easily spread by knives, by first cutting infected vegetative material and then cutting healthy stalks in quick succession (Comstock et al., 1996; Bailey and McFarlane, 1992). Also, the pathogen can be spread between stools along the row by harvesters as they cut the field (Hoy et al., 1999; Damann, 1992). Thus, precautions must be taken to disinfect cane knives and harvesters when going from field to field (see Fig. 2f). Sanitation is particularly important in establishing and operating clean seedcane nurseries.

3.2.3 Disease-free seedcane nurseries

Vegetative propagation of sugarcane by seedcanes in commercial fields poses a risk of transmitting diseases. Systemically diseased seedcane will germinate and lead to primary disease incidence. Disease escalation may then occur in subsequent ratoon crops, not only affecting yield but also providing additional inoculum enabling disease spread. Thus, systemic diseases such as ratoon stunt, leaf scald, various forms of mosaic, smut, chlorotic streak, red stripe, dry top rot, yellow leaf and other diseases (see Fig. 1a–w) must be considered in establishing healthy nurseries and disease-free seedcane programmes.

An initial source of disease-free plants is required to establish nurseries (Bakker, 1999). Proper HWTs (see Fig. 3a–f) will eliminate bacterial and fungal diseases (ratoon stunt, leaf scald, smut and chlorotic streak) but will not effectively eliminate viral pathogens. Pathogens such as the viruses causing sugarcane yellow leaf and mosaic (see Fig. 1s and t) may be eliminated by meristem tissue culture. The initial disease-free material is used to establish small primary nurseries that are used to increase and expand the amount of seedcane in large secondary nurseries. During this process it is critical to prevent any infection and reduce the spread of systemic pathogens that are transmitted via infected seedcane. Thus, primary seedcane fields are often planted in isolation away from commercial plantings. This is to eliminate or at least minimize the risk of introduction of systemic pathogens, such as smut and leaf scald. Nursery fields are monitored; surveys, with associated specific assays, are undertaken to ensure the level of infection is low enough for commercial planting.

Practices that reduce or eliminate inoculum of systemic pathogens greatly improve the success of disease-free seedcane programmes. Varieties that are either resistant or have partial resistance assist with the minimization of disease incidence. Varieties with some resistance to smut limit the amount of spores produced, reducing disease buildup. Most sugarcane variety development programmes do not select for ratoon stunt resistance; however, partial resistance reduces the speed and distance of spread down the harvest row via cutting implements. Roguing smut-infected plants reduces the inoculum that can spread aerially in the seedcane nurseries.

3.3 Therapy

Once a disease is established in an area, the only option to limit the yield losses is to apply therapeutic measures in the form of biological, chemical or physical control practices. Fungicidal control in sugarcane has limited application. Fungicides are useful when the disease is non-systemic and the economic threat is of short duration. Sugarcane is grown throughout the year and if the disease occurs over an extended period, frequent multiple fungicide applications are likely to be unprofitable. The use of fungicides has a role in controlling pineapple sett rot caused by *Ceratocystis paradoxa* (Bhuiyan et al., 2014; Comstock et al., 1984). Fungicides are used at planting to control pineapple sett rot and other fungal pathogens that attack germinating seedcane; control is needed for the short period between planting and when shoots and roots develop, making the young plant independent of the planting material. In Florida, fungicides are also used to control sugarcane rusts, especially brown rust (caused by *Puccinia melanocephala*) since the disease has a shorter duration than orange rust (caused by *Puccinia kuehnii*). There are three different classes of fungicides that are registered for control of orange rust in the United States: two strobilurin class fungicides (pyraclostrobin and azoxystrobin), two triazole class fungicides (metconazole and propiconazole) and a third fungicide class, the carboxamides. It is generally recommended that orange rust fungicide applications be made at 3- to 4-weeks interval (Raid and Comstock, 2013). Only one or two fungicide applications may be needed to control brown rust, whereas more applications are required to control orange rust in Florida. Knowledge of the relationship between environmental factors, crop development and pathogen inoculum with disease severity is important to gain full advantage of fungicides.

In the last thirty years, much research has been conducted to understand and incorporate ecologically friendly, low-cost disease management strategies into IDM in various crops. However, there are not many products commercially available for biological control of sugarcane diseases. In recent years, antagonistic biological controls have been a preferred sustainable disease management method for some sugarcane diseases, especially for red rot caused by *Glomerella tucumanensis*. *Trichoderma* has been demonstrated to control red rot and pineapple sett rot caused by *C. paradoxa* in research trials (Singh et al., 2008; Talukeder et al., 2007). Hassan et al. (2012) reported 45–56% reduction in red rot incidence by using three strains of *Bacillus* spp. under field conditions. *Pantoea dispersa* (syn. *Erwinia herbicola*) provided biocontrol against leaf scald disease (Zhang and Birch, 1996). A bacterial isolate reduced smut disease by 77% in a greenhouse trial (Liu et al., 2017). Also, some growth-promoting bacteria have been shown to suppress pathogens by promoting growth and at the same time by eliciting systemic resistance in the host. Viswanathan (1999) demonstrated induced systemic resistance in sugarcane by plant growth–promoting bacteria for the management of red rot.

4 Managing diseases through the host

4.1 Host resistance

Resistance is the most commonly implemented and effective disease control strategy used in sugarcane production, since it is easy to apply and is cost-effective for growers. If a sugarcane variety has sufficient resistance for all diseases present in the production area, no other control practices may be needed. Unfortunately, this is rare. With multiple diseases and the need for high-yielding varieties, the development of new commercial varieties with adequate resistance is essential. However, an IDM strategy is also often used to minimize losses and maximize yields.

Several factors determine whether resistance for a particular disease is the major emphasis in the IDM programme. First, is the disease a major threat to production and does the disease cause significant economic losses? Simply, is selection for resistance really needed or is there sufficient resistance present in commercial and potential varieties being developed? Second, are sources of resistance available and what is their frequency in the parental breeding population? Third, is there a screening procedure and how easy is it to conduct? Is the screening passive, such as with a natural infection test or is it active, requiring field screening and laboratory assays to evaluate resistance? Can numerous clones be screened or, because of the need for replicated field plots, only a limited number of clones can be screened? Lastly, how effective are the other practices in the IDM programme and what are their ease of application and cost? Answers to these questions will determine the effort required to develop adequate resistance. If other control practices are effective, easier and less costly, screening for resistance may not be the best choice.

Most variety development programmes have active resistance screening and selection programmes for smut, rust (brown and/or orange rust), leaf scald and mosaic (sugarcane and/or sorghum) resistance. These diseases have almost a worldwide distribution and are potential economic threats in most countries where they are present. Resistance screening for these diseases is conducted via natural infection, artificial inoculations or a combination of both for evaluating the level of resistance in clones. Screening for disease resistance ideally is best early in the variety development programme. In which selection stage of the breeding programme resistance screening takes place is determined by many of the following factors: importance of the disease, frequency of resistance in the cultivar development programme, precision of the rating, number of plants required for disease reaction determination (or in other words, the quantity of seedcane of a clone needed), or the fact if the pathogen is systemic within the plant. If the pathogen is systemic a separate planting is required to ensure infected seedcane is not used to propagate the clone in the variety development programme, thus delaying screening. Either if large replicated field plots, greenhouse screening, labour-intensive diagnostic assays are necessary, or if screening needs repeating for rating verification, the resistance testing usually will occur later in the programme.

Smut resistance screening tends to have a standard approach, but this varies between countries and may be altered when specific situations require a different approach. Most smut resistance screening uses a dip-inoculation technique where seedcanes are dipped in a spore suspension of *Sporisorium scitamineum*. Inoculum concentration is often around 10^6 spores per mL; setts are dipped for 30 minutes in the spore suspension (Heinz et al., 1973), then incubated overnight and planted in replicated field plots. Assessments are based on disease incidence (number of smut sori, commonly called whips); plant and first ratoon crop assessments are undertaken and compared to varieties of known ratings. This procedure generally works well in evaluating the smut reaction of clones in the variety development programme.

However, outcomes have to be carefully considered. The screening process is more severe than normal natural infection conditions in the field. Both in Florida and Hawaii, it was observed that some clones that could be grown commercially would be discarded based on ratings in artificial inoculation tests as compared to smut susceptibility in commercial crops. Hawaii used natural infection tests to decide whether clones identified as susceptible were in fact safe and could be grown commercially without significant yield losses. In Florida, CP 73-1547 and CP 78-1628 were successfully grown commercially despite the fact they were rated moderately susceptible in inoculated trials. The level of smut in these two commercial varieties was low even in a hot dry spring that was conducive to smut development. CP 73-1547 is also grown without smut losses in Central America. Discarding these two varieties based only on their ratings in artificial trials would have been a costly mistake, because they have been economically productive.

Smut in Australia presented some unique problems after its arrival. First the incidence of smut susceptibility was extremely high both in varieties in commercial fields and in the parental clones used in breeding. An example is the severe smut development in Q 158 when grown in the field at the USDA-ARS Sugarcane Field Station at Canal Point, Florida; smut essentially destroyed the plot (see Fig. 1d). Smut screening programme is often conducted late in the variety development programme and it usually assumes there is a high percentage of resistance in the breeding population. A late evaluation in Australia's unique situation of high susceptibility would have drastic consequences and eliminate the majority of clones, leaving a very small number of potential commercial possibilities. This would severely impact the sugarcane breeding programme, reducing the selection for all other characteristics, in particular yield. Thus, a smut seedling screening programme was initiated to eliminate the most susceptible clones early in the programme (Cox et al., 2011; Croft et al., 2000a). This essentially created a pool of potential clones in the variety development programme with a much higher frequency of smut resistance. Thus, clones selected for superior yield characteristics in the programme would have less probability of being smut-susceptible. The impact of smut on the parental population was also devastating, because the best parents were smut-susceptible. Eliminating all smut-susceptible parental clones would have set back the development of superior varieties because the breeding pool would be dramatically reduced. This was addressed by establishing a small programme using the best smut-susceptible parents and then screening heavily for smut resistance with the selection of high-yielding (but low frequency) smut-resistant clones. The programme resulted in smut-resistant but superior varieties and enabled other characteristics to be incorporated into the breeding population. Early selection for smut resistance in sugarcane can be achieved by TaqMan qRT-PCR detection of pathogen proliferation and changes in physiological and biochemical indices (peroxidase, ascorbate peroxidase, catalase, superoxide dismutase, β-1,3-glucanase and malondialdehyde) after artificial inoculation (Su et al., 2016).

Varietal resistance is an economical management method for brown and orange rust control in many sugarcane-producing industries. Differences in varietal resistance are evident in adjacent fields of two varieties that differ in their reaction to brown rust (see Fig. 5a and b). Sugarcane rust reactions can be evaluated based on natural infection, but

Figure 5 Sugarcane brown rust causing different severities due to varietal resistance: (a) North Coast, South Africa (Sharon McFarlane, South Africa Sugarcane Research Institute (SASRI), Mount Edgecombe, South Africa). (b) Everglades Agricultural Area, Florida (Jack Comstock, retired, formerly of the USDA-ARS Sugarcane Field Station, Canal Point, Florida).

Figure 6 Screening for rust resistance. (a) Urediniospore collection from rust infected leaves using a vacuum apparatus. (b) Flask with massive amount of urediniospores. (c) Rust spore suspension. (d) Determining spore concentration. (e) Inoculation of plants by placing spores in the whorl. (f) Incubation of inoculated plants in a growth chamber (a–f). Werner Ovalle, CENGICAÑA, Guatemala. (g) and (h) Rust whorl inoculation of plants in the field. (i) Brown rust pustules. (j) Orange rust pustules (Sushma Sood, USDA-ARS Sugarcane Field Station, Canal Point, Florida).

if conditions are not conducive for rust development, valid ratings may not be obtained. A leaf whorl inoculation technique was developed to determine rust reactions (Sood et al., 2009). Figure 6a–j demonstrate the screening method from collection of spores, inoculation and symptom expression on plants grown in growth chambers, and Fig. 6g–h illustrate the screening process in the field.

Leaf scald screening varies between institutions but usually the pathogen, *X. albilineans*, is isolated from 'pencil-line' foliar symptoms, grown for six days and a standard bacterial suspension is prepared and applied to cut or wounded surfaces (Bechet et al., 1992) as shown in Fig. 1q. Ratings are based on incidence and severity of symptom expression.

Several sugarcane diseases are economically important only in specific countries (locations) and have screening programmes to screen and develop resistance. Red rot (primarily in India and Pakistan), Fiji leaf gall (primarily in Australia and Fiji), chlorotic streak (primarily in Australia) and Pachymetra root rot (only in Australia) are examples.

Red rot caused by *G. tucumanensis* infects sugarcane stalks causing a patchy reddish-white discolouration in internal stalk tissues, a reduction in stalk growth and deterioration of stalk tissues that inhibits sugar recovery. Disease screening usually relies on inserting mycelial mats of the pathogen into holes in the stalk and rating based on the extent of tissue colonization by the pathogen in comparison to clones of known resistance (Mohanraj et al., 2012; Viswanathan and Samiyappan, 1999). The selection of resistant clones is complicated by the presence of pathogen strains. Thus, a clone may be resistant to a specific strain but susceptible to other strains that are present, either in local or distant regions.

Fiji leaf gall is caused by Fiji disease virus and is present in Australia, Fiji, and the Philippines. Disease screening relies on the *Perkinsiella saccharicida* leafhoppers as vectors of the Fiji disease virus. The reaction to the virus is assessed by symptom development (the time it takes for leaf galls to develop on the underside of affected leaves) and severity.

Sugarcane varieties are screened for Pachymetra root rot resistance by exposing their root system to *Pachymetra chaunorhiza* under optimum glasshouse conditions for infection (Croft, 1989).

Red stripe in Florida is observed rarely on a few plants during the growing season and is not a concern; thus, resistance and other control practices for red stripe are not required. However, in Honduras near the Pacific coast, foliar red stripe symptoms are abundant with over 5% of the stalks having internal deterioration (see Fig. 1g, h and i), even for variety CP 72-2086 that originated in Florida. Unfortunately, in Honduras there is no screening programme for red stripe resistance. Stalk deterioration causes some commercial yield losses.

Sugarcane mosaic virus (SCMV) is transmitted naturally in the field through an insect vector, but also mechanically via a leaf abrasion technique (Dean, 1971). This makes resistance screening easy and progress has been made in developing SCMV-resistant varieties.

Sugarcane yellow leaf is a serious problem in CP varieties in Florida and resistance in the CP breeding population is limited. The sugarcane yellow leaf virus is aphid transmitted and an efficient resistance screening procedure to handle large numbers of individual clones is yet to be developed. Without a screening technique, development of a resistant clone is largely by chance. Resistant varieties also need to have commercially viable yields; they are of little use if their commercial yield capability is limited. Alternative control practices such as eradication, protection and therapy may be sufficient in an IDM strategy, without developing resistance.

4.2 Protection

Plant nutrition has a significant impact on sugarcane yield, by either reducing or increasing disease severity. Silicon has been reported as a key element in the management of a wide range of diseases in many crops. Silicon significantly reduced brown rust severity in glass house experiments (Ramouthar, et al., 2016); it was also effective in increasing resistance to eye spot, caused by *Bipolaris sacchari* (Raid et al., 1992). The effect of silicon on smut incidence in highly susceptible varieties was not significant in Australia (Bhuiyan and Croft, 2015). Johnson and Grisham (2014) showed that a foliar spray of copper (Cu) and nickel (Ni) fertilizers improves both cane and sugar yields in brown rust–infected varieties. Cadet et al. (2003) found that brown rust appeared to favour uptake of potassium to the detriment of calcium, magnesium and silicon. They hypothesized that early application of calcium before rust infection may reduce disease levels by limiting the uptake of potassium. Ling (1962) found that white leaf disease incidence was not influenced by plant nutrition. However, according to Chen and Kusalwong (2000) white leaf is under control in infected sugarcane areas in Thailand by using various soil amendments (bagasse, filter mud, green manure and waste water of alcohol factories as a nitrogen source) along with other disease management strategies (quarantine, routine use of healthy nursery, HWT of seed pieces and micropropagation of disease-free plants). Severity of sugarcane root rot caused by *Pythium arrhenomanes* was reduced by amendments of some organic materials (Dissanayake and Hoy, 1999).

5 Managing diseases by controlling environment

5.1 Avoidance

Avoiding disease by avoiding the pathogen or by altering the environment can be achieved via cultural practices targeting either one or more components (host, pathogen or environment). The interaction between a pathogen and a host can be avoided by changing the time of planting and/or harvest. Planting and harvesting crops when conditions are not conducive to pathogens and vectors can reduce disease development. Croft et al. (2000b) documented that an early spring and autumn planting of the noble cane Badila (*Saccharum officinarum*) significantly reduced losses from red stripe compared to crops planted in late spring. Similarly, Fiji leaf gall spreads more rapidly in a spring-planted crop due to the vulnerability of the young crop to high populations of the vector planthopper (Croft et al., 2000b). In South Africa, avoiding the planting of susceptible varieties in the summer months reduced the incidence of SCMV by eliminating exposure of the young plants to high aphid populations (*Rhopalosiphum maidis* and *Heteroneura setariae*), the vectors of SCMV (Harborne, 1988). Avoiding the planting of red rot-susceptible varieties during excessively cool and wet weather has been effective in several countries (Raid and Rott 2015). Raid and Rott (2015) also suggested prompt harvest of red rot-infected, or -susceptible crops as an effective IDM strategy. Avoiding deep planting of vegetative material also reduces pineapple sett rot incidence. In contrast, the population of planthopper (*Matsumaratettix hiroglyphicus*) vector and multiplication of the SCWL disease phytoplasma are limited by low temperature during the winter season. Excellent control of SCWL disease has been achieved by planting sugarcane crop in spring rather than in autumn (Chen and Kusalwong, 2000; Pan and Yang, 1971).

Field irrigation and drainage may be very important for disease management of soil-borne diseases. Some pathogens need water to disperse and transmission can be avoided by managing irrigation and drainage of the field (Croft et al., 2000b). Well-drained field is one of the components of the integrated management of the red rot (Raid and Rott, 2015). Leaf wetness is very important for many foliar diseases (Raid and Comstock, 2000) and this can be managed through well-designed irrigation practices (Ryan and Egan, 1989).

Crop rotation and fallowing can be helpful in disease management of soil-borne pathogens. A short fallow was not able to significantly reduce population of the pachymetra root rot fungus (Croft et al., 2000b) and chlorotic streak (Egan, 1989).

6 Factors considered in integrated disease management strategies

Several factors that need to be considered in IDM strategies are presented in Table 2. The process involves considering whether the disease is a current or potential threat and causes economic losses. The next step is to determine the easiest, most effective and profitable control practice(s) and if they can be used together in an integrated programme.

Table 2 Factors considered in IDM strategies

Factor		Control options
Disease wide-spread presence	No	Exclusion/quarantine
	Yes	Everything except exclusion/quarantine
Disease incidence	Low	May not merit control practices, if losses per plant are low
	High	Definitely requires control if losses per plant is high
Losses per plant	Low	May not merit control practices
	Unknown	Requires monitoring and measures to restrict pathogen spread
	High	Definitely requires measures to restrict disease incidence
Pathogen systemic	No	Everything except thermal treatment or nurseries for disease-free seedcane
	Yes	Seedcane programmes to eradicate the pathogen to produce disease-free seedcane, coupled with sanitation and cultural practices to prevent re-infection may be required
Resistance available and easy to develop	No	Must use all suitable control practices other than resistance
	Yes	Resistance may be primary mechanism of control
Transmission	Wind	All factors but exclusion may not be effective
	Seedcane	Eradication via thermal treatment may be possible and disease-free seedcane
	Vectors	All factors but exclusion may not be effective
	Soil-borne	Preventive sanitary and exclusion measures to restrict movement of pathogen infested soil is required.
Fungicidal control	No	Everything except fungicides
	Yes	Option if effective and economical
Composition of varieties	Single or low	May need an increase in the number of resistant varieties
	High	Acceptable if most of the varieties have resistance to the major pathogens

7 Future trends

Research has steadily improved the knowledge of disease aetiology, ecology, pathogen detection and screening methods to identify resistance and disease management practices and this has contributed to limited disease losses in crop production systems.

7.1 Disease detection

Accurate and timely detection of the existing and/or exotic pathogen is important for the management of any disease. There are direct and indirect methods for pathogen detection. A tissue blot immunoassay has been regularly used to detect leaf scald, RSD, SCMV and sugarcane yellow leaf virus (Comstock et al.,1998; Schenck et al., 1997). Flow cytometry has been used to differentiate spores of seven isolates of smut (Rodriguez et al., 2001). A hyper spectral technique has been used in the Mackay region of Australia to detect orange rust (Grisham, et al., 2010; Apan et al., 2004).

Progress in disease diagnosis has been dramatic in the last 20 years with PCR tests for pathogens of sugarcane yellow leaf, chlorotic streak and sugarcane rusts (brown, orange and tawny rusts caused by *Macruropyxis fulva*) (Martin et al., 2017, 2015; Braithwaite and Croft, 2013; Glynn et al., 2010). The initiation of metagenomics has allowed the characterization of the microbiome and pathogen populations in sugarcane and the soil (Souza, et al., 2016); NGS (for next-generation sequencing) technology has also led to pathogen DNA characterization (Mollov and Malapi-Wight, 2016). Since 1990, seven virus species in the genus *Mastrevirus* of the family Geminiviridae have been studied with improved metagenomics and next-generation or high-throughput sequencing; more pathogens may be detected using these techniques. It is likely that these relatively new technologies will be applied more widely. The diagnostic technology will improve quarantine procedures, making disease exclusion for non-windblown pathogens more successful.

7.2 Disease forecasting

The use of fungicides has had a role in controlling pineapple sett rot (Bhuiyan et al., 2014; Comstock et al., 1984), but more recently they have been used to control brown and orange rusts. Durable resistance has been difficult to develop in commercial varieties and strategic fungicide application has been economical. Disease forecasting, by monitoring weather conditions, relative aerial spore concentration or sentinel plots, has been used to predict disease outbreaks and to recommend fungicide application. This has been critical in preventing economic losses in Florida crops. Forecasting optimizes fungicide management and may lead to reduced chemical usage.

Surveys and scouting are the most widely used form of disease monitoring. The Florida sugarcane industry has regularly surveyed crops for orange rust and brown rust over the last several years. However, disease predictive models and computer-generated disease simulators also play important roles in IDM. There are many advanced technologies that have been developed in recent years to remotely monitor disease. Remote sensing has been used in yield forecasting for decades but has not been used in sugarcane disease forecasting. Drones and satellite images can also be used in disease monitoring and forecasting. The use of GPS technology to relate severity of disease to soil nutritional parameters has been investigated in brown rust of sugarcane (Grisham, et al., 2008). Fertilizer applications could be applied using GPS technology, to correct the conditions previously conducive to particular diseases.

7.3 Resistance

Disease screening procedures have been developed for a wide range of diseases: Pachymetra root rot (Croft, 1989); ratoon stunt (Comstock et al., 2000); brown and orange rusts (Sood et al., 2009) and chlorotic streak (Braithwaite et al., 2016). With improvements in screening technology, selection of resistant varieties will be more efficient.

A brown rust resistance gene, *Bru*1, has been identified (Costet et al. 2012a; Asnaghi et al., 2004; Daugrois et al., 1996) and is used in variety development programmes (Glynn et al., 2013). Other potential molecular markers show potential for identifying resistance to sugarcane yellow leaf (Debibakas et al., 2014; Izquierdo, et al., 2013; Costet et al., 2012b) which is needed since high-throughput screening techniques for yellow leaf are not available. More molecular markers for resistance are likely to be developed in the future. With the molecular knowledge of resistance genes, such genes may be used in developing resistant transgenic plants. Also, the recent CRISPR (Clustered Regularly Interspaced Short Palindromic Repeats) technology offers an opportunity for genome editing (Chakravarthi, 2016). This technology may also be used in future in modifying disease reactions in plants. Transgenic lines resistant against the sugarcane yellow leaf virus showed 1000-fold less viral titer and were comparable to natural resistance (Zhu et al., 2011).

7.4 Clean seedcane

The use of disease-free seedcane nurseries plays an important role in controlling sugarcane diseases. HWT of vegetative seedpieces or setts eliminates bacterial pathogens, *L. xyli* subsp. *xyli* and *X. albilineans,* the smut fungal pathogen, *Sporisorium scitamineum,* to some extent and chlorotic streak. HWT schemes have been playing a major role in maintaining disease-free nurseries. However, virus elimination is not effective by HWT, and meristem apical tip culture is used to eliminate the viruses (Viswanathan et al., 2014; Flynn et al., 2005; Fitch et al., 2001). Propagation of virus-free shoots is used to increase material for rapid propagation (Ramgareeb et al., 2010). More industries are using tissue culture to produce disease-free seedcane.

8 Where to look for further information

Information on sugarcane diseases is available on multiple websites of scientific societies, universities and sugarcane research institutes (see below). The International Society of Sugar Cane Technologists (ISSCT), established in 1924, stages congresses every three years and has subject area workshops in pathology between the congresses. The pathology workshops are attended by sugarcane pathologists from the top research organizations and are an excellent way to learn of the latest research. The ISSCT proceedings provide an historical record of sugarcane research as well as the latest developments.

A newer sugar organization established in 2004 is the International Association of Professionals in Sugar and Integrated Technologies (IAPSIT). IAPSIT conferences are held every three years and focus on technology and scientific advancements in developing countries. It also publishes a newsletter and manuscripts in the *Sugar Tech* Journal.

8.1 Sources of sugarcane information

Scientific Societies publications:

International Society of Sugar Cane Technologists
Proceedings http://www.issct.org/proceedings

American Society of Sugar Cane Technologists
Journal http://www.assct.org/

Australian Society of Sugar Cane Technologists
Proceedings http://www.assct.com.au/

South Africa Sugar Technologists Association
Proceedings http://www.sasta.co.za/publications/congress-proceedings

Sugar Tech
Journal http://www.springer.com/life+sciences/agriculture/journal/12355

US Universities:

Louisiana State University
http://www.lsuagcenter.com/portals/communications/publications/management_guides

University of Florida
Sugarcane Handbook http://edis.ifas.ufl.edu/topic_book_sugarcane_handbook
http://edis.ifas.ufl.edu/topic_sugarcane_diseases

Sugarcane Research Organizations:

Sugarcane Production Research
https://www.ars.usda.gov/southeast-area/canal-point-fl/sugarcane-field-station/

Sugarcane research
https://www.ars.usda.gov/southeast-area/houma-la/sugarcane-research/

Sugar Research Australia
http://www.sugarresearch.com.au/

Centro de investigación de la caña de azucar de Colombia
(Colombian Sugarcane Research) http://www.cenicana.org/

Centro Guatemalteco de Investigación y Capacitación de la Caña de Azúcar
(Guatemalan Sugarcane Research) http://www.cengicana.org/

Centro de Investigación de la Caña de Azúcar del Ecuador
http://cincae.org/

CTC - Centro de Tecnologia Canavieira (Sugarcane Research Institute in Brazil)
http://www.bionity.com/en/research-institutes/42296/ctc-centro-de-tecnologia-canavieira.html

CIRAD (French Agricultural Research Centre for International Development)
http://www.cirad.fr/en/research-operations/tropical-supply-chains/sugarcane/context-and-issues

Mauritius Sugar Industry Research Institute
http://www.msiri.mu/

RIDESA (An organization of ten federal universities in Brazil that conducts sugarcane research and development of RB varieties) http://www.cetem.gov.br/

South African Sugarcane Research Institute
www.sasa.org.za/sasri

9 References

Abbott, E. V. (1964), 'Dry top rot', In Hughes, C. G., Abbott, E. V. and Wismer, C. A. (Eds), *Sugar Cane Diseases of the World*, Elsevier, Amsterdam, The Netherlands, pp. 110–13.

Afghan, S., Hussain, Z. U., Ali, K. and Shahzad, A. (2013), 'Performance of promising sugarcane clones development from imported fuzz (true seed) in Pakistan', In *XXVIII Congress Proc.*, Sao Paulo, Brazil.

Apan, A., Held, A., Phinn, S. and Markley, J. (2004), 'Detecting sugarcane orange rust disease using EO-1 Hyperion hyperspectral imagery', *Int. J. Remote Sens.*, 25, 489–98.

Asnaghi, C., Roques, D., Ruffel, S., Kaye, C., Hoarau, J. Y., Télismart, H., Girard, J. C., Raboin, L. M., Risterucci, A. M., Grivet, L. and D'Hont, A. (2004), 'Targeted mapping of a sugarcane rust resistance gene (*Bru*1) using bulked segregant analysis and AFLP markers', *Theor. Appl. Genet.*, 108, 759–64.

Bailey, R. A. and McFarlane, S. M. (1992), 'Ratoon stunting disease: Survival of *Clavibacter xyli* subsp *xyli* in field soil and its spread to newly planted sugarcane', *Proc. S. Afr. Sugar Technol. Assoc.*, 66, 75–7.

Bakker, H. (1999), *Sugar Cane Cultivation and Management*, Kluwer Academic/Plenum Publishers, New York, USA.

Bechet, G. R., Bailey, R. A. and Nuss, K. J. (1992), 'Screening of new sugarcane clones for resistance to leaf scald (*Xanthomonas albilineans*)', *Proc. S. Afr. Sugar Technol. Assoc.*, 57, 99–104.

Bhuiyan, S. A. and Croft, B. J. (2015), 'Effect of silicon fertilizers on sugarcane smut in Australia', *Proc. Aust. Soc. Sugar Cane Technol.*, 37, 53–60.

Bhuiyan S. A., Croft, B. J. and Tucker, G. R. (2014), 'Efficacy of the fungicide flutriafol for the control of pineapple sett rot of sugarcane in Australia', *Aust. Plant Pathol.*, 43, 413–19.

Braithwaite, K. S. and Croft, B. J. (2013), 'A diagnostic test for chlorotic streak disease', *Proc. Aust. Soc. Sugar Cane Technol*, 35, 1–8.

Braithwaite, K., Ngo, C., Croft, B. J., Magarey, R. and Young, A. (2016), 'Progress in understanding and managing chlorotic streak of sugarcane', *Proc. Int. Soc. Sugar Cane Technol.*, 29, 1768–75.

Byther, R. S. and Steiner, G. W. (1973), 'Hot water control of smut', *Experiment Station Hawaiian Sugar Planters Association Annual Report*, pp. 31–2.

Cadet, P., McFarlane, S. A. and Meyer, J. H. (2003), 'Association between nutrients and rust in sugarcane in Kwazulu-Natal', *Proc. S. Afr. Sug. Technol. Ass.*, 7, 223–9.

Chakravarthi, M. (2016), 'Genome editing in sugarcane: challenges ahead', *Front. Plant Sci.*, 7, 1–5.

Chen, C. T. and Kusalwong, A. (2000). 'White leaf', *in* Rott, P., Bailey, R. A., Comstock, J. C., Croft, B. J. and Saumtally, A. S. (Eds), *A Guide to Sugarcane Diseases*, Centre de cooperation internationale en recherche agronomique pour le developpement (CIRAD) and International Society of Sugar Cane Technologists (ISSCT), Montpellier, France, pp. 231–6.

Comstock, J. C., Ferreira, S. A. and Tew, T. L. (1983), 'Hawaii's approach to control sugarcane smut', *Plant Dis.*, 67, 452–7.

Comstock, J. C., Ferreira, S. A., Ching, S. A. and Hilton, H. W. (1984), 'Control of pineapple disease of sugarcane with propiconazole', *Plant Dis.*, 68, 1072–5.

Comstock, J. C., Miller, J. D. and Farr, D. F. (1994), 'First report of dry top rot of sugarcane in Florida: Symptomatology, cultivar reaction and effect on stalk water flow rate', *Plant Dis.*, 78, 428–31.

Comstock, J. C., Shine, J. M., Davis, M. J. and Dean, J. L. (1996), 'Relationship between resistance to *Clavibacter xyli* subsp *xyli* colonization in sugarcane and spread of ratoon stunting disease in the field', *Plant Dis.*, 80, 704–8.

Comstock, J. C., Irey, M. S., Lockhart, B. E. L. and Wang, Z. K. (1998), 'Incidence of yellow leaf syndrome in CP cultivars based on polymerase chain reaction and serological techniques', *Sugar Cane*, 4, 21–4.

Comstock, J. C., Davis, M. J., Tai, P. Y. P. and Miller, J. D. (2000), 'Selecting ratoon stunting disease resistant cultivars for the 21st century', In *1998 Inter-American Sugar Cane Seminar*, pp. 38–45.

Costet, L., Cunff, L. L., Royaert, S., Raboin, L. M., Hervouet, C., Toubi, L., Telismart, H., Garsmeur, O., Rousselle, Y., Pauquet, J., Nibouche, S., Glaszmann, J. C., Hoarau, J. Y. and D'Hont, A. (2012a),

'Haplotype structure around *Bru1* reveals a narrow genetic basis for brown rust resistance in modern sugarcane cultivars', *Theor. Appl. Genet.*, 125, 825–36.

Costet, L, Raboin, L. M., Payet, M., D'Hont, A. and Nibouche, S. (2012b), 'A major quantitative trait allele for resistance to the *Sugarcane yellow leaf virus* (*Luteoviridae*)', *Plant Breed.*, 131, 637–40.

Cox, M., Croft, B. and Bonnett, G. (2011), 'Smutbuster: Accelerated breeding for smut-resistant sugarcane varieties', *BSES Project Report SD11007.*

Croft, B. J. (1989), 'A technique for screening sugarcane cultivars for resistance to Pachymetra root rot', *Plant Dis.*, 73, 651–4.

Croft, B. J., Irawan and Berding, N. (2000a), 'Screening Australian sugarcane clones for smut reaction in Indonesia: Initial results', *Proc. Aust. Soc. Sugar Cane Technol*, 22, 170–7.

Croft, B. J., Magarey, R. C. and Whittle, P. (2000b), 'Disease management', *in* Hogarth, M. and Allsopp, P. (Eds), *Manual of Cane Growing*, Bureau of Sugar Experiment Station, Indooroopilly, Australia, pp. 263–89.

Croft, B. J., Magarey, R. C., Quinn, B., Royal, A. and Kerkwyk, R. (2008), 'Sugarcane smut in Queensland', *Proc. Aus. Soc. Sugar Cane Technol.*, 30, 52–62.

Damann, K. E. (1992), 'Effect of sugarcane cultivar susceptibility on spread of ratoon stunting disease by the mechanical harvester', *Plant Dis.*, 76, 1148–9.

Daugrois, J. H., Grivet, L., Roques, D., Hoarau, J. Y., Lombard, H., Glaszmann, J. C. and D'Hont, A. (1996), 'A putative major gene for rust resistance linked with RFLP marker in sugarcane cultivar R570', *Theor. Appl. Gen.*, 92, 1059–64.

Dean, J. L. (1971), 'Systemic-host assay of sugarcane mosaic virus', *Phytopathology*, 61, 526–31.

Debibakas, S., Rocher, S. and Garsmeur, O. (2014), 'Prospecting sugarcane resistance to *Sugarcane yellow leaf virus* by genome-wide association', *Theor. Appl. Gen.*, 127, 1719–32.

Dissanayake, N. and Hoy, J. W. (1999), 'Organic material soil amendment effects on root rot and sugarcane growth and characterization of materials', *Plant Dis.*, 83, 1039–46.

Egan, B. T. (1989), 'Chlorotic streak', *in* Ricaud, C., Egan, B. T., Gillaspie Jr., A. G. and Hughes, C. G. (Eds), *Diseases of Sugarcane Major Diseases*, Elsevier, Amsterdam, The Netherlands, pp. 247–62.

Fitch, M. M. M., Lehrer, A. T., Komor, E. and Moore, P. H. (2001), 'Elimination of sugarcane yellow leaf virus from infected sugarcane plants by meristem tip culture visualized by tissue blot immunoassay', *Plant Pathol.*, 50, 676–80.

Flynn, J., Powell, G., Perdomo, R., Monres, G., Quebedeaux, K. and Comstock, J. C. (2005), 'Comparison of sugarcane disease incidence and yield of field-run, heat-treated, and tissue-culture based seedcane', *J. Am. Soc. Sugar Cane Tech.*, 25, 88–100.

Gillaspie, A. G. and Davis, M. J. (1992), 'Ratoon stunting of sugarcane', *in* Mukhopadhyay, A. N., Kanar, J., Chaube, H. S. and Singh, U. S. (Eds), *Plant Diseases of International Importance*, Volume 4, Prentice Hall, New Jersey, USA, pp. 41–61.

Glynn, N. C., Dixon, L. J., Castlebury, L. A., Szabo, L. J. and Comstock, J. C. (2010), 'PCR assays for the sugarcane rust pathogens *Puccinia kuehnii* and *P. melanocephala* and detection of a SNP associated with geographical distribution in *P. kuehnii*', *Plant Pathol.*, 59, 703–11.

Glynn, N. C., LaBorde, C., Davidson, R. W., Irey, M., Glaz, B. S., Comstock, J. C. and D'Hont, A. (2013), 'Utilization of a major brown rust resistance gene in sugarcane breeding', *Mol. Breed.*, 30, 321–3.

Grisham, M. P., Johnson, R. M., Viator, R. P. and Zimba, P. V. (2008), 'Use of precision agriculture techniques for sugarcane pathology studies', *Sugar J.*, 71, 24–7.

Grisham, M. P., Johnson, R. M. and Zimba, P. V. (2010), 'Detecting sugarcane yellow leaf virus in asymptomatic sugarcane leaves with hyperspectral remote sensing and associated leaf pigment changes', *J. Virol. Methods*, 167, 140–5.

Harborne, K. M. (1988), 'Population dynamics of the main aphid vectors of sugarcane mosaic virus in Natal', *Proc. S. Afr. Technol. Assoc.*, 62, 195–8.

Hassan, M. N., Afghan, S. and Hafeez, F. Y. (2012), 'Biological suppression of sugarcane red rot by *Bacillus* spp. under field conditions', *J. Plant Pathol.*, 94, 325–9.

Heinz, D. J., Ladd, S. L. and Meyer, H. K. (1973), 'Short note on smut disease in Hawaii', *Sugarcane Breed. Newslett.*, 31, 17–19.

Hoy, J. W., Grisham, M. P. and Damann, K. E. (1999), 'Spread and increase of ratoon stunting disease of sugarcane and comparison of disease detection methods', *Plant Dis.*, 83, 1170–5.

Hughes, C. G. and Robinson, P. E. (1961), 'Downy mildew disease', In J. P. Martin, E. V. Abbott and C. G. Hughes (Eds), *Sugarcane diseases of the world*, Elsevier, Amsterdam, The Netherlands, pp. 140–59.

Izquierdo, P., Gutiérrez, A., Victoria, J. I., Ángel, J. C., Avellaneda, M. C. and López, J. (2013), 'Molecular markers associated with resistance to sugarcane yellow leaf virus', *Proc. Int. Soc. Sugar Cane Tech.*, 28, 10.

Johnson, R. M. and Grisham, M. P. (2014), 'Influence of foliar copper and nickel applications on sugarcane yields in brown rust infested fields in Louisiana', *J. Am. Soc. Sugar Cane Tech.*, 34, 57–8.

Leu, L. S. (1972), 'Freeing sugarcane from mosaic virus by apical meristem and tissue culture', *Taiwan Sugar Exp. Sta. Rep.*, 57, 57–63.

Ling, K. C. (1962), 'White leaf disease of sugarcane', *Taiwan Sugar*, 9, 28–32.

Llu, S., Lın, N., Chen, Y., Liang, J., Liao, L, Lv, M., Chen, Y., Tang, Y., He, F., Chen, S., Zhou, J. and Zhang, L. (2017), 'Biocontrol of sugarcane smut disease by interference of fungal sexual mating and hyphal growth using bacterial isolate', *Front. Microbiol.*, 8, 778.

Magarey, R. C. (1994), 'Effect of Pachymetra root rot on sugarcane yield', *Plant Dis.*, 78, 475–7.

Magarey, R. C., Denney, D., Sheahan, T., Fowell, L., Croft, B. J., Lonie, K. J., Bull, J. I., Bhuiyan, S. and Willcox, T. G. (2009), 'Results from smut epidemiology studies in the Herbert, Mackay and Bundaberg-Isis areas in 2007–2008', *Proc. Aust. Soc. Sugar Cane Technol.*, 31, 145–57.

Maroon-Lango, C. J., Hoy, J. W., Comstock, J. C., Grisham, M. P., Mock, R., Hale, A., Afghan, S., Croft, B. J., de la Cueva, F., Hoffmann, H., Kennedy, A., Orozco, H., Saumtally, S., Victoria, J. and Viswanathan, R. (2013), 'Current knowledge and practices related to seed transmission of sugarcane pathogens and movement of seed', *J. Am. Soc. Sugar Cane Tech.*, 33, 20–9.

Martin, L. A., Lloyd Evans, D. Rutherford, R. S. and McFarlane, S. A. (2015), 'Tawny rust, An update on the new species of rust infecting sugarcane in southern Africa', *Proc. S. Afr. Sugar. Technol.*, 88, 309–13.

Martin, L. A., Evans, D. L., Castlebury, L. A., Sifundza, J. T., Comstock, J. C. Rutherford, R. S. and McFarlane, S. A. (2017), '*Macruropyxis fulva* sp. *nov.*, a new rust (Pucciniales) infecting sugarcane in southern Africa', *Australas. Plant Pathol.*, 46, 63–74.

Mohanraj, D., Padmanaban, P. and Viswanathan, R. (2012), 'Screening of red rot resistance in sugarcane', *Funct. Plant Sci. Biotechnol.*, 6, 51–62.

Mollov, D. and Malapi-Wight, M. (2016), 'Next generation sequencing: A useful tool for detection of sugarcane viruses in quarantine programs', *Proc. Int. Soc. Sugar Cane Technol.*, 29, 1631–5.

NAS (National Academy of Science) (1968), 'Plant-disease development and control'. *Principles of Plant and Animal Pest Control*, Volume 1, Publication 1596, National Academy of Sciences, Washington DC, USA, p. 205.

Pan, Y. S. and Yang, S. L. (1971), 'Studies on the bionomics of *Matsumaratettix hiroglyphicus* (Matsumura), an insect vector of sugarcane white leaf disease. I. Seasonal abundance and dispersal of the adult', *Rep. Repub. Taiwan Sugar Exp. Stat.*, 50, 65–71.

Raid, R. N. and Comstock, J. C. (2000), 'Common rust', *in* Rott, P., Bailey, R. A., Comstock, J. C., Croft, B. J. and Saumtally, A. S. (Eds), *A Guide to Sugarcane Diseases*, Centre de cooperation internationale en recherche agronomique pour le developpement (CIRAD) and International Society of Sugar Cane Technologists (ISSCT), Montpellier, France, pp. 85–9.

Raid, R. N. and Comstock, J. C. (2013), 'A review of fungicide evaluations for the control of orange rust of sugarcane in Florida', *Proc. Int. Soc. Sugar Cane Technol.*, 28, 1–9.

Raid, R. N. and Rott, P. (2015), 'Sugarcane red rot disease', In Sandhu, H. S. (Ed.), *Florida Sugarcane Handbook*, University of Florida Extension, IFAS. http://edis.ifas.ufl.edu/MENU_SC (accessed 8 August 2017).

Raid, R. N., Anderson, D. L. and Ulloa, M. F. (1992), 'Influence of cultivar and amendment of soil with calcium silicate slag on foliar disease development and yield of sugarcane', *Crop Prot.*, 11, 84–8.

Ramgareeb, S., Snyman, S. J., van Antwerpen, T. and Rutherford, R. S. (2010), 'Elimination of virus and rapid propagation of disease-free sugarcane (*Saccharum* spp. cultivar NCo376) using apical meristem culture', *Plant Cell Tiss. Organ Cult.*, 100, 175–81.

Ramouthar, P. V., Caldwell, P. M. and McFarlane, S. A. (2016), 'Effect of silicon on the severity of brown rust of sugarcane in South Africa', *Eur. J. Plant Pathol.*, 145, 53–60.

Rodriguez, E., Lao, M., Gago, S., Espino, A., Acevedo, R. and Muñiz, Y. (2001), 'Razas patogénicas del carbón de la caña de azúcar en Cuba', *Cuba Azúcar*, 31, 34–8.

Rott, P., Bailey, R. A., Comstock, J. C., Croft, B. J. and Saumtally, A. S. (2000), *A Guide to Sugarcane Diseases*, CIRAD/ISSCT, Montpellier, France.

Ryan, C. C. and Egan, B. T. (1989), 'Rust', *in* Ricaud, C., Egan, B. T., Gillaspie Jr., A. G. and Hughes, C. G. (Eds), *Diseases of Sugarcane Major Diseases*, Elsevier, Amsterdam, The Netherlands, pp. 189–210.

Schenck, S., Hu, J. S. and Lockhart, B. E. (1997), 'Use of a tissue blot immunoassay to determine the distribution of sugarcane yellow leaf virus in Hawaii', *Sugar Cane*, 4, 5–9.

Singh, V., Srivastava, S. N., Lal, R. J., Awasthi, S. K. and Joshi, B. B. (2008), 'Biological control of red rot disease of sugarcane through *Trichoderma harzianum and Trichoderma viride, Indian Phytopath.*, 61, 486–91.

Sood, S. G., Comstock, J. C. and Glynn, N. C. (2009), 'Leaf whorl inoculation for screening sugarcane rust resistance', *Plant Dis.*, 93, 1335–40.

Souza, R. S. C., Okura, V. K., Armanhi, J. S. L, Jorrin, B., Lozano, N., Silva, M. J., Gonzales-Guerrero, M., Araujo, L. M., Verza, N. C., Bagheri, H. C., Imperial, J. and Arruda, P. (2016), 'Unlocking the bacterial and fungal communities assemblages of sugarcane microbiome', *Sci. Rep.* 6, 28774.

Steindl, D. R. L. (1972), 'The elimination of leaf scald from infected planting material, *Proc. Int. Soc. Sugar Cane Technol.*, 14, 925–9.

Su, Y., Wang, Z., Xu, L., Peng, Q., Liu, F., Li, Z. and Que, Y. (2016), 'Early selection for smut resistance in sugarcane using pathogen proliferation and changes in physiological and biochemical indices', *Front. Plant Sci.*, 7, 1133.

Talukeder, M. I., Begum, F. and Azad, M. K. (2007), 'Management of pineapple disease of sugarcane through biological means', *Agri. Rural Dev.*, 5, 79–83.

Viswanathan, R. (1999), 'Induction of systemic resistance against red rot disease of sugarcane by plant growth promoting rhizobacteria', PhD thesis, Tamil Nadu Agriculture University, Coimbatore, India.

Viswanathan, R. (2012), '*Handbook on Sugarcane Diseases and their Management*', Sugarcane Breeding Institute, Coimbatore, India.

Viswanathan, R. and Samiyappan, R. (1999), 'Induction of systemic resistance by plant growth promoting rhizobacteria against red rot disease in sugarcane', *Sugar Tech.*, 1, 67–76.

Viswanathan, R., Chinnaraja, C., Malathi, P., Gomathi, R., Rakkiyappan, P., Neelamathi, D. and Ravichandran, V. (2014), 'Impact of Sugarcane yellow leaf virus (SCYLV) infection on physiological efficiency and growth parameters of sugarcane under tropical climatic conditions in India', *Acta. Physiol. Plant*, 26, 1805–22.

Wagih, M. E., Gordon, G. H., Ryan, C. C. and Adkins, S. W. (1995), 'Development of an axillary bud culture technique for Fiji disease virus elimination in sugar cane. *Aust. J. Bot.*, 43, 135–43.

Wongkaew, P. and Fletcher, J. (2004), 'Sugarcane white leaf phytoplasma in tissue culture: Long-term maintenance, transmission, and oxytetracycline remission', *Plant Cell Rep.*, 23, 426–34.

Zhang, L. and Birch, R. G. (1996), 'Biocontrol of sugar cane leaf scald disease by an isolate of *Pantonea dispersa* which detoxify albicidin phytotoxin', *Lett. Appl. Microbial.*, 22, 132–6.

Zhu, Y. J., McCafferty, H., Osterman, G., Lim, S., Agbayani, R., Lehrer, A., Schenck, S. and Komor, E. (2011), 'Genetic transformation with untranslatable coat protein gene of sugarcane yellow leaf virus reduces virus titers in sugarcane', *Transgenic Res.*, 20, 503–12.

Chapter 17

Integrated weed management in sugarcane cultivation

Calvin Odero, University of Florida, USA

1 Introduction

Weeds are not benign in sugarcane cropping systems and become important biotic constraints on production. They reduce sugarcane yield by competing for water, nutrients and light, thereby potentially causing considerable economic loss. Although sugarcane is neither a short stature nor a short-season row crop, its wide row spacing,[1,2] and slow emergence and growth rate early in the season allow for an upsurge in weed pressure,[3] making the crop vulnerable to weed competition before canopy closure shades and suppresses weed growth.[4–8] Competitiveness of weeds with sugarcane is attributed to their ability to grow and develop rapidly, shade emerging and developing shoots and reduce the formation of tillers.[6] Early in the season in resource-rich sugarcane production systems with ideal temperatures, high precipitation, full or supplementary irrigation[3,9,10] and high inorganic fertilizer inputs,[11–16] weed infestation and competition is accentuated by rapid weed growth as compared to sugarcane which utilizes much of its carbohydrate reserves for the production of shoot-roots which are used for the absorption of nutrients and water.[3] In these resource-rich environments, weed competition is also likely for light because more individuals develop leaves that reduce light available for the crop.[17,18]

http://dx.doi.org/10.19103/AS.2017.0035.28

Competition from numerous weed species with sugarcane can result in significant yield losses. Total crop failure can occur when no adequate and timely control measures are taken. Perennial grass species are the most competitive and difficult to selectively control in sugarcane because of similarities between the species. Sugarcane yield losses of up to 91% caused by competition from individual weed species or mixed weed populations under field conditions have been reported in different sugarcane production regions.[4,19–32] The degree of sugarcane yield loss from weed competition depends on the weed species, weed density, weed distribution, duration of weed–sugarcane competition, sugarcane cultivar, climatic condition, agronomic practices and the life history stage of the weed–sugarcane system at which the interaction occurs.[6,19,25,33–39]

Weeds infesting sugarcane also serve as alternative hosts for disease (Table 1) and insect pests (Table 2), increase production cost and lower land value. Combination of weeds and other pests can also favour occurrence of other pests. Showler et al.[40] showed that weed and disease stress factors favoured higher sugarcane nematode populations. Weeds can also create physical hindrance of sugarcane growth and interfere with harvesting operations.[27,41,42] For example, vines of *Ipomea* species climb and wrap sugarcane plants and increase the need for frequent removal from mechanical harvesters, thus interfering with harvesting.[26] Weedy fields can also harbour rats that feed on sugarcane stalks by providing them with seeds that become an important source of protein required for their reproduction.[43–45] The diet of the canefield rat (*Rattus sordidus*) during breeding in Australia predominantly consists of summer grasses in sugarcane fields followed by a switch to a diet of predominantly sugarcane at the end of the breeding season.[46] Furthermore, weeds that mature and produce seed become sources of seed bank replenishment and reinfestation in subsequent years.

The critical period of weed control (CPWC) is important for weed management in sugarcane. It represents two separately measured components of weed-crop competition, namely the critical timing of weed control (CTWC) and the critical weed-free period (CWFP).[39,47] The CTWC is the maximum amount of early-season weed competition that a crop can tolerate before it suffers irrecoverable yield reduction, and the CWFP is the minimum weed-free period from the time of planting to prevent unacceptable yield reductions.[39,47] In sugarcane, the CPWC ranges from 3 to 12 weeks after emergence depending on weed species composition and abundance, cultivar, agronomic practices and prevailing environmental conditions.[4,6,24,38,48,49] The CPWC coincides with the slow early-season sugarcane growth when roots are in transition from sett-roots to shoot-roots, and when tillering occurs.[41] Profuse tillering is important for good sugarcane establishment and is directly related to the final number of stalks at harvest. Weeds exert the most competitive influence during this period of slower sugarcane establishment, making weed control timing very important.[41]

The CPWC is important in sugarcane cultivation for determination of appropriate timing of weed control so that integrated weed management (IWM) strategies can be developed to maximize control measures. Implementation of IWM in sugarcane is essential to reduce the negative impact of weeds to economically acceptable levels. This chapter deals with an integrated approach involving different methods for effective weed management in sugarcane that can be adopted for sustainable production.

Table 1 Examples of sugarcane diseases harboured by weed species*

Sugarcane disease	Pathogen	Weed host	Reference
White leaf	White leaf phytoplasma	*Dactyloctenium aegyptium*	127–130
		Cynodon dactylon	127–131
		Brachiaria distachya	127–130
Leaf yellows	Sugarcane leaf yellows phytoplasma	*Cynodon dactylon*	132
		Conyza canadensis	132
		Sorghum halepense	132
		Macroptilium lathyroides	132
Yellow leaf	*Sugarcane yellow leaf virus*	*Sorghum almum*	133
Root rot	*Pythium arrhenomanes*	*Cynodon dactylon*	134
		Brachiaria platyphylla	134
		Echinochloa crus-galli	134
		Eleusine indica	134
		Rottboellia cochinchinensis	134
		Sorghum halepense	134
		Lolium multiflorum	134
		Cyperus rotundus	134
Ratoon stunting	*Leifsonia xyli* subsp. *xyli*	*Sorghum halepense*	131
		Pennisetum purpureum	131
Eye spot	*Bipolaris sacchari*	*Pennisetum purpureum*	131
		Pennisetum typhoides	131
Banded scletorial disease	*Thanatephorus sasakii, T. cucumeris*	*Sorghum halepense*	131
		Cynodon dactylon	131
		Cyperus rotundus	131
Mosaic	*Sugarcane mosaic virus, Sorghum mosaic virus*	*Commelina benghalensis*	135
		Sorghum arundinaceum	135
		Arundo donax	135

*The above citations are examples of sugarcane diseases harboured by weeds; however, more information can be found in the literature. For further information refer to Ricaud et al.[136] and Rott et al.[137] on epidemiology of sugarcane diseases.

Table 2 Sugarcane insects harboured by weed species in the United States

Sugarcane insect	Weed host	Reference
Diaprepes abbreviatus (L.)	*Sesbania herbacea*	138
	Senna obtusifolia	138
	Cajanus cajan	138
	Schinus terebinthifolius	138,139
	Chenopodium album	138
	Amaranthus spinosus	138
	Portulaca oleracea	138
Diatraea saccharalis (F.)	*Paspalum* spp.	138,141
	Sorghum halepense	138,141–145
Eoreuma loftini (Dyar)	*Sorghum halepense*	143–145
	Echinochloa crus-galli	143

2 Integrated weed management: overview and cultural control

2.1 Overview of integrated weed management

Weed management in leading sugarcane producing countries primarily depends on chemical control practices. However, cost-effective, sustainable and environmentally friendly weed management requires integration of multiple approaches to keep weeds at a competitive disadvantage and below economic threshold levels. Such integrated approaches must diversify selection pressure on weed populations, use resources more efficiently and provide growers with different weed management options.[50] The use of numerous alternative weed control measures including cultural, mechanical, biological and chemical options while maintaining economical yields and reducing reliance on a single control option constitute IWM.[51] Significant weed control is achieved when various components of IWM are implemented in a systematic manner so that the competitive ability of the crop is enhanced while still providing adequate weed control.[51] Biological weed control options as part of IWM are not dealt with in this chapter because there is not much literature on use of biological weed control agents in sugarcane and other cropping systems.

IWM puts emphasis on combining control methods with knowledge of weed biology, agroecology and the influence of agricultural practices that consider causes of weed infestation rather than reaction to infestations after they occur.[50,52] Effective IWM programmes focus on identifying the most vulnerable stages of weed development in order to strategically target weeds so that weed survival and their damaging effects can be diminished. Furthermore, IWM reduces selection pressure imposed on weed species primarily by one control method and ensures that weeds not controlled by one method are controlled by other methods.

Therefore, the goal of effective IWM programmes should be to create cropping systems that are unfavourable for weeds and able to minimize the effect of weeds that survive.[50] Because no single weed control technique is effective in sugarcane cultivation, an IWM programme that utilizes a combination of cultural, mechanical and chemical methods is important for successful weed management. Such programmes should be effective in preventing and suppressing growth, seed production and the spread of weeds in addition to reducing seed bank reserves.[52] For ease of adoption, all components of IWM programmes in sugarcane must be practical, sustainable, cost-effective and environmentally safe.[52,53]

2.2 Cultural weed control

Cultural weed control practices are important components of IWM in sugarcane. Cultural practices reduce the impact of weeds by manipulating the crop-weed relationship in favour of the crop at the expense of the weed. The goal of cultural weed control is to reduce weed emergence, survival and competition through crop rotation, clean tillage, field sanitation and enhance the competitive ability of the crop through appropriate choice of crop genotype, planting date, row spacing and fertilization strategy.[54]

Land preparation to kill emerging weeds and field sanitation are important first steps to limit weed competition with sugarcane. Sanitation can prevent weeds from contaminating and spreading across sugarcane fields. Practices such as using cleaning equipment between fields, using seed cane from weed-free fields and controlling volunteer weeds along field edges, fence lines, canals and ditchbanks can further reduce the spread and establishment of weeds in sugarcane fields. This is because uncontrolled weeds along field margins become seed reservoirs for future infestation. Tillage, spraying, fertilizer distribution and harvest equipment should also be free from weeds to minimize the spread of weeds from field to field.

Crop rotation is important in breaking life cycles of weed species adapted to a particular cropping system by utilizing different weed control programmes and agronomic practices. Therefore, crop rotation changes cultural conditions that a weed must tolerate.[55] The multi-year crop cycle of sugarcane enables rhizomatous perennial grasses, such as *Cynodon dactylon*, *Sorghum halepense* and *Panicum repens*, and annual grasses such as *Rottboellia cochinchinensis* and *Panicum dichotomiflorum* to become well established and difficult to control. Sugarcane growers have historically relied heavily on crop rotation during the fallow renovation period to manage problematic grasses. The bare fallow period provides the most options or flexibility to reduce grass seeds and vegetative propagules that have established over the long crop cycle before establishing the next crop by using tillage, non-selective herbicides and graminicides.[56,57] For example, control of *C. dactylon* and *S. halepense* in fallowed Louisiana sugarcane fields is increased by repeated glyphosate applications and tillage operations throughout the fallow period, illustrating its importance in management of problematic perennial grasses.[56]

Legumes including soybean (*Glycine max* (L.) Merr.), peanut (*Arachis hypogea* L.) and green beans (*Phaseolus vulgaris* L.) are widely used as cover or cash crops for renovation of fallow sugarcane fields for six to nine months.[58–63] These leguminous fallow crops are mainly planted to enhance yields of the following sugarcane crop by providing nitrogen input and improving soil aggregation.[63] Planting legume cover crops has been shown to reduce weed problems in the following plant cane crop.[64] The success of these crops in limiting weed infestation in the subsequent sugarcane crop depends on the ability to control problematic weeds, particularly grasses in the rotational crop. Graminicides (fluazifop, quizalofop, clethodim and sethoxydim) used in these leguminous crops for selective control of several annual and perennial grasses[65,66] provide the opportunity for efficacious control of problematic grasses in fallowed fields before planting the next sugarcane crop. The availability of glyphosate-tolerant soybean as a rotational crop choice also allows use of glyphosate to manage perennial grasses during the fallow period.[60] For problem grasses such as *Panicum maximum* that can grow through a legume cover crop, a pre-emergence application of trifluralin at planting of the crop can be used for its control.[64]

Radish (*Raphanus sativus L.*) and leafy vegetables such as lettuce (*Lactuca sativa* L.) are also grown in rotation with sugarcane.[67,68] Grass control in these crops is achieved using graminicides which aid in managing problematic grasses during the sugarcane fallow period. Herbicides not registered for sugarcane are used for management of problematic broadleaf weeds in leafy vegetables. For example, Dusky and Stall[67] reported control of *Chenopodium album*, *Amaranthus* species and *Portulaca oleracea* with imazethapyr in lettuce grown in rotation with sugarcane. Although cultivation of rotational crops during the sugarcane fallow allows for management of problematic weeds in the crop, rotation requires growers to have knowledge of agronomic practices of the rotational crop and access to additional equipment suited for these crops.

Row spacing, planting density and sugarcane cultivar influence weed-sugarcane competition dynamics by affecting canopy closure and solar radiation interception early during the growing season. Sugarcane is conventionally planted in 0.6 to 1.8 m row spacing[69–73] depending on the level of mechanization used for field operations. Wider sugarcane row spacing can result in significant and gradual increase in weed infestation compared to narrow spacing.[74] Narrow row spaced sugarcane with high plant density can potentially reduce the level of weed infestation during the period between planting and canopy closure. This is attributed to the ability of high plant densities to generate more ground cover, thereby increasing light interception and access to water and nutrients by the crop[75] while reducing light reaching the soil surface for weed emergence and growth. Bull and Bull[76] reported quicker canopy closure when sugarcane row spacing was reduced from 1.5 to 0.5 m. However, altered or new equipment is required to suit field operations for sugarcane planted in narrow spacing with high plant density. Similarly, planting sugarcane cultivars that close canopy fast enhances their competitiveness by promoting efficient use of light, water and nutrients. For example, Inman-Bamber[77] recommended planting 'NCo376' in South Africa for more rapid canopy closure in the summer than 'N12' to enhance solar radiation interception by the crop, thereby limiting light access to weeds.

Sugarcane and weeds generally use and compete for similar nutrients, making fertilizer management an integral component of IWM. Therefore, it is important to develop fertilization strategies that enhance the competitive ability of the crop and reduce interference from weeds.[78] Strategies such as deep band fertilizer application to the crop row rather than broadcasting can favourably place the crop at a competitive advantage over weeds in accessing applied nutrients.[55,78] In sugarcane, fertilizers are typically applied at the base of the planting furrow in close proximity to where the crop absorbs them[79–81] instead of weeds. Proper placement of fertilizers can promote sugarcane growth and supplement other weed control measures by limiting accessibility of applied nutrients to weeds. Although cultural practices can enhance the competitive ability of sugarcane against weeds, other control approaches must also be used in conjunction with these measures to provide season-long weed control.

3 Integrated weed management: mechanical weed control

Mechanical weed control has been a backbone of weed management programmes in sugarcane. It consists of methods that physically destroy already emerged weeds or place barriers that make the environment less favourable for emergence of weed seedlings or survival of existing weeds. Mechanical control methods range in complexity from simple hand weeding to complex multicomponent tillage machines. Other mechanical weed control methods used in sugarcane to create physical barriers that block and impede weed growth include trash mulching and flooding.

Hand weeding is widely used for weed control by many smallholder sugarcane growers in several developing countries in the Asia Pacific, Central America, South America and Africa.[3] It involves hand pulling and hand hoeing for selective weed control, making it very environmentally safe.[18,55] Hand weeding is labour intensive, time consuming and mostly effective for controlling annual and biennial weeds, and not as effective for perennial weed control. Smallholder sugarcane growers normally use hoes and hand pulling for inter- and intra-row weeding followed by collection of removed weeds in heaps along the inter-row to dry out in the sun.[82] Hand weeding around the perimeter of in-field patches (approximately 10–20 m^2 in size) effectively contains and prevents spread of stolons and rhizomes of perennial grass species such as *C. dactylon* and *Digitaria abyssinica* into sugarcane fields. These patches are subsequently treated with herbicides once they green up. During hand weeding, special attention must be paid to the intra-rows to ensure that they are weed-free. More persistent weeds such as *P. maximum* are typically chipped out and removed from the field or placed in field edges to dry out.[82] In highly mechanized sugarcane cropping systems that rely on chemical weed control, hand weeding can be used to destroy weeds that escape chemical control especially in situations where weeds are evolving herbicide resistance.[33,55]

Tillage is the most common method of mechanical weed control in sugarcane to reduce competition and prevent weeds from flowering and setting seed. In sugarcane, tillage is also used to destroy residues from the previous crop, incorporate fertilizer and aid in soil aeration and water infiltration.[83] Tillage involves preplant disking or ploughing for land preparation to destroy weeds present in the field prior to sugarcane planting. Both ploughing and disking bury weed seeds and thereby inhibit or delay their germination.[3,82] Disking can also chop vegetative propagules of perennial weeds, bring them to the surface and subject them to desiccation in the sun. Weed control with these operations can only occur under dry and not wet conditions where chopped pieces of vegetative propagules can easily germinate.[82] However, ploughing and disking may also increase soil aeration and stimulate weed emergence, whereas buried dormant seeds can become sources of infestation in the future.[55,82] Disking can also separate chains of tubers of *Cyperus rotundus*, which can allow easier control. Caution must be taken because disking also spreads tubers throughout the field, so it should only be considered in fields with very dense coverage of *C. rotundus*.

Tillage should always begin as early as possible and continue as along as sugarcane has a height differential advantage over weeds.[84] This is because tillage will only be effective when sugarcane plants are growing taller than competing weeds.[84] Tillage is not recommended when weeds are not present in the field as it can encourage germination of additional weed seeds. Tillage can change patch biomass and spatial structure of *C. dactylon*, spreading propagules throughout the field[85] and can bury non-treated rhizomes and stolons, thereby increasing propagules in the soil. However, tillage can also be beneficial in rejuvenating moribund *C. dactylon*, and regenerated plants can easily be controlled with systemic herbicides because of more leaf area for herbicide absorption. Caution must also be taken before a decision is made for tillage as the operation can remove the layer of herbicides present when soil-applied herbicides have been used.[84] Tillage equipment used in sugarcane includes ox-drawn ploughs or rakes, disk harrows, tine rippers, grubbers and cultivars. In Australia, cultivators with adjustable spring-toothed rakes that follow contours of the soil in a floating manner are used for weed control because of their limited damage to sugarcane roots.[3]

In conservation tillage systems (i.e. reduced and no-tillage programmes) where weed control is efficiently achieved with other control measures, conventional tillage may not be necessary. Reduced and no-tillage programmes in combination with one or two timely applications of glyphosate have been shown to be effective in controlling perennial weeds during sugarcane fallow.[57] These conservation tillage programmes can be viable options for sugarcane growers for plant and ratoon cane fields not rutted in previous harvest and when efficient weed control with herbicides has been achieved.[83]

Green harvesting of sugarcane results in a trash blanket on the surface that can suppress weed emergence and growth depending on the amount of crop residue.[86–88] The trash blanket covering the soil surface intercepts solar radiation needed for weed seed germination and acts as a physical barrier to weed growth and development. In Brazil, sugarcane trash has been shown to provide control of small-seeded broadleaf weeds (*Sida rhombifolia* and *Bidens pilosa*) and not large-seeded broadleaf weeds (*Euphorbia heterophylla* and *Ipomoea grandiflora*).[89] Similarly, acceptable control of grasses such as *Brachiaria decumbens*, *Brachiaria plantaginea*, *P. maximum* and *Digitaria horizontalis* by sugarcane trash has been reported in Brazil.[90,91] However, microclimatic factors (air and soil temperatures) that control sugarcane emergence and growth may be significantly affected by trash blanket on the surface in places such as Florida where sugarcane harvesting coincides with cooler winter months.[92,93]

The efficacy of mechanical weed control in sugarcane can only be achieved when it is integrated with other control measures including chemical and cultural control. Its effectiveness is also influenced by weed species composition, abundance and emergence patterns in addition to timing of control practices.

4 Chemical weed control as part of integrated weed management: overview and preplant and pre-emergence herbicides

4.1 Overview

Chemical weed control in sugarcane probably began in Hawaii in 1913 with the use of sodium arsenite, which lasted until the mid-1940s.[94] This was followed in 1945 by development of phenoxy acid herbicide 2,4 dichlorophenoxy acetic acid (2,4-D) for post-emergence weed control in many crops including sugarcane.[33] Since then, 2,4-D and other herbicides applied alone or in mixtures have become essential components of IWM programmes in sugarcane in many countries. Herbicide mixtures are used because they provide broader weed control spectrum, more consistent control, use lower rates and reduce carryover problems and the risks of development of herbicide-resistant weeds. In the United States, herbicide mixing is approved as long as the mixtures are not incompatible or prohibited by use restrictions on the product label.[95]

Herbicides are usually more efficient than other control methods when used appropriately, and are more energy and labour efficient compared to hand labour and cultivation. Furthermore, herbicides can selectively control weeds in sugarcane rows where cultivation is not possible. With increasing labour cost and scarcity of labour, herbicide use has become more widespread in sugarcane cultivation in many countries. Caution must be taken as herbicide use increases in sugarcane cultivation to avoid food and environmental contamination, and development of herbicide-resistant weeds associated with repeated use of herbicides with the same mode of action.

In sugarcane, herbicides are applied preplant, pre-emergence, and post-emergence for weed control. Accurate herbicide application timing and proper calibration of application equipment are extremely important to maximize weed control and herbicide selectivity. Herbicides are also expensive and must be used appropriately to maximize weed control. Therefore, knowledge of appropriate herbicide application timing, selectivity and weed biology and ecology is essential for herbicide use. The overall goal of any herbicide programme in combination with other weed control practices in sugarcane must be to give the crop a competitive advantage over weeds from early in the season before canopy closure when the crop is most vulnerable to weed competition. Therefore, selection of an efficacious herbicide programme must be based on the weed abundance and spectrum, sugarcane growth stage, herbicide cost, agronomic practices, rotational crop choice, edaphic and prevailing environmental conditions.

4.2 Preplant and pre-emergence herbicides

Preplant herbicides are applied before the planting of sugarcane as burndown treatment to prevent establishment of weeds not controlled during the fallow period or prior to sugarcane planting in successively planted cane. These herbicides are broadcast applied to the entire field. In sugarcane, foliar-based contact or systemic herbicides with no soil residual activity such as glyphosate, paraquat and 2,4-D are commonly used for control of emerged annual and perennial weeds in and around sugarcane fields before planting.[84,96] These herbicides must be applied to actively growing weeds and any need for additional surfactants to improve absorption must be based on the formulation and labelling requirements.

Several pre-emergence herbicides applied after planting and before crop emergence or as soon as possible after harvesting usually when moisture is available are used for weed control in sugarcane. The herbicides can either be applied in a band over the row particularly in ratoon sugarcane or broadcasted over the entire field. Band applications are typically supplemented with at least one inter-row cultivation to control weeds between the rows. The choice of application method depends on weed spectrum and abundance; field condition and herbicide, application, and cultivation costs. Pre-emergence herbicides are applied before sugarcane emergence or soil-applied before emergence of weeds. Foliar-applied herbicides include glyphosate and paraquat, and must be applied prior to sugarcane emergence or with shielded applications. These foliar-applied herbicides have no residual activity and only control weeds that emerge before sugarcane emergence.

Soil-applied herbicides provide residual control of several broadleaf, annual and seedling perennial grasses such as *S. halepense* in sugarcane. Rainfall or irrigation is usually required to incorporate soil-applied herbicides to make them available for uptake by germinating weed seeds. Some herbicides (e.g. amicarbazone, hexazinone, and clomazone) are less water dependent and more tolerant to drier conditions. Mechanical incorporation in the top 5–7.6 cm of soil can be used where herbicide movement into the weed germination zone in the soil is not feasible by rainfall or irrigation and to incorporate volatile dinitroaniline herbicides such as trifluralin. For example, trifluralin mechanically incorporated immediately after treatment using rolling cultivators provides *R. cochinchinensis* control in Louisiana.[97] However, deeper incorporation of herbicides into the soil can reduce their efficacy because of the diluting effect of soil and care must be taken to minimize sugarcane injury associated with shaving during mechanical incorporation. Incorporation is important because herbicides that remain on the soil surface become subject to several degradation processes, thereby reducing their efficacy and residual activity.

Triazine herbicides are the most widely used for pre-emergence weed control in sugarcane. These herbicides include atrazine, simazine, hexazinone and metribuzin.[49,66,84,96,98,99] Efficacious residual weed control by these herbicides depends on their persistence in soil.[99] Atrazine, an *s*-triazine, is the most widely used herbicide for control of broadleaf weeds and some grasses because of its affordability, consistent performance, residual activity, flexibility in time and method of application, compatibility with other herbicides and crop safety.[49,66,84,96,100] Tank-mixes of atrazine with other herbicides provide broad-spectrum pre-emergence weed control. For example, atrazine in combination with pendimethalin and prodiamine provides excellent control of seedling *S. halepense.*[101] The half-life of atrazine under field conditions has been reported to be 60 days.[66] However, enhanced atrazine degradation has been reported in sugarcane fields with previous atrazine use history, resulting in shorter residual activity.[100] This is attributed to adaptation of bacterial communities able to use atrazine as carbon and nitrogen sources in soils with repeated atrazine use, thereby enabling rapid degradation.[102] Atrazine can be applied in combination with acetochlor, *S*-metolachlor, pendimethalin, or alachlor to enhance grass control and provide longer residual activity.[84,96,98] Paraquat or glyphosate can be added to atrazine for burndown of small emerged weeds before sugarcane emergence especially when application is delayed in plant cane.[96,98]

Metribuzin, an *as*-triazine, is used for control of annual broadleaf weeds and certain grasses in sugarcane.[66,84,96,98] It can be applied in combination with sulfentrazone, clomazone, hexazinone, terbacil or pendimethalin to enhance management of problematic grasses such as *C. dactylon* and *S. halepense.*[965,103] Programmes that include metribuzin for control of seedling *S. halepense* can also minimize *C. dactylon* interference in sugarcane.[103] Combination of metribuzin with pendimethalin is also generally more effective in controlling *S. halepense* than mixtures with atrazine.[101] Because metribuzin is not cross-adapted with atrazine-adapted sugarcane production soils,[104] it is a better option for weed control in sugarcane grown on soils exhibiting enhanced atrazine degradation.[100] Cross-adaptation is a phenomenon where microorganisms are adapted to degrading structurally similar herbicides.[102] Majumdar and Singh[105] reported that sorption of metribuzin increases with increasing soil organic matter, implying that metribuzin should not be used in mineral soils with low organic matter because of leaching concerns.[95,98]

Hexazinone can be used alone or as a premix with diuron for broad-spectrum residual weed control in sugarcane.[96,98,99,106] For example, hexazinone plus diuron provided more than 97% control of *Digitaria sanguinalis*, *Colopogonio* spp., *Paspalum distichum* and *Wedelia paludosa* 100 days after treatment.[106] The efficacy of hexazinone is enhanced when rainfall or irrigation moves the herbicide into the root zone within 3–4 days after application.[98,99] Hexazinone must be applied before sugarcane emergence only on clay or high organic matter soils with reduced leaching potential because of phytotoxicity on sugarcane.[107,108] The other triazine, simazine, was widely used for selective weed control in sugarcane following its introduction in 1960, but today it is only used in a few countries where it is still registered for use in sugarcane.[49,109]

Dinitroaniline herbicides pendimethalin and trifluralin are applied pre-emergence for control of many annual grasses in sugarcane.[96–99,101,110,111] These herbicides require incorporation by cultivation, rainfall or irrigation to minimize loss by vaporization and photodecomposition.[112] Odero and Shaner[113] reported rapid dissipation of pendimethalin in sugarcane fields irrespective of formulation when applied under dry soil conditions with no incorporation into the soil. Millhollon[101] reported that pendimethalin and trifluralin provided excellent control of *R. cochinchinensis* (97–99% control) in sugarcane when they were incorporated.[106] These dinitroaniline herbicides are applied in combination with triazines and terbacil to enhance and broaden the spectrum of weed control in sugarcane.[49,101] However, mixtures of pendimethalin with metribuzin or terbacil generally provide more effective control of grasses such as *S. halepense* than mixtures with atrazine.[101]

Chloroacetamide herbicides alachlor, acetochlor, *S*-metolachlor and metazachlor are used for control of grasses and certain broadleaf weeds in sugarcane.[98] *S*-metolachlor applied pre-emergence has been shown to be effective in controlling *B. decumbens*, *D. horizontalis* and *P. maximum*.[114] The chloroacetamides are applied in combination with ametryn, atrazine or diuron to broaden the weed control spectrum.[98] For instance, combination of acetochlor with diuron provides excellent grass (*D. sanguinalis* and *Panicum schinzii*) and broadleaf (*Amaranthus* spp. and *Bidens pilosa*) control.[115] A premix of *S*-metolachlor, mesotrione and terbuthylazine is also used for broad-spectrum weed control.[98] However, rainfall within 7–10 days after application is required to move the premix into the weed germinating zone for better efficacy.

Other pre-emergence herbicides used in sugarcane include diuron, clomazone, isoxaflutole, amicarbazone, mesotrione, sulfentrazone, terbacil, flumioxazin, azafeniden, EPTC and tebuthiuron.[84,96,98,116] These herbicides can be applied alone or be tank-mixed with other pre-emergence or post-emergence herbicides to broaden the weed control spectrum. Pre-emergence clomazone can provide comparable *R. cochinchinensis* control to that of pendimethalin and trifluralin when incorporated by rainfall.[97] However, transient bleaching of emerging sugarcane foliage can occur following application of clomazone in fine sand soils where it has moderate mobility.[55,66]

A tank-mix of isoxaflutole with diuron or ametryn can control several broadleaf and grass weeds including *P. maximum*.[98] Pre-emergence herbicides in conjunction with mechanical cultivation are important in giving sugarcane the early-season competitive advantage when the crop is most vulnerable to weed competition; however, the use of pre-emergence herbicides in sugarcane will vary depending on agronomic practices, soil type and prevailing environmental conditions.

5 Chemical weed control as part of integrated weed management: post-emergence herbicides

Post-emergence herbicides are applied after sugarcane emergence to control established weeds. These herbicides are often broadcasted over the entire field, banded to narrow strips over the sugarcane rows or as spot treatment depending on the cropping practice and prevalent weed species. Similar to banded pre-emergence herbicides, at least one inter-row cultivation is necessary with banded post-emergence applications to control weeds in the sugarcane row middles. Over-the-top post-emergence herbicides used in sugarcane will selectively kill weeds with insignificant to no phytotoxic effect on the crop.

Sugarcane herbicides applied over-the-top of the crop are either contact or systemic. Contact herbicides only affect parts of the weed contacted by the herbicide with little or no movement inside the plant, whereas systemic herbicides translocate throughout the plant following absorption by foliage or roots of weeds. Contact herbicides are most active on small, actively growing annual and young perennial weeds, while systemic herbicides are effective in controlling both annual and established perennial weeds. Good coverage is absolutely necessary for maximum activity of contact herbicides. Because systemic herbicides translocate in the plant, good coverage is desirable but not as critical as for contact herbicides.

Unlike preplant and pre-emergence herbicide applications, post-emergence applications are not strongly influenced by the soil environment.[55] However, the efficacy of herbicides applied post-emergence to sugarcane, but pre-emergence to weeds may be influenced by soil texture and soil organic matter. For example, the efficacy of herbicides including metribuzin, pendimethalin, sulfentrazone, trifluralin and flumioxazin applied pre-emergence as lay-by treatments between May and June in Louisiana to provide residual weed control[96] are influenced by the soil environment. Lay-by treatments are directed applications commonly made after the last cultivation underneath the sugarcane canopy to minimize exposure of the crop to herbicides. Other directed applications on weeds especially with contact herbicides are generally only effective and safe when sugarcane grows taller than weeds.[84]

Triazine herbicides are widely used for post-emergence weed control in sugarcane. Atrazine is the most widely used herbicide primarily for post-emergence broadleaf weed control in sugarcane.[84,96] In the United States it is used in approximately 89% of the sugarcane hectarage.[43] Atrazine is usually tank-mixed with other herbicides to broaden the weed control spectrum. Best results with atrazine are achieved when applied early post-emergence to broadleaf weeds <5 cm in size.[99] It is also used after lay-by cultivation mainly for vine control. However, inconsistent control of red morning glory (*Ipomea coccinea* L.) with lay-by atrazine and other herbicides including diuron, metribuzin or terbacil in Louisiana attributed to environmental conditions such as rainfall and soil factors has been reported.[116] More consistent control of red morning glory compared to atrazine can be achieved with lay-by sulfentrazone and azafeniden with minimal environmental effects from sulfentrazone.[26,116,117] However, azafeniden is not used in the United States because of sugarcane injury concerns from over-the-top applications.

Ametryn and metribuzin are other triazines used for post-emergence weed control in sugarcane. Both herbicides are tank-mixed with other herbicides to broaden weed control spectrum. Ametryn has excellent activity on small grasses and small-seeded broadleaf weeds and offers acceptable, short-term residual activity.[43] However, phytotoxicity in sugarcane can occur at higher rates of ametryn applied under relatively warm or hot conditions.[43,84,99] Ametryn is commonly mixed with atrazine primarily for grass and broadleaf weed control. Unlike ametryn, metribuzin is relatively nonphytotoxic to sugarcane, but should not be used on mineral soils low in organic matter and clay content because of concerns regarding leaching[84] and reaction with nematicides.[98] Metribuzin tank-mixed with pendimethalin can be applied underneath the sugarcane canopy following lay-by cultivation to control emerged weeds or provide residual control.[118]

Growth regulator herbicides 2,4-D and dicamba are used for broadleaf weed control in sugarcane. These herbicides are tank-mixed with other herbicides to broaden weed control spectrum. 2,4-D is the most widely used herbicide for control of annual and perennial broadleaf weeds. It is also applied aerially to control vines that escape post-emergence lay-by applications. However, 2,4-D applications pose considerable injury risk to nontarget susceptible crops and can also reduce shoot emergence of treated sugarcane used for seed.[84,99,119] Other herbicides used for broadleaf weed control in sugarcane include mesotrione, trifloxysulfuron, sulfentrazone, ioxynil plus bromoxynil, hexazinone, hexazinone plus diuron, carfentrazone, halosulfuron, amicarbazone and isoxaflutole.[84,96,98] These herbicides are either applied over-the-top of sugarcane or directed applied underneath sugarcane canopy to minimize phytotoxicity.

Asulam, a carbamate herbicide, and trifloxysulfuron, a sulfonyl urea, are used for post-emergence control of annual and certain perennial grasses in sugarcane. Inclusion of surfactants with asulam ensures maximum retention and generally enhances absorption rate, thereby ensuring consistent performance.[120,121] Sequential applications of asulam are most effective in controlling problematic grasses such as *S. halepense*.[122,123] Phytotoxicity from asulam application can occur when applied under high temperature and moisture stress, but varies depending on sugarcane variety and stage of growth.[123] Tank-mix of trifloxysulfuron asulam enhances grass control in sugarcane.[124] Combination of asulam and trifloxysulfuron has been shown to slow the spread of rhizome *S. halepense* enough to allow an increased number of ratoon crops.[125] However, combinations of asulam with residual herbicides atrazine, metribuzin, pendimethalin and terbacil do not improve on *S. halepense* control.[122] Other herbicides used for post-emergence grass control in sugarcane include isoxaflutole and MSMA and must be directed underneath the sugarcane canopy to minimize injury to the crop.[98]

Problematic nutsedges in sugarcane, *C. rotundus* and *C. esculentus*, are controlled with sulfonyl urea herbicides, halosulfuron and trifloxysulfuron. Nutsedge control programmes should be first implemented during the sugarcane fallow period using glyphosate to reduce tuber populations that can reinfest the subsequent plant cane.[126] Halosulfuron generally provides better nutsedge control compared to trifloxysulfuron and other herbicides such as sulfentrazone.[126] All the aforementioned post-emergence herbicide weed control programmes in sugarcane must be integrated with other weed control measures, particularly weeds that escape pre-emergence weed control and mechanical control measures.

6 Conclusion

Currently, herbicide-resistant weeds are not a significant problem in sugarcane cultivation in spite of present heavy reliance on herbicides for weed management. This is likely attributed to use of multiple herbicides applied in combination with other weed control measures. Because weeds infesting sugarcane are very diverse, any single weed control strategy will not provide effective season-long control required for economical production. Therefore, an IWM approach that uses different weed control methods (i.e. cultural, mechanical and chemical) is required for sustainable sugarcane production to meet growing demands for sugar for human consumption and industrial uses. The greatest challenge in sugarcane cultivation is management of perennial grasses in both plant and ratoon crops because of a limited number of selective grass herbicides and limited control from other measures such as in-crop tillage.

7 Where to look for further information

The sugarcane industry is switching its focus to reduced tillage systems and is adopting green cane harvesting by moving from the long-standing practice of burning the crop prior to harvest because of environmental concerns and regulations. These production systems will be more reliant on herbicide use, reducing mechanical weed control measures, and will likely result in shifts in weed species composition. Research to mitigate herbicide resistance development with increased herbicide use in these systems and to ensure continued use of IWM programmes even with the reduction of mechanical weed control measures will be required. Because management of perennial grasses in sugarcane is the most challenging, the ability to produce high yielding cultivars with herbicide-resistant traits is an area of research that needs to be pursued to ensure continued profitability of the crop in the future. Organizations currently involved in the aforementioned research areas include Centro de Technologia Canavierira (CTC), South African Sugarcane Research Institute, the University of Florida-IFAS, Louisiana State University AgCenter, United States Department of Agriculture-Agricultural Research Service, Sugar Research Australia, CIRAD (French Agricultural Research Centre for International Development), and the Indian Institute of Sugarcane Research.

8 References

1. Ridge, D. R. and Hurney, A. P. (1994), 'A review of row spacing research in the Australian sugar industry', *Proc. Aust. Soc. Sugar Cane Technol.*, 16, 63–9.
2. Garside, A. L. and Bell, M. J. (2009), 'Row spacing and planting density effects on the growth and yield of sugarcane. 3. Responses with different cultivars', *Crop Pasture Sci.*, 60(6), 555–65.
3. Peng, S. Y. (1984), *The Biology and Control of Weeds in Sugarcane*. Elsevier, Amsterdam.
4. Yirefu, F., Tana, T., Tafesee, A. and Zekarias, Y. (2012), 'Competitive ability of sugarcane (*Saccharum officinarum* L.) cultivars to weed interference in sugarcane plantations of Ethiopia', *Crop Prot.*, 32, 138–43.
5. Miller, J. D., Gilbert, R. A. and Odero, D. C. (2012), *Sugarcane Botany: A Brief View*. IFAS, Florida Coop. Ext. Serv., University of Florida, EDIS SS-AGR-234.
6. Odero, D. C., Duchrow, M. and Havranek, N. (2015), 'Critical timing of fall panicum (*Panicum dichotomiflorum*) removal in sugarcane', *Weed Technol.*, 30(1), 13–20.
7. Suganthi, M., Muthukrishnan, P. and Chinnusamy, C. (2013), 'Influence of early post emergence sulfonylurea herbicides on growth, yield parameters, yield and weed control efficiency in sugarcane', *J. Agron.*,12(1), 59–63.
8. Singles, A. and Donaldson, R. A. (2002), 'A simple model of unstressed sugarcane canopy development', *Proc. S. Afr. Sug. Technol. Ass.*, 74, 151–4.

9. Silva, V. P. R, Silva, B. B., Albuquerque, W. G., Borges, C. J. R, Sousa, I. F. and Neto, J. D. (2013), 'Crop coefficient, water requirements, yield and water use efficiency of sugarcane growth in Brazil', *Agric. Water Manag.*, 128, 102–9.
10. Silva, V. P. R., Sousa, F. S. S., Cavalcanti, E. P., Souza, E. P. and Silva, B. B. (2006), 'Teleconnections between sea-surface temperature anomalies and air temperature in northeast Brazil', *J. Atmos. Sol. Terr. Phys.*, 68, 781–92.
11. Wiedenfeld, B. and Enciso, J. (2008), 'Sugarcane responses to irrigation and nitrogen in semiarid south Texas', *Agron. J.*, 100(3), 665–71.
12. Franco, H. C. J., Otto, R., Faroni, C. E., Vitti, A. C., Almeida, D. O. and Trivelin, P. C. O. (2011), 'Nitrogen in sugarcane derived from fertilizer under Brazilian field conditions', *Field Crops Res.*, 121(1), 29–41.
13. de Sousa, R. T. X., Korndörfer, G. H., Brem Soares, R. A. and Fontoura, P. R. (2015), 'Phosphate fertilizers for sugarcane used at pre-planting (phosphorus fertilizer application)', *J. Plant Nutr.*, 38(9), 1444–55.
14. McCray, J. M., Rice, R. W., Luo, Y. and Ji, S. (2010), 'Sugarcane response to phosphorus fertilizer on everglades histosols', *Agron. J.*, 102(5), 1468–77.
15. Lofton, J. and Tubaña, B. (2015), 'Effect of nitrogen rates and application time on sugarcane yield and quality', *J. Plant Nutr.*, 38(2), 161–76.
16. Cerrizuela, E. A., Arevalo, R. A. and Soldati, A. A. (1974), 'Effects of weed competition on nitrogen fertilization in sugarcane', *Rev. Agron. N. O. Argent.*, 11(3–4), 179–91.
17. Goldberg, D. E. (1990), Components of resource competition. In: *Perspectives on Plant Competition* (Grace, J. B. and Tilman, D. (Eds)), Academic Press, San Diego, CA, pp. 27–49.
18. Booth, B. D., Murphy, S. D. and Swanton, C. J. (2003), *Weed Ecology in Natural and Agricultural Systems*. CAB International.
19. Lall, M. C. (1977), 'Weed management can raise yields in sugarcane', *Ind. Farm.*, 26, 25–37.
20. Kuntohartono, T. and Sasongko, D. (1978), 'The economic importance of weeds and weeding methods in sugarcane nurseries', *Perusahaan Gula* 14, 1–14.
21. Ibrahim, A. A. S. (1984), 'Weed competition and control in sugarcane', *Weed Res.*, 24(4), 227–31.
22. Taye, E. (1991), Survey of weed flora and evaluation of some foliage applied herbicides in the sugarcane plantation of Wonji-Shoa and Metahara. MSc. thesis, Alemaya University of Agriculture, Alemaya, Ethiopia.
23. Millholon, R. W. (1992), 'Effect of itchgrass (Rottboellia cochinchinensis) interference on growth and yield of sugarcane (Saccharum spp. hybrids)', *Weed Sci.*, 40, 48–53.
24. Mani, V. S., Gautam, K. C. and Chakraborty, K. C. (1968), 'Losses in crop yield in India due to weed growth', *Pans(C)*, 14(2), 142–56.
25. Etheredge Jr., L. M., Griffin, J. L. and Boudreaux, J. M. (2010), 'Purple nutsedge (*Cyperus rotundus*) competition with sugarcane and response to shade', *J. Am. Soc. Sugar Cane Tech.*, 30, 89–103.
26. Jones, C. A. and Griffin, J. L. (2009), 'Red morningglory (*Ipomoea coccinea*) control and competition in sugarcane', *J. Am. Soc. Sugar Cane Tech.*, 29, 25–35.
27. Arevalo, R. (1977), 'Competition of specific weeds in sugarcane II. *Sorghum halepense* (L.) Pers', *Rev. Agron. N. O. Argent.*, 14(1–4), 39–51.
28. Richard Jr., E. P. and Dalley C. D. (2007), 'Sugarcane response to bermudagrass interference', *Weed Technol.*, 21(4), 941–6.
29. Richard Jr., E. P. (1997), 'Effects of fallow bermudagrass (*Cynodon dactylon*) control programs on newly planted sugarcane (*Saccharum* spp. hybrids)', *Weed Technol.* 11(4), 677–82.
30. McMahon, G. G., Williams, R. C. and McGuire, P. J. (1989), 'The effects of weed competition on ratoon sugar cane yield', *Proc. Aust. Soc. Sugar Cane Technol.*, 11, 88–92.
31. Millholon, R. W. (1986), 'Control of itchgrass (*Rottboellia cochinchinensis* Lour. Clayton) in sugarcane with postemergence herbicide treatments', *Proc. Int. Soc. Sugar Cane Technol.*, 19, 80–91.
32. Kadam, B. S., Suryavanshi, M. M., Veer, D. M., Patil, S. M. and Khot, R. B. (2011), 'Influence of weed management practices on cane yield and weed intensity of ratoon crop of sugarcane (CO86032)', *Co-op. Sug.*, 42, 41–6.

33. Zimdahl, R. L. (2007), *Fundamentals of Weed Science*. Academic Press, San Diego, CA.
34. Vila, M., Williamson, M. and Lonsdale, M. (2004), 'Competition experiments on alien weeds with crops: Lessons from measuring plant invasion impact?' *Biol. Invasions*, 6(1), 59–69.
35. Alström, S. (1990), Fundamentals of weed management in hot climate peasant agriculture. In: *Crop Production Science*, vol. 11, pp. 271. Department of Crop Production Science, Swedish University of Agricultural Sciences, Uppsala, Sweden.
36. Lamusse, M. J. M. (1965), 'The effects of weed competition on the sugar content and yield of sugar cane', *Trop. Agric.*, 42(1), 31–7.
37. Millholon, R. W. (1992), 'Effect of Itchgrass (*Rottboellia cochinchinensis*) interference on growth and yield of sugarcane (*Saccharum* spp. hybrids)', *Weed Sci.*, 40(1), 48–52.
38. Cerrizuela, E. (1977), 'Interaction of planting rate and weed competition in sugarcane', *Rev. Agron. N. O. Argent.*, 14(1–4), 53–9.
39. Knezevic, S. Z. and Datta, A. (2015), 'The critical period for weed control: Revisiting data analysis', *Weed Sci.*, 63(1), 188–202.
40. Showler, A. T., Reagan, T. E. and Knaus, R. M. (1991), 'Sugar-cane weed community interactions with arthropods and pathogens,', *Int. J. Trop. Insect Sci.*, 11(1), 1–11.
41. McMahon, G. G., Lawrence, P. and O'Grady, T. (2000), Weed control in sugarcane. In: *Manual of Cane Growing* (Hogarth, D. M. and Allsopp, P. (Eds)), Bureau of Sugarcane Experiment Stations. Fergies Printers, Brisbane, Australia, pp. 241–61.
42. Thakar, C. and Singh, H. N. (1954), 'Nilkalamine (*Ipomoea hederacea)*, a menace to sugarcane', *Hort. Abstr.*, 24, 530.
43. Smith, D. T., Richard Jr., E. P. and Santo, L. T. (2002), Weed control in sugarcane and the role of triazine herbicides. In: *The Triazine Herbicides 50 Years Revolutionizing Agriculture* (LeBaron, H., McFarland, J. and Burnside, O. (Eds)), Elsevier, San Diego, CA, pp. 185–98.
44. Smith, H., Staples, L., Dyer, B. and Hunt, W. (2002), 'Incorporation of a zinc phosphide rodenticide into integrated management of rats in sugarcane crops', *Proc. Aust. Soc. Sugar Cane Technol.*, 24, 228–373.
45. Robertson, L. N., Story, P. G. and Wilson, J. (1995), 'Integrated pest management for rodents in sugarcane', *Proc. Aust. Soc. Sugar Cane Technol.*, 17, 70–4.
46. Wilson, J. and Whisson, D. (1993), *The Management of Rodents in North Queensland Canefields*. CBPM SLS Queensland University of Technology, Brisbane, Australia.
47. Knezevic, S. Z., Evans, S. P., Van Acker, R. C. and Lindquist, J. L. (2002), 'Critical period for weed control: The concept and data analysis', *Weed Sci.*, 50, 773–86.
48. Lamusse, M. J. M. (1965), 'The effects of weed competition on the sugar content and yield of sugar cane', *Trop. Agric.*, 42(1), 31–7.
49. Srivastava, T. K., Singh, A. K. and Srivastava, S. N. (2003), 'Critical period of weed competition in sugarcane ratoon,' *Indian J. Weed Sci.*, 34(3–4), 320–1.
50. Buhler, D. D. (2002), '50th anniversary – invited: Challenges and opportunities for integrated weed management', *Weed Sci.*, 50(3), 273–80.
51. Swanton, C. J. and Weise, S. F. (1991), 'Integrated weed management: The rationale and approach', *Weed Technol.*, 5(3), 657–63.
52. Sanyal, D., Bhowmik, P. C., Anderson, R. L. and Shrestha, A. (2008), 'Revisiting the perspective and progress of integrated weed management', *Weed Sci.*, 56(1), 161–7.
53. Bhowmik, P. C. and Inderjit (2003), 'Challenges and opportunities in implementing allelopathy for natural weed management', *Crop Prot.*, 22(4), 661–71.
54. Bàrberi, P. (2002), 'Weed management in organic agriculture: Are we addressing the right issues?' *Weed Res.*, 42(3), 177–93.
55. Monaco, T. J., Weller, S. C. and Ashton, F. M. (2002), *Weed Science: Principles and Practices*, John Wiley & Sons, Inc, New York.
56. Griffin, J. L., Viator, B. J., Clay, P. A., Ellis, J. M, Miller, D. K., Bruff, S. A., Lanie, A. J. and Lencse, R. J. (2001), 'Weed control: Essential to sugarcane production', *Louisiana Agriculture*, 44(4), 12–14.

57. Etheredge Jr., L. M., Griffin, J. L. and Salassi, M. E. (2010), 'Economics of summer fallow conventional and reduced-tillage programs for sugarcane', *Weed Technol.*, 23(2), 274–9.
58. Garside, A. L. and Bell, M. J. (2001), 'Fallow legumes in the Australian sugar industry: Review of recent research findings and implications for the sugarcane cropping system', *Proc. Aust. Soc. Sugar Cane Technol.*, 23, 230–5.
59. Garside, A. L., Watters, T. S., Berthelsen, J. E., Sing, N. J., Robotham, B. G. and Bell, M. J. (2004), 'Comparisons between conventional and alternative sugarcane farming systems which incorporate permanent beds, minimum tillage, controlled traffic and legume fallows', *Proc. Aust. Soc. Sugar Cane Technol.*, 26, 1–12.
60. Viator, H. P. and Griffin, J. L. (2001), 'Fallow period cropping to soybeans can provide benefits', *Louisiana Agric.*, 44(4), 15.
61. Borrero, M. A. V., Pereira, J. T. V. and Miranda, E. E. (2003) 'An environmental management method for sugar cane alcohol production in Brazil', *Biomass Bioenerg.*, 25(3), 287–99.
62. Mkhize, N. D. (2013), *Investigating Crop Rotational Benefits of a Soybean and Sugarcane Cropping System in South Africa*. Master's Thesis, University of Kwa-Zulu Natal, Pietermaritzburg, South Africa.
63. Whan, B. M., Scott, C. H. and Jefferson, T. R. (1976), 'Scheduling sugar cane plant and ratoon crops and a fallow – a constrained Markov model', *J. Agric. Engng. Res.*, 21, 281–9.
64. Sturgess, O. W. (1989), 'Legume cover crops provide benefits later', *BSES Bull.*, 25, 8–9.
65. Vidrine, P. R., Reynolds, D. B. and Blouin, D. C. (1995), 'Grass control in soybean (*Glycine max*) with graminicides applied alone and in mixtures', *Weed Technol.*, 9(1), 68–72.
66. Shaner, D. L. (2014), *Herbicide Handbook*. Weed Science Society of America, Lawrence, KS.
67. Dusky, J. A. and Stall, W. M. (1996), 'Evaluation of imazethapyr for weed control in leafy vegetable crops', *Weed Technol.*, 10(2), 253–7.
68. Odero, D. C., Fernandez, J. V. and Havranek, N. (2016). 'Weed control and radish (*Raphanus sativus*) response to *S*-metolachlor', *HortScience*, 51(1), 79–83.
69. Devi, T. C., Lakshmi, M. B. and Naidu, N. V. (2005), 'Response of new sugarcane genotypes to wide row spacing', *Sugar Tech.*, 7(4), 154–6.
70. Ali, H. H., Peerzada, A. M., Hanif, Z., Hashim, S. and Chauhan, B. S. (2017), 'Weed management using crop competition in Pakistan: A review', *Crop Prot.*, 95, 22–30.
71. Singels, A. and Smit, M. A. (2009), 'Sugarcane response to row spacing-induced competition for light', *Field Crop Res.*, 113, 149–55.
72. Garside, A. L. and Bell, M. J. (2009), 'Row spacing and planting density effects on the growth and yield of sugarcane. 1. Responses in fumigated and non-fumigated soil', *Crop Pasture Sci.*, 60(6), 532–43.
73. Richard Jr., E. P, Dunckelman, J. W. and Carter, C. E. (1991), 'Productivity of sugarcane on narrow rows, as affected by mechanical harvesting', *Field Crop Res.*, 26, 375–86.
74. El-Shafai, A. M. A., Fakkar, A. A. O. and Bekheet, M. A. (2010), 'Effect of row spacing and some weed control treatments on growth, quality and yield sugarcane', *Int. J. Acad. Res.*, 2(4), 297–306.
75. Bull, T. A. and Bull, J. K. (2000), 'High density planting as an economic production strategy: (b) theory and trial results', *Proc. Aust. Soc. Sugar Cane Technol.*, 22, 104–12.
76. Bull, T. A. and Bull, J. K. (1996), Increasing sugarcane yields through higher planting density - preliminary results. In: *Sugarcane: Research Towards Efficient and Sustainable Production* (Wilson, J. R., Hogarth, D. M., Campbell, J. A. and Garside, A. L. (Eds)), CSIRO Division of Tropical Crops and Pastures, Brisbane, Australia, pp. 166–8.
77. Inman-Bamber, N. G. (1994), 'Temperature and seasonal effects on canopy development and light interception of sugarcane', *Field Crops Res.*, 36(1), 41–51.
78. Di Tomaso, J. M. (1995), 'Approaches for improving crop competitiveness through the manipulation of fertilization strategies', *Weed Sci.*, 43(3), 491–7.
79. Jones, C. A. (1982), 'The role of phosphorus in the Hawaii sugar industry', *Hawaii Plant Rec.*, 59(11), 229–64.

80. Franco, H. C. J., Otto, R., Faroni, C. E., Vitti, A. C., de Oliveira, E. C. A. and Trivelin, P. C. O. (2011), 'Nitrogen in sugarcane derived from fertilizer under Brazilian field conditions', *Field Crops Res.*, 121(1), 29–41.
81. McCray, J. M., Rice, R. W., Luo, Y. and Ji, S. (2010), 'Sugarcane response to phosphorus fertilizer on Everglades histosols', *Agron. J.*, 102(5), 1468–77.
82. Bakker, H. (1999), *Sugar Cane Cultivation and Management*. Kluwer Academic/Plenum Publishers, New York.
83. Judice, W. E., Griffin, J. L., Jones, C. A., Etheredge Jr., L. M. and Salassi, M. E. (2006), 'Weed control and economics using reduced tillage programs in sugarcane', *Weed Technol.*, 20(2), 319–25.
84. Odero, D. C. and Dusky, J. A. (2014), *Weed Management in Sugarcane*. IFAS, Florida Coop. Ext. Serv., University of Florida, EDIS SS-AGR-09.
85. Guglielmini, A. C. and Satorre, E. H. (2004), 'The effect of non-inversion tillage and light availability on dispersal and spatial growth of *Cynodon dactylon*', *Weed Res.*, 44(5), 366–74.
86. Wood, A. W. (1991), 'Management of crop residues following green harvesting of sugarcane in North Queensland', *Soil Till. Res.*, 20(1), 69–85.
87. Braunbeck, A., Bauen, A., Rosillo-Calle, F. and Cortez, L. (1999), 'Prospects for green cane harvesting and cane residue use in Brazil', *Biomass Bioenerg.*, 17(6), 495–506.
88. Meyer, E., Norris, C. P., Jacquin, E., Richard, C. and Scandaliaris, J. (2005), 'The impact of green cane production systems on manual and mechanical farming operations', *Proc. Int. Soc. Sugar Cane Technol.*, 25, 500–10.
89. Martins, D., Velini, E. D., Martins, C. C. and de Souza, L. S. (1999), 'Broadleaf weed emergence in soil covered with sugar cane straw', *Planta Daninha*, 17(1), 151–61.
90. Velini, E. D. and Martins, D. (1998), *Efeito da Palha da Cana-de-açúcar Sobre a Germinação das Principais Espécies de Plantas Daninhas Desta Cultura*. FCA/UNESP, Relatório Técnico, Botucatu.
91. Medeiros, D. (2001), Efeitos da palha de cana-de-açúcar (*Saccharum* spp.) sobre o manejo de plantas daninhas e diâmica do banco de sementes. Piracicaba, Dissertação Mestrado, Escola Superior de Agricultura 'Luiz de Queiroz'.
92. Richard Jr., E. P. (1999), 'Management of chopper harvester-generated green cane trash blankets: A new concern for Louisiana', *Proc. Int. Soc. Sugarcane Technol.*, 23(2), 52–62.
93. Sandhu, H. S., Gilbert, R. A., Kingston, G., Subiros, J. F., Morgan, K., Rice, R. W., Baucum, L., Shine Jr., J. M. and Davis, L. (2013), 'Effects of sugarcane harvest method on microclimate in Florida and Costa Rica', *Agric. For. Meteorol.*, 177, 101–9.
94. Hanson, N. S. (1962), 'Weed control practices and research for sugar cane in Hawaii', *Weeds*, 10(3), 192–200.
95. United States Environmental Protection Agency (1982). PRN 82-1: Revised Policy on Label Claims for Tank Mixing. Accessed13 February 2017. https://www.epa.gov/pesticide-registration/prn-82-1-revised-policy-label-claims-tank-mixing.
96. Orgeron, A. (2016), Sugarcane weed management. In: *Louisiana Suggested Weed Management Guide*, LSU AgCenter, Baton Rouge, LA, pp. 67–93.
97. Griffin, J. L. and Lencse, R. J. (1992), 'Preemergence control of itchgrass (*Rottboellia cochinchinensis*) in sugarcane', *J. Am. Soc. Sugar Cane Technol.*, 12, 65–70.
98. Anonymous (2014), *Herbicide Guide*, SASRI, Mount Edgecombe, KwaZulu-Natal, South Africa.
99. Anonymous (1989), *Weeds in Australian Cane Fields. Part B A Guide to the Selection of Chemical Control Measures*. BSES Bulletin No. 28, October 1989, Bureau of Sugar Experiment Stations, Indooroopilly, Qld.
100. Odero, D. C. and Shaner, D. L. (2014), 'Field dissipation of atrazine and metribuzin in organic soils in Florida', *Weed Technol.*, 28(4), 578–86.
101. Millhollon, R. W. (1993), 'Preemergence control of itchgrass (*Rottboellia cochinchinensis*) and johnsongrass (*Sorghum halepense*) in sugarcane (*Saccharum* spp hybrids) with pendimethalin and prodiamine', *Weed Sci.*, 41(4), 621–6.

102. Krutz, L. J., Shaner, D. L., Weaver, M. A., Webb, R. M. T., Zablotowicz, R. M., Reddy, K. N., Huang, Y. and Thomson, S. J. (2010), 'Agronomic and environmental implications of enhanced s-triazine degradation', *Pest Manag. Sci.*, 66(5), 461–81.
103. Richard, E. P. Jr. (1993), 'Preemergence herbicide effects on bermudagrass (*Cynodon dactylon*) interference in sugarcane (*Saccharum* spp. hybrids)', *Weed Technol.*, 7(3), 578–84.
104. Shaner, D. L., Krutz, L. J., Henry, W., Hanson, B. D., Poteet, M. D. and Rainbolt, C. R. (2010), 'Sugarcane soils exhibit enhanced atrazine degradation and cross adaptation to other s-triazines', *J. Am. Soc. Sugar Cane Technol.*, 30, 1–10.
105. Majumdar, K. and Singh, N. (2007), 'Effect of soil amendments on sorption and mobility of metribuzin in soils', *Chemosphere*, 66(4), 630–7.
106. Clement, A. A., Lammel, J. S., Filho, J. A. and Barbosa, J. C. (1989), 'Weed control in sugar cane with hexazinone and its mixtures with diuron in pre-emergence', *Planta Daninha*, 2(2), 85–8.
107. Millhollon, R. W. (1980), 'Johnsongrass (Sorghum halepense) control and sugarcane tolerance from preemergence treatments with hexazinone', *Proc. Int. Soc. Sugar Cane Technol.*, 17, 63–74.
108. Richard Jr., E. P. (1989), 'Response of sugarcane (*Saccharum* sp.) cultivars to preemergence herbicides', *Weed Technol.*, 3(2), 358–63.
109. Sund, K. A. (1964), 'An evaluation of atrazine, simazine, monuron and diuron on ten Hawaiian sugar cane plantations', *Weeds*, 12(3), 215–19.
110. Richard Jr., E. P. (1997), 'Johnsongrass (*Sorghum halepense*) control in fallow sugarcane (*Saccharum* spp. hybrids) fields', *Weed Technol.*, 11(3), 410–16.
111. Bhullar, M. S., Kamboj, A. and Singh, G. P. (2006), 'Weed management in spring-planted sugarcane (*Saccharum officinarum*)-based intercropping systems', *Indian J. Agron.*, 51(3), 183–5.
112. Weber, J. B. (1990), 'Behavior of dinitroaniline herbicides in soils', *Weed Technol.*, 4(2), 394–406.
113. Odero, D. C. and Shaner, D. L. (2014), 'Dissipation of pendimethalin in organic soils in Florida', *Weed Technol.*, 28(1), 82–8.
114. Correia, N. M., Perussi, F. J. and Gomes, L. J. P. (2012), '*S*-metolachlor efficacy on the control of *Brachiaria decumbens, Digitaria horizontalis*, and *Panicum maximum* in mechanically green harvested sugarcane', *Planta Daninha, Viçosa-MG*, 30(4), 861–70.
115. Chedzey, J. and Findlay, J. B. R. (1986), 'The use of acetochlor for weed control in sugarcane', *Proc. S. Afr. Sug. Technol. Ass.*, 60, 183–90.
116. Viator, B. J., Griffin, J. L. and Ellis, J. M. (2002), 'Sugarcane (*Saccharum* spp.) response to azafeniden applied preemergence and postemergence', *Weed Technol.*, 16(2), 444–51.
117. Richard Jr., E. P. and Griffin, J. L. (1993), 'Johnsongrass (*Sorghum halepense*) control in sugarcane with selected preemergence and postemergence herbicides', *J. Am. Soc. Sugar Cane Technol.*, 13, 58–72.
118. Dalley, C. D., Viator, R. P. and Richard Jr., E. P. (2013), 'Integrated management of bermudagrass (*Cynodon dactylon*) in sugarcane', *Weed Technol.*, 61(3), 482–90.
119. Griffin, J. L., Hook, B. J., Peregoy, R. S. and Kitchen, L. M. (1990),'Emergence and yield of 2,4-D treated seed cane', *J. Am. Soc. Sugar Cane Technol.*, 10, 56–60.
120. Richard Jr., E. P. (1986), 'Influence of surfactants on the toxicity of asulam to johnsongrass (*Sorghum halepense*) and sugarcane (*Saccharum* sp.)', *Weed Sci.*, 34(2), 299–303.
121. Richard Jr., E. P. (1991), 'Optimizing diluent volume improves johnsongrass control in sugarcane (*Saccharum* sp.) with asulam', *Weed Technol.*, 5(2), 363–8.
122. Richard Jr., E. P. and Griffin, J. L. (1993), 'Johnsongrass (*Sorghum halepense*) control in sugarcane (*Saccharum* sp.) with asulam', *Weed Technol.*, 7(3), 657–62.
123. Richard Jr., E. P. (1990), 'Timing effects on johnsongrass (*Sorghum halepense*) control with asulam in sugarcane (*Saccharum* sp.)', *Weed Technol.*, 4(1), 81–6.
124. Odero, D. C., Sellers, B., Baucum, L. and Rainbolt, C. (2014), *Fall Panicum: Biology and Control in Sugarcane*. IFAS, Florida Coop. Ext. Serv., University of Florida, EDIS SS-AGR-367.
125. Dalley, C. D. and Richard Jr., E. P. (2008), 'Control of rhizome johnsongrass (*Sorghum halepense*) in sugarcane with trifloxysulfuron and asulam', *Weed Technol.*, 22(3), 397–401.

126. Etheredge, L. M., Griffin, J. L. and Boudreaux, J. M. (2010), 'Nutsedge (*Cyperus* spp.) control programs in sugarcane', *J. Am. Soc. Sugar Cane Technol.*, 30, 67–80.
127. Wongkaew, P., Hanboonsong, Y., Sirithorn, P., Choosai, C., Boonkrong, S., Tinnangwattana, T., Kitchareonpanya, R. and Damak, S. (1997), 'Differentiation of phytoplasmas associated with sugarcane and gramineous weed white leaf disease and sugarcane grassy shoot disease by RFLP and sequencing', *Theor. Appl. Genet.*, 95, 660–3.
128. Chen, C. T., Lee, C. S. and Chen, M. J. (1972), 'Mycoplasma like organisms in *Cynodon dactylon* and *Brachiaria distachya* affected by white leaf diseases', *Report Taiwan Sugar Exp. Stn.*, 56, 49–55.
129. Rishi, N. and Chen, C. T. (1989), Grassy shoot and white leaf disease. In: *Diseases of Sugarcane: Major Diseases* (Ricaud, B. C. and Egan, B. T. (Eds)), Elsevier Science, Amsterdam, pp. 289–300.
130. Nakashima, K., Chaleeprom, W., Wongkaew, P. and Sirithorn, P. (1994), 'Detection of mycoplasma-like organisms associated with white leaf disease of sugarcane in Thailand using DNA probes', *JIRCAS*, J1, 57–67.
131. Sathe, T. V., Shinde, K. P. and Shaikh, A. L. (2009), *Sugarcane Pests and Diseases*, Manglam Publications, Dehli, India.
132. Arocha, Y., Lopez, M., Pinol, B., Fernandez, M., Picornell, B., Almeida, R., Palenzuela, I., Wilson, M. R. and Jones, P. (2005), "*Candidatus* Phytoplasma graminis' and '*Candidatus* Phytoplasma caricae', two novel phytoplasmas associated with diseases of sugarcane, weeds and papaya in Cuba', *Int. J. Syst. Evol. Microbiol.*, 55, 2451–63.
133. Espinoza Delgado, H. V., Kaye, C., Hincapie, M., Boukari, W., Wei, C., Fernandez, J. V., Mollov, D., Comstock, J. C. and Rott, P. (2016), 'First report of *Sugarcane yellow leaf virus* infecting Columbus grass (*Sorghum almum*) in Florida', *Plant Dis.* 100, 1027.
134. Dissanayake, N., Hoy, J. W., and Griffin, J. L. (1997), 'Weed hosts of the sugarcane root rot pathogen, *Pythium arrhenomanes*', *Plant Dis.*, 81(6), 587–91.
135. Yahaya, A., Dangora, D. B., Khan, A. U. and Zangoma, M. A. (2014), 'Detection of Sugarcane Mosaic Disease (SCMD) in crops and weeds associated with sugarcane fields in Makarfi and Sabon Gari local government areas of Kaduna state, Nigeria', *Int. J. Curr. Sci.*, 11, E99–104.
136. Ricaud, C., Egan, B. T., Gillaspie Jr., A. G. and Hughes, C. G. (Eds) (1989), *Diseases of Sugarcane: Major Diseases*, Elsevier Science Publishers, Amsterdam,.
137. Rott, P., Bailey, R. A., Comstock, J. C., Croft B. J. and Saumtally, A. S. (Eds) (2000), *A Guide to Sugarcane Diseases*, CIRAD/ ISSCT Publication Service, Montpellier, France.
138. Odero, D. C., Cherry, R. H. and Hall, D. G. (2013), 'Weedy host plants of the sugarcane root weevil (Coleoptera: Curculionidae) in Florida sugarcane', *J. Entomol. Sci.*, 48(2), 81–9.
139. Odero, D. C., Cherry, R. H., Wright, A. L., Sandhu, H. S. and Luo, Y. (2015), 'Woody host plants of the sugarcane root weevil (Coleoptera: Curculionidae) in Florida sugarcane', *J. Entomol. Sci.*, 50(1), 3–13.
140. Kirst, L. D. and Hensley, S. D. (1974), 'A study of overwintering populations of *Diatraea saccharalis* (F.) in Louisiana', *Proc. Int. Sugar Cane Tech.*, 15, 258–65.
141. Reagan, T. E., Hensley, S. D., Huffman, F. R. and Fuchs, T. W. (1979), 'Response to insecticides of the sugarcane borer in Louisiana and Texas', *J. Econ. Entomol.*, 72, 94–6.
142. Bessin, R. T., and Reagan, T. E. (1990), 'Fecundity of the sugarcane borer (Lepidoptera: Pyralidae) as affected by larval development on gramineous host plants', *Environ. Entomol.*, 19(3), 635–9.
143. Showler, A. T., Beuzelin, J. M., and Reagan, T. E. (2011), 'Alternate crop and weed host plant oviposition preferences by the Mexican rice borer (Lepidoptera: Crambidae)', *Crop Prot.*, 30(7), 895–901.
144. Beuzelin, J. M., Reagan, T. E., Way, M. O., Mészáros, A., Akbar, W. and Wilson, L. T. (2011), 'Potential impact Mexican rice borer non-crop hosts on sugarcane IPM', *Int. Sugar J.*, 113(1153), 660–5.
145. Beuzelin, J. M., Mészáros, A., Reagan, T. E., Wilson, L. T., Way, M. O., Blouin, D. C. and Showler, A. T. (2011), 'Seasonal infestations of two stem borers (Lepidoptera: Crambidae) in non-crop grasses of Gulf Coast rice agroecosystems', *Environ. Entomol.*, 40(5), 1036–50.

Index

Lightning Source UK Ltd.
Milton Keynes UK
UKHW02n2205070618
323883UK00001B/5/P

9 781786 761484